PROBABILISTIC RISK ASSESSMENT

Reliability Engineering, Design, and Analysis

Ernest J. Henley
Department of Chemical Engineering
University of Houston

Hiromitsu Kumamoto
Department of Precision Mechanics
Kyoto University

IEEE
PRESS

The Institute of Electrical and Electronics Engineers, Inc., New York

IEEE PRESS
445 Hoes Lane, PO Box 1331
Piscataway, NJ 08855-1331

ISBN 0-87942-290-4

IEEE Order Number: PP0285-7

Printed in the United States of America

10 9 8 7 6 5 4 3 2 1

This is the IEEE PRESS edition of a book previously published by Prentice-Hall, Inc. under the title *Reliability Engineering and Risk Assessment.*

Library of Congress Cataloging-in-Publication Data

Henley, Ernest J.
[Reliability engineering and risk assessment]
Probabilistic risk assessment : reliability engineering, design, and analysis / Ernest J. Henley, Hiromitsu Kumamoto.
p. cm.
Originally published as: Reliability engineering and risk assessment. Englewood Cliffs, N.J. : Prentice Hall, 1981.
Includes bibliographical references and index.
ISBN 0-87942-290-4
1. Reliability (Engineering) 2. Health risk assessment.
I. Kumamoto, Hiromitsu. II. Title.
Reliability engineering and risk assessment.
TS173.H47 1991
620'.00452—dc20 91-28802

CONTENTS

PREFACE

With the possible exception of environmental and computer technology, no other branch of applied science has developed and broadened so dramatically during this decade as safety, risk, and reliability analysis. In the early 1960's safety analyses were empirically based, the term *risk analysis* was virtually unknown, and the word *reliability* used only in isolated sectors of the aerospace and weapons industries. In the literature of the world's largest manufacturing industry, the chemical, there were no articles on reliability until 1966, and only a few before 1970.

Since 1970, problems associated with product liability, environmental constraints, and massive governmental intrusion into plant design, construction, and operating procedures, particularly in Europe, have spawned an entirely new technology. Dissemination of that technology has been slow and difficult, since the literature is complex and frustrating and some of the mathematical techniques unfamiliar to many engineers. The diversity of applications and the bewildering array of literature and nomenclature typical of a newly emerging, broadly based technology represent obstacles for novitiates as well as aspiring authors.

This book originated from a set of lecture notes for a two-day, continuing education course for practicing engineers sponsored by the American Institute of Chemical Engineers. This course has been given sixteen times since 1974 and we are grateful to the many students who contributed material and ideas. In 1975, the Sloan Foundation provided a small grant, which permitted the development of further material, including the first drafts of Chapters 7, 11, and 13, which were written in 1976 at the Computer-Aided Design Centre in Cambridge, England, with the help of Mr. Hiromitsu Hoshino. This material was used first in a short course sponsored jointly by the Computer-Aided Design Centre and the United Kingdom Atomic Energy Authority's National Centre of Systems Reliability.

A two-week, NATO-sponsored Advanced Study Institute on Synthesis and Analysis Techniques in Safety and Reliability Analysis held in Sogesta, Italy, in 1978 and co-directed by Giuseppi Volta provided the impetus for development of much of Chapters 4, 6, and 13. Leonardo Caldarolla, Bob Taylor, Eric Green, and Jerry Fussell, who served on the steering committee of the ASI, all contributed ideas which led to the final reorganization of this material. Indeed, Bob Taylor, Jerry Fussell, as well as Gary Powers were, at one time or another, potential or actual coauthors. In retrospect, it seems as if so much was happening and so many things were changing from 1975 to 1979 that we had to discard more material than we used.

This book is intended to serve two kinds of readers: the practicing engineer and the advanced undergraduate or graduate student. After a brief history of the field we introduce, in Chapter 1, the various risk study methodologies such as preliminary hazards analysis, failure modes and effects analysis, and event trees. Chapters 2 and 3 deal with methods for constructing fault trees and decision tables and with their qualitative analysis, in terms of cut and path sets, common-cause failures, and prime implicants.

In Chapter 4 we begin our discussion of quantitative methods by showing how component failure characteristics are obtained from failure data. At first reading, this chapter may seem unnecessarily complicated because we develop the full complement of conditional and unconditional failure and repair parameters. However, a clear understanding of the relationships between these parameters is required for understanding kinetic tree theory, which, despite its shortcomings, is currently the single most widely used technique for obtaining system failure parameters.

In Chapter 5 we introduce the subject of error bounds and confidence limits for failure data. In both Chapter 4 and 5 we assume that the reader has some undergraduate training in statistics. In general, we do not feel that advanced statistics are an important tool for practicing safety engineers, since they seldom have sufficient failure data for the application of sophisticated statistical methods. (We recognize of course that statistics are the lifeblood of quality control engineering, but we see that as a different discipline.)

In Chapter 6 we present something which we know that practicing engineers will welcome: a comprehensive survey of failure data and data banks.

Chapter 7, which was perhaps the most difficult to write, develops kinetic tree theory. Even though it is not completely self-consistent, this theory correctly calculates the expected number of failures (ENF) for both cut sets and systems. We believe the ENF to be, by far, the single most useful system parameter; far more useful than reliability. In subsequent chapters, incidentally, we compare kinetic tree theory with exact Markov results and show that the approximations in the theory are not at all restrictive.

In Chapter 8 we develop Markov methods and in Chapter 9 show how the system reliability, which cannot be obtained from kinetic tree theory, can be calculated by using Markov theory. Here, as elsewhere in the book, we stress bounding theorems, short-cut methods, and engineering approximations.

In Chapters 10 and 11 we describe some less general material: importance parameters, protective systems, cold standby, and storage tanks. Monte Carlo methods, the most flexible of all analytical simulation techniques, are the subject of Chapter 12. We hope the reader will find some of the newly developed, dagger-sampling, and state-transition methods interesting. In Chapter 13, we conclude with a number of "case-studies," to illustrate either a useful methodology or result.

To make the book self-contained we include, as appendices, a variety of specialized mathematical techniques that may be unfamiliar to some. In developing this book, a number of computer programs were written. These, as well as other available programs that have come to the attention of the authors, are listed and described in the section following the Acknowledgments. A solutions manual for all end-of-chapter problems is available.

ACKNOWLEDGMENTS

Other than the authors, the person most responsible for this book was Jerry Fussell. He was a "silent partner" during the five years of its development. Many pages contain Jerry's ideas. Another silent partner has been the National Science Foundation. Both of us, at one time or another, received sustenance from grant ENG-75-16713 AOI.

As stated in the Preface, Hiromitsu Hoshino wrote the first drafts of Chapters 7 and 11. Most of the computations in Chapters 7, 10, and 11 were done by Hiro at the Computer-Aided Design Centre at Cambridge, England, in 1976, and we are grateful to the Centre's staff for their generous forbearance. Most of the computer programs for Chapter 12 were written by Mr. Kazuo Tanaka.

Richard Barlow was kind enough to read one of the original versions of the manuscript and to help us stay on what we hope was the right track. Chuck Donaghey was an early contributor. His chapters on Markov and

Monte Carlo methods were not ultimately used in the book, but the fact that we had a "failsafe position" gave us confidence to go on. We owe the same kind of debt to Bob Taylor and Jans Rasmussen, who contributed material that we finally decided not to use but whose ideas we surely used consciously and unconsciously.

Leonardo Caldarolla, Eric Green, Gieuseppi Volta, Sergio Garriba, Gary Powers, George Apostolakis, Jack Lynn, Annick Carnino, Mitch Locks, and Ralph Evans, each of whom have seen at least parts of this manuscript during the five years it was written were kind enough to be supportive, and we are deeply appreciative.

It is a pleasure to also express our gratitude to the fine people at Prentice-Hall: Lori Opre who did an outstanding editing job, and Hank Kennedy, who made it all happen.

If this book were to have a dedication, it would be to Professor Koichi Inoue, whose friendship and scientific insight has been an inspiration to both of us.

ERNEST J. HENLEY
Houston, Texas

HIROMITSU KUMAMOTO
Kyoto, Japan

COMPUTER PROGRAMS FOR RELIABILITY AND RISK STUDIES

The following eleven computer programs are available for a modest price from either of the authors of this book (in print form only):

Program title: KITT-IT
Abstract: Top-event parameters, Q_S, W_S, Λ_S for systems using kinetic tree theory (augmented KITT program which handles storage tanks and standby redundancy).
Computer used: Honeywell 66/60
Minimum core: 50K words
Language: FORTRAN
Number of source statements: 1361
Environment: Batch
User manual: 24 pages
Section references to book: Example 8 of Chap. 13

Program title: PATH-CUT
Abstract: Conversion of min path sets into min cut sets, and vice versa, using a classification method.
Computer used: FACOM M-190 (compatible with IBM 3033)
Minimum core: 45K words
Language: FORTRAN
Number of source statements: 250
Environment: Interactive or batch
User manual: 15 pages
Section references to book: Sec. 6 of Chap. 3

Program title: PROTECT
Abstract: Based on a fault tree for a plant and two fault trees for a protective system, the computer code produces time profiles of expected numbers of normal trips, spurious trips, and destructive hazards.
Computer used: FACOM M-190 (compatible with IBM 3033)
Minimum core: 25K words
Language: FORTRAN
Number of source statements: 340
Environment: Interactive or batch
User manual: 75 pages
Section references to book: Sec. 2 of Chap. 11

Program title: PITE
Abstract: Simplifying decision tables, using the Quine's consensus theory.
Computer used: Honeywell 66/60
Minimum core: 60K words
Language: FORTRAN
Number of source statements: 940
Environment: Batch
User manual: 200 pages
Section reference to book: Sec. 9 and 10 of Chap. 3

Program title: PRIME
Abstract: Generating prime implicants for non-coherent fault trees, by using a classification method.
Computer used: FACOM M-190 (compatible with IBM 3033)
Minimum core: 60K words
Language: FORTRAN
Number of source statements: 980
Environment: Interactive or batch
User manual: 40 pages
Section reference to book: Sec. 6 of Chap. 3

Program title: MARKOV
Abstract: This code calculates time profiles of state probabilities for a Markov transition diagram. State transition matrix is used for numerical integration of linear differential equations.
Computer used: FACOM M-190 (compatible with IBM 3033)
Minimum core: 40K words
Language: FORTRAN
Number of source statements: 310
Environment: Interactive or batch
User manual: 15 pages
Section references to book: Appendix 1 to Chap. 8

Program title: NLB (New Lawler & Bell)
Abstract: Redundancy optimization using Lawler & Bell's integer programming.
Computer used: FACOM M-190 (compatible with IBM 3033)
Minimum core: 10K words
Language: FORTRAN
Number of source statements: 320
Environment: Batch

User manual: 15 pages
Section references to book: Sec. 3 of Chap. 11

Program title: HEUR
Abstract: Reliability optimization under constraints on cost, weight, etc., by using heuristic approach.
Computer used: FACOM M-190 (compatible with IBM 3033)
Language: FORTRAN
Minimum core: 20K words
Number of source statements: 310
Environment: Interactive or batch
User manual: 20 pages
Section references to book: Sec. 3 of Chap. 11

Program title: SCHE
Abstract: Conversion of reliability block diagrams into fault trees.
Computer used: Honeywell 66/60
Language: FORTRAN
Minimum core: 60K words
Number of source statements: 970
Environment: Batch
User manual: 40 pages
Section reference to book: Chap. 2

Program title: CONVERSION
Abstract: Obtaining min cut sets through expansion of product of sum expression of top event, given min path sets.
Computer used: FACOM M-190 (compatible with IBM 3033)
Minimum core: 60K words
Language: FORTRAN
Number of source statements: 220
Environment: Interactive or batch
User manual: 20 pages
Section references to book: Chap. 3

Program title: FAMULS (Fault Tree for Multi-Loop Systems)
Abstract: Generating cut sets for systems with multiple control loops, given signal flow graph representation of plant.
Computer used: FACOM M-190 (compatible with IBM 3033)

Minimum core: 40K words
Language: FORTRAN
Number of source statements: 620
Environment: Interactive or batch
User manual: 40 pages
Section references to book: Sec. 7 of Chap. 2

The following eight computer programs are available as card decks from JBF Associates, 10700 Dutchtown Drive, Knoxville, Tennessee 37922.

*Program title: MOCUS
Abstract: To obtain minimal cut sets or path sets for fault trees with AND/OR and INHIBIT logic.
Computer used: IBM 360, 370, CDC 7600
Minimum core: 228K (IBM)
Number of source statements: 1800
Environment: Batch

*Program title: PREP
Abstract: Obtains cut sets or path sets from fault trees with AND/OR and INHIBIT logic using combinatorial testing.
Computer used: IBM 360, 370, CDC 7600
Minimum core: 336K (IBM)
Number of source statements: 1200

Program title: BACFIRE
Abstract: Aids in common cause failure analysis by commonality searches.
Computer used: IBM 360, 370, CDC 7600
Minimum core: 128K (IBM)
Number of source statements: 800

*Program title: KITT-1
Abstract: For calculating top-event parameters (Q_S, W_S, Λ_S) given cut sets and failure and repair rates for components. Uses kinetic tree theory.

*The original version of these programs can also be obtained from the Argonne National Laboratory, Code Center, Argonne, Illinois, 60439.

Computer used: IBM 360, 370, CDC 7600
Minimum core: 372K (IBM)
Number of source statements: 1800

*Program title: KITT-2
Abstract: A version of KITT-1 which permits the input of time-varying failure and repair rates.
Computer used: IBM 360, 370, CDC 7600
Minimum core: 448K (IBM)
Number of source statements: 1700

†Program title: SAMPLE
Abstract: Uses Monte Carlo techniques to obtain confidence limits for top-event, given confidence limits for component failure and repair rates.
Computer used: IBM 360, 370, CDC 7600
Minimum core: 94K (IBM)
Number of source statements: 400

Program title: SUPERPOCUS
Abstract: A simplified KITT which uses bounding theorems to approximate top event probabilities. Also calculates Fussell-Vesely importance.
Computer used: IBM 360, 370, CDC 7600
Minimum core: 430K (IBM)
Number of source statements: 600

Program title: TREDA
Abstract: Draws report-quality fault trees using a CALCOMP plotter.
Computer used: IBM 360, 370, CDC 7600
Minimum core: 300K (IBM)
Number of source statements: 3700

Two other codes mentioned in this text, IMPORTANCE and F-TAP are available from Dr. Howard Lambert, TERA Corporation, 2150 Shattuck Ave., Berkeley, CA 94704.

†The original program appears in Wash 1400. A random number generator is required.

NOTATION*

Roman Letters

$\mathbf{A}$	An $n \times n$ matrix (Appendix 9.1)
A	Event (4.1)
$A(t)$	Availability (Appendix 4.6)
a	Constant in Laplace transform (4.93)
B	Function (12.65)
$\mathbf{b}$	Binary vector (12.6)
b_i	Component i of $\mathbf{b}$ (12.33)
C	Consequence (Fig. 13.1)
C	Capacity of units (11.34)
C	Event (4.1)
$\mathbf{c}$	Sample vector (12.11)
C_r	Criticality number for system component (Section 1.11)
D	Plant failure rate without protective instrument (11.39)
D	Determinant of graph (2.1)
d_i	An event in the ith minimal cut set (7.74)
$E(x)$	Expected value (Appendix 4.2)
e_i	The event of the ith minimal cut set failure (7.107)
$F(r)$	Poisson probability (Appendix 4.3)
$f(t)$	Failure density (Appendix 4.6)
$F(t)$	Unreliability (Appendix 4.6)
G_k	Transmittance (Fig. 2.37)
$G(t)$	Repair distribution (Appendix 4.6)
$g(t)$	Repair density (Appendix 4.6)
$\mathbf{g}(t)$	Importance function (10.5)
$g[\mathbf{Q}]$	Probability-of-top-event function (10.5)
h	Function (12.33)
$h(t)$	Function (4.92)
$\mathbf{I}$	Identity matrix (Appendix 8.1)
I^{BP}	Barlow-Proschan importance (Table 10.1)
I^{CR}	Criticality importance (Table 10.1)
I^{SC}	Sequential contributory importance (Table 10.1)
I^{UF}	Upgrading function (Table 10.1)
I^{FV}	Fussell-Vesely importance (Table 10.1)
K	Probability of no upstream repair (11.2)
K	Event of cut set (9.25)
K	Cut set (12.32)
K_A	Operational factor in criticality number expression (Section 1.11)
K_E	Environmental factor in criticality number expression (Section 1.11)

*Parentheses indicate equation numbers, Sections, or Tables where variable is defined.

L	Statistically independent factors (12.33)
L	Numerator to denominator ratio (2.3)
L	Level (2.5)
L	Laplace transform (4.92)
$L(\mathbf{y})$	Lower Bayesian confidence limit (5.43)
$L(t)$	Number of living at time t (Table 4.1)
M	Number of failures (12.2)
$M(t)$	Expected value of tank mass (11.30)
$m(t)$	Repair rate (Appendix 4.6)
N	Total sample (4.10)
N	Total units available (11.67)
N	Total number of samples (12.11)
N	Last critical failure mode in criticality number expression (Section 1.11)
N_c	Total number of minimal cuts (7.73)
$n(t)$	Number of deaths (4.10)
P	Path sets (12.35)
P	Path transmittance (2.2)
P	Transition probability (4.116)
Pr	Probability (4.1)
$Q(t)$	Unavailability (Appendix 4.6)
r	Number of failures (Appendix 4.4)
r	Total failures (5.4.1)
r	Number of units required for operation in (11.67)
$r(t)$	Failure rate (Appendix 4.6)
$R(t)$	Reliability (Appendix 4.6)
S	Life length of protective system (11.37)
S	Number of critical states (11.72)
S	Measured characteristic of a sample (Section 5.2.1)
S	Top-event occurrence (7.24)
s	Laplace transform variable (4.92)
$\mathbf{s}$	Random samples (12.29)
$\mathbf{T}$	An $n \times n$ matrix (Appendix 9.1)
$\mathbf{T}^{-1}$	Inverse of matrix $\mathbf{T}$ (Appendix 9.1)
T	Inspection interval of the protective system (Section 11.2.2)
T	Time to empty tank (11.1)
t	Time
U	Life length of plant (11.38)
$U(\mathbf{y})$	Upper Bayesian confidence limit (5.44)
$V(t)$	Expected number of repairs (Appendix 4.6)
$v(t)$	Unconditional repair intensity (Appendix 4.6)
$\mathbf{v}_i$	Eigenvectors (Appendix 9.1)
W	Event (4.3)
W	Expected number of failures (Appendix 4.6)

X	Source variable (2.1)
$\mathbf{x}$	An n-dimensional column vector (Appendix 9.1)
$\mathbf{x}$	Basic event state vector (12.6)
x_i	Basic event state (12.6)
$\mathbf{x}^*$	A linear transformation of $\mathbf{x}$ (Appendix 9.1)
$x(t)$	An indicator variable (4.65)
Y	Sink variable (2.1)
Y	Unconditional failure intensity (11.73)
Y	A binary indicator variable for basic events (7.30)
$\mathbf{y}$	Observations (5.39)
$\mathbf{z}$	Sample states generated for dagger sampling (12.48)
Z_i	Normalized variable (2.5)

Greek Letters

α	Level of significance (Fig. 5.1)
α	Failure mode ratio of critical failure mode (Section 1.11)
β	Shape parameter of Weibull distribution (4.128)
β	Parameter for beta probability density (Problem 5.8)
β	Conditional probability in criticality number expression (Section 1.11)
γ	Parameter in Weibull distribution (4.128)
γ	Time when the component begins to fail (4.128)
γ	Constant transition rate (12.55)
γ_i	Eigenvalues (Appendix 9.1)
Γ	Parameter (12.59)
Θ	Random variable (5.12)
θ	Parameter (8.48)
θ	True characteristic of a population (5.1)
$\lambda(t)$	Conditional failure intensity (Appendix 4.6)
λ_G	Generic failure frequency used in criticality number expression (Section 1.11)
$\mu(t)$	Conditional repair intensity (Appendix 4.6)
μ	Parameter in lognormal distribution (4.122)
μ	Kth moment about mean (Appendix 4.3)
π	Parameter (8.45)
ρ	Minimal path structure function (7.64)
σ	Characteristic life (4.128)
σ	Standard derivation (Appendix 4.3)
σ	Parameter in lognormal distribution (4.122)
$\bar{\sigma}$	Sample standard deviation (Problem 5.1)

σ^2	Variance (Appendix 4.3)
ψ	Binary function (12.8)
$\psi(\mathbf{Y})$	Top-event indicator variable, structure function (Section 7.4)
$\psi_L(\mathbf{Y})$	Lower bound structure function (7.80)
$\psi_U(\mathbf{Y})$	Upper bound structure function (7.81)
κ	Minimal-cut structure function (7.57)
χ^2	Percentiles of the chi-square distribution (5.22)

Subscripts

C	Direct sampling (12.11)
c	Cold standby (11.68)
c	Minimal cut sets (7.80)
D	Dagger (12.48)
L	Lower bound (7.80)
m	Mode (Fig. 12.16)
N	Total parallel units (11.66)
o	Median (Fig. 12.16)
p	Minimal path sets (7.81)
R	Restricted sampling (12.29)
SD	System downstream (Section 11.1)
SU	System upstream (Section 11.1)
TE	Empty tank (11.1)
TF	Full tank (Section 11.1.7)
U	Upper bound (7.81)

Superscripts

*	Cut set parameter
—	Standby (Section 8.2.1))
—	Event negation
.	Derivative

Abbreviations

ENF	Expected number of failures (Section 4.2.5)
FAFR	Fatal accident frequency rate (Section 1.3)
MTBF	Mean time between failures (Appendix 4.6)
MTBR	Mean time between repairs (Appendix 4.6)
MTTF	Mean time to failure (Appendix 4.6)

MTTF	Mean time to failure (Appendix 4.6)
MTTR	Mean time to repair (Appendix 4.6)
WSUM	Expected number of failures (Table 7.8)
TTF	Time to fail (Appendix 4.6)
TTR	Time to repair (Appendix 4.6)

Mathematical Conventions

Π	Product
[ ]	Square brackets enclose definition of a function, e.g., $R \equiv \frac{C}{T}$ is expressed as $R\left[\frac{C}{T}\right]$
$\int_1^2$	Integration over the interval 1 to 2
$L[f(t)]$	Laplace transform of function $f(t)$ (4.92)
$\binom{n}{m}$	$\equiv \frac{n!}{(n-m)!m!}$
Σ	Summation
$\vee$	Boolean summation
$\wedge$	Boolean product
$\cup$	Union
$\cap$	Intersection
∞	Infinite

HISTORICAL PERSPECTIVE

For obvious reasons, the earliest impetus for reliability quantification came from the aircraft industry. After World War I, as air traffic and air crashes increased, reliability criteria and necessary safety levels for aircraft performance emerged. Comparisons of one and multi-engine aircraft from the point of view of successful flights were made, and requirements in terms of accident rates per hours of flying time developed. By 1960, for instance, it had been deduced that fatal accidents occurred in approximately one out of one million landings. Thus, for automatic landing systems, design criteria for a fatal landing risk less than one per 10^7 landings could be established.[1] R. H. Jennings[2] has chronicled the development of the reliability engineering in the 1940–1970 decades.

THE 1940'S

The early development of mathematical reliability models began during World War II in Germany, where a group led by Wernher Von Braun was developing the V-1 missile. The first series of ten missiles was totally unreliable; they all blew up on the launching pads or fell into the English

[1]Green, A. E. and A. J. Bourne, *Reliability Technology*, John Wiley & Sons, Inc., New York, p. 3, 1972.

[2]Jennings, R. H., "Historic and Modern Practices in Reliability Engineering," Paper Presented at the AIChE Meeting, Washington, Dec., 1974.

Channel. Robert Lusser, a mathematician, was called in as a consultant. He determined that the old saw, "A chain is no stronger than its weakest link," was not applicable to a series system since it failed to account for the random failure. Lusser then produced the product law of series components, namely that the reliability of a series system is equal to the product of the reliabilities of the components: $R_s = R_1 R_2 R_3 \ldots R_n$, or $R_s = \prod_{i=1}^{n} R_1$ as we now use it. Thus, in a series system, the reliability of the individual components must be much higher than the system reliability for satisfactory system performance.

In the United States, efforts to improve reliability during the 1940's were focused on an extension of quality. Better design, stronger materials, harder and smoother wearing surfaces, advanced inspection instruments, etc.—all were emphasized to extend the useful life of a part or an assembly. The Electro-Motive Division of General Motors Corporation, for example, extended the useful life of traction motors used in locomotives from 250,000 miles to 1 million miles by the use of better insulation, high temperature testing and improved tapered-spherical roller bearings. Life of the diesel engine was greatly extended through the development of Tocco hardening of the bearing surfaces of crankshafts and camshafts. Advances were made in designing for maintenance accessibility and for providing plans, facilities, techniques, and schedules for preventive maintenance. Other noticeable progress was displayed during the 1940's by management interest and enthusiasm in sampling plans for inspection, control charts for high production machine tools, levels of evaluation, and purchasing incentives for quality products. This marked the entry of industrial engineers into the field and, as a result, most "reliability" texts and courses deal only with quality control and inspection and associated statistical techniques.

THE 1950'S

Increased importance was attached to safety—most prominently in the aerospace and nuclear fields. This decade saw the beginnings of the use of component reliability in terms of failure rate, life expectancy, design adequacy, and success prediction.

During the Korean War, the Department of Defense found that unreliable equipment required a tremendous amount of maintenance. It found that the cost to the Armed Services was $2 per year to maintain every dollar's worth of electronic equipment: for an equipment life of ten years, it cost 20 million dollars to maintain every million dollars of purchase value of the equipment. It was, thereby, demonstrated to the Government that it is wiser to design for reliability than it is to wait and repair equipment after failure.

It was in the early 1950's that efforts were applied towards understanding and correcting human errors which contribute to system failures. One of the first quantitative estimates of human performance was done at the Sandia Laboratories in 1952. A classified study of an aircraft nuclear weapon system was accomplished, using as an approach the compilation of estimates of human error rates per task. The task assignment was further subdivided into the environment; i.e., a 0.01 error rate was somewhat subjectively assigned to "on ground" operation, and a 0.02 error rate "in the air." These figures were incorporated into operational reliability equations in the same manner as were other system events.

THE 1960'S

The 1960's saw the emergence of new reliability techniques and a wider variety of specialized applications. Spreading from the earlier concentration on the ways that components behaved, whether mechanical, electrical, or hydraulic, the emphasis broadened to studies of the effect component failures had on the system of which they were a part. The era of the Intercontinental Ballistic Missiles and the subsequent man-rated rocket development such as the Mercury and Gemini programs accelerated the "demand-for-success." This was prompted by the "one-shot" requirements, culminating with the countdown of the rocket engines and systems on the launch pad.

Considerable effort was applied to both component and system functional testing during the "aerospace years." Records were kept of each failure, its analysis, and the inspection records of the deficiencies that turned up in the investigations. Each component mode of failure, mechanism, and cause, and its failure effect on the system was evaluated for application of corrective action to preclude recurrence. Systems analysis, utilizing reliability block diagrams, became highly developed and extensively used as success models to help in attaining safety and reliability goals.

With the increased complexity of the more sophisticated block diagrams, another approach was required. In 1961, the concept of Fault Tree Analysis was originated by H. A. Watson of Bell Telephone Laboratories as a plan to evaluate the safety of the Minuteman Launch Control System. Later the Boeing Company modified the concept for computer utilization. In 1965, D. F. Haasl further developed the technique of fault tree construction and its application to a wide variety of industrial safety and reliability problems.

The emergence of a system safety study as an independent and separate activity was first mandated by the Air Force in 1962, following disastrous accidents at four ICBM missle/silo complexes. In 1966, the Department of

Defense (DOD) adopted the Air Force standards and began requiring system safety studies in all phases of system development, for all defense contracts. These standards were continually broadened and revised and, in 1969, DOD adopted MIL-STD-882, "System Safety Programs for Systems and Associated Subsystems and Equipment: Requirements For," as a standard ukase for all defense contractors.

Parallel DOD requirements relating reliability, availability, and maintainability for individual hardware items were evolving. Standards such as MIL-STD 471, "Maintainability/Verification/Demonstration/Evaluation," and MIL-STD 781, "Reliability Tests: Exponential Distribution," are documents which serve to maintain a high rate of employment among reliability engineers and consultants as well as civil servants.

The 1960's also witnessed the beginning of a still-burgeoning production of books and journals. The seminal text of Igor Bazovsky, *Reliability Theory and Practice*, was published in 1961 by Prentice-Hall, Inc., and at least 15 other books were available by the end of the decade. (A bibliography is provided at the end of this chapter.) This period also saw the birth of the *IEEE Transactions on Reliability* which, under the leadership of Dr. Ralph Evans, has become the premier journal in its field. Distinguished mathematicians such as Z. W. Birnbaum, R. Barlow, F. Proschan, J. Esary, and W. Weibull led the way in developing advanced statistical techniques generic to reliability and maintainability problems.

An enterprise that started in 1950 and accelerated in the 1960's was the collection and archiving of component, system, and human failure data. Developments in this field are the subject of Chapter 6.

THE 1970'S

The extensive risk assessment of nuclear power plants sponsored by the United States Atomic Energy Commission and completed in 1974, "WASH 1400, The Reactor Safety Study," has been literally epoch-making. Professor N. Rasmussen and his multi-million-dollar team analyzed a vast spectrum of nuclear accidents, numerically ranked them in order of their probability of occurrence, and then assessed their potential consequence to the public. The event tree, fault tree, and risk-consequence techniques used in this study are being widely adopted by the chemical and other industries. Rasmussen-like studies are proliferating in Europe, Asia, and the United States.

Rising public clamor regarding industrial hazards, coupled with strident consumerism and environmentalism, has had a profound impact on this decade. In Europe, following the serious industrial accidents at Flixborough, England and Cervesa, Italy, there has been a rash of new legislation requiring major risk studies prior to all new plant construction. In Britain, the new Toxic Substances Act could similarly affect every plant that has as much as a single cylinder of compressed gas. In the United States, we have given birth to OSHA and the product liability lawsuit: the cost of the latter to the chemical industry in 1977 was estimated to have been 2 billion dollars.

REFERENCES

Henney, K., et al.: *Reliability Factors for Ground Electronic Equipment*, McGraw-Hill Book Company, New York, 1956.

Reliability and Maintainability Symposia Proceedings (known by different names starting in 1954, annually since 1956), Institute of Electrical and Electronic Engineers, New York.

IRE, Inc.: *Reliability Training Text*, New York, 1959.

Chorofas, Dimitri N., *Statistical Processes and Reliability Engineering*, Van Nostrand Reinhold Company, New York, 1960.

Gryna, F. M., N. J. Ryerson, and S. Swerling (eds.), *Reliability Training Text*, 2nd ed., 1 East 79th Street, Institute of Radio Engineers, New York, 1960.

Dummer, G., and N. Griffin, *Electronic Equipment Reliability*, John Wiley & Sons, Inc., New York, 1960.

Bazovsky, Igor, *Reliability Theory and Practice*, Prentice-Hall, Inc., Englewood Cliffs, N. J., 1961.

Lloyd, David K., and Myron Lipow, *Reliability: Management, Methods and Mathematics*, Prentice-Hall, Inc., Englewood Cliffs, N. J., 1962.

Zelen, Marvin S., *Statistical Theory of Reliability*, University of Wisconsin Press, Madison, 1962.

Calabro, S. R., *Reliability Principles and Practice*, McGraw-Hill Book Company, New York, 1962.

Barlow, W. R., L. Hunter, and F. Proschan, *Probabilistic Models in Reliability Theory*, John Wiley & Sons, Inc., New York, 1962.

Cox, D. R., *Renewal Theory*, John Wiley & Sons, Inc., New York, 1962.

Wilcox, M. H., and W. C. Mann (eds.), *Redundancy Techniques for Computing Systems*, Spartan Press, Washington, D. C., 1962.

Pieruschka, E., *Principles of Reliability*, Prentice-Hall, Inc., Englewood Cliffs, N. J. 1963.

Sandler, G., *System Reliability Engineering*, Prentice-Hall, Inc., Englewood Cliffs, N. J., 1963.

Ankenbrandt, F. L., *Maintainability Design*, Engineering Publishers, Division of A. C. Book Company, Inc., Elizabeth, N. J., 1963.

Landers, R., *Reliability and Product Assurance*, Prentice-Hall, Inc., Englewood Cliffs, N. J., 1963.

Lipson, C., J. Kerawalla, and L. Mitchell, *Engineering Applications of Reliability*, The University of Michigan Press, Ann Arbor, 1963.

Reliability Control in Aerospace Equipment Development, Society of Automotive Engineers, Inc., SAE Technical Progress Series, Vol. 4, 1963.

Goldman, A., and T. Slattery, *Maintainability: A Major Element of System Effectiveness*, John Wiley & Sons, Inc., New York, 1964.

Handbook of Reliability Engineering, Bureau of Naval Weapons Handbook, NAVWEPS 100-65-502, 1964.

Haviland, R. P., *Engineering Reliability and Long Life Design*, Van Nostrand Reinhold Company, New York, 1964.

Myers, R., et al., *Reliability Engineering for Electronic Systems*, John Wiley & Sons, Inc., New York, 1964.

Roberts, N., *Mathematical Methods in Reliability Engineering*, McGraw-Hill Book Company, New York, 1964.

Reliability Engineering Handbook, NAVAIR 00-65-502/NAVORD OD-41146, 1 June, 1964.

Buckland, William R., *Statistical Assessment of the Life Characteristic*, Griffin, 1964 (available from NTIS, Springfield, VA, 22151).

Aeronautical Radio, Inc., *Reliability Engineering*, Prentice-Hall, Inc., Englewood Cliffs, N. J., 1964.

ARINC Research Technical Staff, *Reliability Engineering*, Prentice-Hall, Inc., Englewood Cliffs, N. J., 1964.

Levenback, G. J., "System Reliability and Engineering Statistical Aspects," *Am. Scientist*, September, 1965.

Häckler, Jurgen, *Methoden für die Untersuchung der Zuverlassigkeit*, Institut f̈urDatenverarbeitung, Dresden, 1965.

Barlow, Richard E., and Frank H. Proschan, *Mathematical Theory of Reliability*, John Wiley & Sons, Inc., New York, 1965.

Gryna, F. M., Jr. (ed.), *Reliability Theory and Practice*, Sixth Annual Workshop, Chicago, June 22–24, 1965. Sponsored by Electrical Engineering Division of ASEE at the Illinois Institute of Technology (available from ASEE Headquarters).

Pierce, W., *Failure Tolerant Computer Design*, Van Nostrand Reinhold Company, New York, 1966.

Stewart, Donald Arnott, *Probability, Statistics and Reliability*, Draughtsmen's and Allied Technicians Association, Richmond, Surrey, 1966.

Dummer, Geoffrey W., and Norman B. Griffin, *Electronic Reliability: Calculation and Design*, Pergamon Press, Inc., Elmsford, N. Y., 1966.

Enrick, Norbert Lloyd, *Quality Control and Reliability*, Industrial Press, New York, 1966.

Ireson, W. G., *Reliability Handbook*, McGraw-Hill Book Company, New York, 1966.

Kenney, D. P., *Application of Reliability Techniques*, Argyle Publishing Corporation, New York, 1966. (A self-instructional programmed training course book.)

Gedye, Gordon R., *A Manager's Guide to Quality and Reliability*, John Wiley & Sons, Inc., New York, 1968.

Haugen, Edward B., *Probabilistic Approaches to Design*, John Wiley & Sons, Inc., New York, 1968.

Polovko, A. M., *Fundamentals of Reliability Theory*, translation edited by William H. Pierce, Academic Press, Inc., New York, 1968.

Research Triangle Institute, *Practical Reliability* (in five volumes), National Aeronautics and Space Administration, 1968.

Shooman, Martin, *Probabilistic Reliability: An Engineering Approach*, McGraw-Hill Book Company, New York, 1968.

Dummer, Geoffrey W., and R. C. Winton, *An Elementary Guide to Reliability*, Pergamon Press, Inc., Elmsford, N. Y., 1968.

Polovko, A. M., *Fundamentals of Reliability Theory*, Academic Press, Inc., New York, 1968.

Chapouille, Pierre, *Fiabilité des Systèmes*, Masson et cie, Paris, 1968.

Gnedenko, B. V., Yu K. Belyayev, and A. D. Solovyev, *Mathematical Methods of Reliability Theory*, translation edited by Richard E. Barlow, Academic Press, New York, 1969.

International Electrotechnical Commission, *Managerial Aspects of Reliability*, Geneva, 1969.

Smith, Charles Stanley, *Quality and Reliability: An Integrated Approach*, Pitman Press, New York, 1969.

Thomason, Roy, *An Introduction to Quality and Reliability*, Machinery Publishing Company, Brighton, England, 1969.

Barlow, Richard E., and Ernest M. Scheuer, *An Introduction to Reliability Theory*, CEIR, Inc. 1969.

Blanchard, Benjamin S., Jr., and Edward Lowery, *Maintainability, Principles and Practice*, McGraw-Hill Book Company, New York, 1969.

Störmer, Horand, *Mathematisch Theorie der Zuverlässigkeit*, R. Oldenborgh, Munich, Germany, 1970.

Breipohl, A. M., *Probabilistic Systems Analysis*, John Wiley & Sons, Inc., New York, 1970.

Amstadter, Bertram, *Reliability Mathematics*, McGraw-Hill Book Company, New York, 1971.

Green, A. E., and A. J. Bourne, *Reliability Technology*, John Wiley & Sons, Inc., New York, 1972.

Hamner, W., *Handbook of System and Product Safety*, Prentice-Hall, Inc., 1972.

O'Connell, E. P. (ed.), *Handbook of Product Maintainability*, Reliability Division, American Society for Quality Control, 1973.

Locks, M. O., *Reliability, Maintainability, and Availability Assessment*, Hayden Book Co., Inc., Rochelle Park, N. J., 1973.

Smith, C. O., *Introduction to Reliability in Design*, McGraw-Hill Book Company, 1976.

Henley, E. J., and J. Lynn, *Generic Techniques in Reliability Assessment*, Noordhoff International, Leyden, Holland, 1976.

Barlow, Richard E., J. B. Fussell, and N. D. Singpurwalla, "Reliability and Fault Tree Analysis," *SIAM*, Philadelphia, 1975.

Fussell, J. B., and G. R. Burdick, "Nuclear Systems Reliability Engineering and Risk Assessment," *SIAM*, Philadelphia, 1977.

1

RISK ANALYSIS

1.1 RELIABILITY, RISK, AND SAFETY

There is considerable overlap (and often confusion) between the terms *Reliability*, *Safety*, *Hazard*, and *Risk*.

In this text, the terms *safety* or *hazard analysis* are viewed interchangeably, and they, as well as *reliability analysis*, refer to studies of process or equipment failure or operability. If the purpose of the study is to determine safety parameters, it is necessary to consider, in addition to equipment failure and operability, the possibility of damage by (or to) the system. If this phase of the safety study suggests that there could be system failures, then a *risk study* is done to determine the consequence of the failure in terms of possible damage to property or people.

An example of a reliability study would be an analysis of how frequently a chemical reactor might overheat due to malfunctioning pumps, heat exchangers, human operators, control systems, and other plant equipment and utilities. If this study were extended to include an assessment of how frequently a temperature excursion results in an explosion, we would be looking at the safety (hazard) problem. To conclude the safety study, we must also verify that the chemical reactor will not overheat, given no hardware or utility failures, due to factors outside the design envelope.

If we now extend the chemical reactor explosion analysis to include the array of consequences, and their associated frequency, and the damage in terms of human and property losses, a risk analysis will have been accomplished. For example, the consequence of an explosion following a reactor temperature excursion might be minor injuries due to flying shrapnel or a major disaster due to fire. One of the purposes of a risk analysis would be to assign a frequency (probability) to these and other possible consequences of system failure.

The outcome of a risk study might be a statement (or series of statements) such as: "The number of people expected to be killed per year due to a reactor explosion is 10^{-4}." In effect, for every 10,000 hours we predict one death. It is of societal interest, therefore, to compare a number such as one death per 10,000 hours with the risks involved in everyday living, to obtain some idea of what might constitute a reasonable outcome for a risk analysis and as a basis for decision making.

1.2 DEFINITION AND MEASUREMENT OF RISK

A dictionary definition of risk is "the *possibility* of loss or injury to people and property." If reliability engineers wrote dictionaries, the entry might have read "the *probability* of loss or injury to people and property." In engineering terms, for example, the individual risk (probability) of death of any of the 200 million people in the United States, per year, from automobile accidents is

$$\frac{50{,}000 \text{ deaths/year}}{200{,}000{,}000 \text{ persons}} = 2.5 \times 10^{-4} \frac{\text{deaths}}{\text{person-year}}$$

since there is a yearly total of 50,000 automobile fatalities in the United States. A risk can also be a consequence other than death, a more general expression being

$$\text{risk}\left[\frac{\text{consequence}}{\text{time}}\right] = \text{frequency}\left[\frac{\text{events}}{\text{unit time}}\right]\text{magnitude}\left[\frac{\text{consequence}}{\text{event}}\right]$$

In the case of automobile accidents, if there are 50 million car accidents in the U. S. every year, the consequence (death/accident) is 10^{-3}, since

$$50{,}000 \frac{\text{deaths}}{\text{year}} = \left(50 \times 10^{6} \frac{\text{accidents}}{\text{year}}\right)\left(10^{-3} \frac{\text{deaths}}{\text{accidents}}\right)$$

The societal risk of property loss from automobile accidents would be

$$\text{risk}\left[\frac{\text{loss}}{\text{time}}\right] = \text{frequency}\left[\frac{\text{accidents}}{\text{time}}\right]\text{magnitude}\left[\frac{\text{loss}}{\text{accident}}\right]$$

The probabilistic figure 2.5×10^{-4} deaths/person-year means that if all persons in the U. S. have equal probabilities of being killed in an automobile accident then, if there were no other causes of death, the entire population of the United States would die of car accidents in a period of four thousand years. It is only that we are dealing with a large data base which makes this a valid conclusion. Any single driver could say, "This is meaningless to me, I might be killed in a crash tomorrow," and he would be correct.

In applying this probabilistic criteria to risk of a death from a train wreck, there is a major difference in whether a risk of, for example, 0.1 fatalities per year refers to 100 persons killed in a single accident every 1000 years, or 1 person killed every 10 years. In general, the public ignores single-death accidents, but there is grave public concern about potential accidents in which hundreds of people might die.

The risk analysis approach described in the foregoing paragraphs invoke the classical concept of long-run relative frequencies. However, when a risk analysis of a yet-unbuilt nuclear reactor predicts a public risk of 10^{-6} fatalities per year, it can be argued that we are dealing not with long-run relative frequencies derived from data, but with "rare events" to which the classical probabilistic approach of statistical inference cannot be applied. There is always the example of the statistician who drowned in a stream having an average depth of five centimeters.

An alternative approach to the rare-event problem is based on subjectivistic logic. This rejects the notion of a true probability and embraces the idea of a probability as a measure of opinions and beliefs. The methods for converting beliefs and opinions into risk criteria involve non-trivial and, at times, controversial exercises in modifying probabilities, using expert opinion in conjunction with Bayes's theorem. The interested reader is referred to footnotes 1 and 2 below.

1.3 PUBLIC RISK

Safety cannot be guaranteed any individual, whatever his mode of life. Each of us survives from one day to the next by avoiding or overcoming risks such as those quoted in Table 1.1.[3] As the risk diminishes to less than 10^{-6} year^{-1}, the average individual does not show undue concern, and so

[1]Apostolakis, G., "Probabilistic Risk Assessment: The Subjectivistic Viewpoint and Some Suggestions," *Nuclear Safety*, **19**, (3) June, 1978.

[2]De Finetti, B., *Theory of Probability*, John Wiley & Sons, Inc., New York, 1974.

[3]WASH 1400, as cited by Wall, I. A., "Some Insights from the Reactor Safety Study," Office of Nuclear Regulatory Research, U. S. NRC.

elaborate precautions against this risk level are seldom taken—we do not pass our lives in fear of being struck down by lightning. Based on this premise, risk levels of 10^{-6} have been cited by many as target figures for risks posed by industrial activities.

An interesting perspective on the precarious nature of our daily activities was developed by Brian Bulloch, Mond Division, Imperial Chemical Industries Ltd. The ordinate of Fig. 1.1 is the *fatal accident frequency rate*, the average number of deaths by accidents in 10^8 hours of a particular activity. The chemical plant, contrary to public belief generated by tabloid headlines, which insinuate that the chemical industry is, at best a nuisance and at worst a creator of mass poisons, is an extremely safe workplace, the average risk lying at the low end of the spectrum. Moreover, about half of the 3.5 FAFR is due to road accidents, tripping, falling, i.e., non-process related incidents.

TABLE 1.1 INDIVIDUAL RISK OF EARLY FATALITY BY VARIOUS CAUSES

Accident Type	Total Number for 1969	Approximate Individual Risk Early Fatality Probability/Yr[a]
Motor vehicle	55,791	3×10^{-4}
Falls	17,827	9×10^{-5}
Fires and hot substance	7,451	4×10^{-5}
Drowning	6,181	3×10^{-5}
Poison	4,516	2×10^{-5}
Firearms	2,309	1×10^{-5}
Machinery (1968)	2,054	1×10^{-5}
Water transport	1,743	9×10^{-6}
Air travel	1,778	9×10^{-6}
Falling objects	1,271	6×10^{-6}
Electrocution	1,148	6×10^{-6}
Railway	884	4×10^{-6}
Lightning	160	5×10^{-7}
Tornadoes	118[b]	4×10^{-7}
Hurricanes	90[c]	4×10^{-7}
All others	8,695	4×10^{-5}
All accidents	115,000	6×10^{-4}
Nuclear accidents (100 reactors)	—	2×10^{-10}[d]

[a]Based on total U. S. population, except as noted.
[b](1953–1971 avg.).
[c](1901–1972 avg.).
[d]Based on a population at risk of 15×10^6.

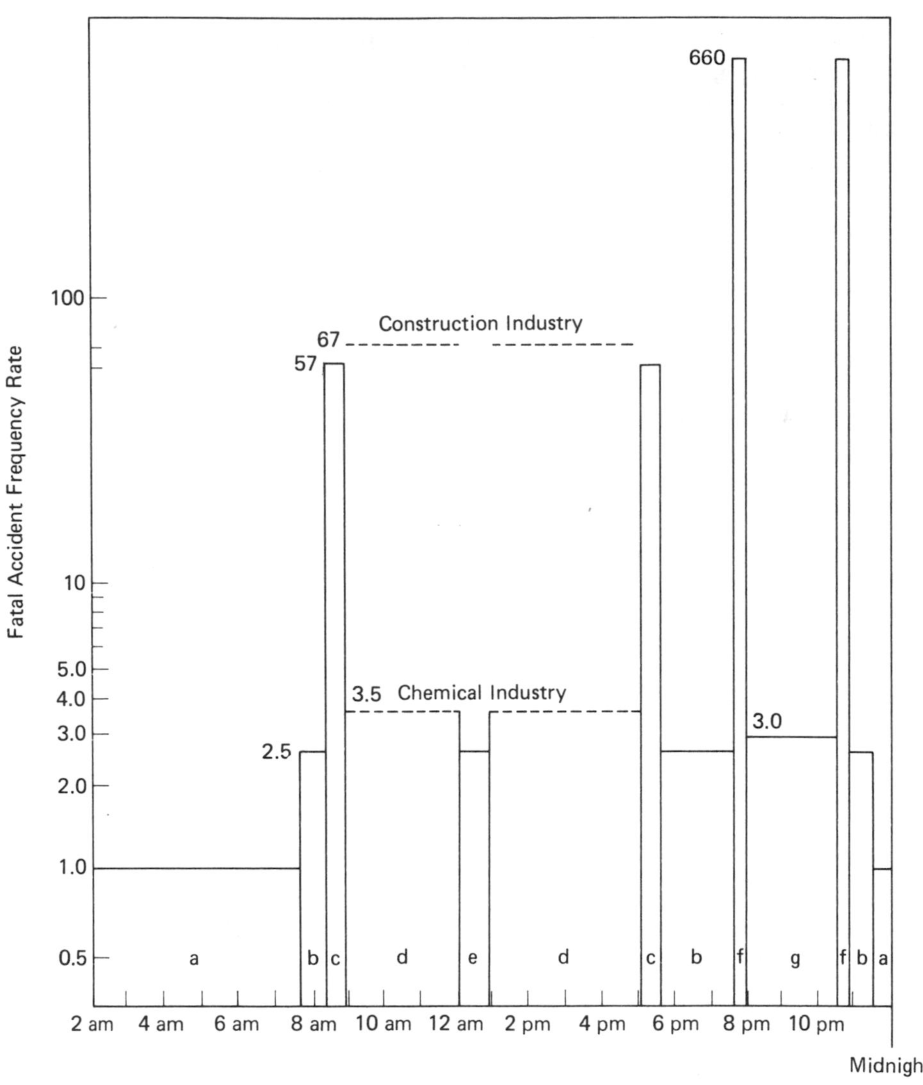

KEY

a Sleeping time
b Eating, washing, dressing, etc. at home
c Driving to or from work by car
d The day's work
e The lunch break
f Motor-cycling
g Communal entertainment (e.g., pub)

Figure 1.1. Daily hazards.

1.4 THE FARMER CURVE

A key concept in risk analysis is the idea introduced by Farmer in 1967[4] of an arbitrary but carefully chosen relationship between the estimated size of a radioactive release to the atmosphere caused by a nuclear reactor accident and the probability (long-term average frequency per year, or the reciprocal of the average time period between such events) that the specific accident described might occur. This defines an *accident release frequency limit line*, which could be used primarily as a guide for designers of a new plant and those who are required to assess the safety of a plant. The current form of the accident release frequency limit line in use in the United Kingdom Atomic Energy Authority is shown in Fig. 1.2. The line is regarded as separating an upper area of unacceptably high risk from one of lower and acceptable risk beneath and to the left of the line. The line may thus be used as a safety criterion defining an upper boundary of permissible probability. If so, a general intention is satisfied, which one feels instinctively to be correct, namely that while accidents which give rise

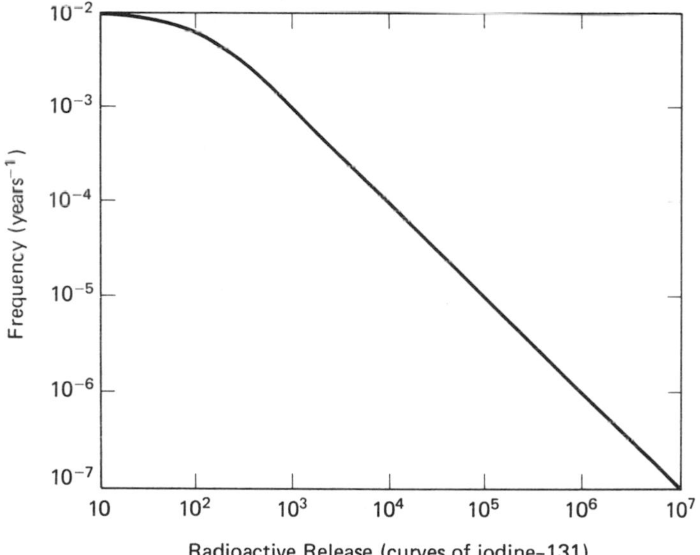

Figure 1.2. Accident release frequency limit line.[5]

[4]Farmer, F. R., "Siting Criteria—New Approach," "Containment Siting of Nuclear Power Plants," *Proc. Symp. Vienna Pap.*, SM-89/34 (1967).

[5]Farmer, F. R., and J. R. Beattie, "Evaluation of Population Hazards from An Escape of Fission Products," *Advances in Nuclear Science and Technology*, Vol. 10, Henley, E. J., and J. Lewins (eds.), Academic Press, Inc., New York, 1976.

to small releases of activity with minor consequences to public health and the environment could occur relatively frequently (say every 10 or 100 years on average for one reactor), the larger the release, the lower should be the probability or frequency of its occurrence; and for very large releases, the probability should be very low indeed. In Example 1, Chapter 13, we construct a curve of this type.

1.5 PUBLIC ATTITUDE AND SOCIETAL RISK MANAGEMENT

In these days of "full disclosure," risk identification and quantification studies conducted by skilled technicians are subject to public debate. The flow of information is illustrated in Fig. 1.3, which includes the headings "psychological" and "social aspects," i.e., the psychological well-being of individuals due to their perception of social risks, and their impact on

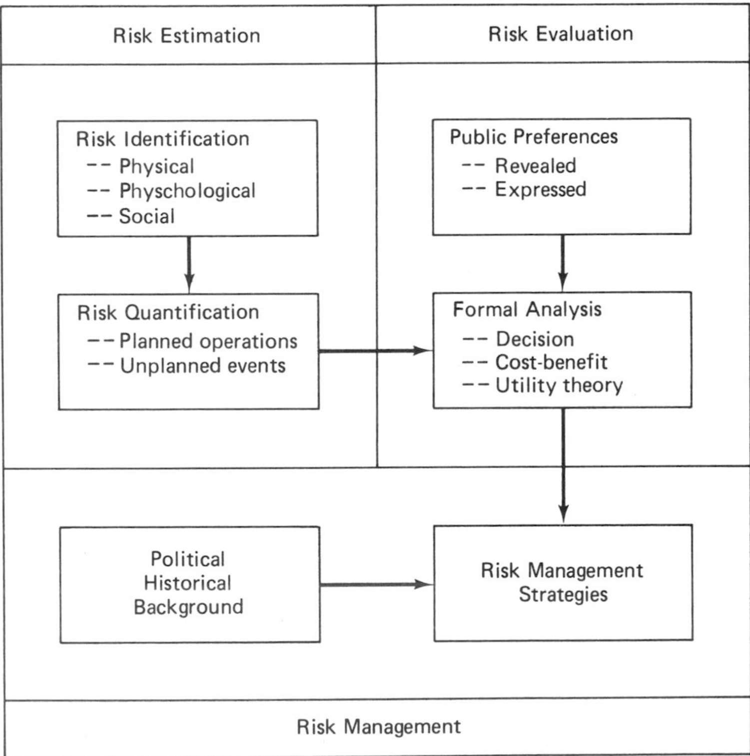

Figure 1.3. A risk assessment framework.

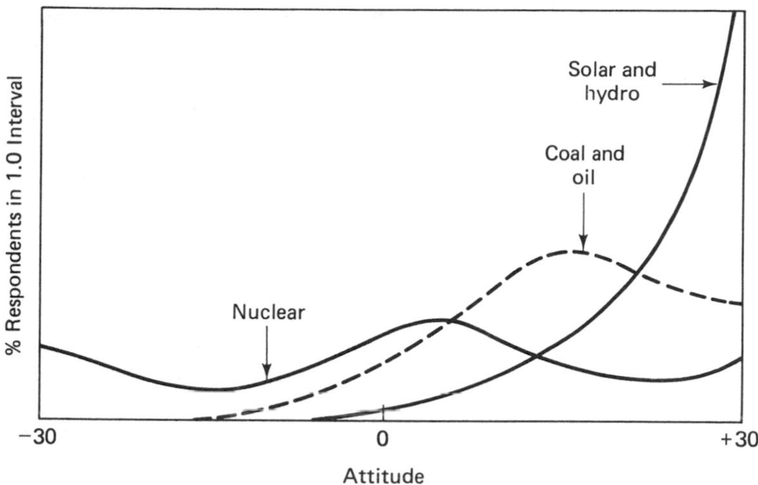

Figure 1.4. Frequency distribution of attitudes toward energy sources.

society and individuals. A well-known observation of public risk acceptance is that the acceptable level of mortality risk for voluntary exposure is a factor of 10^3 higher than for involuntary exposure.[6] This implies that railroad equipment is expected to be 1000 times "safer" than mountain climbing gear. Also, it is generally agreed that society views single, large-consequence events less favorably than the total of small events having the same risk. These are examples of "public preferences" in Fig. 1.3.

Observations of this sort are the result of attitude-based methodology used to identify the underlying determinants of public perceptions of technological systems. In general, a favorable attitude being in the direction of risk acceptance, an unfavorable one corresponding to risk aversion.

Figure 1.4 shows the frequency distribution of attitudes toward five energy systems.[7] Only three types of distribution were found: for hydropower and solar, attitudes were highly favorable; attitudes toward coal and oil were moderately favorable; however, attitudes toward nuclear energy were essentially normally distributed with the addition of clusters of highly positive and highly negative attitudes. This latter distribution reflects the degree of polarization on the nuclear issue. Studies like this are interesting but far from conclusive; indeed, they are suspect and may reflect the bias of interviewers. In the United States, whenever the public was given the opportunity to vote on the nuclear issue, they voted overwhelmingly in

[6]Starr, C., "Social Benefits vs. Technological Risk," *Science*, **165**, 1232–1238, 1969.

[7]Otway, H. J., "Towards the Development of Societal Risk Criteria," RSA/9/78, CEC, Joint Research Center, ISPRA, Italy.

TABLE 1.2 Life Valuation for Purposes of Cost-Benefit Analysis

Approaches	Typical Values	Some Limitations
(1) Implicit value	\$9,000–\$9,000,000	Assumes past decisions are optimal
(2) Human capital	\$100,000–\$400,000	Based solely on lifetime income Ignores individual preferences Discriminates against unproductive members of society
(3) Insurance premiums	Wide range	Does not take into account individual's interest in protecting his own life
(4) Court awards	\$250,000	Based on lost income
(5) Willingness to pay	\$180,000–\$1,000,000	Difficult to estimate Depends on risk situation

Summary: All measures depend to some extent on the lifetime earning potential of the individuals at risk and ignore perception of seriousness.

Conclusion: Cannot be rigorously determined. Choose value (say \$300,000) weight according to personal variables, third party interests and psychological factors.

favor of nuclear power. Opponents of nuclear power are apparently successful only in intimidating courts and bureaucratic agencies. In this context it is interesting to note that a recent public-opinion poll in America showed that the public had more confidence in engineers than in lawyers or doctors. Of course, the publicity given the Three Mile Island nuclear power plant accident may change this picture dramatically.

In Fig. 1.3, the second entry, under Risk Evaluation, "formal analyses," suggests that utility theory[8] or cost-benefit approaches can be used.[9] When the program being evaluated involves the potential saving or loss of human life then the quantification problem becomes one of placing a monetary value on human life.

[8]Garriba, S., and A. Ovi, "Statistical Utility Theory for Comparison of Nuclear Power Versus Fossil Power Plant Alternatives," *Nuclear Technology*, **34**, 18–37, 1977.

[9]Mishan, E. J., *Cost Benefit Analysis*, Allan and Unwin, London, 1971.

TABLE 1.3 ESTIMATED DEATHS PER GWy OUTPUT FROM VARIOUS ENERGY SOURCES

Fuel Source	Type of Energy Output[a]	Deaths per GWy Occupational	+	Public	=	Total
Methanol/ biomass	(m)	110	+	0	=	110
Windpower	(e)	(20 to 30)	+	(2 to 40)	=	(22 to 70)
Solar, photo-voltaic	(e)	(16 to 21)	+	(1 to 40)	=	(17 to 61)
Coal	(e)	(2 to 10)	+	(3 to 150)	=	(5 to 160)
Solar, thermal	(e)	(7 to 10)	+	(1 to 40)	=	(8 to 50)
Oil	(e)	(0.2 to 2)	+	(1.4 to 140)	=	(2 to 140)
Solar, space heating	(th)	(9 to 10)	+	0.4	=	(9 to 10)
Hydroelectric	(e)	(2 to 4)	+	(1 to 2)	=	(3 to 6)
Ocean, thermal	(e)	(2 to 3)	+	0.1	=	(2 to 3)
Nuclear	(e)	(0.2 to 1.3)	+	(0.04 to 0.24)	=	(0.25 to 1.5)
Natural gas	(e)	(0.1 to 0.4)	+	0	=	(0.1 to 0.4)

[a] Energy outputs as electrical (e), mechanical (m), or thermal (th).

There are basically five approaches which have been put forward for placing a monetary value on life-saving for use in cost-benefit analysis. They are (as summarized by Otway[10]):

Implicit Value: Life is valued according to the values implied by post-policy decisions to reduce mortality risk.

Human Capital: Life is evaluated as the discounted future earnings of those at risk.

Insurance: Life evaluated on the basis of individual life insurance decisions.

Court Awards: Awards made by courts to compensate for loss of life are used as the basis for life value figures.

Willingness to Pay: Values risk reduction by the public's willingness to pay for it.

The monetary values derived by using these methods are summarized in Table 1.2, along with limitations to the methods. All these methods, in one way or another, depend on the income of those at risk and the legal system. To derive a valid number for use in cost-benefit analysis, attention should also be given to the mode and timing of death and associated

[10] Otway, H. J., "An Interdisciplinary Approach to the Management of Technological Risk," RSA/2/78, CEC, Joint Research Center, ISPRA, Italy, 1978.

anxieties. In the United States, courts usually award smaller claim settlements for death than for incapacitation. The desirability of killing someone rather than maiming them was first taught by Machiavelli, so our courts have not advanced much during the last 500 years.

The last entry in Fig. 1.3, "risk management strategies," involves political, historical, and background information. It is the organizational aspect of the integration of technical and social systems. "Political judgment" is the key term, and the implications are disquieting. Studies by Ponchin and others have resulted in Table 1.3. This table shows that natural gas and nuclear power, the two power-generating modes subject to the most political action, are considerably more risk-free than other methods of power production.[11]

1.6 RISK LEGISLATION

Like the environmental impact studies required of industry by the government, most safety and risk studies are now done for the purpose of satisfying the public (i.e., government agencies) not for the purpose of reducing risk.* The current trend in risk legislation is to require quantitative assessments. Nuclear Regulatory Commission (NRC) RG 1.115 *Protection Against Low-Trajectory Turbine Missiles* (page 1.115-3) states "NRC staff considers 10^{-7} per year are acceptable risk rates for the loss of an essential system for a single event." Another interesting quantitative risk concept is proposed in RG 1.110, *Cost-Benefit Analysis for Radwaste Systems for LWR Nuclear Power Reactors* (p. 1.110-6): "Each applicant for a permit to construct a light-water cooled nuclear power reactor should demonstrate by means of a cost-benefit analysis that further reductions to the cumulative (collective) dose to the population within a 50-mile radius of the reactor site cannot be effected at an annual cost of $1000 per man-rem† or $1000 per man-thyroid-rem (or such less cost as demonstrated to be suitable for a particular case)."

The United States has pioneered the application of this "visionary" legislation to nuclear power and environmental matters.[12] Europe, however, is rapidly catching up in terms of legislation based on quantitative

[11]Ponchin, E. E., "The Risk Involved in Different Methods of Power Production," RSA/4/78, CEC, Joint Research Center, ISPRA, Italy, June 30, 1978.

*As proof of this statement we offer the observation that perhaps as few as 10% of the plants for which risk (or environmental impact) studies are made are eventually constructed.

†The abbreviation rem is for roentgen equivalent man, a measure of absorbed nuclear radiation.

[12]The term *visionary* was first applied to this type of legislation by Farmer, F. R., "Risk Quantification and Acceptability" *Nuclear Safety*, **17**, No. 4, July–August, 1976.

criteria and promises to outstrip the United States in extending such concepts to the licensing and inspection of chemical plants. Holland has already enacted such legislation, with England and France soon to follow.[13]

Editorially speaking, we do not believe that legislation will stop accidents any more than legislative prohibition of drinking, sex, or abortions were effective deterrents in a democracy: they merely raise the cost to the public. Indeed, if the only motivation for doing risk analysis is to satisfy a government regulation, then it represents an abrogation of personal liability on the part of the analyst, rather than a constructive, personal effort for which the analyst himself feels accountable.

1.7 RISK STUDY METHODOLOGY—PHASE I

System Definition. Risk arises because of an uncontrolled release of energy or toxic materials. Usually, certain parts of a plant are more likely to pose hazards than others; therefore, an early step in the analysis is to decompose the plant into subsystems to identify the sections or components which are likely sources of an uncontrolled release. Hence, the first two steps are:

Step 1. Identify the *hazard*(*s*). (Is it a toxic release, an explosion, a fire...?)

Step 2. Identify the *parts of the system which give rise to the hazard*(*s*). (Does it involve the chemical reactor, the storage tank, the power plant...?)

In identifying subsystems of the plant which give rise to a hazard it has been found useful to list *guide words* which stimulate the exercise of creative thinking. Robinson suggests looking at a process to see how it might deviate from design intent by applying the guide words:[14]

More of	Other than
Less of	As well as
None of	Reverse
Part of	Later than
	Sooner than

The only guideposts in achieving an understanding of system hazards are engineering judgment and a detailed grasp of the environment, the

[13]Vinck, W., "Status and Trends in the European Practices and Legislation for Safety and Environment," CEC, D. G. XII, Brussels, BSA 78/13, June 26, 1978.

[14]Robinson, B. W., "Risk Assessment in the Chemical Industry," RSA 5/78, CEC, Joint Research Center, ISPRA, Italy, 1978.

Hazardous Energy Sources

1. Fuels
2. Propellants
3. Initiators
4. Explosive charges
5. Charged electrical capacitors
6. Storage batteries
7. Static electrical charges
8. Pressure containers
9. Spring-loaded devices
10. Suspension systems
11. Gas generators
12. Electrical generators
13. rf energy sources
14. Radioactive energy sources
15. Falling objects
16. Catapulted objects
17. Heating devices
18. Pumps, blowers, fans
19. Rotating machinery
20. Actuating devices
21. Nuclear devices, etc.

Hazardous Processes and Events

1. Acceleration
2. Contamination
3. Corrosion
4. Chemical dissociation
5. Electrical
   - shock
   - thermal
   - inadvertent activation
   - power source failure
   - electromagnetic radiation
6. Explosion
7. Fire
8. Heat and temperature
   - high temperature
   - low temperature
   - temperature variations
9. Leakage
10. Moisture
    - high humidity
    - low humidity
11. Oxidation
12. Pressure
    - high pressure
    - low pressure
    - rapid pressure changes
13. Radiation
    - thermal
    - electromagnetic
    - ionizing
    - ultraviolet
14. Chemical replacement
15. Mechanical shock,
    etc.

Figure 1.5. Checklists of hazardous sources.

process, and the equipment. A knowledge of toxicity, safety regulations, explosive conditions, reactivity, corrosiveness, and flammabilities is fundamental. Checklists such as the one used by Boeing Aircraft and shown in Fig. 1.5 are a basic tool in identifying hazards.

It is necessary to put some boundary on the technical system and the environment under study. It is not reasonable, for example, to study in detail, risk parameters for the event of an airplane crashing into a distillation column. However, airplane crashes, seismic risk, and other low-probability hazards do enter into calculations for nuclear power plant risks because (a) one can protect against them, (b) theoretically, a nuclear power plant can kill more people than a distillation column. Therefore:

Step 3. Bound the study. (Will it include detailed studies of risks from sabotage, adversary action, war, public utility failures, lightning, earthquakes...?)

The objective of Phase I of a risk study is to *define the system* and to identify, in broad terms, the potential hazards.

Preliminary Hazards Analysis (PHA). Frequently, the Phase I study will involve more than a preliminary identification of the system elements or events that lead to hazards. If the analysis is extended to a more formal (qualitative) manner to include consideration of the event sequences which transform a hazard into an accident, as well as corrective measures and consequences of the accident, the study is called a *preliminary hazards analysis*.

In the aerospace industry, for example, the hazards, after they are identified, are characterized according to their effects. A common ranking scheme is:*

Class I Hazards: Negligible effects
Class II Hazards: Marginal effects
Class III Hazards: Critical effects
Class IV Hazards: Catastrophic effects

The next step is to decide on accident prevention measures, if any, that must be taken to eliminate Class IV, and possibly Class III and II hazards. The decisions to be considered are shown in the form of a *decision tree* in Fig. 1.6. One can decide to take corrective action in the form of equipment design changes, or redirection of goals or functions, and/or to provide contingency action in the form of protective and alarm systems, fire walls, etc.

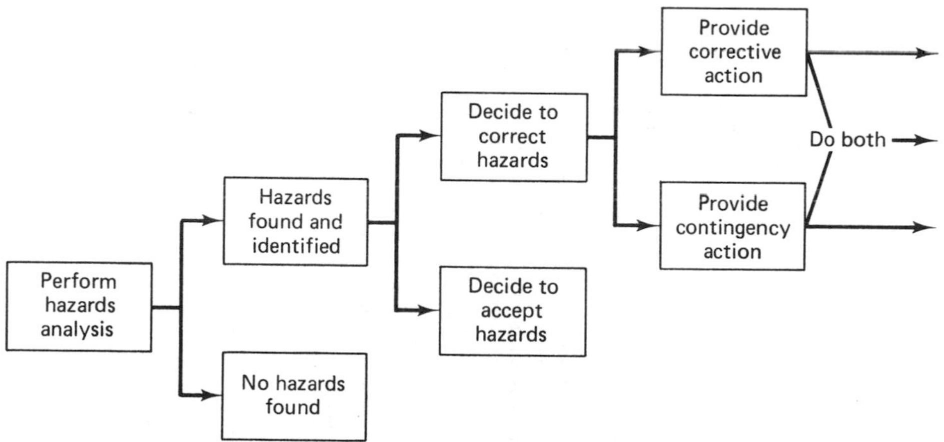

Figure 1.6. Hazards analysis decision tree.

*Similar categorizations, when applied to specific components, are called *criticality categorizations* (see Section 1.11 of this chapter).

Boeing Company Format

1. Subsystem or function	2. Mode	3. Hazardous element	4. Event causing hazardous element	5. Hazardous condition	6. Event causing hazardous condition	7. Potential accident	8. Effect	9. Hazard class	10. Accident prevention measures			11. Validation
									10A1 hardware	10A2 procedures	10A3 personnel	

1. Hardware or functional element being analyzed.
2. Applicable system phases or modes of operation.
3. Elements in the hardware or function being analyzed that are inherently hazardous.
4. Conditions, undesired events, or faults that could cause the hazardous element to become the identified hazardous condition.
5. Hazardous conditions that could result from the interaction of the system and each hazardous element in the system.
6. Undesired events or faults that could cause the hazardous condition to become the identified potential accident.
7. Any potential accidents that could result from the identified hazardous conditions.
8. Possible effects of the potential accident, should it occur.
9. Qualitative measure of significance for the potential effect on each identified hazardous condition, according to the following criteria: Class I - Safe - Condition(s) such that personnel error, deficiency/inadequacy of design, or malfunction will not result in major degradation and will not produce equipment damage or personnel injury. Class II - Marginal - Condition(s) such that personnel error, deficiency/inadequacy of design, or malfunction will degrade performance. Can be counteracted or controlled without major damage or any injury to personnel. Class III - Critical - Condition(s) such that personnel error, deficiency/inadequacy of design, or malfunction will degrade performance, damage equipment or result in a hazard requiring immediate corrective action for personnel or equipment survival. Class IV - Catastrophic - Condition(s) such that personnel error, deficiency/inadequacy of design, or malfunction will severely degrade performance and cause subsequent equipment loss and/or death or multiple injuries to personnel.
10. Recommended preventive measures to eliminate or control identified hazardous conditions and/or potential accidents. Preventive measures to be recommended should be hardware design requirements, incorporation of safety devices, hardware design changes, special procedures, personnel requirements.
11. Record validated preventive measures and keep aware of the status of the remaining recommended preventive measures. Complete by answering (1) has the recommended solution been incorporated? and (2) is the solution effective?

Figure 1.7.(a). Suggested format for preliminary hazard analysis: Boeing Company format.

Hazardous element	Triggering event 1	Hazardous condition	Triggering event 2	Potential accident	Effect	Corrective measures
1. Strong oxidizer	Alkali metal perchlorate is contaminated with lube oil	Potential to initiate strong redox reaction	Sufficient energy present to initiate reaction	Explosion	Personnel injury; damage to surrounding structures	Keep metal perchlorate at a suitable distance from all possible contaminants
2. Corrosion	Contents of steel tank contaminated with water vapor	Rust forms inside pressure tank	Operating pressure not reduced	Pressure tank rupture	Personnel injury; damage to surrounding structures	Use stainless steel pressure tank; locate tank at a suitable distance from equipment and personnel

Figure 1.7.(b). Format for preliminary hazards analysis: 1. Hazardous situation: alkali metal perchlorate is contaminated by a spill of lube oil; 2. Hazardous situation: moisture inside pressurized steel tank.

A common format for a PHA are entry formulations such as shown in Figs. 1.7(a) and 1.7(b). These are partially narrative in nature, listing both the events and the corrective actions that might be taken.

Of particular importance in a PHA are equipment and subsystem interface conditions. Lambert[15] cites a classic example that occurred in the early stages of ballistic missile development in the United States. Four major accidents occurred as the result of numerous interface problems. In each accident, the loss of a multimillion dollar missile/silo launch complex resulted.

The failure of Apollo 13 was due to a subtle interface condition. During prelaunch, improper voltage was applied to the thermostatic switches to the heater of oxygen tank #2. This caused Teflon on the wires leading to a fan inside the tank to crack. During flight, the switch to the fan was turned on, a short circuit resulted that caused Teflon to ignite and, in turn, caused the oxygen tank to explode.

In this section, we have used the terms *system definition* and *preliminary hazards analysis* in a somewhat arbitrary manner. Safety-study methodology varies between industries and among companies. Unless specific government regulations dictate the procedures to be used, industrial practice and terminologies relating to Phase I studies will vary widely. In general, the PHA represents a first attempt to identify the (gross) system hardware and events which can lead to hazards, while the system is still in a preliminary design stage. Detailed event analysis is commonly done by

[15]Lambert, H. E., "Fault Tree in Decision Making in Systems Analysis," Lawrence Livermore Laboratory, UCRL-51829, 1975.

fault tree methodology (Section 1.8, and Chapter 2) after the system is fully defined, and detailed hardware fault analysis techniques such as *failure modes* and *effects analysis* (Section 1.10) are also applied at a later design stage. *Fault hazards analysis* is a term somewhat analogous to PHA and is applied to PHA-type studies performed on subsystems.

1.8 RISK STUDY METHODOLOGY—PHASE II: IDENTIFYING ACCIDENT SEQUENCE, EVENT TREES, FAULT TREES

A Phase II study is usually started after a choice of hardware and a system configuration have been made. Two of the common analytical techniques, event trees and fault trees, are described in this section. Two other Phase II methodologies, failure modes and effects analysis and criticality analysis, are considered in Sections 1.10 and 1.11.

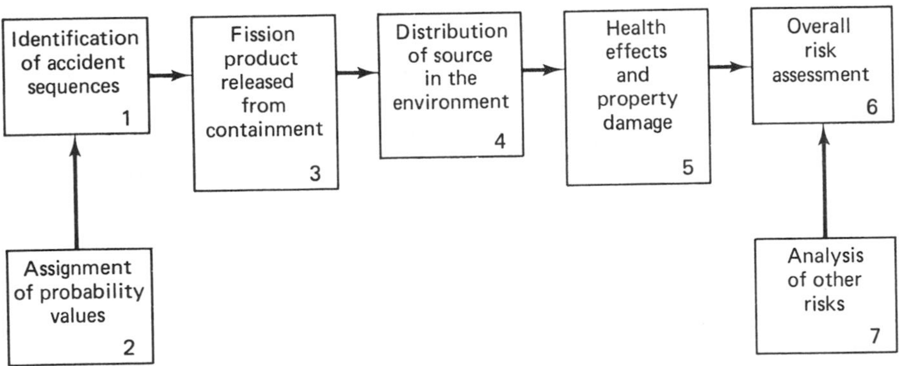

Figure 1.8. Seven basic tasks in reactor safety study.

Consider as an example the reactor safety study WASH 1400. In Phase I of the study it was determined that the overriding risk was that of a radioactive (toxic) fission product release. The second phase begins, as shown in Fig. 1.8, with the task of identifying the *accident sequences*: the different ways in which a release might occur.[16] The Phase I study indicated that the critical portion of the nuclear reactor system, i.e., the subsystem whose failure initiates the risk, is the reactor cooling system, and so the risk study begins by following the potential course of events beginning with (coolant) "pipe breaks," the *initiating event* having a probability of P_A in Figure 1.9.

[16]Wall, Ian A., "Some Insights from the Reactor Safety Study," Office of Nuclear Regulatory Research, U. S. NRC.

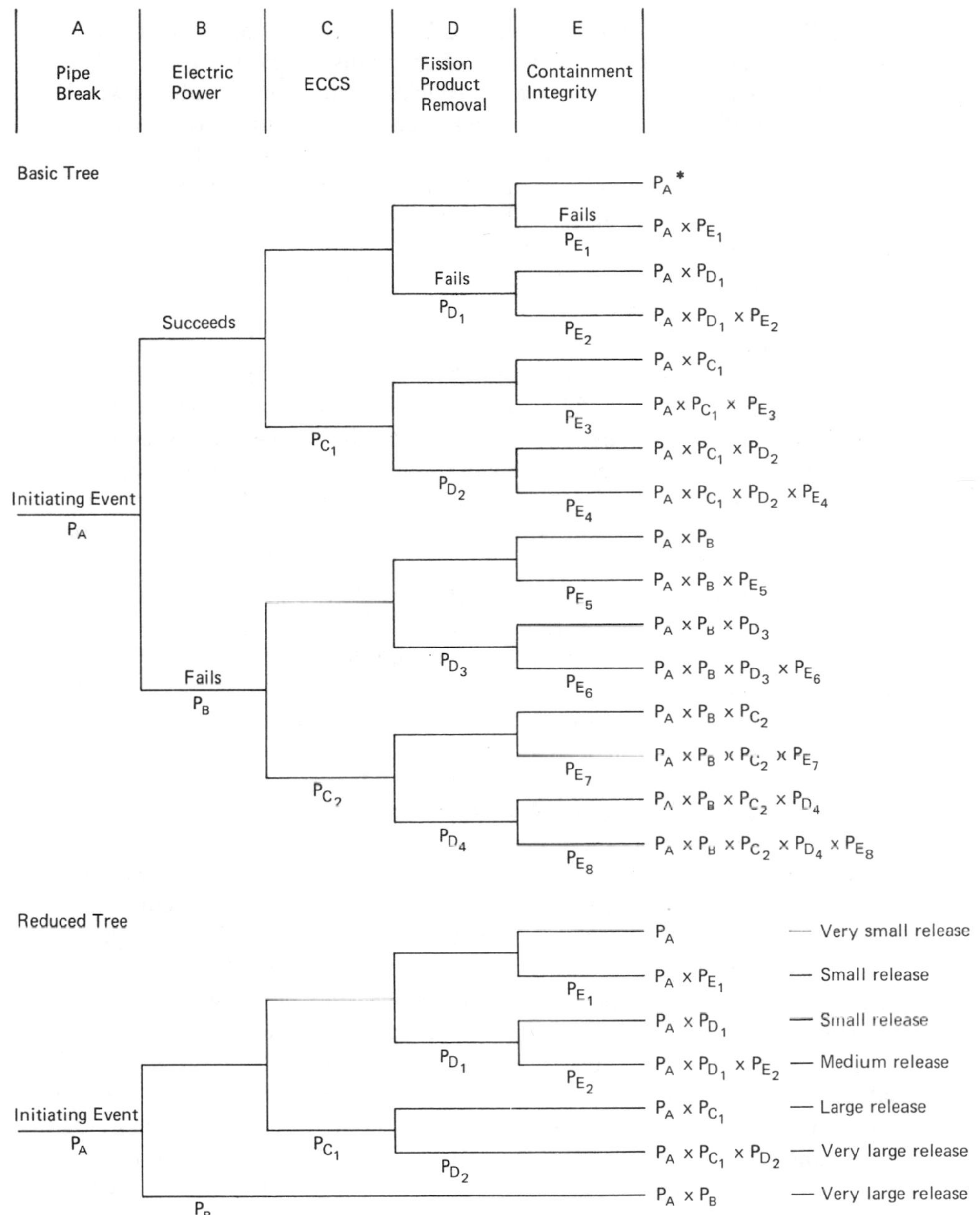

Figure 1.9. Pruning an event tree.

* This is an approximation to $P_A(1-P_B)\,(1-P_{C_1})\,(1-P_{D_1})\,(1-P_{E_1})$. Note that all P's are very small. Similar approximations are made throughout.

The WASH 1400 study has obtained by "backward logic," using *fault tree* techniques, the numerical value of P_A. This methodology, which is described in Chapter 2, seeks out the equipment or human failures which could result in the "top event," the loss of coolant. Failure rates, based on experience data for components, human error, and contribution from testing and maintenance are combined appropriately by means of the fault trees to determine the unavailability and unreliability of the engineered systems. This procedure is identified as task 2 in Fig. 1.8.

Now, let us return to box 1, by considering the *event tree* for a loss of coolant accident (LOCA) in a typical nuclear power plant (Fig. 1.9). The accident starts with a pipe break having a probability of occurrence, P_A. The potential course of events that might follow such a pipe break are then examined. The upper part of Fig. 1.9 shows the basic tree, which shows all possible alternatives. At the first branch, the status of the electric power is considered. If it is available, the next-in-line system, the emergency core cooling system (ECCS), is studied. Failure of the ECCS results in fuel meltdown and varying amounts of fission product releases, depending on the containment integrity.

For a binary analysis where a system either succeeds or fails, the number of potential accident sequences is 2^{N-1}, where N is the number of systems being considered. In practice, as shown in the following discussion, the basic tree may be pruned by simple engineering logic to the reduced tree shown in the lower part of Fig. 1.9. One of the first things of interest is the availability of electrical power. The question is, what is the probability, P_B, of electrical power failing, and how would it affect other safety systems? If there is no electric power, the emergency core cooling pumps, the sprays, in fact, none of the post accident functions can be performed. Thus, no choices are shown in the simplified event tree when electric power is unavailable and a very large release can occur whose probability is $P_A \times P_B$. In the event that the unavailability of electric power is dependent on the pipe that broke, the probability P_B should be calculated as a conditional probability to reflect such a dependency. It is important to recognize that event trees are used to define accident sequences that involve the complex interrelationships among engineered safety systems. They are constructed by using *forward logic*: We ask the question "What happens if the pipe breaks?" The probability, such as P_B, is obtained by *backward* (*fault tree*) *logic* by asking the question "How could the electric power fail?" and constructing a fault tree for the electric power subsystem. Forward logic, such as was used in event tree construction and failure modes and effects analysis (FMEA) is often referred to as *inductive* logic, whereas the type of logic used in fault tree analysis is *deductive*. A PHA involves a mixture of inductive and deductive logic.

If electric power is available, the next choice for study is the availability of the emergency core cooling system. It can work, or it can fail, and its unavailability, P_{C1}, would lead to the sequence shown in Fig. 1.9. Notice that there are still choices available which can affect the course of the accident. If the radioactivity removal systems operate, a smaller radioactive release would result than if they were to fail. Of course, their failure would in general produce a lower probability accident sequence than one in which they operated. By working through the entire event tree, we produce a spectrum of release magnitudes and their probabilities for the various accident sequences. The top line of the tree is the conventional design basis LOCA that is analyzed in the licensing process for each reactor. In this sequence, the pipe is assumed to break but each of the safety systems are assumed to operate.

Decision trees are a special case of event tree models. In event trees, working states are not considered, so the sum of all probabilities do not add up to one. In decision trees, the system outcomes are expressed in terms of component states, so the outcomes must be coherent; i.e., they must add up to one. Decision trees can be used if the probabilities of component states are independent or if there are multiple component states or unilateral (one-way) dependencies. They cannot be used in the case of two-way dependencies, and provide no logical method of choosing initiating event.

Example .[17]

Figure 1.10 shows a series system comprising a pump and a valve having successful working probabilities of 0.98 and 0.95, respectively, and the associated decision tree. Note that, by convention, desirable outcomes branch upward and undesirable outcomes downward. The tree is read from left to right.

If the pump is not working, the system has failed, regardless of the valve state. If the pump is working, we examine the question of whether the valve is working at the second nodal point.

The probability of system success is $(0.98)\times(0.95)=0.931$. The probability of failure is $(0.98)\times(0.05)+(0.02)=(0.069)$; the total probability of the two system states adding up to one.

Another way of obtaining this result is via a *truth table* (see Chapter 7). For the pump and valve:

Pump State	Valve State	Probability of System Working	Probability of System Failed
Working	Working	$(0.98)\times(0.95)$	
Failed	Working		$(0.02)\times(0.95)$
Working	Failed		$(0.48)\times(0.05)$
Failed	Failed		$(0.02)\times(0.05)$
		Total: 0.931	0.069

[17]After Lambert, H. E., "Systems Safety Analysis and Fault Tree Analysis," Lawrence Livermore Laboratory, RLPT UCID-16238, 1973.

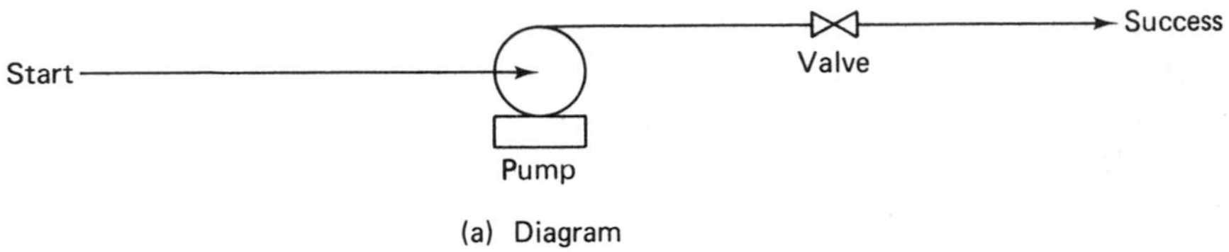

(a) Diagram

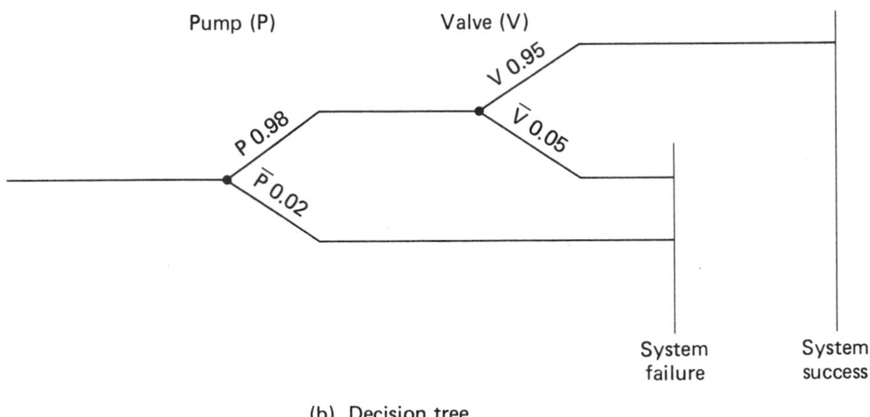

(b) Decision tree

Figure 1.10. Diagram and decision tree for two components.

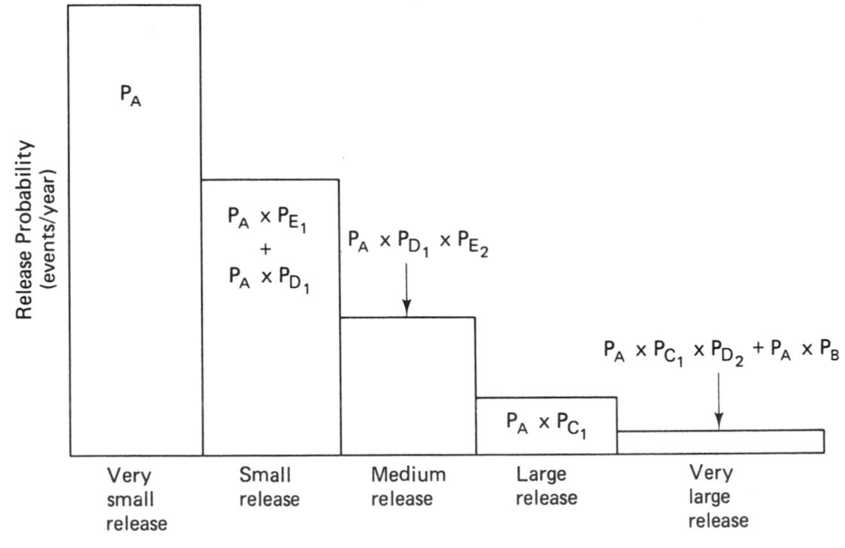

Figure 1.11. Frequency histogram for release magnitude.

1.9 RISK STUDY METHODOLOGY—PHASE III: CONSEQUENCE ANALYSIS

Taking as an example the WASH 1400 study, in this final phase of the risk study we: 1. calculate the amount of toxic material, or energy released by each of the accident paths (task 3, Fig. 1.8); 2. follow the trajectory of the lethal toxins, shock waves, or flame fronts (task 4, Fig. 1.8); 3. assess the health effects and property damage (task 5, Fig. 1.8); 4. make some overall

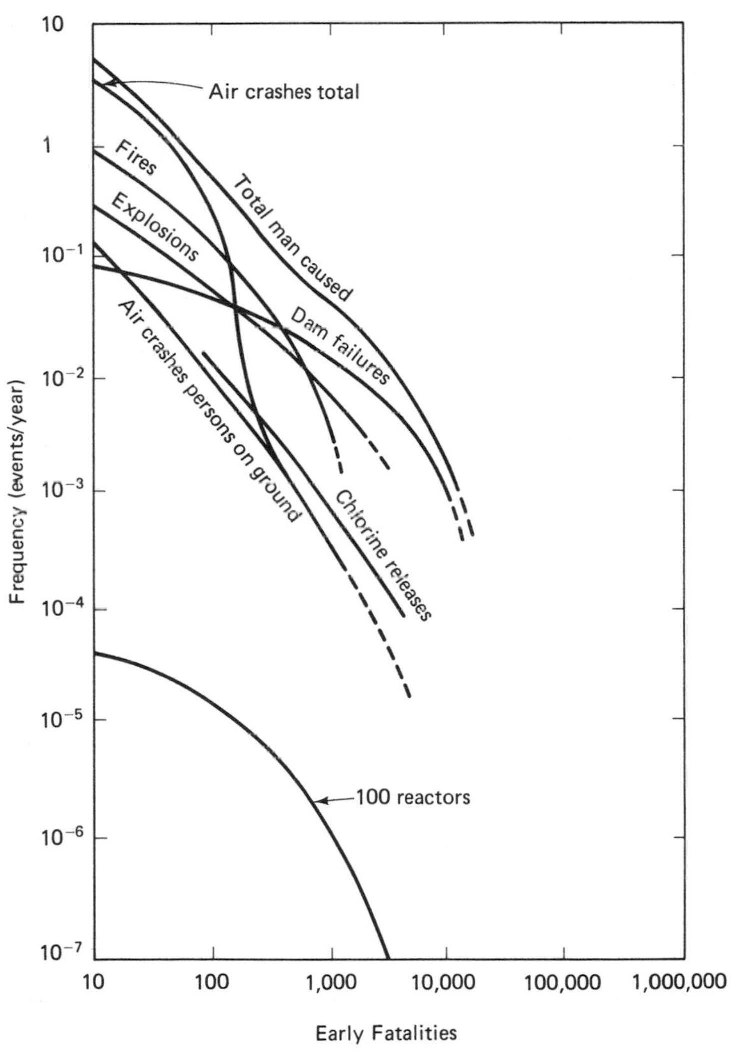

Figure 1.12. Frequency of man-caused events involving fatalities.

judgment of the technology (task 6, Fig. 1.8) based on a comparison with other societal risks (task 7, Fig. 1.8).

The result of the task 3 effort is the histogram of probability versus release magnitude shown in Fig. 1.11. Conceptually, this histogram is a Farmer curve, which represents an assessment of the reactor design and is the key input into the consequence model. All other inputs are site-specific, involving meteorlogical and demographic data.

We see the outcome of task 6, in Figs. 1.12 and 1.13. Here the data in Table 1.1 was used to compare the risk from nuclear reactors (bottom curve) with other man-caused events involving fatalities and property damage.

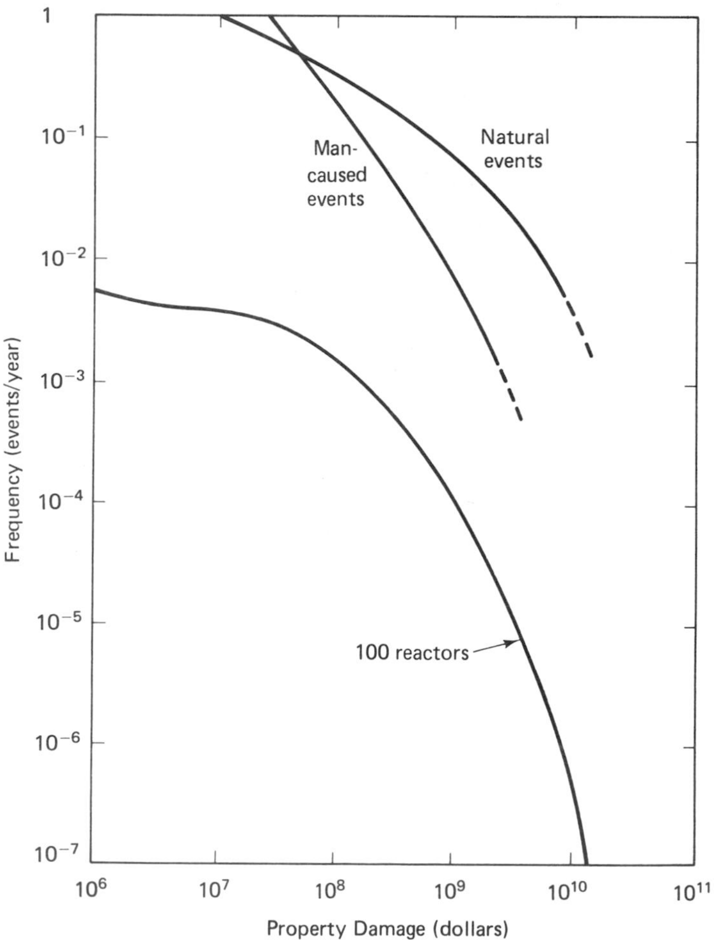

Figure 1.13. Frequency of accidents involving property damage.

Although these nuclear risks are very low, we see daily substantiation in terms of the safe operation of the hundreds of reactors in operation throughout the world.

1.10 OTHER RISK ANALYSIS TECHNIQUES: FAILURE MODES AND EFFECTS ANALYSIS (FMEA)

FMEA is an inductive analysis that systematically details, on a component-by-component basis, all possible failure modes and identifies their resulting effects on the system. Possible single modes of failure or malfunction of each component in a system are identified and analyzed to determine the effect on surrounding components and the system.

This technique is used to perform single random failure analysis as required by IEEE 279-1971, 10 CFR 50, Appendix K, and regulatory guide 1.70, Revision 2. FMEA can be much more detailed than a fault tree, since every mode of failure of every component must be considered. A relay, for example, can fail by:[18]

Failure modes for relay:

- —contacts stuck closed
- —contacts slow in opening
- —contacts stuck open
- —contacts slow in closing
- —contact short circuit
    - —to ground
    - —to supply
    - —between contacts
    - —to signal lines
- —contacts chattering, intermittent contact
- —contact arcing, generating noise
- —coil open circuit
- —coil short circuit
- —coil resistance
    - —low
    - —high
- —coil overheating
- —coil short circuit
    - —to supply
    - —to contacts
    - —to ground
    - —to signal lines
- —overmagnetized, or excessive hysteresis (same effect as contacts stuck closed or slow in opening)

[18] R. Taylor, RISØ National Laboratory, Roskilde, Denmark, private communication.

Functional level				Diagram	Nuclear Services Corporation		Program TMI Unit 1 ECCS	
System Reactor core flooding				Gilbert Associates, Inc. Piping Flow Diagram C-302-711-REV 12 core flooding			Report No. NSC-LS&R-GPU-0104-1	
Sub-system							Prepared by	
Equipment Mechanical equipment							Date ___ Rev ___ Reviewed ___ Date ___ Approved ___ Date ___	

No.	Name	Failure mode	Cause	Symptoms and local effects including dependent failures	Method of detection	Inherent compensating provision	Effect upon system	Remarks and other effects
500	CF-V1A or CF-V1B CF headers isolation valves	Open (FO)	Short, operator error, testing error	No effect on the operation of the system.	Status lights and annun-ciators during shutdown.	Check valves prevent backflow from reactor.	None.	D-1 type failure
		Closed (FC)	Open circuit, etc.	One leg of the Core Flooding becomes inoperable. System fails to meet the minimum ECCS requirements.	Status lights and annunciators.	Operator action is required to close these normally locked open valves.	CF portion of ECCS fails.	A-2 type failure
501	CF-V4A or CF-V4B CF headers check valves	Open (FO)	Dirt, metal burrs, etc.	No effect on the system. Reactor coolant may enter into the flooding tanks.	High level and pressure alarms at the flooding tanks.	Downstream check valves are operable.	None	D-1 type failure
		Closed (FC)	Dirt, metal burrs, etc.	One leg of the Core Flooding becomes inoperable. System fails to meet the minimum ECCS requirements.	Tank level and pressure indicator only after a LOCA.	None	CF portion of ECCS fails	A-2 type failure
502	CF-V5A or CF-V5B injection header check valves	Open (FO)	Dirt, metal burrs, etc.	None. Increased possibility that Reactor coolant will flow into other systems.	Surveillance	Upstream check valves are operable.	None	D-1 type failure
		Closed (FC)	Dirt, metal burrs, etc.	One leg of the Core Flooding and Decay Heat Removal Systems becomes inoperable. System fails to meet the minimum ECCS requirements.	Surveillance or alarms during LOCA.	None	CF portion of ECCS fails	A-2 type failure
503	CF-V2A or CF-V2B or CF-V20A or CF-V20B sampling valves	Open (FO)	Short circuit, operator error, etc.	Valve remains open.	Sample run, valve position indicators.	Redundant valves close on ES signal.	None	D-1 type failure
		Closed (FC)	Open circuit, etc.	None. Sampling becomes inoperable	Lack of sample or valve position indicators.	Manual grab sampling	None	D-2 type failure

Figure 1.14. Failure mode and effects analysis. (*Courtesy of Nuclear Services Corp.*)

FAILURE MODES AND EFFECTS ANALYSIS

1. SUBSYSTEM ________ 2. DWG. NR. ________ 3. PREPARED BY ________ 4. DATE ________

ITEM	FAILURE MODES	CAUSE OF FAILURE	POSSIBLE EFFECTS	PROBABILITY OF OCCURRENCE	CRITICALITY	POSSIBLE ACTION TO REDUCE FAILURE RATE OR EFFECTS
Motor case	Rupture	a. Poor workmanship b. Defective materials c. Damage during transportation d. Damage during handling e. Overpressurization	Destruction of missile	0.0006	Critical	Close control of manufacturing processes to ensure that workmanship meets prescribed standards. Rigid quality control of basic materials to eliminate defectives. Inspection and pressure testing of completed cases. Provision of suitable packaging to protect motor during transportation.
Propellant grain	a. Cracking b. Voids c. Bond separation	a. Abnormal stresses from cure b. Excessively low temperatures c. Aging effects	Excessive burning rate; overpressurization; motor case rupture during otherwise normal operation	0.0001	Critical	Carefully controlled production. Storage and operation only within prescribed temperature limits. Suitable formulation to resist effects of aging.
Liner	a. Separation from motor case b. Separation from motor grain or insulation	a. Inadequate cleaning of motor case after fabrication b. Use of unsuitable bonding material c. Failure to control bonding process properly	Excessive burning rate Overpressurization Case rupture during operation	0.0001	Critical	Strict observance of proper cleaning procedures. Strict inspection after cleaning of motor case to ensure that all contaminants have been removed.

Figure 1.15. Failure modes and effects analysis. (Hammer, W., *Handbook of System and Product Safety*, Prentice-Hall, Inc., p. 153, 1972. By permission.)

In addition, checklists for each category of equipment must be devised. For tanks, vessels, and pipe sections, a possible check list is:

Variables: Flow, quantity, temperature, pressure, pH, saturation, etc.
Services: Heating, cooling, electricity, water, air, control, N_2, etc.
Special states: Maintenance, start-up, shut-down, catalyst change, etc.
Changes: Too much, too little, none, water hammer, non-mixing, deposit, drift, oscillation, pulse, fire, drop, crash, corrosion, rupture, leak, explosion, wear, opening by operator, overfull with liquid.
Instrument: Sensitivity, placing, response time.

Figure 1.14 is part of a FMEA carried out by the Nuclear Services Corporation, Campbell, California, for the Three Mile Island Unit 1 emergency cooling system, for the GPU Service Corporation. Figure 1.15 offers a slightly different format for the FMEA. Both formats are similar to those used in a preliminary hazards analysis, the primary difference being the greater specificity and degree of resolution of the FMEA.

1.11 OTHER RISK ANALYSIS TECHNIQUES: CRITICALITY ANALYSIS

In both Figs. 1.14 and 1.15, each component is labeled with respect to its critical importance to mission operation. *Criticality* is rated in more than one way and for more than one purpose, as was illustrated in Section 1.7, where a criticality ranking for system hazards was given. The Society of Automotive Engineers (SAE) in *Aerospace Recommended Practice (ARP) 926* categorizes criticality of component failure modes as:

Category 1: Failure resulting in potential life.
Category 2: Failure resulting in potential mission failure.
Category 3: Failure resulting in potential delay or loss of operational availability.
Category 4: Failure resulting in excessive unscheduled maintenance.

Criticality categorization is an obvious "next step" after a FMEA, and is described in this way in ARP 926, the combination being termed a *FMEA CA—failure modes and effects and criticality analysis*. Component rankings can be achieved by computing a *criticality number* C_r:

$$C_r = \sum_{n=1}^{N} \beta\alpha K_E K_A \lambda_G t \times 10^6, \qquad n = 1, 2, \ldots, N$$

where C_r = criticality number for the system component in losses per million trials,
n = critical failure modes in the system component that fall under a particular loss statement,

N = last critical failure mode in the system component under loss statement,

λ_G = generic failure frequency of the component in failures per hour or cycle,

t = operating time in hours or number of operating cycles of the component per mission,

K_A = operational factor that adjusts λ_G for the difference between operating stresses when λ_G was measured and the operating stresses under which the component is going to be used,

K_E = environmental factor that adjusts λ_G for difference between environmental stresses when λ_G was measured and the environmental stresses under which the component is going to be used.

Note: For simplified uses, omit K_E, K_A, and use λ_G as the estimated failure rate for the given failure mode and operating condition.

α = failure mode ratio of critical failure mode. The failure mode ratio is that fraction of λ_G attributable to the critical failure mode,

β = conditional probability that the failure effects of the critical failure mode will occur, given that the critical failure mode has occurred. Values of β should be selected from an established set of ranges:

Failure Effects	Typical Value of Beta
Actual loss	100%
Probable loss	>10% to <100%
Possible loss	>0% to 10%
None	0%

10^6 = factor that transforms C_r, from losses per trial to losses per million trials, so C_r will normally be greater than one.

Note that this ranking method places no "value" on possible consequences or damage. Its primary usefulness is for achieving system upgrades by identifying:[19]

—Which items should be given more intensive study for elimination of the hazard that could cause the failure and for fail-safe design, failure-rate reduction, or damage containment.

—Which items require special attention during production, require tight quality control, and need protective handling at all times.

[19]Hammer, W., *Handbook of System and Product Safety*, Prentice-Hall, Inc., Englewood Cliffs, N. J., p. 157, 1972.

— Special requirements to be included in specifications for suppliers concerning design, performance, reliability, safety, or quality assurance.
— Acceptance standards to be established for components received at a plant from subcontractors and for parameters that should be tested most intensively.
— Where special procedures, safeguards, protective equipment, monitoring devices, or warning systems should be provided.
— Where accident prevention efforts and funds could be applied most effectively. This is especially important, since every program is generally limited by the availability of funds.

1.12 OTHER RISK ANALYSIS TECHNIQUES: HAZARDS AND OPERABILITY STUDIES

In a sense, a *hazards and operability study* is an extended FMEA technique, the extension being in the direction of including operability factors in addition to equipment fault modes.

Robinson[20] describes the use of this technique at the Imperial Chemical Industries, Ltd. In their terminology, this is a "Hazard Study III," Hazard Study I being similar to the preliminary, Phase I Risk Study (system definition) described in Section 1.6 of this chapter, and Phase II being a fault tree analysis carried out on a subsystem level.

> Hazard Study III is a detailed failure mode and effect analysis of the Piping and Instrument (P & I) line diagram. A team of four or five people study the P & I line diagram in a formal and systematical manner. The team includes the process engineer responsible for the chemical engineering design; the project engineer responsible for the mechanical engineering design and having control of the budget; the commissioning manager who has the greatest commitment to making the plant a good one and who is usually appointed at a very early stage of the project design; a hazard analyst who guides the team through the hazard study and quantifies any risks as necessary.
>
> This team studies each individual pipe and vessel, in turn, using a series of guide words to stimulate creative thinking about what would happen if the fluid in the pipe were to deviate from the design intention in any way. The guide words which we use for continuous chemical plants include high flow, low flow, no flow, reverse flow, high and low temperature and pressure and any other deviation of a parameter of importance. Maintenance, commissioning, testing, start-up, shutdown and failure of services are also considered for each pipe and vessel.

[20]Robinson, B. W., "Risk Assessment," RSA 5/78, CEC.

This in-depth investigation of the line diagram is a key feature of the whole project and obviously takes a lot of time—about 200 man hours per $2,000,000 capital. It is very demanding and studies, each lasting about $2\frac{1}{2}$ hours, can only be carried out at a rate of about two or three per week. On a multimillion dollar project, therefore, the studies could extend over many weeks or months. Problems identified by the hazard study team are referred to appropriate members of the team or to experts in support groups. If, during the course of this study, we uncover a major hazard which necessitates some fundamental redesign or change in design concept, the study will be repeated on the redesigned line diagram. This iterative process can lose us much time in the design of the project. By carrying out the Hazard Study II at a much earlier stage, we ensure that major hazards are taken care of before the detailed design stage. As a result the Study III can be done on frozen line diagrams and problems identified rarely disrupt the project to any significant degree. Detailed design and procurement can, therefore, continue with confidence while at the same time many operability, maintenance, start-up and shutdown problems are identified and dealt with satisfactorily.

1.13 OTHER RISK ANALYSIS TECHNIQUES: CAUSE-CONSEQUENCE ANALYSIS

*Cause-consequence diagrams** were invented at the RISØ Laboratories in Denmark. Construction starts with a choice of a *critical event*. Critical events are chosen so as to be convenient starting points for analysis, with most sequential problems following the critical event. Suitable critical events leading to a hazard might be:

— Disturbance of a major plant variable in some vessel or container.
— Breadth of a pressure or containment boundary.
— Start of a batch process or start of a startup or shutdown procedure.
— An event which activates a safety system.

The "consequence tracing" part of the cause-consequence analysis involves taking the initial event, and following the resulting chains of events through the system. At various steps, the chains may branch, taking two paths. For example, the start of a fire may lead to two chains of events—gradual destruction of the plant, and at the same time to activation of fire alarms, and to calling the fire departments. The chains of events may take alternative forms, depending on different conditions. For example, the progress of a fire may depend on whether there is a traffic jam to prevent the fire department from reaching the fire.

*Figure 13.1 shows an example of a cause-consequence diagram.

The procedure for construction of a consequence diagram is to take first the initiating event, and later each event so far developed and ask:

— Under what conditions does this event lead to further events?
— What are the alternative plant conditions which lead to different events?
— What other components does the event affect? Does it affect more than one component?
— What further event does this event cause?

Cause-consequence technology is a marriage of fault trees (to show causes) and event trees (to show consequences), all taken in their natural sequence of occurrence. Like fault trees, their construction will be described in detail in subsequent chapters.

1.14 OPTIMAL ALLOCATION OF RISK-REDUCTION RESOURCES

The optimal allocation of limited national or industrial funds to reduce various risks to human life and health is a difficult task. One way of evaluating a risk-reduction program is to compare estimated costs with expected benefits, both measured in dollars. As discussed previously in this chapter, this kind of analysis is controversial, since it requires placing a price on life itself.

A new approach developed at the General Motors Research Laboratory (Chem. and Eng. News, p. 82, July 23, 1979), avoids this problem by focusing on longevity, and rests on the simple premise that since all life inevitably ends, no amount of risk-reduction can *save* lives—only *lengthen* them. The method involves using data for all categories of mortality risks and determining the effect on longevity of each category independently. The results can be summarized for each risk by the equation

$$\text{Average years of longer life} = 0.2 \times \text{Annual deaths per million population}$$

This provides a perspective of days or years gained from risk-reduction programs and, combined with cost estimates, it helps rate the effectiveness of programs.

Figure 1.16, which was constructed by General Motors scientists, compares the cost-effectiveness of several medical, environmental, and safety programs presently under consideration by the government. The chart demonstrates the extreme variations in the cost of extending life by implementing the various options.

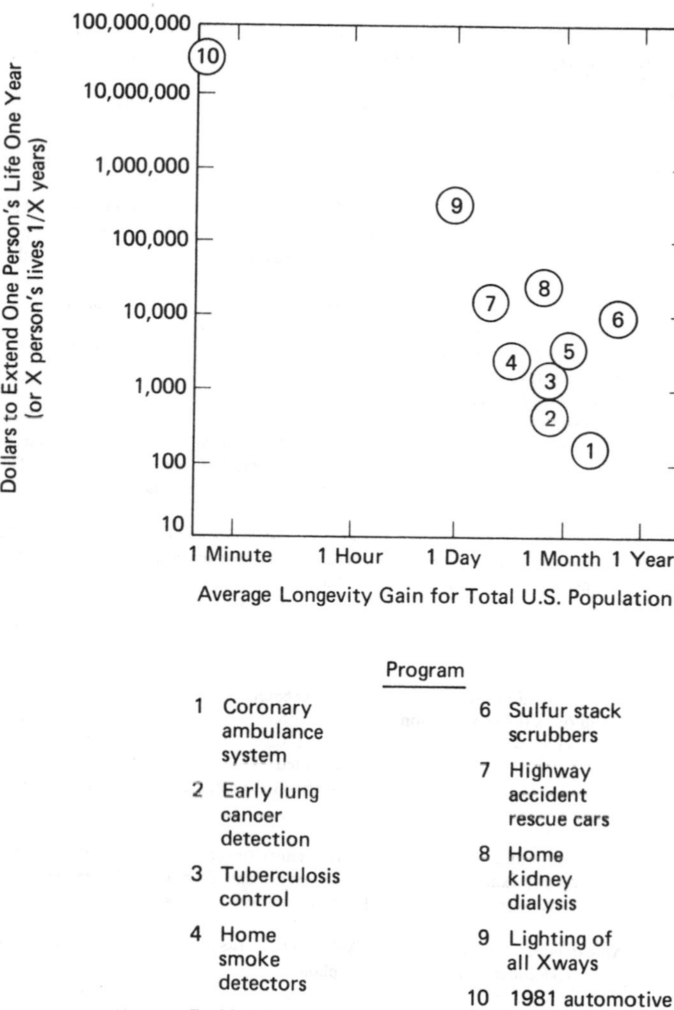

Figure 1.16. Cost of extending human life.

1.15 SUMMARY AND COMPARISON OF TECHNIQUES

Table 1.4 is a summary of the various risk analysis techniques described in this chapter. It deserves careful study, and we recommend that the reader restudy this table (and this chapter) after mastering the technical material developed in this text.

TABLE 1.4 Summary and Risk Study Comparison of Techniques

Method	Characteristics	Advantages	Disadvantages
Preliminary hazards analysis	Defines the system hazards and identifies elements for FMEA and fault tree analysis. Overlaps with FMEA and criticality analysis.	A required first step.	None.
Failure modes and effects analysis	Examines all failure modes of every component. Hardware oriented.	Easily understood. Well accepted, standardized approach; non-controversial, non-mathematical.	Examines non-dangerous failures. Time consuming. Often, combinations of failures and human factors not considered.
Criticality analysis	Identifies and ranks components for system upgrades.	Well standardized technique. Easy to apply and understand. Nonmathematical.	Follows FMEA. Frequently does not take into account human factors, common cause failures, system interactions.
Fault tree analysis	Starts with initiating event and finds the combination of failures which cause it.	Well accepted technique. Very good for finding failure relationships. Fault oriented: we look for ways system can fail.	Large fault trees are difficult to understand, bear no resemblance to system flow sheet, and are not mathematically unique. Complex logic is involved.
Event tree analysis	Starts with initiating events and examines alternative event sequences.	Can identify (gross) effect sequences and alternative consequences of failure.	Fails in case of parallel sequences. Not suitable for detailed analysis.
Hazards and operability studies	An extended FMEA which includes cause and effect of changes in major plant variables.	Suitable for large chemical plants.	Technique is not well standardized or described in the literature.
Cause-consequence analysis	Starts at a critical event and works forward, using consequence tree; backward, using fault tree.	Extremely flexible. All-encompassing. Well documented. Sequential paths clearly shown.	Cause-consequence diagrams can become too large very quickly. They have many of the disadvantages of fault trees.

It is important to note that failure modes and effects analysis, criticality analysis, and fault tree analysis are well-standardized and documented techniques. They are listed, in the previous sentence, in their order of sophistication. A PHA, FMEA, or CA can be carried out by hardware-oriented, non-mathematically sophisticated engineers. The same is true of event tree analysis and hazards and operability studies, even though the ground rules for doing these kinds of analyses are not firmly established.

Fault trees and cause-consequence diagrams are complex logic structures and their construction and quantitative analysis involves, minimally, a sound knowledge of Boolean algebra, set theory, and other advanced mathematical topics which will be fully developed in this text.

PROBLEMS

1.1. (a) Draw a decision tree for a system consisting of three pumps, A, B, and C, in parallel, where the system is considered operable if one of the three pumps works. (b) If $P=0.01$ is the probability of a pump failure, what is the probability of the system's working, assuming that B and C are switched on if A and B, respectively, fail?

1.2. The following sequence of protection is provided in case of a plant fire: (1) sprinkler system, A; (2) fire alarm, B; (3) plant fireman, C; (4) city fire department, D. Draw a decision tree which can be used to determine the probability of a major, uncontrolled fire, assuming statistically independent events.

1.3. The block diagram of Fig. P1.3 represents a series-parallel system, where a power supply furnishes power for two radio receivers. The reliability of each component is shown on the diagram. Draw the decision tree and calculate the probability of system success (Lambert, H. E., UCID-16238, May, 1973).

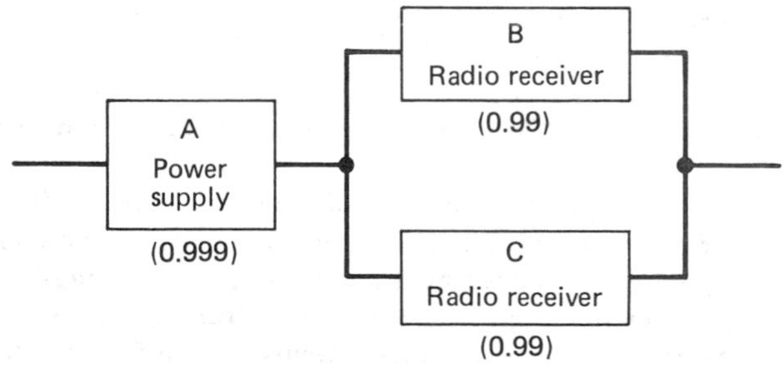

Figure P1.3.

1.4. The (bridge) system of Fig. P1.4 is successful if either pump A or B operates and if the proper valves can be opened. The success combinations are $\{A, C\}$, $\{B, D\}$, $\{A, E, D\}$, and $\{B, E, C\}$. Draw a decision tree for this system and obtain an expression for the probability of system success in terms of component reliabilities (Lambert, H. E., UCID-16238, May, 1973).

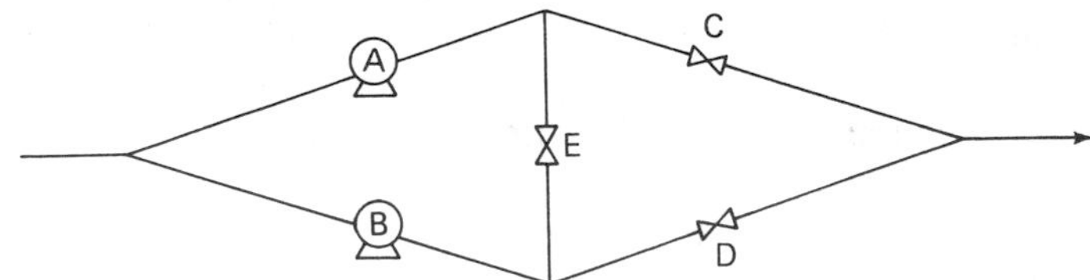

Figure P1.4. A bridge system.

1.5. Consider the system of Fig. P1.5, where the same component appears in more than one path. A and B are power supplies, and C, D, and E are pumps. Pump C requires the combined power capacity of both power supplies, whereas pump D will work from power supply A alone and pump E from power supply B alone. Any one of the three paths lead to success. If power supply A is failed in the A, B, C path, assume it is failed in the A, D path. Draw the decision tree and obtain an expression for the probability of system success in terms of component reliabilities. (Lambert, H. E., UCID-16238, May, 1973).

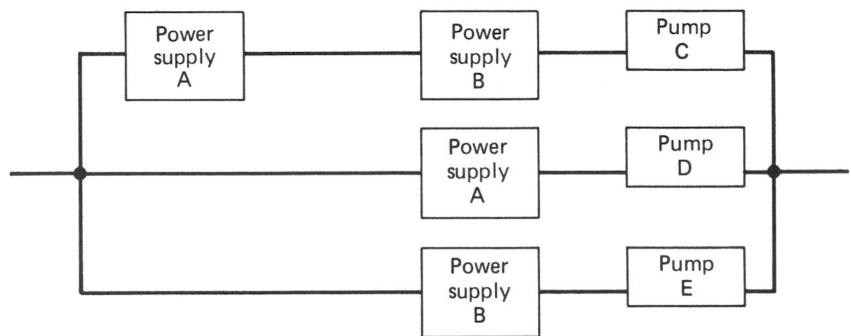

Figure P1.5. Block diagram consisting of power supplies and pumps.

1.6. Figure P1.6 is a diagram of a domestic hot-water system (Lambert, H. E., UCID-16238, May, 1973). The gas valve is operated by the controller which, in turn, is operated by the temperature measuring and comparing device. The gas valve operates the main burner in full-on/full-off modes.

The check valve in the water inlet prevents reverse flow due to overpressure in the hot-water system, and the relief valve opens when the system pressure exceeds 100 psi.

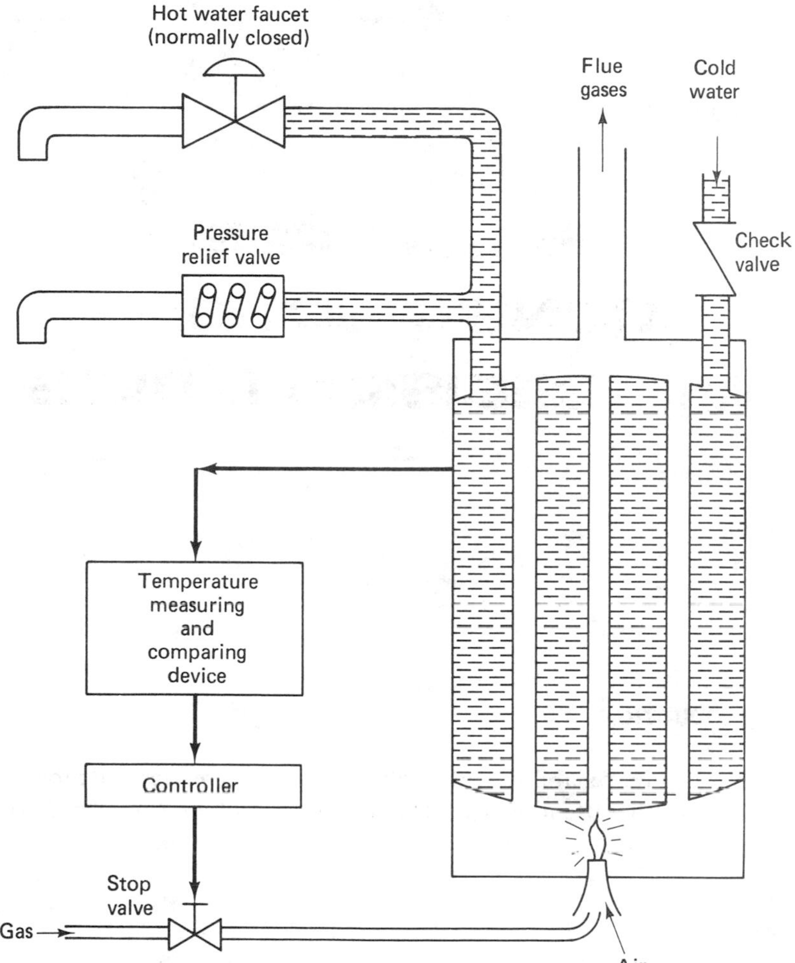

Figure P1.6. Schematic of domestic hot water system.

Control of the temperature is achieved by the controller opening and closing the main gas valve when the water temperature goes outside the preset limits (140°F–180°F). The pilot light is always on.

(a) Formulate a list of undesired safety, reliability, and security events.
(b) Do a preliminary hazards analysis on the system.
(c) Do a failure modes and effects analysis.
(d) Do a criticality ranking.

2

FAULT TREE CONSTRUCTION AND DECISION TABLES

2.1 INTRODUCTION

A major goal of a reliability and safety analysis is to reduce the probability of failure and the attending human, economic, and environmental losses.

The human losses include:

1) Death
2) Injury
3) Sickness or disability

The economic losses are, for example:

1) Production or service shutdown
2) Off-specification products or services
3) Loss of capital equipment

Some typical environmental losses are:

1) Air and water pollution
2) Other degradations of the environment such as odor, vibration, and noise

Losses occur when one or more basic failure events create a system hazard. The three types of basic failure events most commonly encountered are:

1) Events related to human beings:
   a) operator error
   b) design error
   c) maintenance error
2) Events related to hardware, for example:
   a) leakage of toxic fluid from a valve
   b) loss of lubrication in a motor
   c) incorrect measurement by a sensor
3) Events related to the environment such as:
   a) earthquakes or ground subsidence
   b) storm, flood, tornado
   c) ignition caused by sparks or lightning

System hazards are frequently caused by a combination of failure events, i.e., hardware failures plus human error and/or environmental fault events. Some typical policies used to minimize hazards and risk include:

1) equipment redundancies
2) inspection and maintenance
3) protective systems such as sprinklers, fire walls, relief valves, and emergency cooling systems
4) alarm displays

A primary purpose of a system hazards study is to identity the *causal relationships* between the basic human, hardware, and environmental events which result in system failures and to find ways of ameliorating their impact by system redesign and upgrades.

The causal relations can be developed by fault trees, which are then analyzed both qualitatively and quantitatively. After the combination of the basic failure events which lead to system hazards are identified, the system can be improved and the hazards reduced.

2.2 FAULT TREES AND DECISION TABLES

Fault tree analysis was developed by H. A. Watson of the Bell Telephone Laboratories in 1961–62 during an Air Force study contract for the Minuteman Launch Control System. The first published papers were presented at the 1965 Safety Symposium sponsored by the University of Washington and the Boeing Company, where a group including D. F. Haasl, R. J. Schroder, W. R. Jackson, and others had been applying and extending the technique. Despite the widespread use of the methodology, there is a great paucity of tutorial and descriptive literature; the papers of

Haasl,[1] Lambert,[2] and Fussell[3] being the only ones that have come to our attention. Of these, the Lambert paper is the most detailed.

Fussell declares the value of a fault tree to be in:

1) Directing the analysis to ferret out failures.
2) Pointing out the aspects of the system important to the failure of interest.
3) Providing a graphical aid in giving visibility to those in systems management who are removed from systems design changes.
4) Providing options for qualitative and quantitative systems reliability analysis.
5) Allowing the analyst to concentrate on one particular system failure at a time.
6) Providing an insight into system behavior.

To this, one might add that a fault tree, like any other engineering report, is a communication tool and, as such, must be a clear and demonstratable record.

The structure of a tree is shown in Fig. 2.1. The undesired event appears as the *top event*, and this is linked to more basic *fault events* by *event statements* and *logic gates*. The central advantage of the fault tree vis-a-vis other techniques such as FMEA is that the analysis is restricted only to the identification of the system elements and events that lead to one particular undesired failure or accident.

Since the early 1970's when computer-based analysis techniques for fault trees were developed, their use has become very widespread. It is the most widely used method, having replaced block diagram analysis for most purposes. Indeed, the use of fault tree analysis, or its first cousin, cause-consequence analysis (see Chapter 13) is being mandated by a number of governmental agencies responsible for worker and/or public safety.

Decision tables, which are introduced later in this chapter, have been applied to safety analysis problems only in the past few years, primarily as a method for automatic fault tree synthesis. However, we believe that they are more versatile than fault trees and will be used more widely in the future. Their versatility stems from the fact that:

1) More than one component failure state can be shown. Fault trees are Boolean logic diagrams which show only success or failure states, for

[1]Haasl, D. F., "Advanced Concepts in Fault Tree Analysis," System Safety Symposium, June 8–9, 1965, Seattle: the Boeing Company.

[2]Lambert, H. E., "Systems Safety Analysis And Fault Tree Analysis," UCID-16238, **31**, May 9, 1973.

[3]Fussell, J., "Fault Tree Analysis—Concepts and Techniques" in *Generic Techniques in Reliability Assessment*, Henley, E., and J. Lynn, (eds.) Noordhoff, Publishing Co., Leyden, Holland, 1976.

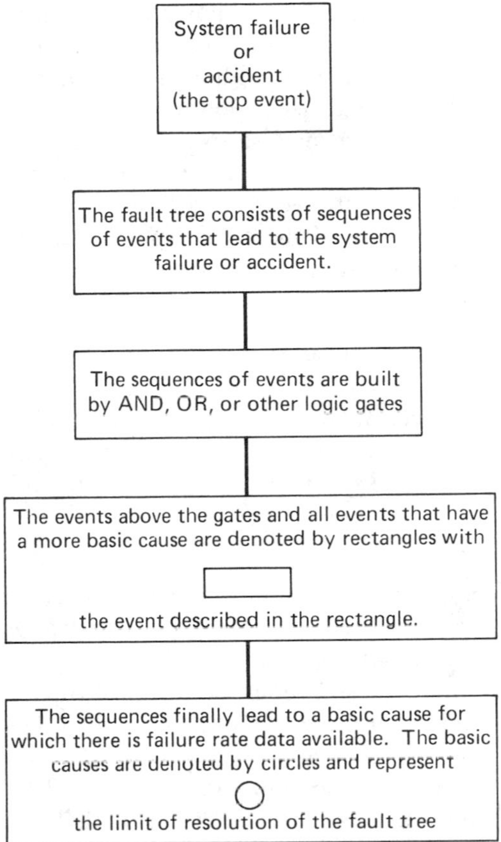

Figure 2.1. Fundamental structure of fault tree.

example, "valve functioning or valve broken." In a decision table one can show additional states such as "valve output reduced," etc.

2) In systems having control loops as well as many other situations, the time and/or sequence of failure events are important. Fault trees describe systems at a certain instant of time (usually the steady state) and sequential events can be shown only with difficulty, if at all, whereas decision table technology can be fairly readily extended to even such complex phenomena as cascaded control loops.[4] Cause-consequence diagrams, potentially, can also handle more complex logic than fault trees.

Decision tables are not the graphic communication tools that fault trees or cause-consequence diagrams are. However, as fault trees become very

[4]Ogunbiyi, E., "Application of Decision Tables to Risk Analysis Studies," Ph.D. Thesis, Department of Chemical Engineering, University of Houston, 1980.

large, mistakes are difficult to find, and the logic becomes difficult to follow or obscured. Professor C. Apostolakis, for example,[5] found a minor error in a WASH 1400 fault tree, although those trees were supposedly checked and rechecked. The construction of fault trees is perhaps as much an art as a science. No two analysts will construct identical fault trees (although the trees should be equivalent in the sense that they yield the same failure modes). If fault tree construction *is* an art, that raises the question as to whether some of the construction rules and procedures we invented in this chapter are at all useful. We don't know. Time will tell.

2.3 FAULT TREE BUILDING BLOCKS

In order to find and visualize causal relations by fault trees, we require building blocks to classify and connect a large number of events. There are two types of building blocks: *gate symbols* and *event symbols*.

2.3.1 Gate Symbols

Gate symbols connect events according to their *causal relations*. The symbols for the gates are listed in Table 2.1. A gate may have one or more *input events* but only one *output event*.

The output events of AND gates occur if all input events occur simultaneously. On the other hand, the output events of OR gates happen if any one of the input events occurs.

Examples of these two gates are shown in Fig. 2.2. The event "fire breaks out" happens when two events, "leakage of flammable fluid" and "ignition source is near the fluid," occur simultaneously. The latter event happens when either one of the two events, "spark exists" or "employee is smoking" occurs.

The causal relation expressed by an AND gate or OR gate is *deterministic* because the occurrence of the output event is completely controlled by the input events. There are causal relations that are not deterministic. Consider the two events: "a person is struck by an automobile" and "a person dies." The causal relation of these two events is not deterministic but *probabilistic* because the accident does not always result in a death.

A hexagon, *the inhibit gate* in row 3 of Table 2.1, is used to represent a probabilistic causal relation. The event at the bottom of the inhibit gate in Fig. 2.3 is called an *input event*, whereas the event to the side of the gate is a *conditional event*. The conditional event takes the form of an event conditioned by the input event. The output event occurs if both the input event and the conditional event occur. In other words, the input event

[5] Personal communication.

TABLE 2.1 GATE SYMBOLS

	Gate Symbol	Gate Name	Causal Relation
1		AND gate	Output event occurs if all input events occur simultaneously.
2		OR gate	Output event occurs if any one of the input events occurs.
3	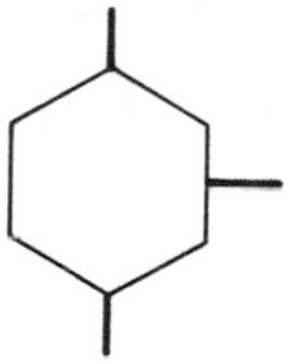	Inhibit gate	Input produces output when conditional event occurs.
4		Priority AND gate	Output event occurs if all input events occur in the order from left to right.
5	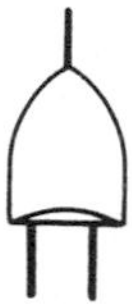	Exclusive OR gate	Output event occurs if one, but not both, of the input events occur.
6	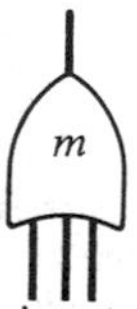 *m* *n* inputs	*m* Out of *n* gate (voting or sample gate)	Output event occurs if *m* out of *n* input events occur.

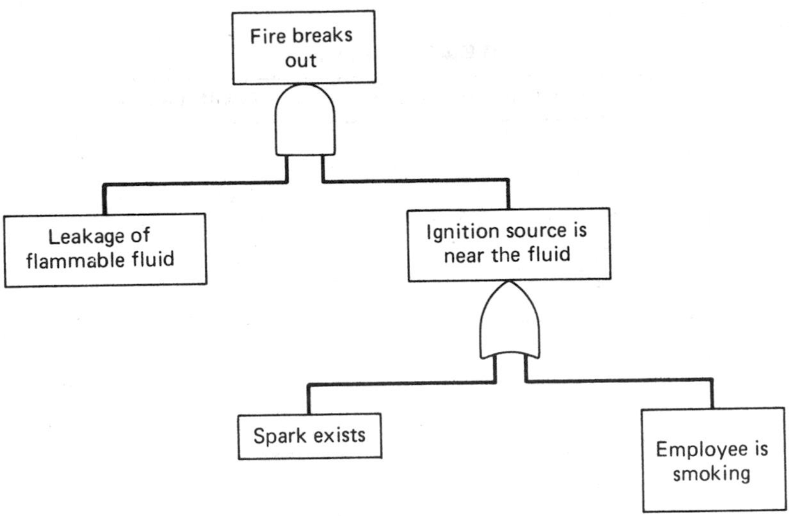

Figure 2.2. Example of AND gate and OR gate.

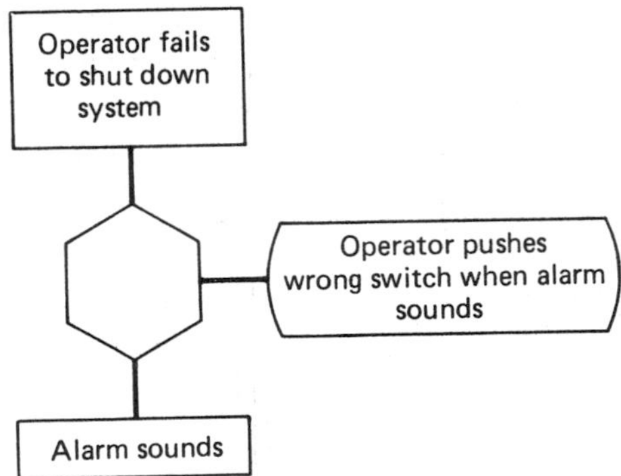

Figure 2.3. Example of inhibit gate.

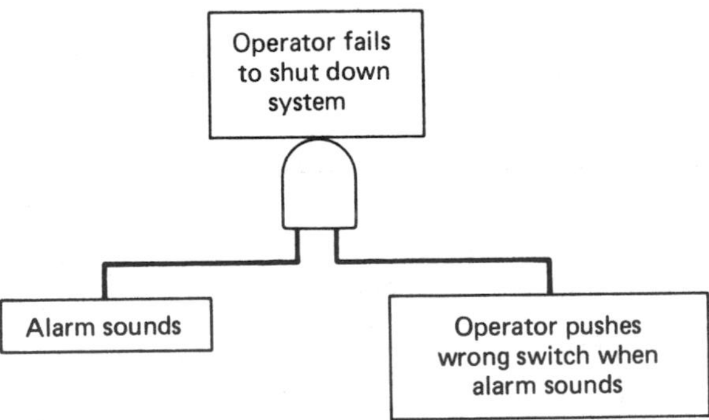

Figure 2.4. Equivalent expression to Fig. 2.3.

causes the output event with the (usually constant) probability of occurrence of the conditional event. The inhibit gate frequently appears when an event occurs according to a demand. It is used primarily for convenience and can be replaced by an AND gate as shown in Fig. 2.4.

The *priority AND gate* in row 4 of Table 2.1 is logically equivalent to an AND gate, with the additional requirement that the input events occur in a specific order.[6] The output event occurs if the input events occur in the order that they appear from left to right. The occurrence of the input events in a different order does not cause the output event. Consider, for example, a system that has a principal power supply and a standby power supply. The standby power supply is switched into operation by an automatic switch when the principal power supply fails. Power is unavailable in the system if:

1) the principal and standby units both fail, or
2) the switch controller fails first and then the principal unit fails.

It is assumed that the failure of the switch controller followed by the failure of the principal unit does not yield a loss of power if the standby unit is functioning. The causal relations in the system are shown in Fig. 2.5. The priority AND gate can be represented by a combination of an AND gate and an inhibit gate and hence is equivalent to an AND gate. The conditional event to the inhibit gate is that the input events to the AND gate occur in the specified order. Representations equivalent to Fig. 2.5 are shown in Figs. 2.6 and 2.7.

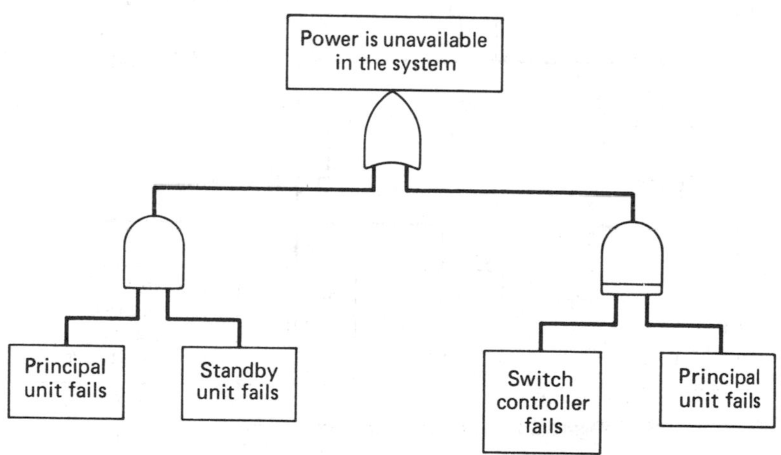

Figure 2.5. Example of priority AND gate.

[6]Fussell, J. B., E. F. Aber, and R. G. Rahl, "On the Quantitative Analysis of Priority AND Failure Logic," *IEEE Trans. on Reliability*, Vol. R-25, No. 5, December, 1976.

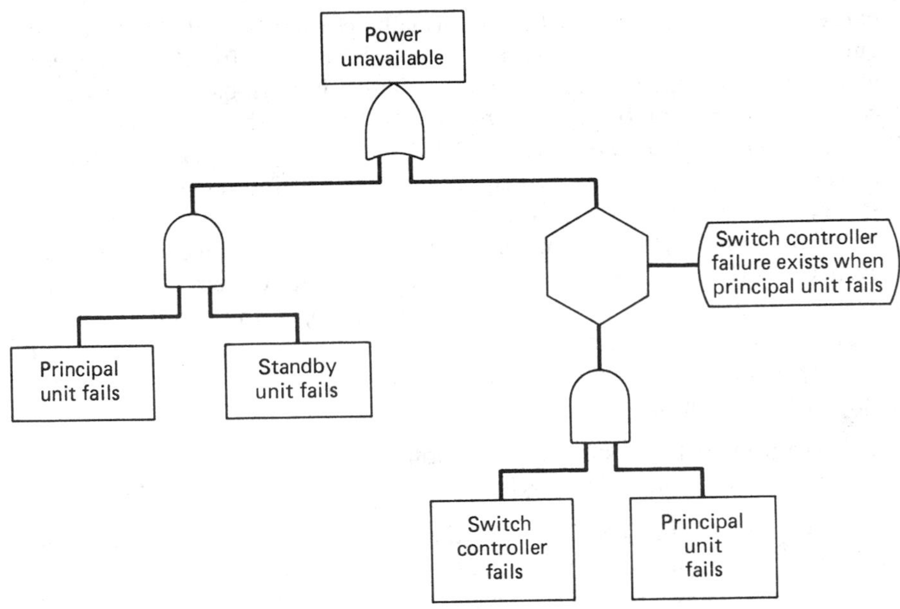

Figure 2.6. Equivalent expression to Fig. 2.5.

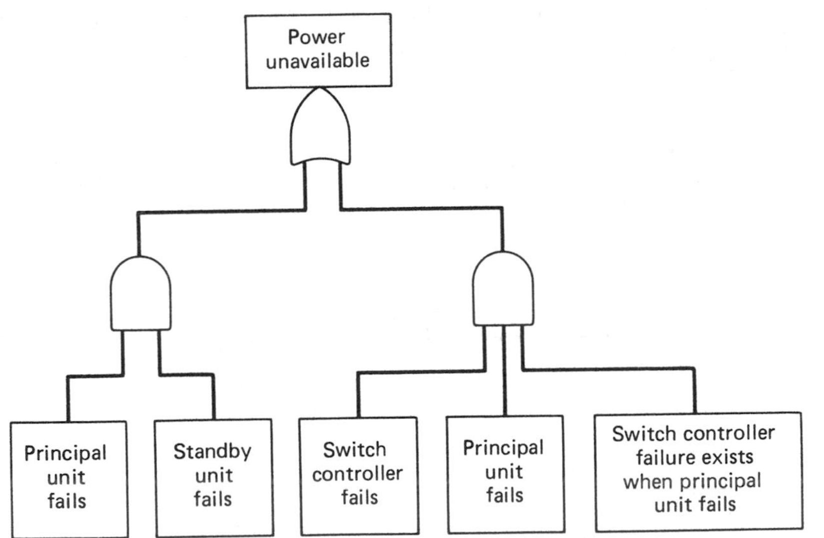

Figure 2.7. Equivalent expression to Fig. 2.5.

Exclusive OR gates (Table 2.1, row 5) describe a situation where the output event occurs if either one, not both, of the two input events occur. Consider a system powered by two generators. A partial loss of power can be represented by the exclusive OR gate shown in Fig. 2.8. The exclusive OR gate can be replaced by a combination of an AND gate and an OR gate, as illustrated in Fig. 2.8.†

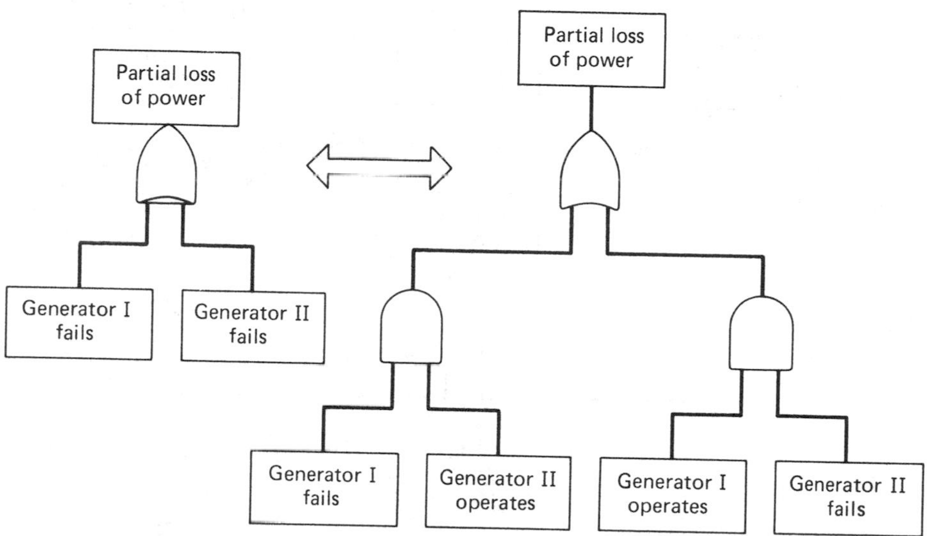

Figure 2.8. Example of exclusive OR gate and its equivalent expression.

An m-out-of-n voting gate (row 6, Table 2.1) has n input events and the output event occurs if at least m out of n input events occur. Consider a shutdown system consisting of three monitors. Assume that system shutdown should occur if and only if two or more monitors generate shutdown signals. Thus, unnecessary shutdowns occur if two or more monitors create spurious signals while the system is in its normal state. This situation can be expressed by a two-out-of-three gate as shown in Fig. 2.9. The voting gate is equivalent to a combination of AND gates and OR gates as illustrated in Fig. 2.10.

New gates can be defined to represent special types of causal relations. We note, however, that most special gates can be rewritten as combinations of AND and OR gates.

†Usually, we avoid having success states such as "generator operating" appearing in fault trees, since these greatly complicate the quantitative analysis. A prudent and conservative policy is to replace exclusive OR gates by OR gates.

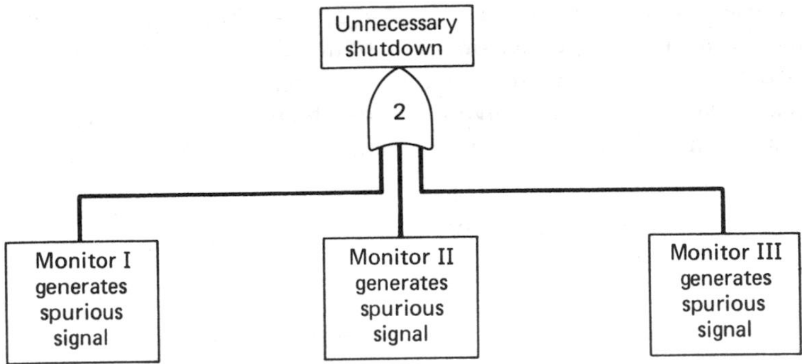

Figure 2.9. Example of two-out-of-three gate.

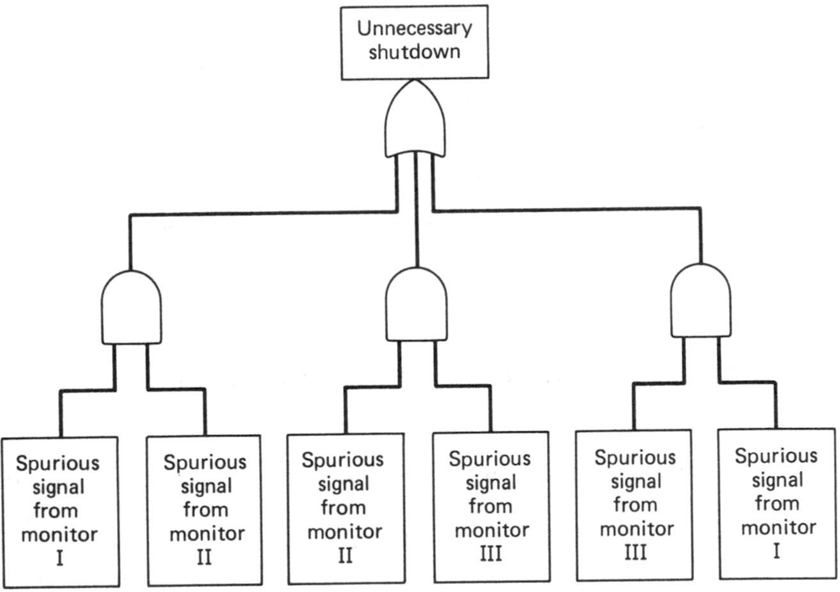

Figure 2.10. Equivalent expression to Fig. 2.9.

2.3.2 Event Symbols

Event symbols are shown in Table 2.2. In the schematic fault tree of Fig. 2.1, a rectangular box denotes a *fault event* resulting from a combination of more basic faults acting through logic gates. The circle designates a *basic component failure* (within the design envelope or environment) that represents the limit of resolution of a fault tree. In order to obtain a quantitative solution for a fault tree, circles must represent events for which reliability information is available.[3] Events that appear as circles are called *basic events*. "Valve failure due to wearout" would be an example of a basic component failure found in a circle. In general, it is an event for which the component itself is responsible and, once it occurs, the component must be repaired or replaced.

TABLE 2.2 EVENT SYMBOLS

	Event Symbol	Meaning of Symbols
1	Circle	Basic event with sufficient data
2	Diamond	Undeveloped event
3	Rectangle	Event represented by a gate
4	Oval	Conditional event used with inhibit gate
5	House	House event. Either occurring or not occurring
6	Triangles	Transfer symbol

In Fig. 2.11 we see that the fault, "excessive current in circuit," is analyzed as being caused either by the basic event, "shorted wire," or the undeveloped event, "line surge."† Had we chosen to develop the event "line surge" more fully, a rectangle would have been used to show that this

†Diamonds are used to signify *undeveloped events*, in the sense that a detailed analysis into the basic faults is not carried out because of lack of information, money, or time. "Failure due to sabotage" is an example of an undeveloped event. Frequently, such events are removed prior to a quantitative analysis. They are included initially because a fault tree is a communication tool, and their presence serves as a reminder of the depth and bounds of the analysis.

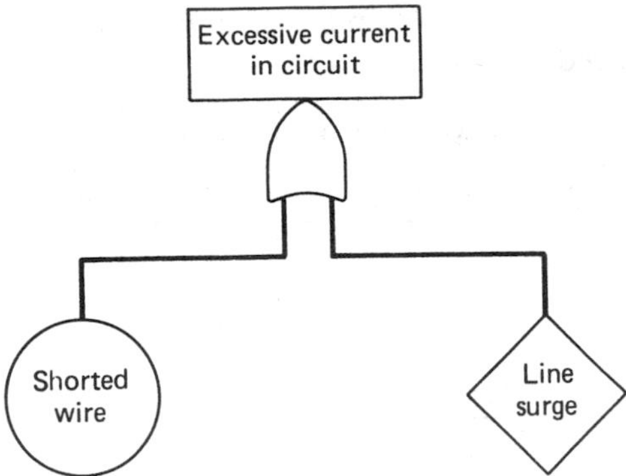

Figure 2.11. Example of event in rectangle.

is developed to more basic events, and then the analysis would have to be carried further back, perhaps to a generator or another in-line hardware component.

Sometimes we wish to examine various special cases of fault trees by forcing some events to occur and other events not to occur. In such instances, we would use the *house event* (row 5, Table 2.2). When we turn on the house event, the fault tree presumes the occurrence of the event and vice versa when we turn it off. We can also delete causal relations below an AND gate by turning off a dummy house event introduced as an input to

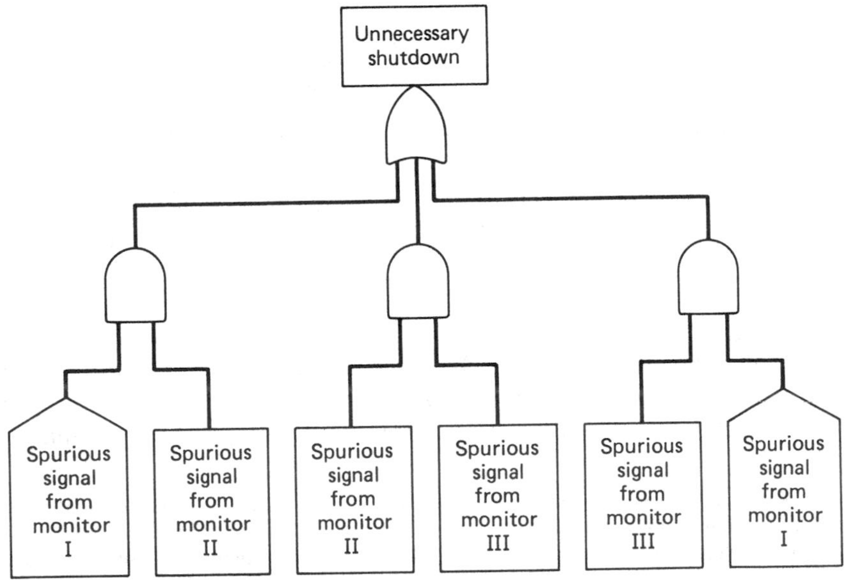

Figure 2.12. Example of house event.

the gate. Similarly, we can erase relations below an OR gate by turning on a house event to the gate.

The house event is illustrated in Fig. 2.12. When we turn on the house event, monitor I is assumed to be generating a spurious signal. Thus, we have a one-out-of-two gate, i.e., a simple OR gate with two inputs, II and III. If we turn off the house event, a simple AND gate is obtained.

In row 6 of Table 2.2 the pair of triangles, a *transfer-out triangle* and a *transfer-in triangle*, cross references two identical parts of the causal relations. The two triangles have the same identification number. The transfer-out triangle has a line to its side from a gate, whereas the transfer-in triangle has a line from its apex to another gate. The triangles are used to simplify the representation of fault trees as illustrated in Fig. 2.13.

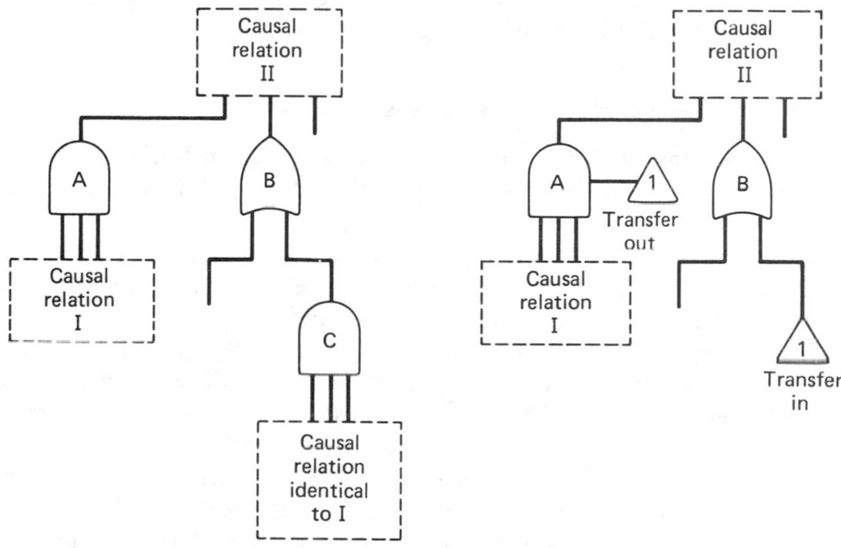

Figure 2.13. Illustration of the use of transfer symbol.

2.4 FINDING TOP EVENTS

There are two approaches for analyzing causal relations: One is *forward analysis*, the other is *backward analysis*. A forward analysis starts with a set of failure events and proceeds forward, seeking possible consequences resulting from the events. A backward analysis begins with a system hazard and traces backward, searching for possible causes of the hazard.

The event tree (ET), failure mode and effect analysis (FMEA), criticality analysis (CA), and preliminary hazards analysis (PHA), use the forward approach (see Chapter 1). The backward approach is typified by fault tree analysis (FTA). As was discussed in Chapter 1, the cooperative use of these two approaches is necessary to attain completeness in a reliability and risk analysis.

The backward analysis, i.e., the fault tree analysis, is used to identify the causal relations leading to a given system hazard. The hazard itself becomes the top event of the fault tree. A particular top event may be only one of many possible system hazards of interest; the fault tree analysis itself does not identify possible hazardous events in the system. Large systems can have many different top events, and fault trees.

The forward analysis assumes sequences of events and writes a number of scenarios ending in the system hazards. Check lists[7] are very helpful in a forward analysis. The information which must be developed in order to write good scenarios are *component interrelations and system topography*, *component failure characteristics*, and *accurate system specifications*. These items are also useful for constructing fault trees.

2.4.1 Component Interrelations and System Topography

A *system* consists of components such as hardware, materials, and plant personnel,[†] is surrounded by its physical and social environment, and suffers from aging.

Hazards are caused by one or a set of system components generating failure events. The environment, plant personnel, and aging can affect the system only through the system components (Fig. 2.14).

Each component in a system is related to the other components in a specific manner, and identical components may have different characteristics in different systems. Therefore, we must clarify component interrelations and system topography. The interrelations and the topography are found by examining plant piping, electrical wiring, mechanical couplings, information flows, and the physical location of components. These can be best expressed by a system schematic; system word models and logic flow charts also help.

For example, a water hammer caused by a quick valve closure, which in turn causes a leak from a flange connection, should be deducible by tracing a piping diagram. Interactions between two closely located tanks are possible in case of fire. Possible changes of component states resulting from other causes have to be described by system word models or logic flow charts.

[7]Wells, G. L., C. J. Seagrave, and R. M. C. Whiteway, "Flowsheeting for Safety," Inst. Chem. Engrs., Warwickshire, England, 1979.

[†]Components are not necessarily the smallest constituents of the system; they may be units or subsystems.

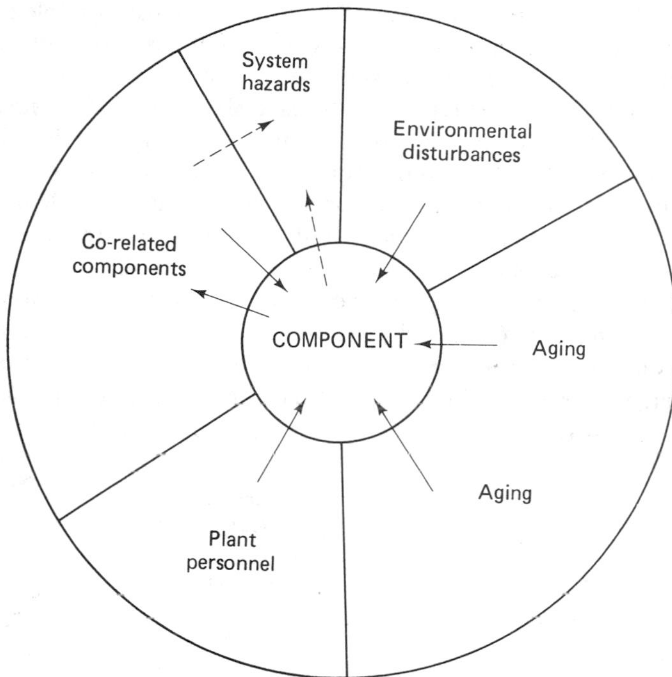

Figure 2.14. Component interrelations.

2.4.2 Component Failure Characteristics

Component failures are fundamental in causal relation analysis. They are classified as either *primary failures*, *secondary failures*, or *command faults*.

A *primary* failure is defined as the component being in the non-working state for which the component is held accountable, and repair action on the component is required to return the component to the working state. The primary failure occurs under inputs within the design envelope, and component natural aging is responsible for the failure. For example, "tank rupture due to metal fatigue" is a primary failure.

A *secondary failure* is the same as a primary failure except that the component is not held accountable for the failure. Past or present excessive stresses placed on the component are responsible for the secondary failure. These stresses involve out-of-tolerance conditions of amplitude, frequency, duration, or polarity, and energy inputs from thermal, mechanical, electrical, chemical, magnetic, or radioactive energy sources. The stresses are caused by neighboring components or the environment, which includes meteorological or geological conditions, and other engineering systems.

Human beings such as operators and inspectors are also possible causes for secondary failures, if they break the component. Examples of secondary failures are "fuse is opened by excessive current" and "earthquake cracks storage tanks." Note that repair of the excessive stresses does not guarantee the working state of the component because the stresses have left damage (memories) in the component which must be repaired. When the exact failure mode for a primary or secondary failure is identified, and failure data are obtained, primary and secondary failure events are the same as basic failures and are shown as circles in a fault tree.

A *command fault* is defined as the component being in the non-working state due to improper control signals or noise and, frequently, repair action is not required to return the component to the working state. Inadvertent control signals or noise often leave no memories (damage) and subsequent normal inputs can operate the component as expected. Examples of

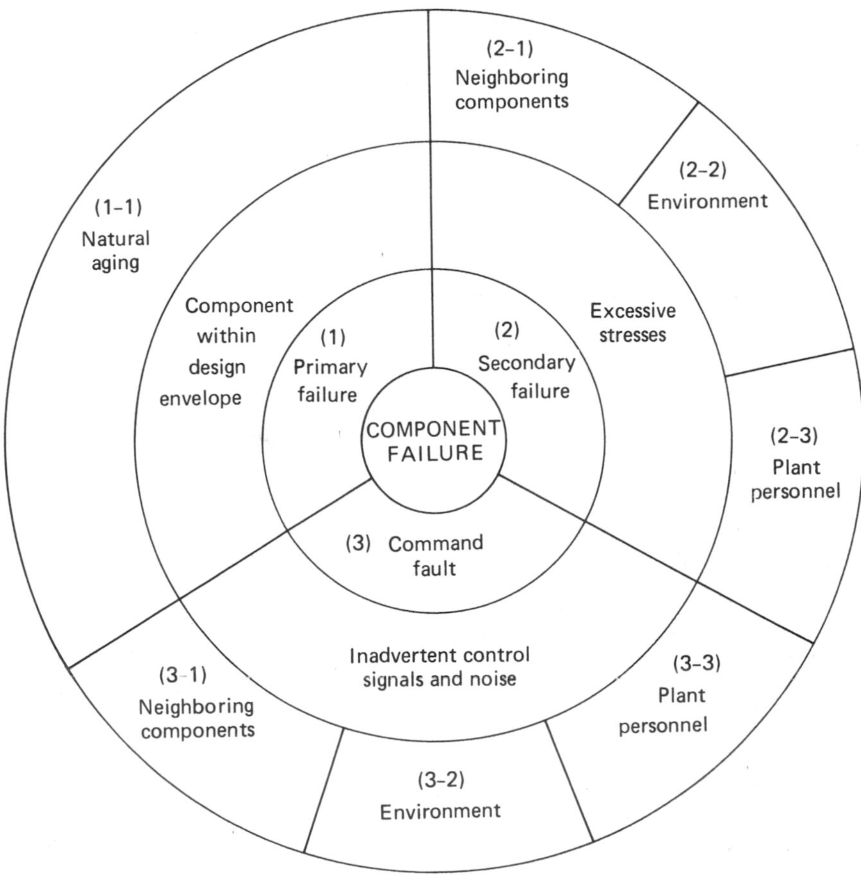

Figure 2.15. Component failure characteristics.

command faults are "power is applied, inadvertently, to the relay coil," "switch randomly fails to open because of noise," " noisy input to safety monitor randomly generates spurious shutdown signals," and "operator fails to push panic button" (command fault for the panic button).

The first concentric circle about "component failure" in Fig. 2.15 shows that failure can result from (1) primary failure, (2) secondary failure, or (3) command faults. These categories have their possible causes shown in the outermost circle.

2.4.3 System Specification

Only major, highly probable, or critical events should be considered in the initial steps of the analysis. Criticality analysis (CA) can be used to identify these events. We can include increasingly rare or less probable events as the analysis proceeds, or choose to ignore them.

The system environment, in principle, includes the entire world outside the system. Thus, an appropriate boundary for the environment is necessary to prevent the event tree analysis (ETA) or the consequence analysis from diverging, since these two techniques extend the initial failure events deeply into the system and its environment.

System identification requires a careful delineation of *component initial conditions*. All components that have more than one operating state generate initial conditions. For example, the initial quantity of fluid in a tank can be specified. The event "tank is full" becomes one initial condition, while "tank is empty" is another. The *time domain* must also be specified; start-up or shut-down conditions, for example, can generate different hazards than a steady state operation.

When enough information on the system have been collected, we can write scenarios and define top events. Causal relations leading to each top event are then found by fault tree analysis.

2.4.4 Example of a Preliminary Forward Analysis

Consider the pumping system in Fig. 2.16.[2,8,9] This schematic gives component relationships which are made clearer by the following word model:

In the operational mode, to start the system pumping, the reset switch $S1$ is closed and then opened immediately. This allows current to flow in the control branch circuit, activating relay coils $K1$ and $K2$; relay $K1$ contacts are closed and latched, while $K2$ contacts close and start the pump motor.

[8]Haasl, D. F., Institute For Systems Sciences, Bellevue, Wash., private communication.

[9]Barlow, R. E., and F. Proschan, *Statistical Theory of Reliability and Life Testing Probability Models*, Holt, Rinehart and Winston, New York, 1975.

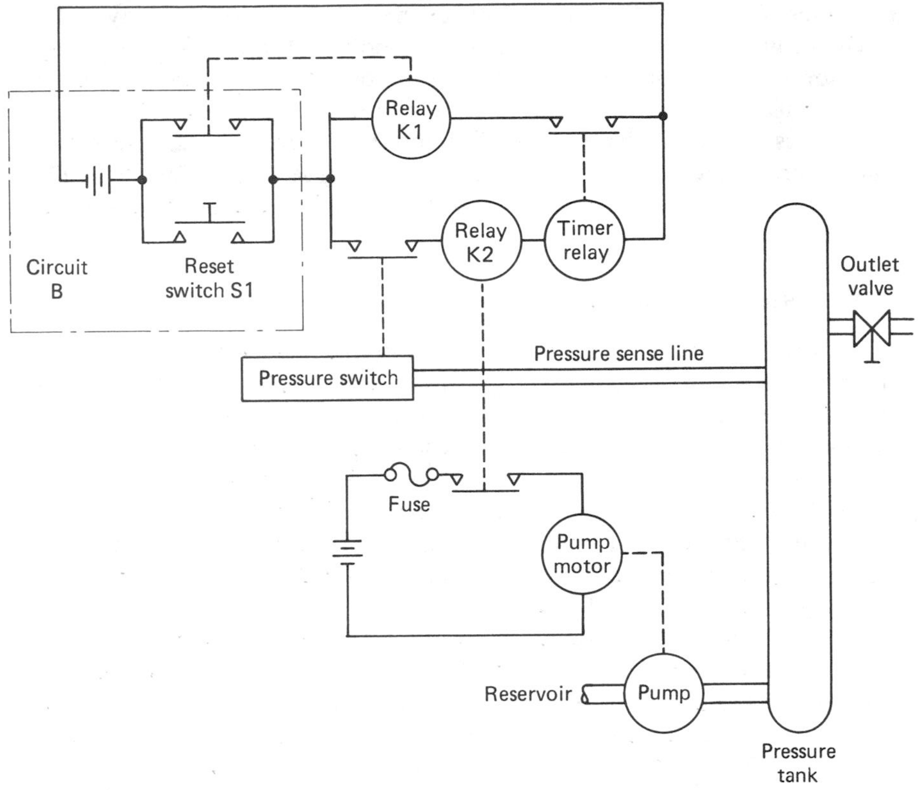

Figure 2.16. System schematic for a pumping system.

In the shutdown mode, after approximately 20 sec, the pressure switch contacts should open (since excess pressure should be detected by the pressure switch), deactivating the control circuit, de-energizing the $K2$ coil, opening the $K2$ contacts, and thereby shutting the motor off. If there is a pressure switch hang-up (emergency shutdown mode), the timer relay contacts should open after 60 sec, de-energizing the $K1$ coil, which in turn de-energizes the $K2$ coil, shutting off the pump. We assume that the timer resets itself automatically after each trial, that the pump operates as specified, and that the tank is emptied of fluid after every run.

We can also introduce the flow chart of Fig. 2.17, showing the sequential functioning of each component in the system with respect to each operational mode.

Forward analyses such as PHA and FMEA are carried out, and we detect sequences of component failure events leading to the system hazards. For the pumping system of Fig. 2.16, for example:

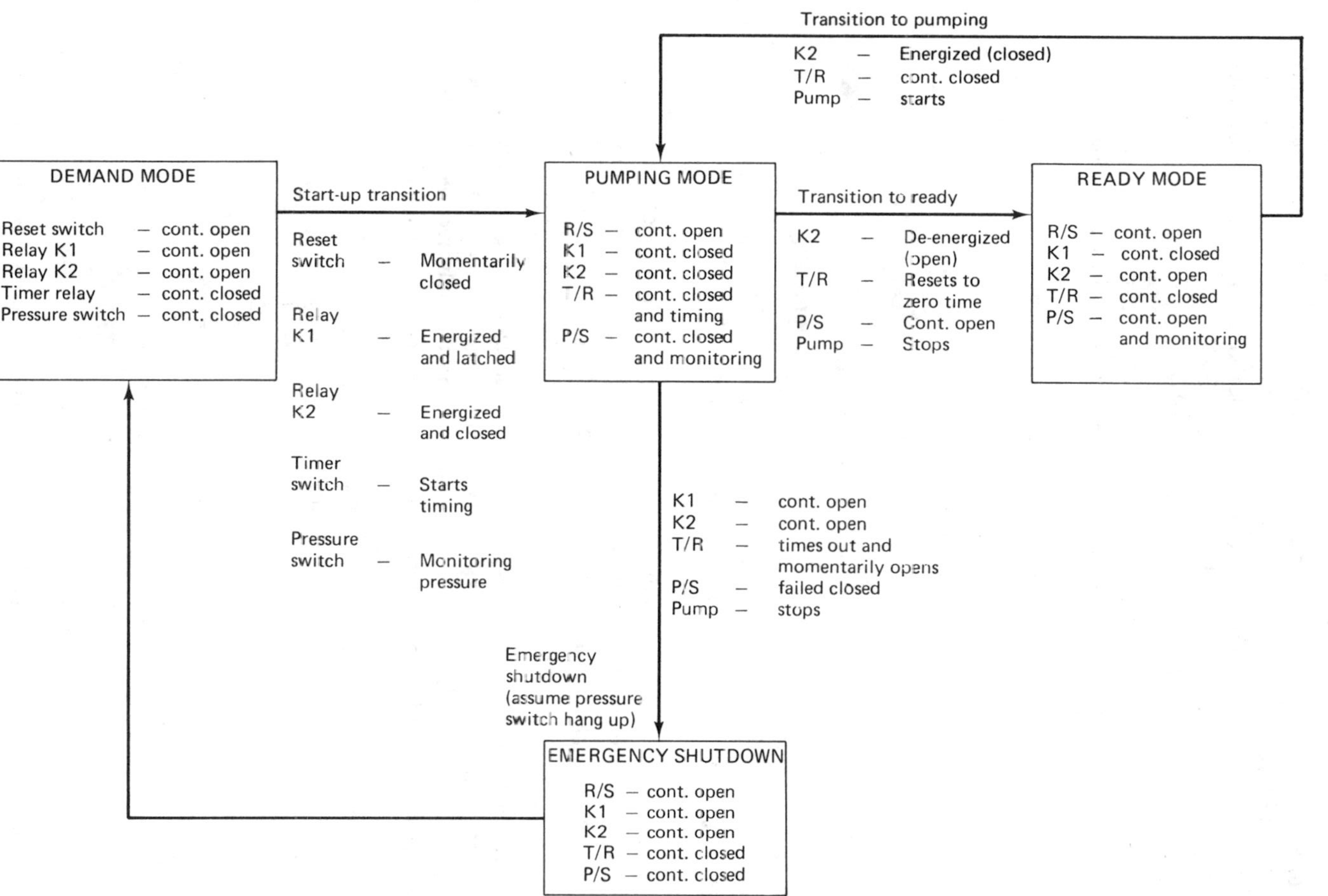

Figure 2.17. Flow chart for the pumping system.

1) Pressure switch fails to open—timer fails to time-out—overpressure —rupture of tank.
2) Reset switch fails to close—pump does not start—fluid becomes unavailable from the tank.
3) Leak of flammable fluid from tank—relay sparks—fire.

By an appropriate choice of the environmental boundary, these system hazards can be traced forward into the system and its environment. Examples are:

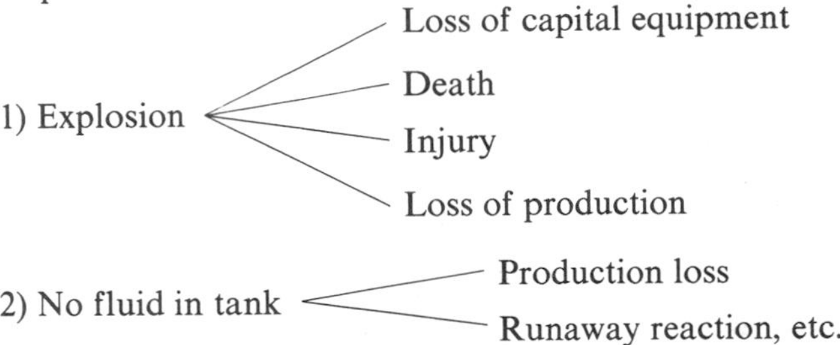

2.5 PROCEDURE FOR FAULT TREE CONSTRUCTION

A fault tree is a graphical representation of causal relations obtained when a system hazard is traced backward to search for its possible causes. The system hazard is then the top event of the fault tree.

Example 1.

As a first example of fault tree construction consider the top event, "motor fails to start," for the system of Fig. 2.18. A clear definition of the top event is necessary even if the event is expressed in abbreviated form in the fault tree. In the present case, the complete top event is "motor fails to start when switch is closed at a given time t."[10]

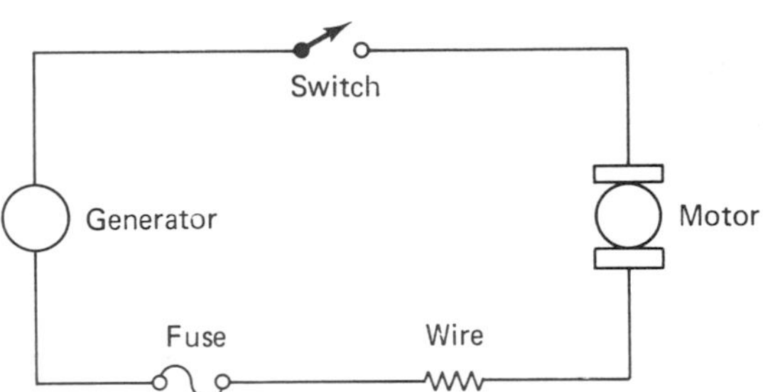

Figure 2.18. System schematic for an electric circuit.

[10]Variable t can be expressed in terms other than time. For example, transport reliability information is usually expressed in terms of mileage. The variable sometimes means "cycles of operation" as indicated by Fothergill, C. D. H., "The Collection, Storage and Use of Equipment Performance Data for the Safety and Reliability Assessment of Nuclear Power Plants," *Proceedings of a Symposium on the Reliability of Nuclear Power Plants*, Int'l. Atomic Energy Agency, Vienna, 1975.

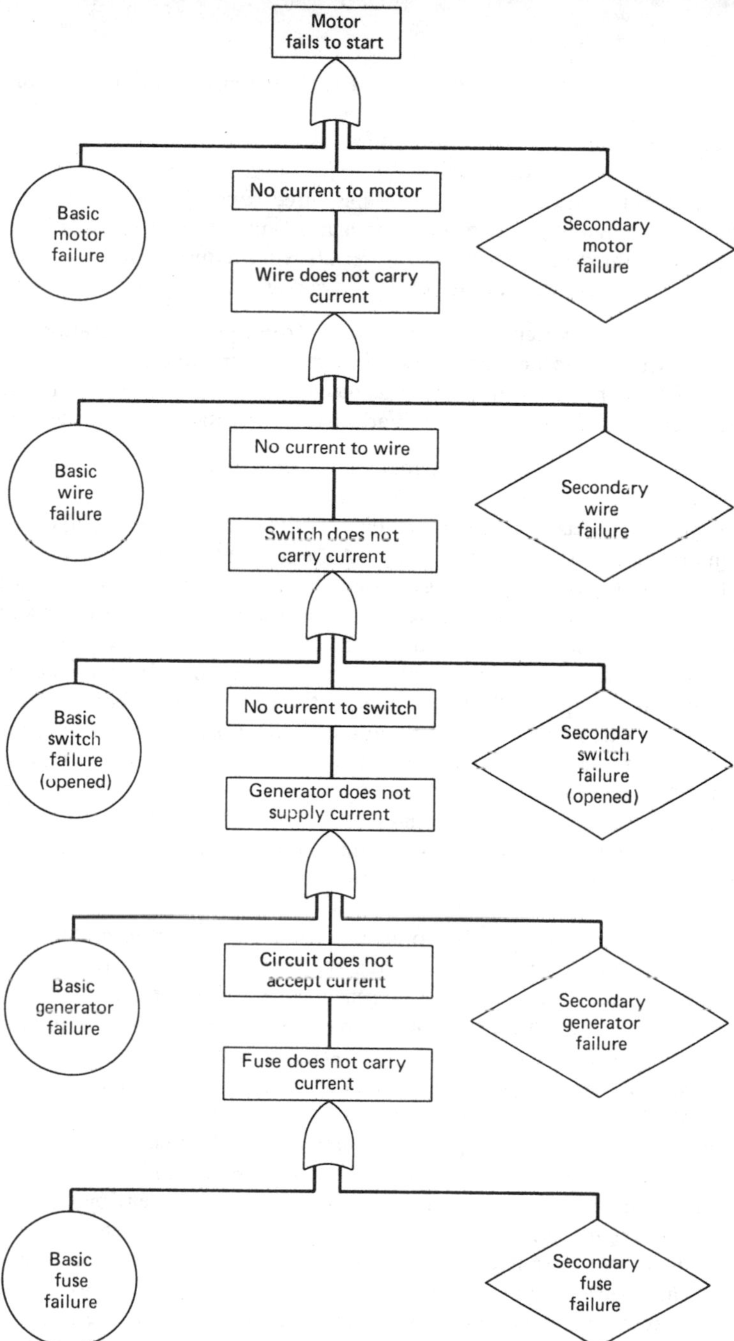

Figure 2.19. Fault tree for system in Fig. 2.18. (Note that the terms *primary event* and *basic failure* become synonymous when the failure mode [and data] are specified and that the secondary events above will ultimately either be removed or become basic events.)

The classification of component failure events in Fig. 2.15 is useful for constructing the fault tree shown in Fig. 2.19.

The top event "motor fails to start" has three causes: primary motor failure, secondary motor failure, and motor command fault. The primary failure is the motor failure in the design envelope and results from natural aging. The secondary failure is due to causes outside the design envelope such as:[3]

a) Over-run, i.e., switch remained closed from previous operation, causing motor windings to heat and then to short or open circuit.
b) Out-of-tolerance conditions such as mechanical vibration, thermal stress, etc.
c) Improper maintenance such as inadequate lubrication of motor bearings.

The command fault is caused by inadvertent control signals or noise, and in the present case the fault is "no current to motor."

Primary or secondary failures are caused by disturbances from the sources shown in the outermost circle of Fig. 2.15. A component can be in the non-working state at time t if past disturbances broke the component and it has not been repaired. The disturbance could have occurred at any time before t. However, we do not go back in time, so the primary or the secondary failures at time t become a terminal event, and further development is not carried out. In other words, fault trees are instant "snapshots" of a system at time t. The disturbances are factors controlling transition from normal component to broken component. More precisely speaking, these disturbances determine transition probabilities of the component. The primary event is enclosed by a circle because it is a basic event for which failure data are available. The secondary failure is an undeveloped event and is enclosed by a diamond. Quantitative failure characteristics of the secondary failure should be estimated by appropriate methods, in which case it becomes a basic event.

As was shown in Fig. 2.15, the command fault "no current to motor" is created by the failure of neighboring components. We have the event "wire does not carry current" in Fig. 2.19. A similar development is possible for this failure, and we finally reach the event "fuse does not carry current." We have the primary fuse failure, "fuse fails open by natural aging," and secondary failure, "fuse opened by excessive current." We might introduce the command fault "no current to fuse" as in category (3) of Fig. 2.15. However, all the components have already been examined, and there is no component failure causing "no current to fuse." Thus, we can neglect this command fault, and the fault tree is completed.

The secondary failure of the fuse may be caused by present or past excessive current from neighboring components. Any excessive current before time t could break the fuse. We cannot develop the event "excessive current occurred before time t" because infinitely many past times are involved. However, we can develop the event "excessive current exists at a specified time t^*," and the resulting fault tree is shown in Fig. 2.20.† Note that the event "generator not dead" exists with a

†Secondary failures are neglected in Fig. 2.20. Note that an inhibit gate is equivalent to an AND gate.

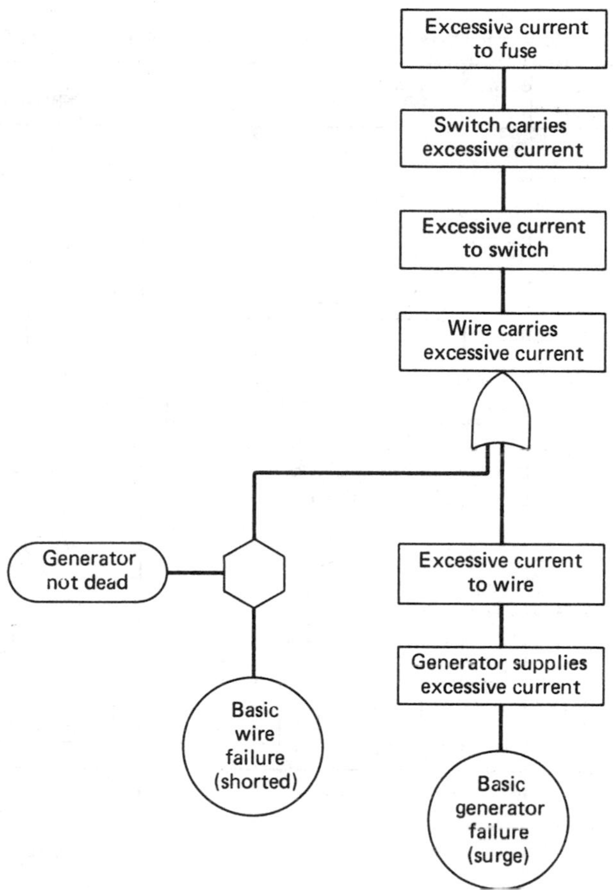

Figure 2.20. Fault tree with the top event "excessive current to fuse" (secondary failures are neglected).

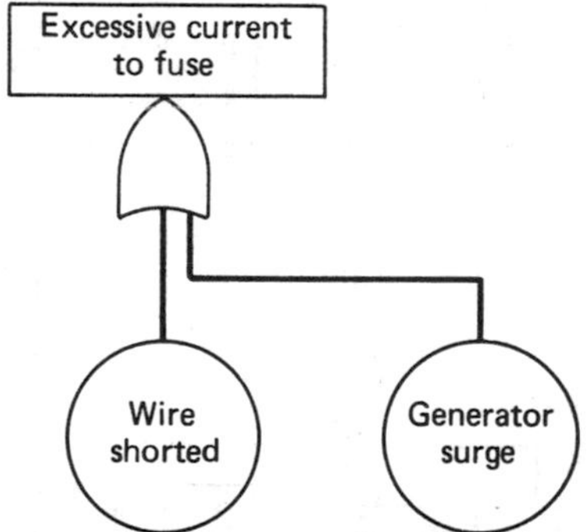

Figure 2.21. Fault tree obtained by neglecting very high probability event, generator not dead.

TABLE 2.3 Simplification by Very High or Very Low Probability Event

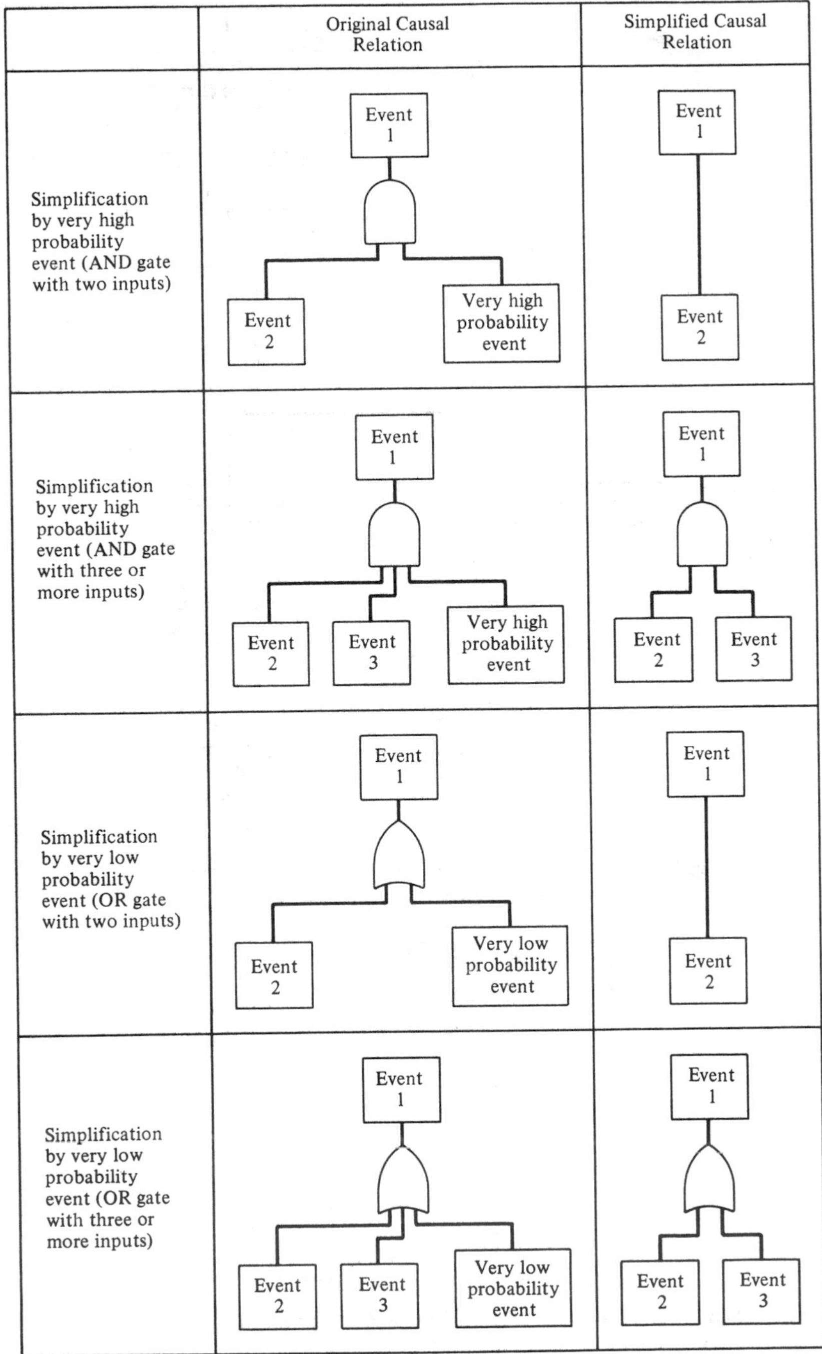

very high probability, say 0.9999. We will call such events "*very high probability events*," and they should be removed from input to AND (or inhibit) gates without any major changes in the top event probabilities.† Only in a rigorous analysis will very high probability events remain in fault trees. We have a simplified fault tree for Fig. 2.21 for the top event, "excessive current," in Fig. 2.21. This fault tree can be quantified, by the methods described in Chapter 7, to determine how many times the excessive current will occur per unit time at a given time t^*. This information, in turn, is used to quantify the secondary fuse failure, and finally, the probability of the occurrence of "motor failing to start" is calculated. Simplification methods for very high or very low probability events are shown in Table 2.3.

2.5.1 Heuristic Guidelines

Some heuristic guidelines are desirable for the construction of the fault trees. These are summarized in Table 2.4 and Fig. 2.22. We have seven guidelines:

1) Replace an abstract event by a less abstract event. Example: "motor operates too long" versus "current to motor too long."
2) Classify an event into more elementary events. Example: "explosion of tank" versus "explosion by overfilling" or "explosion by runaway reaction."
3) Identify distinct causes for an event. Example: "runaway reaction" versus "excessive feed" and "loss of cooling."
4) Couple trigger event with "no protective action." Example "overheating" versus "loss of cooling" coupled with "no system shutdown."
5) Find cooperative causes for an event. Example: "fire" versus "leak of flammable fluid" and "relay sparks."
6) Pinpoint a component failure event. Example: "no current to motor" versus "no current in wire." Another example is "no cooling water" versus "main valve is closed" coupled with "bypass valve is not opened."
7) Develop a component failure via Fig. 2.22. As we trace backward to search for more basic events, we eventually encounter component failure. These events can be developed by using the structure of Fig. 2.22.

If an event in a rectangle can be developed in the form of Fig. 2.22, H. E. Lambert calls it a *state-of-component event*.[2] Otherwise, an event is called a *state-of-system event*. For the state-of-system event, we cannot specify a particular component which is the entire cause of the event. More than one component or subsystems are responsible for a state-of-system event. Such events should be developed by guidelines (1) through (6), until state-of-component events appear. The state-of-component events will

†Very high probability events are typified by component success states which, as emphasized earlier, should not appear in fault trees.

TABLE 2.4 HEURISTIC GUIDELINES FOR FAULT TREE CONSTRUCTION

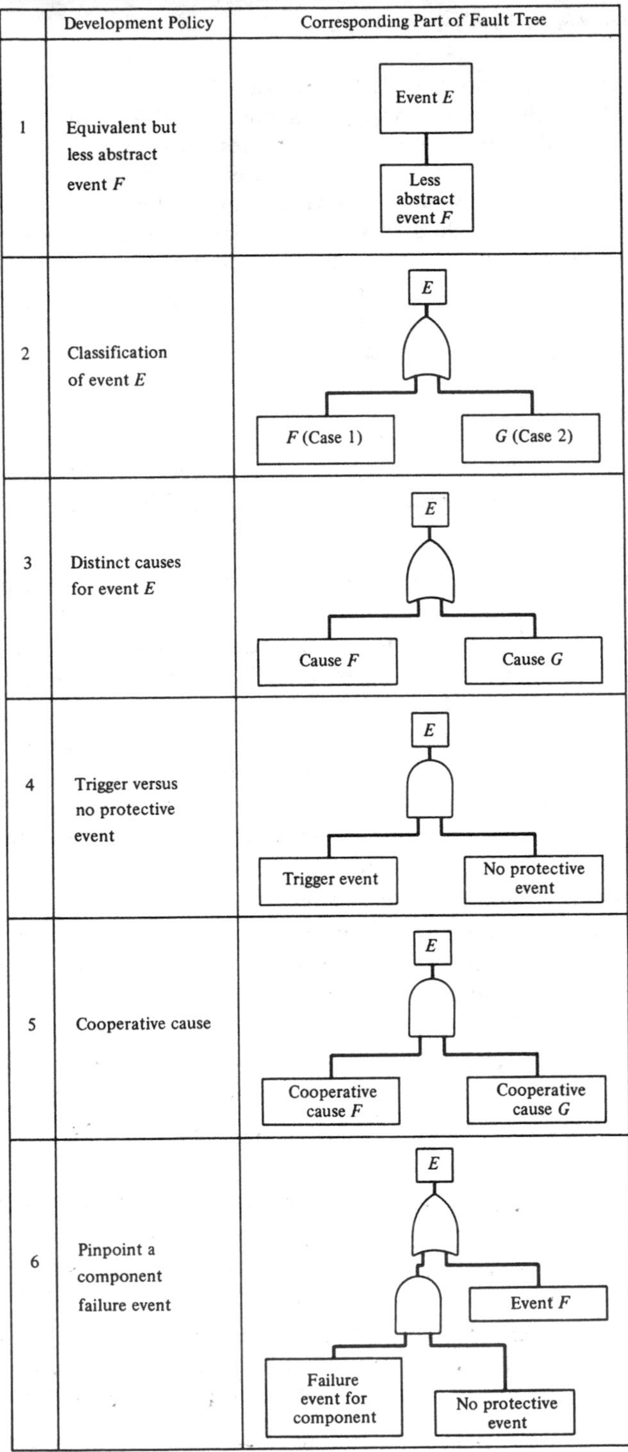

	Development Policy	Corresponding Part of Fault Tree
1	Equivalent but less abstract event F	Event E; Less abstract event F
2	Classification of event E	E; F (Case 1); G (Case 2)
3	Distinct causes for event E	E; Cause F; Cause G
4	Trigger versus no protective event	E; Trigger event; No protective event
5	Cooperative cause	E; Cooperative cause F; Cooperative cause G
6	Pinpoint a component failure event	E; Event F; Failure event for component; No protective event

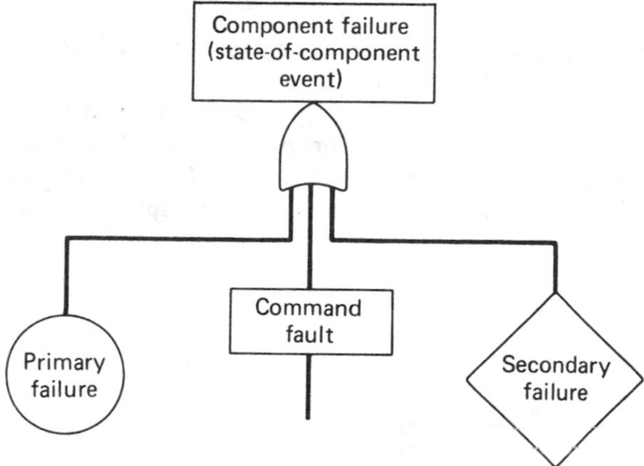

Figure 2.22. Development of a component failure (state-of-component event).

ultimately be developed further in terms of primary failures, secondary failures, and command faults. If the primary or secondary failures are not developed further, they become terminal (basic) events in the fault tree under construction. The command faults are usually state-of-system events which are developed further until relevant state-of-component events are found. The resulting state-of-component events are again developed via Fig. 2.22. The procedure is repeated, and the development is eventually terminated when there is no possibility of command faults.

Top events are usually state-of-system events. A complicated top event should be defined by a so-called *tree top*. The tree top includes the top event and the sub-undesired events, including potential accidents and hazardous conditions that are the immediate causes of the top event. The top event and the sub-events must be carefully defined and all significant causes of the top event identified. Heuristic guidelines (1) through (5) are also useful for constructing the tree top.

To our heuristic guidelines, we can add a few practical considerations by H. E. Lambert[2]:

> Expect no miracles; if the "normal" functioning of a component helps to propagate a fault sequence, it must be assumed that the component functions "normally."† Write complete, detailed fault statements. Avoid direct gate-to-gate relationships. Think locally. Always complete the inputs to a gate. Include notes on the side of the fault tree to explain assumptions not explicit in the fault statements. Repeat fault statements on both sides of the transfer symbols.

†This is equivalent to the removal of very high probability events from the fault tree (see Table 2.3).

Example 2.

This example[11] shows how the heuristic guidelines of Table 2.4 and Fig. 2.22 can be used to construct fault trees.

In the pumping system shown in Fig. 2.23, the tank is filled in 10 min and empties in 50; thus, the cycle time is 1 hr. After the switch is closed, the timer is set to open the contacts in 10 min. If the mechanisms fail, then the alarm horn sounds and the operator opens the switch to prevent a tank rupture due to overfilling.

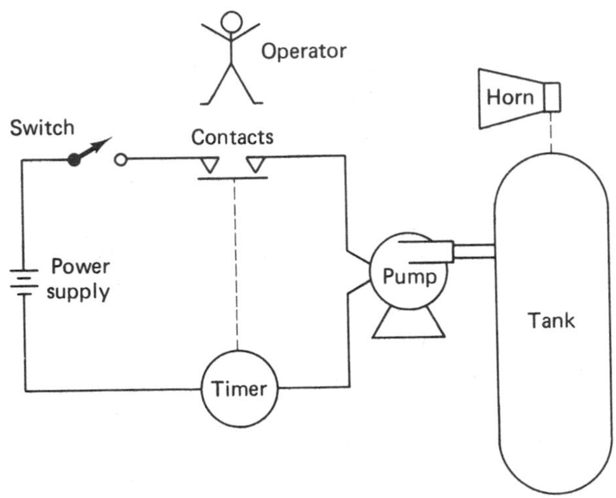

Figure 2.23. Schematic diagram for a pumping system.

A fault tree with the top event of "tank rupture (at time t)" is shown in Fig. 2.24. This tree shows which guidelines are used to develop events in the tree. The operator in this example can be regarded as a system component, and the gate E is developed by using the guidelines of Fig. 2.22. A primary operator failure means that the operator functioning within the design envelope fails to push the panic button when the alarm sounds. The secondary operator failure is, for example, "operator has been killed by a fire when the alarm sounded." The command fault for the operator is "no alarm sounds."

2.5.2 Conditions Induced by OR and AND Gates

We have three applications for OR gates. These are shown in rows 1 through 3 of Table 2.5. Row 1 has two events, A and B, and these two will overlap as shown in the figure by the corresponding *Venn diagram*.† Row 2 subdivides the Venn diagram into two parts: event B plus complement $\bar{B}$ and A. The latter part is equivalent to the *conditional event* $A|\bar{B}$ coupled

[11]From Lambert, H. E.

†Appendix 7.1 provides an explanation of Venn diagrams, and Chapter 4 contains a discussion of conditional probabilities. Readers without this background may wish to skip this section, and return to it later.

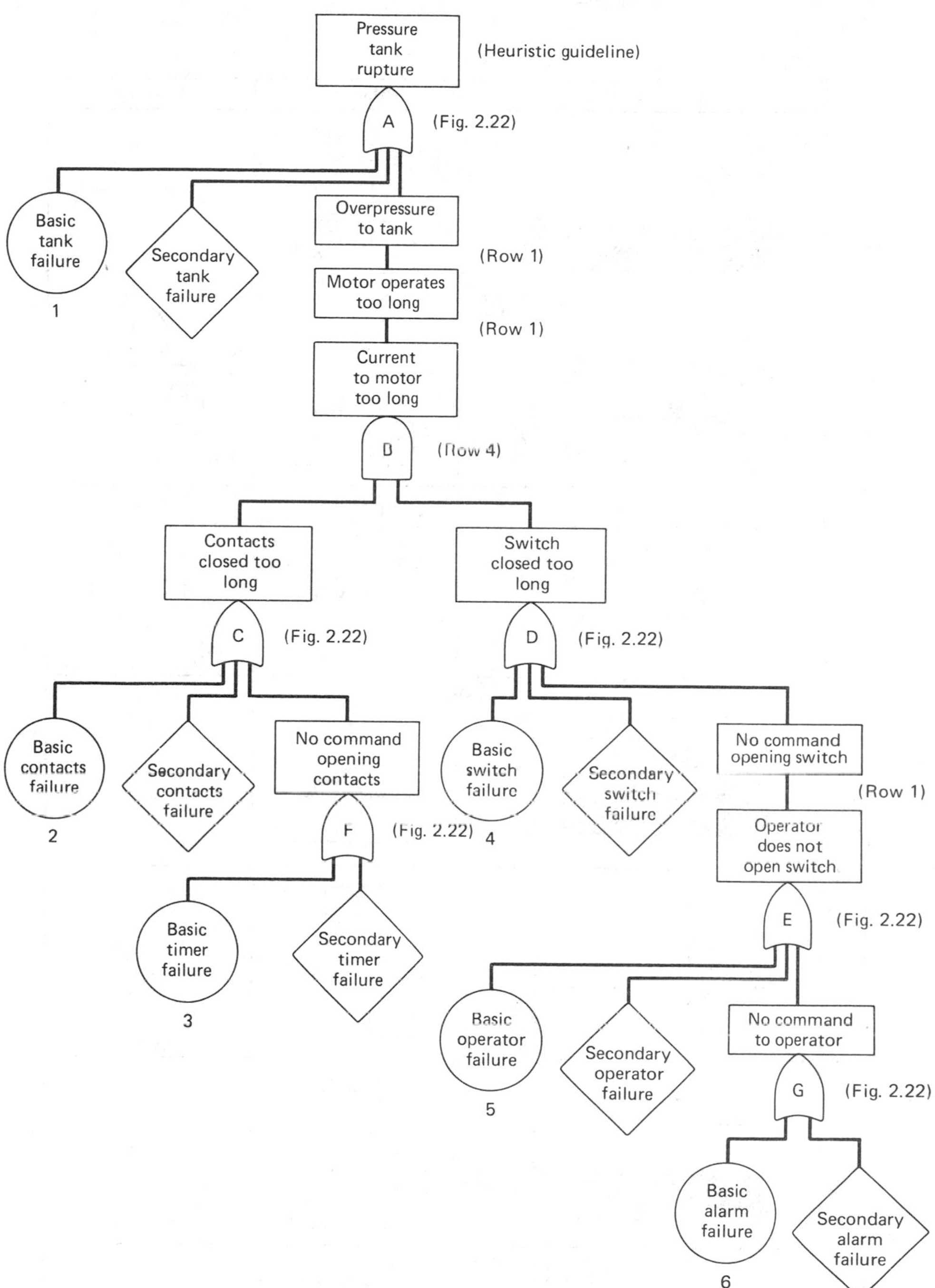

Figure 2.24. Fault tree for pumping system. (The basic event failures are not specified in detail here to save space.)

TABLE 2.5

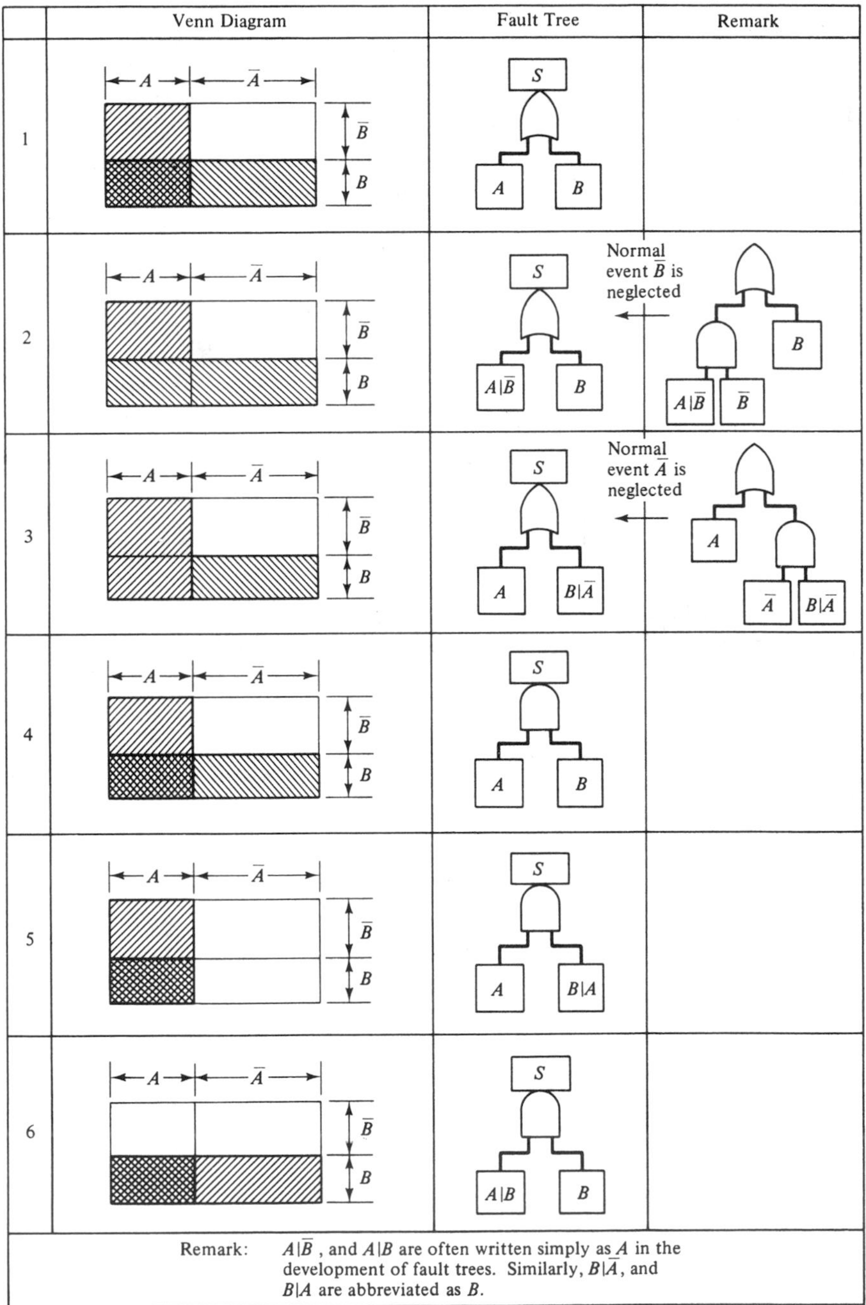

Remark: $A|\bar{B}$, and $A|B$ are often written simply as A in the development of fault trees. Similarly, $B|\bar{A}$, and $B|A$ are abbreviated as B.

with $\bar{B}$ (see the tree in the last column). Conditional event $A|\bar{B}$ means that event A is observed when event $\bar{B}$ is true, i.e., when event B is not occurring. Since $\bar{B}$ is usually a very high probability event, it can be removed from the AND gate and the tree in row 2, column 2 is obtained. An example of this type of OR gate is gate E in Fig. 2.24. Event 5, operator basic failure, is an event conditioned by event $\bar{B}$ meaning "alarm to operator." (event B is "no alarm to operator"). This conditional event implies that the operator (in a normal environment) does not open the switch when there is an alarm. In other words, the operator is careless and neglects the alarm or opens the wrong switch. Considering condition $\bar{B}$ for the primary operator failure, we estimate that this failure has a relatively small probability. On the other hand, the unconditional event, "operator does not open switch," has a very high probability, since he would open it only when the tank is about to explode and the alarm horn sounds. These are quite different probabilities, which depend on whether the event is conditioned. These three uses for OR gates provide a useful background for quantifying primary or secondary failures in fault trees.

Rows 2 and 3 of Table 2.5 introduce conditions for branching of fault trees. They account for why the fault tree of Fig. 2.19 in Example 1 could be terminated by the primary and secondary fuse failures. All OR gates in the tree are used in the sense of row 3. We might have been able to introduce a command fault for the fuse. However, the command fault cannot occur, since at this final stage we have the following conditions:

1) Normal motor (i.e., no primary or secondary failures).
2) Wire is connected (same as above).
3) Switch is closed (same as above).
4) Generator is not dead (same as above).
5) Fuse is connected (same as above).

Three different situations exist for AND gates. They are shown by rows 4 through 6 in Table 2.5.

Example 3.[12]

Figure 2.25 shows a reaction system in which the temperature increases with the feed rate of flow-controlled stream D. Heat is removed by water circulation through a water-cooled exchanger. Normal reactor temperature is 200°F, but a catastrophic runaway will start if this temperature reaches 300°F. In view of this situation:

1) The reactor temperature is monitored.
2) Rising temperature is alarmed at 225°F (see horn).

[12]Browning, R. L., "Human Factors in Fault Trees," *Chem. Eng. Prog.*, **72**, June, 1976.

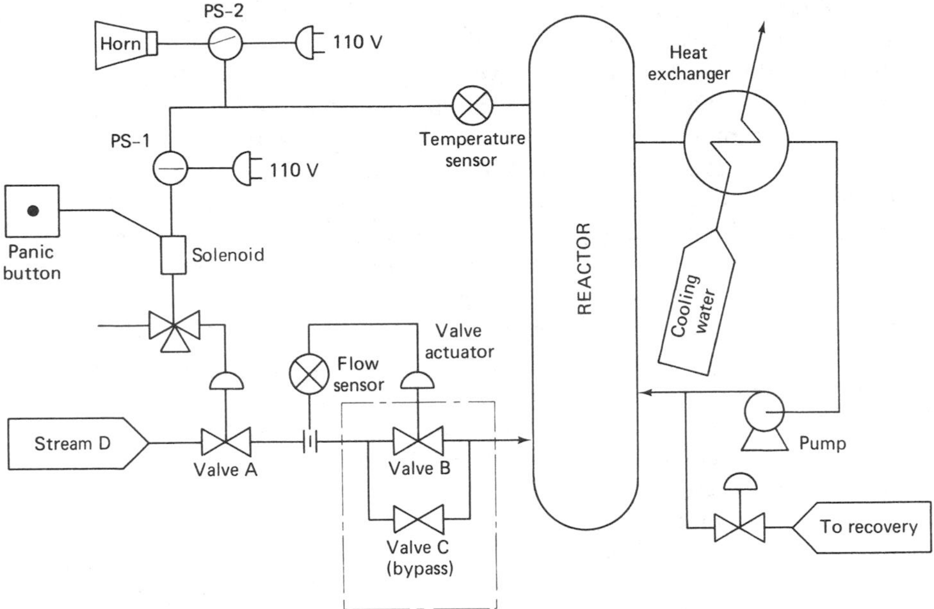

Figure 2.25. Schematic diagram for reactor system.

3) An interlock shuts off stream D at 250°F, stopping the reaction (see solenoid and valve A).
4) The operator can initiate the interlock by punching the panic switch.

Construct a fault tree for this system.

Solution: The fault tree is shown in Fig. 2.26. Secondary failures are neglected. It is further assumed that the alarm signal always reaches the operator whenever the horn sounds, i.e., the alarm has a sufficiently large signal-to-noise ratio. Heuristic guidelines and gate usages are indicated in the fault tree. It is recommended that the reader trace them.

One might think that the event "non-zero feed" to gate C is a very high probability event because the feed is non-zero if the system is in normal operation. This is not true, since this "non-zero feed" is conditioned by "loss of cooling to reactor." Under these circumstances, the shut-down system will operate, and the "non-zero feed" has a small probability of occurrence. This event is an input to an AND gate and should not be neglected.

Table 2.5, if properly applied:

1) Clarifies and quantifies events.
2) Finds very high or very low probability events.
3) Terminates further development of a fault tree under construction.
4) Provides a clear background and useful suggestions at each stage of fault tree construction.

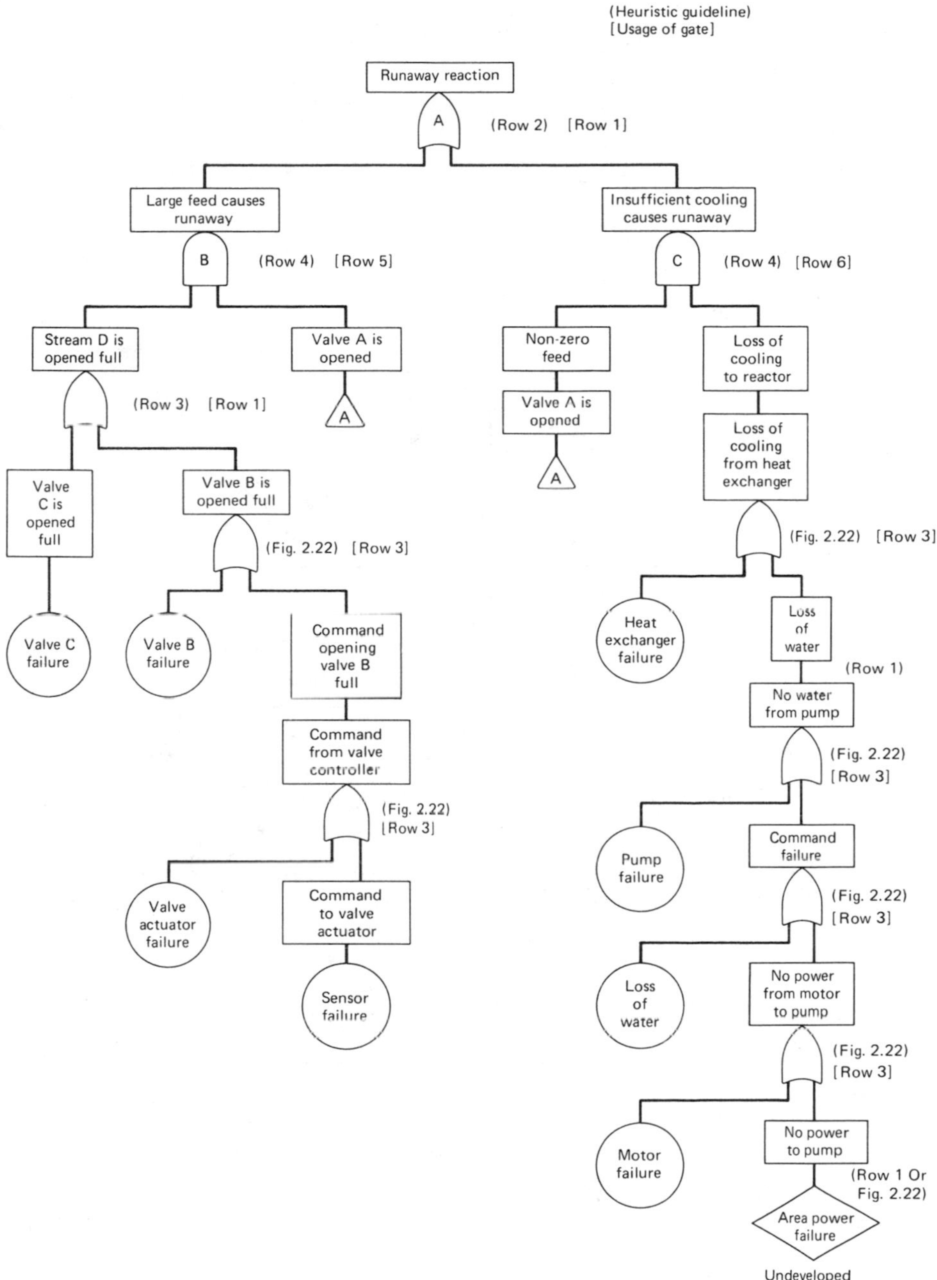

Figure 2.26. Fault tree for reactor system.

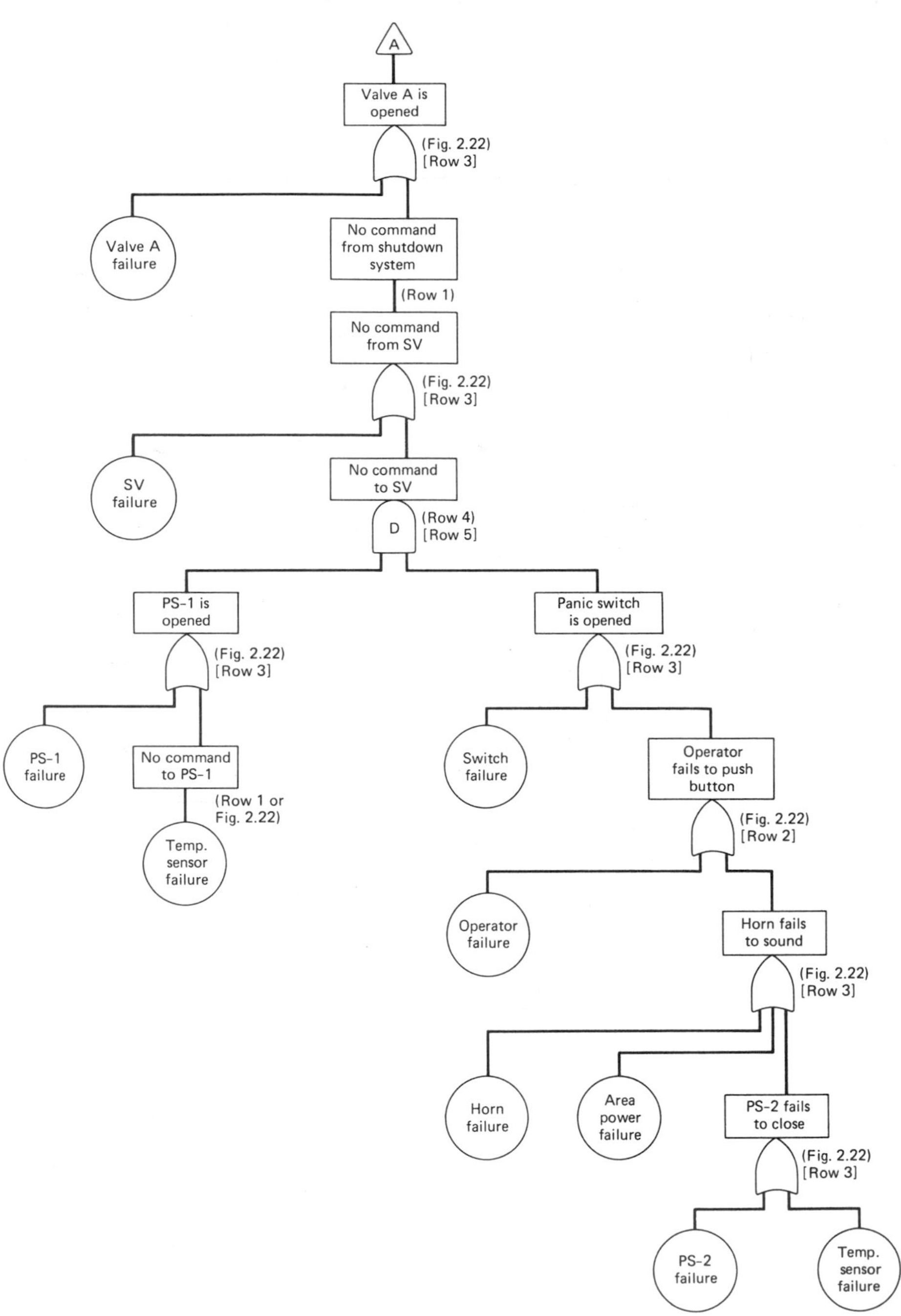

Figure 2.26. (continued).

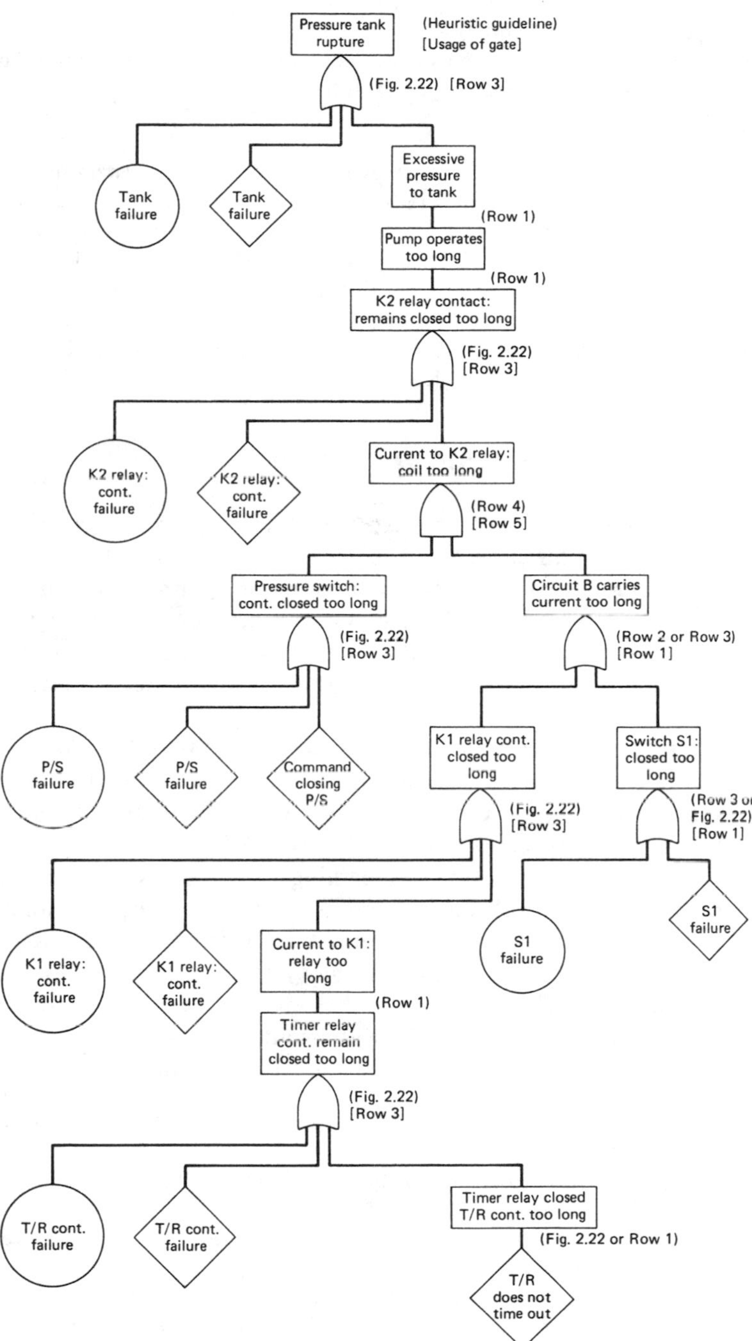

Figure 2.27. Fault tree for pressure tank system.

We note that gates *B*, *C*, and *D* are obtained by the guidelines of row 4 in Table 2.4. Systems should be designed so as to protect them from a single failure event. In such systems, AND gates appear in the fault tree.

Example 4 .

Consider the pressure tank system shown in Fig. 2.16. This system has been a bit of a straw-man since it was first published by Vesely in 1971 (*Trans. Nuclear Sci.*, NS-18, **1**, 472, 1971). It appears also in Barlow and Proschan, *Statistical Theory of Reliability and Life Testing Probability Models*, Holt, Rinehart and Winston, New York, 1975. The fault tree given here in Fig. 2.27 is identical to that given by Lambert[2] except for some terminologies and minor modification. The fault tree can be constructed by the heuristic guidelines of Table 2.4. It also includes the gate usages of Table 2.5.

2.6 FAULT TREE CONSTRUCTION BY DECISION TABLES

The fault tree construction method developed in the previous section is heuristic insofar as human judgment is required to develop state-of-system events. In general, fault tree construction requires time and money for large, complicated systems: at least two man-years for a reactor system.[13] In addition, human errors frequently creep into the construction process, although it is theoretically possible to produce reasonably precise fault trees if one is sufficiently careful.

Various automated approaches have been proposed to avoid errors caused by human tedium and to simplify the construction of fault trees.[14–17] In this section we discuss a systematic fault tree construction method which uses *decision tables*.[18]

Given sufficient information concerning the specific system to be analyzed, and a library of component models, this approach allows a rapid, systematic construction of a fault tree which is as complete and detailed as the component models and the system description supplied. A computer code called *CAT* (Computer Automated Tree) is available.[16]

First, a set of events is assigned to each output from each component. These events are called *output events*. Each of the output events specifies

[13]WASH 1400, as cited by Wall, I. A., "Some Insights from the Reactor Safety Study," Office of Nuclear Regulatory Research, U. S. NRC.

[14]Fussell, J. B., "Synthetic Tree Model. A Formal Methodology for Fault Tree Construction," **UC32**, ANCR-32, 1973.

[15]Powers, G. J., and F. C. Tompkins, Jr., "Fault Tree Synthesis for Chemical Processes," *AIChE Journal*, **20**, 2, 376–387, 1974.

[16]Salem, S. L., G. E. Apostolakis, and D. Okrent, "A New Methodology for the Computer-Aided Construction of Fault Trees," *Annals of Nuclear Energy*, **4**, 417–433, 1977.

[17]Kumamoto, H., and E. J. Henley, "Safety and Reliability Synthesis of Systems with Control Loops," *AIChE Journal*, **25**, No. 1, 108, 1979.

[18]Pollack, S. L., *Decision Tables: Theory and Practice*, Wiley-Interscience, New York, 1971.

the state of the output. For example, the flow rate from a valve may have one of three output events: i.e., high, normal, or zero flow rate. More states could be specified if the complexity can be justified.

Similarly, a set of *input events* is assigned to each input to the component. These specify the states of the inputs. For instance, input pressure to the valve may be assigned one of the three input events: i.e., high, normal, or zero input pressure.

So-called *component internal modes* are regarded as inputs from another component or from the system environment. The opening of a valve may be regarded as an input which has one of the three input events: full, normal, or zero opening. When the valve opening is adjusted by a controller, it is regarded as an input from the controller. When the opening is independent of other components, it is considered an input from the system environment.

Every input from the system environment is regarded as a basic event, while an input event from another component is a state-of-system or state-of-component event. The entire set of inputs events and output events constitute events allowed in the system.

Each component is modelled by a decision table, which is an extended version of a truth table. The decision table describes how each combination of input events specifies the output events, i.e., the state of the output. A component can have plural inputs, but it should have only one output. The assumption of the one output for each component reduces the size of input/output (I/O) entries in the decision table.

When a component has more than one output, a hypothetical component can be defined for each output. For example, when a heat exchanger has two outputs of interest, such as flow rate and temperature, two hypothetical heat exchangers must be used in the system diagram. Two decision tables are then required to model one heat exchanger.

The component connectivity yields the system description. It is obtained by connecting each output from a component with relevant inputs to other components. A top event is, as before, an output event of interest.

To illustrate the decision table approach, consider the cooling system shown in Fig. 2.28. A pump causes cooling water to flow through a heat exchanger and a valve. Acid flows into the exchanger and is cooled by the water.

The component connectivity is expressed by the I/O relation shown in Fig. 2.29. The pump internal mode is regarded as an input from the system environment. This input consists of one of two events: "normal operation" or "pump stops." The cooling water pressure is an output from the pump and an input to the valve. The pressure is either normal or zero. The valve opening is another input to the valve; it is always normal in this example. The cooling water flow rate is an output from the valve. The flow rate is either high, normal, or zero.

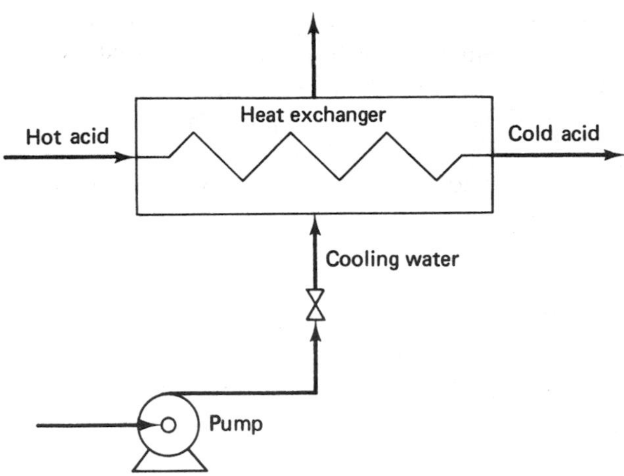

Figure 2.28. Schematic diagram for heat exchanger.

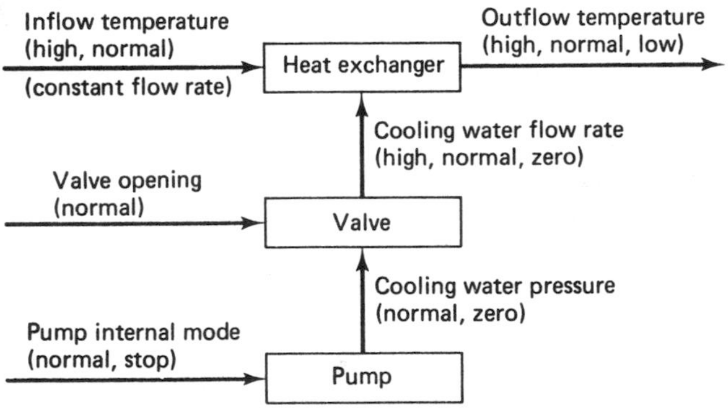

Figure 2.29. I/O relations for cooling system.

The heat exchanger has two inputs, the cooling water and the temperature of the acid. The flow rate of the acid is assumed constant in this example. The temperature is either normal or high. The high temperature is considered as an input from the system environment because it is not in the domain of the cooling system as defined for this problem.

The temperature of the outflow acid is an output of the heat exchanger. The outflow temperature is either high, normal, or low. The high temperature causes a system hazard, since the outflowing acid is used in a subsequent reactor. Hence, the top event is defined as "high temperature of the outflow acid."

The decision tables which model the components are now considered. The I/O relation of the pump is expressed by Table 2.6. Pump stoppage results in zero output pressure, whereas normal pump operation creates normal pressure. Next, the valve is modelled by Table 2.7. The valve opening is excluded from the input columns because the opening is assumed to be normal. The heat exchanger is expressed by decision Table 2.8, which has two input columns. All combinations of the input events appear in the table. Row c, the pair "high flow rate" and "high input temperature," is assumed to result in normal output temperature. Pair d, "high flow rate" and "normal input temperature," creates low output temperature because excess cooling water is available in the heat exchanger. It is important to recognize that the construction of decision tables requires a knowledge of component functions and interrelationships which may be system-specific.

TABLE 2.6 Decision Table for Pump

	INPUT	OUTPUT
	Pump Internal Mode	Output Pressure
h	Normal	Normal
i	Stop	Zero

TABLE 2.7 Decision Table for Valve

	INPUT	OUTPUT
	Cooling Water Pressure to Valve	Cooling Water Flow Rate from Valve
j	Normal	Normal
k	Zero	Zero

TABLE 2.8 Decision Table for Heat Exchanger

	INPUT		OUTPUT
	Cooling Water Flow Rate to Heat Exchanger	Temperature of Inflow Acid	Temperature of Outflow Acid
a	Normal	High	High
b	Normal	Normal	Normal
c	High	High	Normal
d	High	Normal	Low
e	Zero	High	High
f	Zero	Normal	High

Rows e and f in Table 2.8 show that the output temperature is high regardless of the event in the second input column, if the first input column takes the value "zero." We can simplify rows e and f, thus obtaining row g in Table 2.9. The symbol "—" denotes "*don't care*." Table 2.9 is a simplified version of Table 2.8. A general procedure for simplifying a decision table is as follows.[16]

TABLE 2.9 Simplified Decision Table for the Heat Exchanger

	INPUT		OUTPUT
	Cooling Water Flow Rate to Heat Exchanger	Temperature of Inflow Acid	Temperature of Outflow Acid
a	Normal	High	High
b	Normal	Normal	Normal
c	High	High	Normal
d	High	Normal	Low
g	Zero	—	High

1) Search for rows with identical output events.
2) For rows with identical output events, search for those rows which agree in all but one input column.
3) A "don't care" situation then occurs for the inputs in that column if, in the rows obtained in step 2, every possible input event with identical input entries appears in the input column. Eliminate all but one of the rows and replace the input event in that column with a "—."

Let us now analyze the method for constructing the fault tree of Fig. 2.30 for the top event, "high outflow temperature." The row symbols in Fig. 2.30 show how the tree was constructed.

A search is made for rows with the top event in its output column. Rows a and g in Table 2.9 are detected. Since both rows have the correct output, they are connected by an OR gate. Now the two rows must be developed. Row g has a "don't care" event in the input column. This means that zero flow rate of the cooling water can cause the top event regardless of the other input events. Thus, row g is replaced by the zero flow rate event. Row a has two events in the input columns: "high inflow temperature" and "normal cooling water flow rate." The result is an AND gate with two input events. The high temperature event is regarded as a primary event, since it is an input from the system environment. The event is enclosed by a circle, implying that it is a basic event for which data are available.

We now have two undeveloped events: one is "zero cooling water flow rate to heat exchanger" and the other is "normal cooling water flow rate to

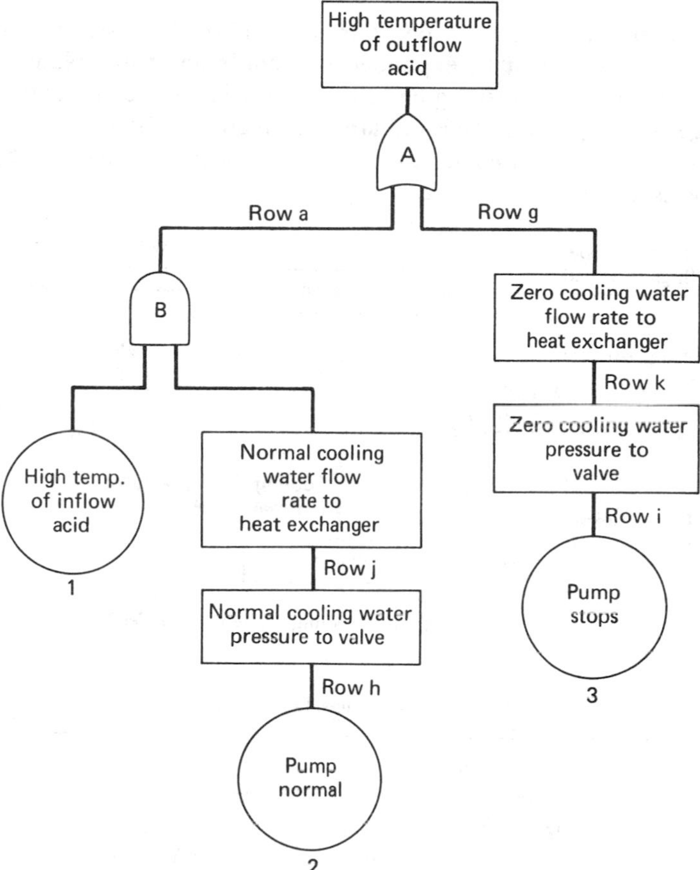

Figure 2.30. Fault tree for cooling system.

the exchanger." A search is made for rows with the undeveloped events in the output column. Row k is found for the zero flow rate and row j is detected for the normal flow rate (Table 2.7). Row k is attached to the zero flow rate, whereas row j is connected with the normal flow rate. Since each row has only an input event in its input column, rows k and j are replaced by the normal pressure event and zero pressure event, respectively.

A search of component decision tables is made for rows containing one of these two events in the output columns. Rows h and i in Table 2.6 are detected and replaced by the corresponding input events. We obtain the fault tree of Fig. 2.30.†

We see from the fault tree that the high temperature of the inflow acid causes the top event even if the cooling water pump is operating. In order to improve the system, let us introduce the feed-forward control loop shown in Fig. 2.31. The valve controller operates so as to enlarge the valve

†The general practice is to avoid listing non-fault events or very high probability events such as "pump normal" in fault trees. It is done here only for pedagogical reasons.

opening when the inflow temperature is high. In this case, the decision table for the valve must be expanded to include an input column, "valve opening," because the opening has two possible states, i.e., normal opening and large opening. This decision table is given by Table 2.10 which, in turn, is simplified, producing Table 2.11. The decision table for the valve controller is given as Table 2.12.

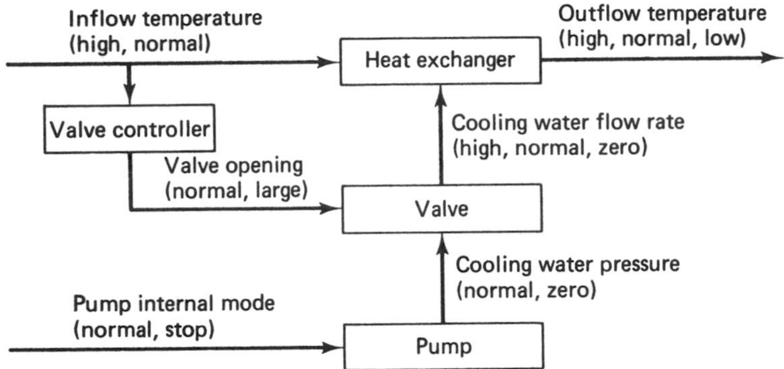

Figure 2.31. I/O relations for cooling system with feed-forward control loop.

TABLE 2.10 DECISION TABLE FOR VALVE

INPUT		OUTPUT
Cooling Water Pressure to Valve	Valve Opening	Cooling Water Flow Rate from Valve
Normal	Normal	Normal
Normal	Large	High
Zero	Normal	Zero
Zero	Large	Zero

TABLE 2.11 SIMPLIFIED DECISION TABLE FOR THE VALVE

	INPUT		OUTPUT
	Cooling Water Pressure to Valve	Valve Opening	Cooling Water Flow Rate from Valve
m	Normal	Normal	Normal
n	Normal	Large	High
o	Zero	—	Zero

TABLE 2.12 Decision Table for Valve Controller

	INPUT	OUTPUT
	Temperature of Inflow Nitric Acid	Valve Opening
p	Normal	Normal
q	High	Large

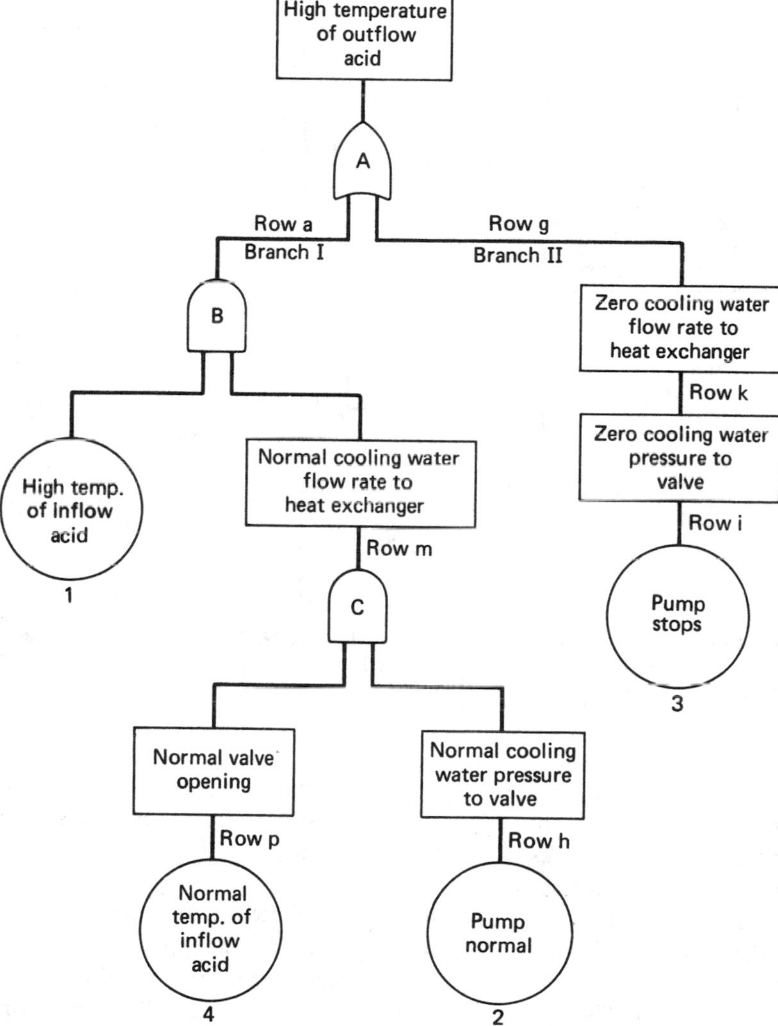

Figure 2.32. Fault tree for cooling system with feed-forward control loop.

The fault tree of Fig. 2.32 is obtained by the procedure of the last example. We see from Fig. 2.32 that there is no possibility of the simultaneous occurrence of the three basic events 1, 2, and 4 because event 1 contradicts event 4. Thus, we can remove the zero-possibility branch I from the OR gate, obtaining Fig. 2.33. We observe that the event "pump stops" is the only fault creating the top event.

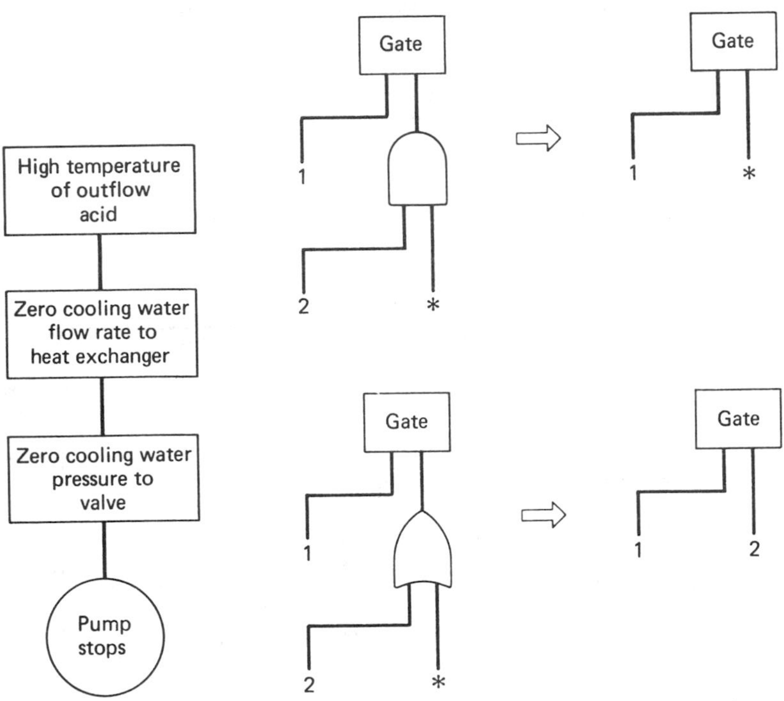

Figure 2.33. Simplified fault tree for cooling system with feed-forward control loop.

Figure 2.34. Fundamental simplification by zero-possibility branch (*).

We can simplify the fault tree either during the process of its construction or after the tree has been constructed. The fundamental simplifications are shown in Figs. 2.34 and 2.35. The simplified fault tree is more easily understood intuitively.

There are only shades of differences between automated fault tree construction by linking decision tables or by linking mini-fault trees, because mini-fault trees must be converted into tabular form prior to computer entry. The synthetic tree method of J. B. Fussell[14] is very similar, if not identical, to the decision table approach used in CAT.

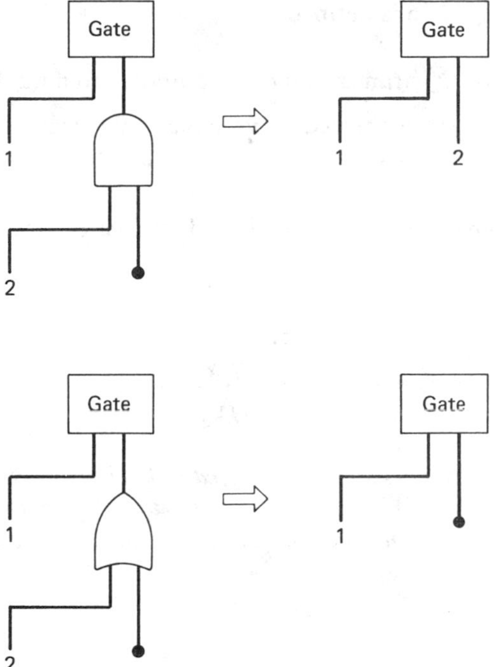

Figure 2.35. Fundamental simplification by certainly-occurring branch (●).

2.7 FAULT TREE CONSTRUCTION FOR SYSTEMS WITH COMPLEX CONTROL LOOPS

Systems such as chemical plants involve large numbers of control loops. In this section we develop a *signal-flow-based* graphical technique to model the system, and apply *Mason's rule* to assess the effect of the control loops. Fault trees for control loop failures are constructed by analyzing the causes for loop transmittance failures.

2.7.1 Signal Flow Graphs

A *signal flow graph* is a graph that consists of *nodes* and directed *branches* connecting them. A node represents a process variable such as flow rate, pressure, level, etc. Each branch has a constant called its *branch transmittance*.

The signal flow graph defines relations among the variables at the nodes; this can best be illustrated by a simple flow graph. For example, variable X_3 of Fig. 2.36 is defined by

$$X_3 = \sum (\text{transmittance on a branch to node } X_3) \cdot (\text{the antecedent variable of the branch})$$
$$= 3X_1 + 5X_2$$

Similarly, the signal flow graph of Fig. 2.37 indicates that

$$X_2 = X_1 - X_5$$
$$X_3 = G_1 X_2$$
$$X_4 = G_2 X_3$$
$$X_5 = H X_4$$

This flow graph models a *negative feedback* control system for positive transmittance H. Variables X_1, X_2, X_3, X_4, and X_5 denote *reference input*, *error*, *control signal*, *controlled variable*, and *measurement signal*, respectively.

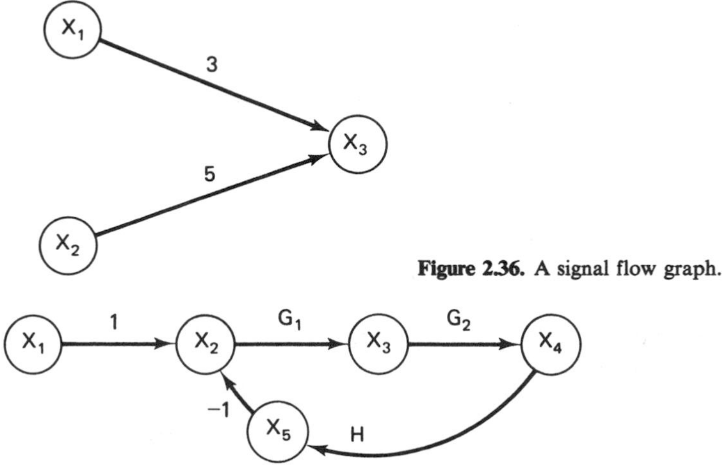

Figure 2.36. A signal flow graph.

Figure 2.37. A signal flow graph containing a loop.

2.7.2 A Signal Flow Graph for a Cascade Level-Control System

Consider the cascade level-control system of Fig. 2.38. Bypass valves B and C are usually closed. In the block diagram for the control system shown in Fig. 2.39, the *outer control loop* monitors the level of the fluid in the tank, and the level controller determines reference input Q_r to the *inner*

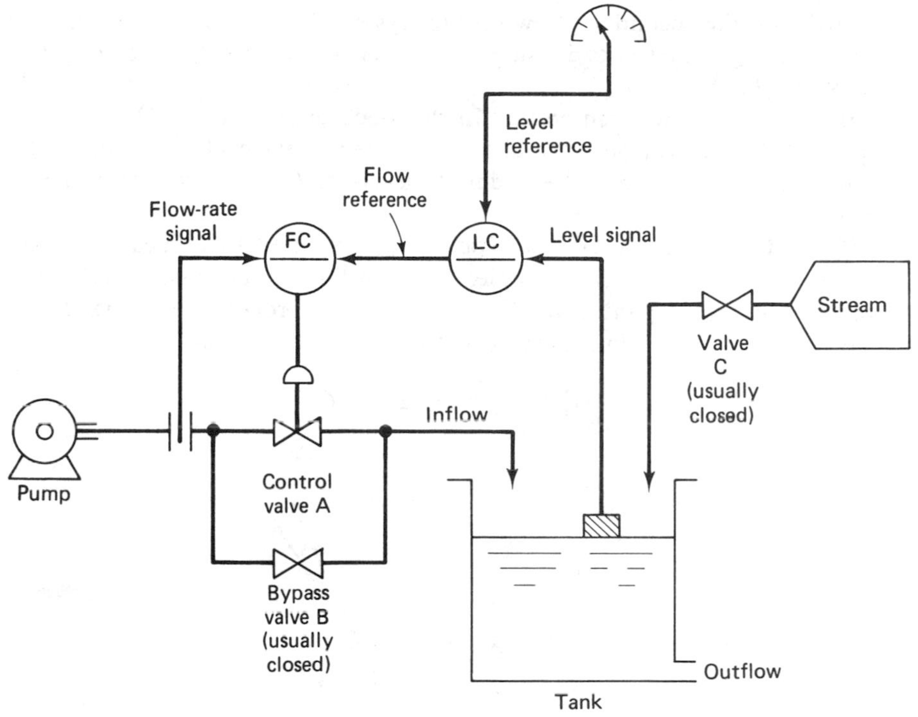

Figure 2.38. A cascade level-control system.

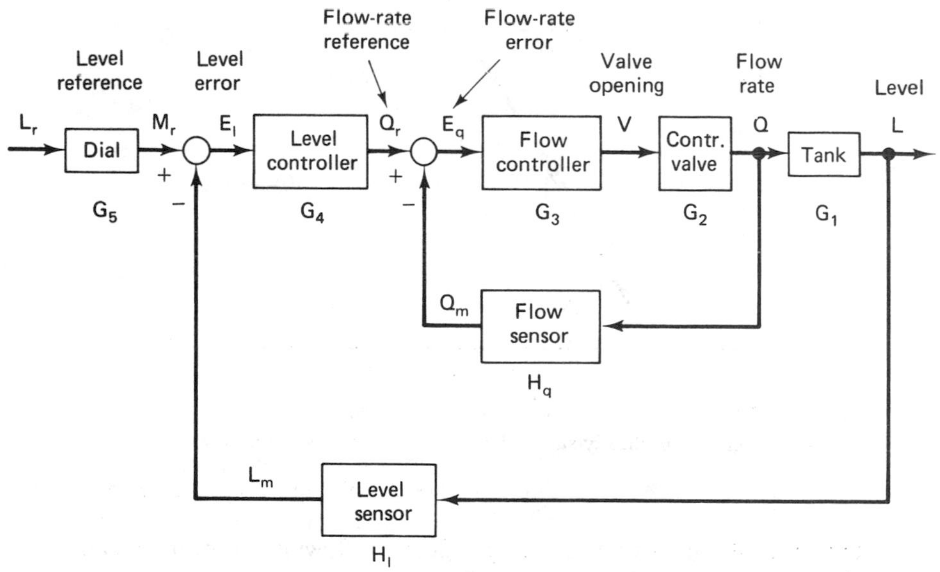

Figure 2.39. A block diagram for the cascade level-control system.

control loop, the secondary flow control system. The inner loop monitors inflow rate Q and works in such a way as to regulate Q according to reference Q_r.

By convention, we assume that in the block diagram the I/O variables represent deviations from normal states. If the normal level of the fluid is 5 meters, then $L=1$ and $L=-1$ denote levels of 6 and 4 meters, respectively.

The I/O relationship at each block is approximated by a linear relation at the steady state. For example, the non-linear, steady-state relation between flow rate Q and level L is shown by the broken line in Fig. 2.40. The linear approximation is given by the straight line. Thus,

$$G_1=32 \quad \text{and} \quad L=32Q$$

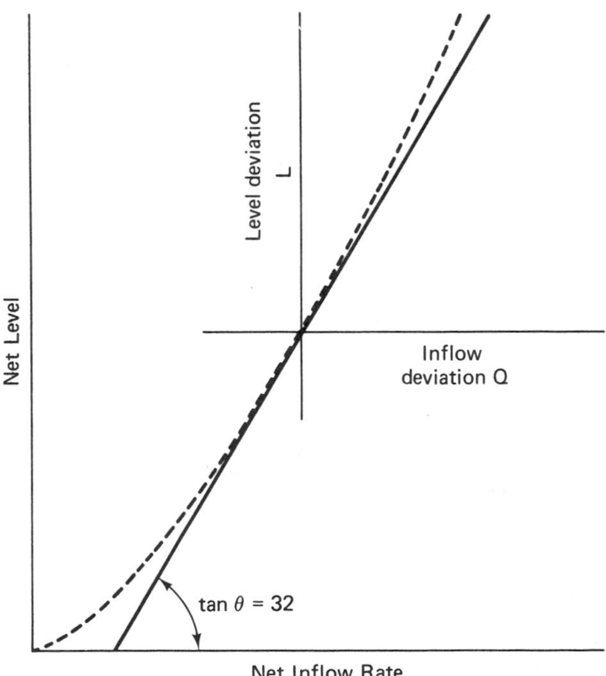

Figure 2.40. Steady-state relation between inflow rate and level.

The block diagram of Fig. 2.39 yields the following algebraic equations, resulting in the signal flow graph of Fig. 2.41.

$$M_r = G_5 L_r, \qquad Q = G_2 V$$
$$E_l = M_r - L_m, \qquad L = G_1 Q$$
$$Q_r = G_4 E_l, \qquad L_m = H_l L$$
$$E_q = Q_r - Q_m, \qquad Q_m = H_q Q$$
$$V = G_3 E_q,$$

The transmittances at the normal operating mode are

$$G_1 = 32 \qquad G_5 = 1$$
$$G_2 = 0.0031 \qquad H_l = 1$$
$$G_3 = 10{,}000 \qquad H_q = 1$$
$$G_4 = 3.2$$

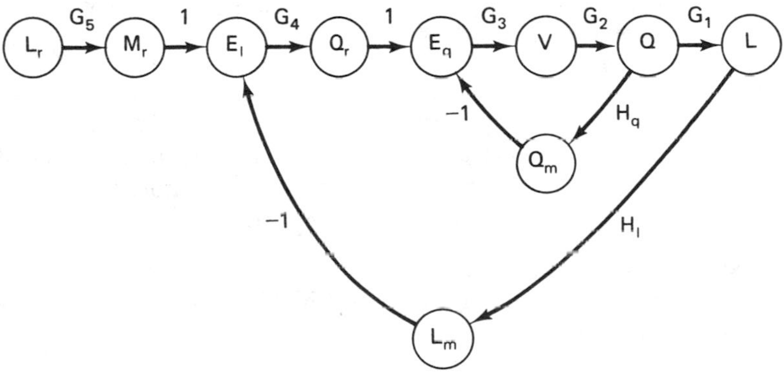

Figure 2.41. A signal flow graph for the cascade level-control system.

2.7.3 Modeling Basic Failures

The signal flow graph of Fig. 2.41 has only one source variable, L_r, the reference input to the outer control loop. We now introduce ten other source variables, X_1 to X_{10}, to model basic failures. The resulting signal flow graph has the source variables shown in Fig. 2.42. If bypass valve B is opened inadvertently, then the inflow rate and level increase, and an overflow of the fluid from the tank results. The bypass valve failure can be modeled by source variable X_2 because the failure yields additional inflow to the tank: $X_2 = 0$ if the bypass valve is closed, and $X_2 > 0$ if the valve is opened.

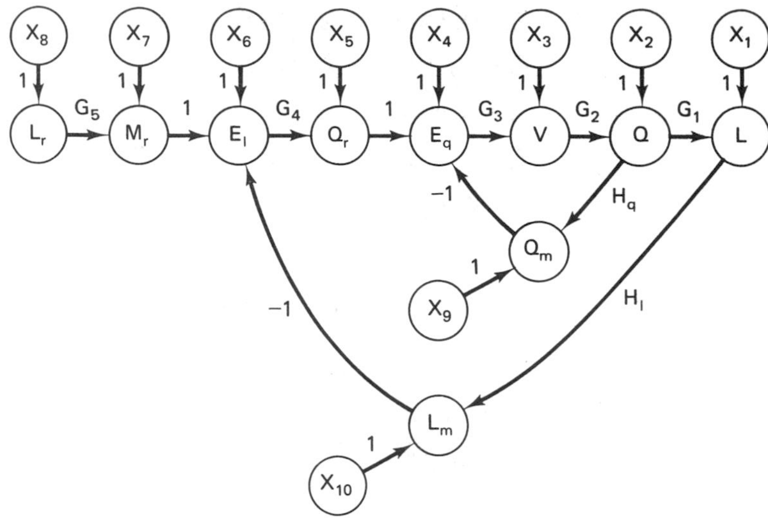

Figure 2.42. A revised signal flow graph including disturbances X_i's to the control system.

Inadvertent opening of valve *C*, on the other hand, is not modeled by variable X_2, since the resulting increase in inflow rate cannot be monitored by the flow sensor. The failure of valve *C*, however, may be represented by source variable X_1, since the additional inflow is monitored by the level sensor indirectly. The implications of the other source variables are:

X_3: Flow controller failure (biased high).
X_4: Flow rate comparator failure (biased high).
X_5: Level controller failure (biased high).
X_6: Level comparator failure (biased high).
X_7: Level reference dial failure (biased high).
X_8: Operator error (high level reference).
X_9: Flow sensor failure (biased low).
X_{10}: Level sensor failure (biased low).

The transmittances also deviate from their norms when basic failures occur. For simplicity we consider only cut-off failures of the control loops, which is equivalent to some transmittances decreasing to zero. For example, $H_l=0$ indicates that the level sensor is stuck, making the measurement signal from the sensor insensitive to the actual level of the fluid.

Combinations of the zero transmittances and source variables can be used to represent a variety of basic failures. For instance, ($H_l=0$, $X_{10}=-6$) shows that the level sensor being stuck at low level results in an increase of inflow rate reference Q_r despite the high level of fluid in the tank. This is a reversed action of the outer control loop.

2.7.4 Mason's Rule

Let X be a source variable. Denote by Y a sink or an intermediate variable. *Mason's rule* states that

$$Y = X\frac{\sum_k R_k D_k}{D} \tag{2.1}$$

where D = determinant of the signal flow graph,

R_k = transmittance of path k from source X to Y.

D_k = determinant of the partial graph which does not touch path k.

The *principle of superposition* is applied when we need an expression for Y in terms of several source variables X_i's.

The determinant of the graph D is defined by

$$D = 1 - \sum_1 P_1 + \sum_2 P_2 - \sum_3 P_3 \ldots \tag{2.2}$$

where P_m is the product of the transmittances of m independent loops, the loops being called *independent* when they have no common nodes. To illustrate Mason's rule we now analyze the signal flow graph of Fig. 2.42. There are 10 sources, X_1 to X_{10}, and we represent sink L in terms of those sources.

The graph has two loops.

$$\text{Loop 1:} \quad E_q - V - Q - Q_m$$
$$\text{Loop 2:} \quad E_l - Q_r - E_q - V - Q - L - L_m$$

The transmittance of each loop is defined by the product of the transmittances in the loop:

$$\text{Loop 1:} \quad G_3 G_2 H_q(-1)$$
$$\text{Loop 2:} \quad G_4(1) G_3 G_2 G_1 H_l(-1)$$

These two loops are not independent because they have common nodes E_q, V, and Q; thus, we neglect the summations over the combinations of two or more independent loops:

$$D = 1 - \sum_1 P_1 = 1 + G_3 G_2 H_q + G_4 G_3 G_2 G_1 H_l$$

There is a single path from X_8 to L; i.e., $X_8 - L_r - M_r - E_l - Q_r - E_q - V - Q - L$. Hence, the path transmittance R_k for the source L_r is

$(1)G_5(1)G_4(1)G_3G_2G_1$. Determinant D_k of the partial graph is obtained by applying (2.2) to the loops that do not touch the path. Since loop 1 and loop 2 touch the path, determinant D_k is $1-0=1$, and the Mason rule result becomes:

$$L=\frac{X_8 \cdot G_5G_4G_3G_2G_1 \cdot 1}{1+G_3G_2H_q+G_4G_3G_2G_1H_l}$$

Source X_1 has a single path to L, i.e., X_1-L. The path transmittance for the source is unity. Loop 1 does not touch the path but loop 2 does. Thus, determinant D_k for this source is $1+G_3G_2H_q$, yielding

$$L=\frac{X_1 \cdot 1 \cdot (1+G_3G_2H_q)}{1+G_3G_2H_q+G_4G_3G_2G_1H_l}$$

In a similar way we can obtain the path transmittances R_k and partial graph determinants D_k for other sources summarized in Table 2.13. Superposing Mason's rule over the 10 sources yields

$$L=\frac{\text{numerator}}{\text{denominator}} \tag{2.3}$$

where

$$\text{denominator}=D=1+G_3G_2H_q+G_4G_3G_2G_1H_l$$

$$\begin{aligned}\text{numerator}=&(1+G_3G_2H_q)X_1+G_1X_2+G_2G_1X_3+G_3G_2G_1X_4\\&+G_3G_2G_1X_5+G_4G_3G_2G_1X_6+G_4G_3G_2G_1X_7\\&+G_5G_4G_3G_2G_1X_8-G_3G_2G_1X_9-G_4G_3G_2G_1X_{10}\end{aligned}$$

TABLE 2.13 PATH TRANSMITTANCES AND PARTIAL GRAPH DETERMINANTS FOR SOURCES

Source	Path	R_k	Loop	D_k
X_1	X_1-L	1	Loop 1	$1+G_3G_2H_q$
X_2	X_2-Q-L	G_1	None	1
X_3	$X_3-V-Q-L$	G_2G_1	None	1
X_4	$X_4-E_q-V-Q-L$	$G_3G_2G_1$	None	1
X_5	$X_5-Q_r-E_q-V-Q-L$	$G_3G_2G_1$	None	1
X_6	$X_6-E_l-Q_r-E_q-V-Q-L$	$G_4G_3G_2G_1$	None	1
X_7	$X_7-M_r-E_l-Q_r-E_q-V-Q-L$	$G_4G_3G_2G_1$	None	1
X_8	$X_8-L_r-M_r-E_l-Q_r-E_q-V-Q-L$	$G_5G_4G_3G_2G_1$	None	1
X_9	$X_9-Q_m-E_q-V-Q-L$	$-G_3G_2G_1$	None	1
X_{10}	$X_{10}-L_m-E_l-Q_r-E_q-V-Q-L$	$-G_4G_3G_2G_1$	None	1

2.7.5 Defining Top Events

The top event of a fault tree can be defined by a set of inequalities on the node variables of the signal flow graph. Assume in the cascade control system that the tank has a height of 10 meters, that the normal level of fluid in the tank is 5 meters, and that the top event is the overflow of fluid. Variable L of Fig. 2.42 is the deviation from the normal level; thus the top event is expressed by the inequality

$$L \geq 5 \tag{2.4}$$

2.7.6 Classifying the Cut-Off Failures of the Loops

The cut-off failure of loop 1 occurs when one of the transmittances in the set $\{G_3, G_2, H_q\}$ takes on the value zero. Similarly, a cut-off failure of loop 2 results if one of the transmittances in the set $\{G_4, G_3, G_2, G_1, H_l\}$ decreases to zero. We have four possible loop states:

1) Both loop 1 and loop 2 are normal. All transmittances have the normal values given in Section 2.7.2. The determinant D of the signal flow graph is

$$D = 1 + G_3G_2H_q + G_4G_3G_2G_1H_l \cong 3200$$

2) Loop 1 is cut off, but loop 2 is normal. The necessary and sufficient condition for this failure is $H_q = 0$, since $G_2 = 0$ or $G_3 = 0$ imply the cut-off failure of loop 2 along with loop 1. The transmittances other than H_q are assumed normal.† The determinant of the graph is

$$D = 1 + G_4G_3G_2G_1H_l \cong 3200$$

3) Loop 1 is normal, but loop 2 is cut off. The causes of this failure are $H_l = 0$ or $G_4 = 0$ or $G_1 = 0$. The determinant of the graph is

$$D = 1 + G_3G_2H_q \cong 32$$

4) Both loop 1 and loop 2 are failed. This can be caused by one of the following zero transmittances:

$$G_3 = 0,\ \ G_2 = 0,\ \ H_q = 0 \text{ and } G_4 = 0,\ \ H_q = 0 \text{ and } G_1 = 0,\ \ H_q = 0 \text{ and } H_l = 0$$

†It can be proven that the failures of other transmittances contribute to the systems success. Thus, a failed system will continue to fail even if we replace the failed transmittances by normal ones. This implies that the transmittance failures are redundant and can be neglected (see Section 2.7.10).

The determinant of the graph is

$$D=1+0=1$$

2.7.7 Representation of Level *L* in Terms of Source Variables

For each of the four cases of Section 2.7.6 we represent level L in terms of source variables, using (2.3). The results are shown in Table 2.14. For example, for the simultaneous cut-off failure of loop 1 and loop 2 caused by $G_2=0$, we have

$$L=X_1+32X_2$$

2.7.8 Discretizing Source Variables

Each source variable is discretized into a set of values specific to the basic failures. When these values are determined, the basic failures have been quantified.

Table 2.15 shows the results of the discretization. At the normal operating mode all transmittances have normal values and all variables at the nodes are zero. Consider level L first. To evaluate the effect of X_1 on L, suppose that the antecedent variable Q remains zero. If the basic failure, valve C open, increases the level by 4 (m), then the failure is characterized by X_1 taking a value of 4.

As another example, consider inflow rate Q. The source variable to Q is X_2. To estimate the effect of opening bypass valve B, suppose that the antecedent variable V remains zero and that opening the bypass valve increases the inflow rate by 0.1 (m^3/s). Then the failure is modeled by $X_2=0.1$ as shown in Table 2.15.

A rather subtle situation is represented by variable X_3. If the flow controller is biased high, the control valve is opened more, resulting in an increase of inflow rate Q. Thus, the flow controller failure is modeled by variable X_3.† Assume that the failure increases the inflow rate by 0.1 (m^3/s), provided that antecedent variable E_q remains zero. Then the failure is characterized by

$$X_3=\frac{0.1}{G_2}=\frac{0.1}{0.0031}=32$$

It is not necessary to substitute positive values for variables X_9 and X_{10} because coefficients a_9 and a_{10} in Table 2.14 are non-positive. Positive

†Pump overheat failure, on the other hand, should be modeled by X_2, since this increases inflow rate Q without additional opening of the control valve.

TABLE 2.14 Representations of Level L in Terms of Source Variables When Cut-Off Failures of Loops Are Specified

Cut-Off Loops	Zero Transmittances	COEFFICIENTS a_i's in $L=\Sigma a_i X_i$									
		a_1	a_2	a_3	a_4	a_5	a_6	a_7	a_8	a_9	a_{10}
None	None	0.01	0.01	0.000031	0.31	0.31	1	1	1	−0.31	−1
Loop 1	$H_q=0$	0.00031	0.01	0.000031	0.31	0.31	1	1	1	−0.31	−1
Loop 2	$H_l=0$	1	1	0.0031	31	31	100	100	100	−31	−100
	$G_1=0$	1	0	0	0	0	0	0	0	0	0
	$G_4=0$	1	1	0.0031	31	31	0	0	0	−31	0
Loop 1 and Loop 2	$G_2=0$	1	32	0	0	0	0	0	0	0	0
	$G_3=0$	1	32	0.099	0	0	0	0	0	0	0
	$H_q=0$ and $G_1=0$	1	0	0	0	0	0	0	0	0	0
	$H_q=0$ and $G_4=0$	1	32	0.099	990	990	0	0	0	−990	0
	$H_q=0$ and $H_l=0$	1	32	0.099	990	990	3200	3200	3200	−990	−3200

TABLE 2.15 DISCRETIZATION OF SOURCE VARIABLES

Variable	Component	Values	Basic Failures	Quantification
X_1	Valve C	0	None	
		4	Opened full	Level is increased by 4 (m)
X_2	Valve B	0	None	
		0.1	Opened full	Inflow is increased by 0.1 (m^3/s)
X_3	Flow-rate controller	0	None	
		32	Biased high	Inflow is increased by 0.1 (m^3/s)
X_4	Flow-rate comparator	0	None	
		0.1	Biased high	Comparator is biased by 0.1 (m^3/s)
X_5	Level controller	0	None	
		0.1	Biased high	Flow-rate reference is biased by 0.1 (m^3/s)
X_6	Level comparator	0	None	
		3	Biased high	Comparator is biased by 3 (m)
X_7	Set point dial	0	None	
		6	Biased high	Dial is biased by 6 (m)
X_8	Operator	0	None	
		3	High set point	Reference is biased by 3 (m)
		6	Too high set point	Reference is biased by 6 (m)
X_9	Flow sensor	0	None	
		−0.1	Biased low	Sensor is biased by −0.1 (m^3/s)
X_{10}	Level sensor	0	None	
		−6	Biased low	Sensor is biased by −6 (m)

values of X_9 or X_{10} contribute to the system success, preventing the overflow. In other words, even if we replace these positive values by normal values (i.e., zeros) the system failure will continue to exist if other failures are occurring.

The final effect on level L by those non-zero source variables is not the same as the ones given in the last column of Table 2.15 because the last column was obtained by assuming that the antecedent variables are zero, in order to estimate the effect of the basic failures. The final effect should be calculated by the equations for L shown in Table 2.14.

2.7.9 Identifying Basic Failures Leading to System Hazards

For the cascade, level-control system, let us first examine the case without the cut-off failures of Table 2.14. The level is expressed as

$$
\begin{aligned}
L &= \sum_k a_k X_k \\
&= 0.01X_1 + 0.01X_2 + 0.000031X_3 + 0.31X_4 + 0.31X_5 + X_6 + X_7 + X_8 \\
&\quad - 0.31X_9 - X_{10}
\end{aligned}
$$

The top event is defined by inequality (2.4). Defining Z_k by

$$Z_1=0.01X_1,\ Z_2=0.01X_2,\ Z_3=0.000031X_3,\ Z_4=0.31X_4$$
$$Z_5=0.31X_5,\ Z_6=X_6,\ Z_7=X_7,\ Z_8=X_8,\ Z_9=-0.31X_9,\ Z_{10}=-X_{10}$$

the top event is represented by

$$L=Z_1+Z_2+Z_3+Z_4+Z_5+Z_6+Z_7+Z_8+Z_9+Z_{10}\geq 5 \qquad (2.5)$$

The values for the X_k's of Table 2.15 are rewritten as

$$\begin{aligned}&Z_1\in\{0,0.04\},\ Z_2\in\{0,0.001\},\ Z_3\in\{0,0.001\},\ Z_4\in\{0,0.031\}\\ &Z_5\in\{0,0.031\},\ Z_6\in\{0,3\},\ Z_7\in\{0,6\},\ Z_8\in\{0,3,6\}\\ &Z_9\in\{0,0.031\},\ Z_{10}\in\{0,6\}\end{aligned} \qquad (2.6)$$

Combinations of Z_k's satisfying (2.5) can be found by the classification tree of Fig. 2.43. It is not necessary to examine all possible combinations. The largest value of Z_k's, i.e., $Z_{10}=6$, is examined first in the classification tree. This satisfies inequality (2.5), and we denote this successful search by a circle, as shown in Fig. 2.43. Then, we assume a value other than $Z_{10}=6$. The largest value among the remaining values is $Z_8=6$. Again, inequality (2.5) is satisfied, and we use a circle as a success symbol for the search.

In a similar way success node C is found. At stage α in the classification tree, we cannot say whether the inequality holds or not. Additional values are searched for and $Z_6=3$ is found. The inequality is now satisfied, and we have success node D.

The inequality cannot be satisfied at node E because, even if we assume the largest value for each remaining variable, level L remains smaller than 5:

$$\begin{aligned}L&=0.04+0.001+0.001+0.031+0.031+0+0+0+0.031+0\\&=0.135<5\end{aligned}$$

We denote by a cross failure node E of the search.

The classification tree of Fig. 2.43 gives four combinations of Z_k's satisfying (2.5) which can be rewritten in terms of X_k's:

Node A:	$\{X_{10}=-6\}$;	level sensor biased low
Node B:	$\{X_8=6\}$;	operator error (high set point)
Node C:	$\{X_7=6\}$;	set point dial biased high
Node D:	$\{X_8=3, X_6=3\}$;	operator error (high set point) and level comparator biased high

The part of the fault tree shown as the left-most branch of Fig. 2.44 is now complete. The other cases shown in Table 2.14 are processed similarly and the combinations of X_k's satisfying (2.4) are as shown in Table 2.16. The system fault tree can now be constructed.

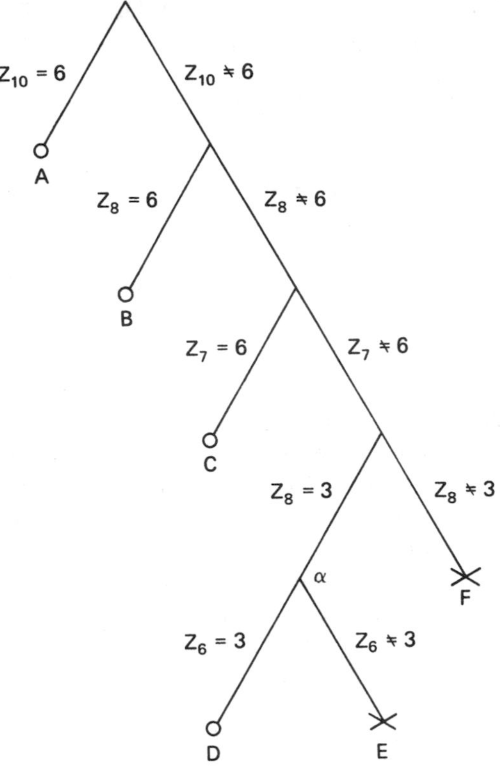

Figure 2.43. A classification tree for finding combinations of Z's satisfying (2.5).

2.7.10 Remarks

In Section 2.7.6, each combination of loop cut-off failure was specified by necessary and sufficient sets of zero transmittances, and those transmittances that are not involved in the sets were assigned normal values. We now justify this assignment.

Assume in Case 2 of Section 2.7.6 the failure of G_5, along with that of H_q. Denote by L_1 the level with $H_q=0$. Then, from Table 2.14,

$$L_1=0.00031X_1+0.01X_2+0.000031X_3+0.31X_4 \\ +0.31X_5+X_6+X_7+X_8-0.31X_9-X_{10} \tag{2.7}$$

Denote by L_2 the level with $G_5=0$, along with $H_q=0$. Then, from (2.3),

$$L_2=0.00031X_1+0.01X_2+0.000031X_3+0.31X_4+0.31X_5 \\ +X_6+X_7+0X_8-0.31X_9-X_{10}$$

Since X_1 to X_8 are non-negative and X_9 and X_{10} are non-positive, the inequality $L_1 \geq L_2$ holds. This inequality shows that the level is lowered by the failure of transmittance G_5. In other words, the failure contributes to

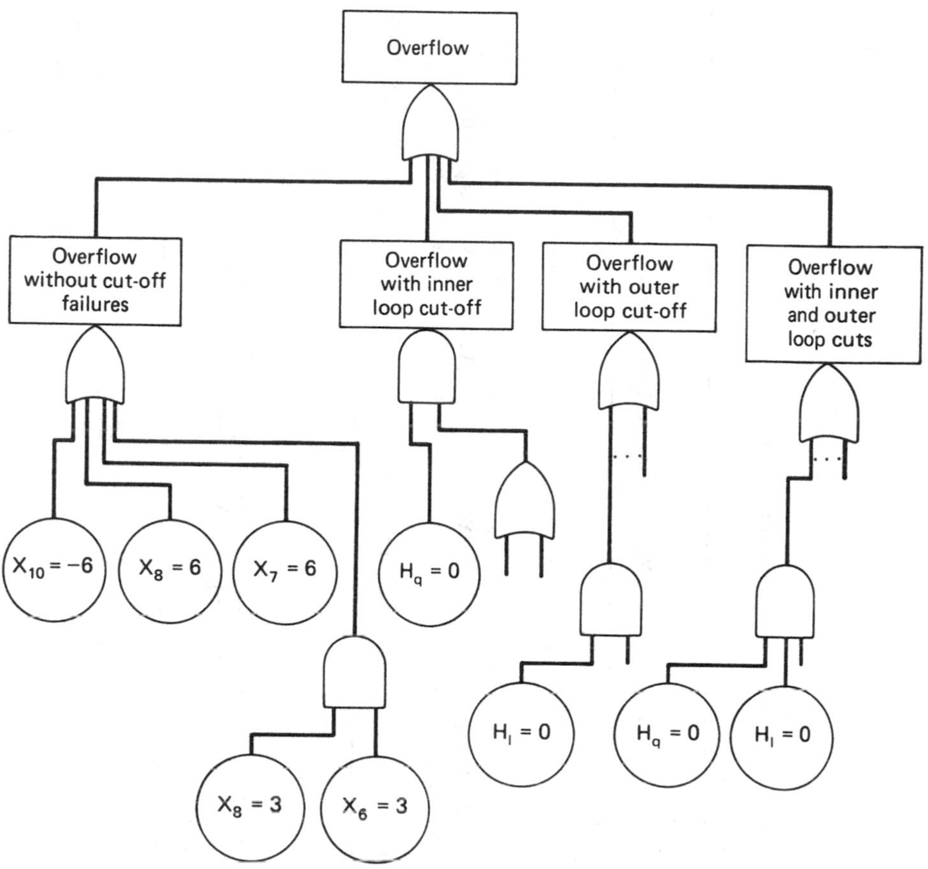

Figure 2.44. Part of fault tree for overflow in cascade level-control system.

system success. Thus, if $\{X_i, X_j, H_q = 0, G_5 = 0\}$ is a sufficient cause of the overflow, then $\{X_i, X_j, H_q = 0, G_5 = \text{normal}\}$ also causes the system hazard. The latter cause can be found by setting H_q to zero and the other transmittances to normal, as was done in Section 2.7.6. Since normal events are assumed to occur with certainty, we can simplify $\{X_i, X_j, H_q = 0, G_5 = \text{normal}\}$ to $\{X_i, X_j, H_q = 0\}$. This simplified cause was used in constructing the fault tree.

In identifying causes of cut-off loop failures, we must concentrate on necessary and sufficient sets of zero transmittances. Other redundant zero transmittances should not be introduced because these contribute to system success; hence, the resulting sets of causes for the system hazard are reduced to the ones which can be obtained without assuming redundant zero transmittances. The redundant failures can be regarded as miracles in the sense that they contribute to system success. In the real world, failures are not negated by miracles.

TABLE 2.16 COMBINATIONS OF SOURCE VARIABLES CAUSING SYSTEM HAZARDS

Cut-Off Loops	Zero Transmittances	Combinations of Source Variables Causing System Failure
None	None	$\{X_{10}=-6\}$, $\{X_8=6\}$, $\{X_7=6\}$, $\{X_8=3, X_6=3\}$
Loop 1	$H_q=0$	$\{X_{10}=-6\}$, $\{X_8=6\}$, $\{X_7=6\}$, $\{X_8=3, X_6=3\}$
Loop 2	$H_l=0$	$\{X_{10}=-6\}$, $\{X_8=6\}$, $\{X_8=3\}$, $\{X_7=6\}$, $\{X_6=3\}$, $\{X_1=4, X_4=0.1\}$, $\{X_1=4, X_5=0.1\}$, $\{X_1=4, X_9=-0.1\}$, $\{X_4=0.1, X_5=0.1\}$, $\{X_4=0.1, X_9=-0.1\}$, $\{X_5=0.1, X_9=-0.1\}$
	$G_1=0$	None
	$G_4=0$	$\{X_1=4, X_4=0.1\}$, $\{X_1=4, X_5=0.1\}$, $\{X_1=4, X_9=-0.1\}$, $\{X_4=0.1, X_5=0.1\}$, $\{X_4=0.1, X_9=-0.1\}$, $\{X_5=0.1, X_9=-0.1\}$
Loop 1 and Loop 2	$G_2=0$	$\{X_1=4, X_2=0.1\}$
	$G_3=0$	$\{X_1=4, X_2=0.1\}$, $\{X_1=4, X_3=32\}$, $\{X_2=0.1, X_3=32\}$
	$H_q=0$, $G_1=0$	None
	$H_q=0$, $G_4=0$	$\{X_9=-0.1\}$, $\{X_5=0.1\}$, $(X_4=0.1\}$, $\{X_1=4, X_2=0.1\}$, $\{X_1=4, X_3=32\}$, $\{X_2=0.1, X_3=32\}$
	$H_q=0$, $H_l=0$	$\{X_{10}=-6\}$, $\{X_9=-0.1\}$, $\{X_8=6\}$, $\{X_8=3\}$, $\{X_7=6\}$, $\{X_6=3\}$, $(X_5=0.1\}$, $\{X_4=0.1\}$, $\{X_1=4, X_2=0.1\}$, $\{X_1=4, X_3=32\}$, $\{X_2=0.1, X_3=32\}$

The method developed is based on a signal flow graph as a system model. Any model is an approximation of an actual system. Thus, the resulting fault tree should be examined again by using past experience and more accurate simulation models. The fault-tree construction method described here should be viewed primarily as a useful approach for uncovering causal relations in complicated systems with control loops.

PROBLEMS

2.1. There are four way stations (Fig. P2.1) on the route of the Dead Eye Stages from Hangman's Hill to Placer Gulch. (Problem courtesy of J. Fussell.) The distances involved are:

Hangman's Hill–Station 1	20 miles
Station 1–Station 2	30 miles
Station 2–Station 3	50 miles
Station 3–Station 4	40 miles
Station 4–Placer Gulch	40 miles

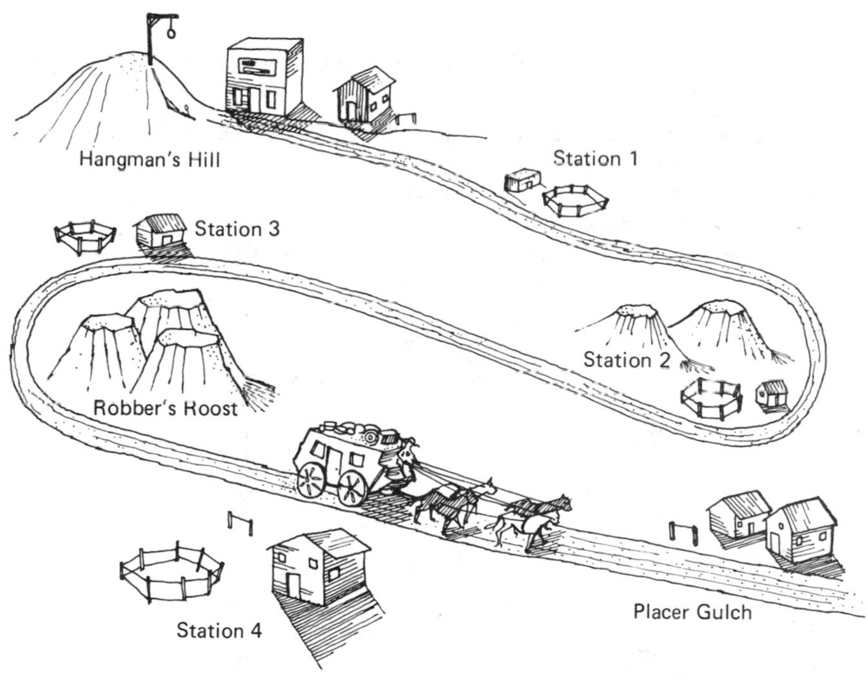

Figure P2.1.

The maximum distance the stage can travel without a change of horses, which can only be accomplished at the way stations, is 85 miles. The stages change horses at every opportunity; however, the stations are raided frequently, and their stock driven off by marauding desperadoes.

Draw a reliability block diagram and a fault tree for the system of stations.

2.2. Construct a fault tree for the circuit in Fig. P2.2, with the top event "no light from bulb" and the boundary conditions.

Initial condition: Switch closed.
Not-allowed events: Failures external to the system.
Existing Events: None.

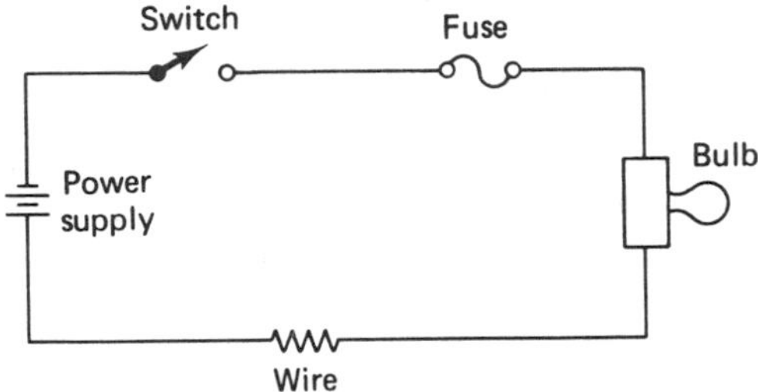

Figure P2.2.

2.3. Construct a fault tree for the dual, hydraulic, automobile braking system shown in Fig. P2.3.

System bounds: Master cylinder assembly, front and rear brake lines, wheel cylinder, and brake shoe assembly.
Top event: Loss of all braking capacity.
Initial condition: Brakes released.
Not-allowed events: Failures external to system bounds.
Existing events: Parking brake inoperable.

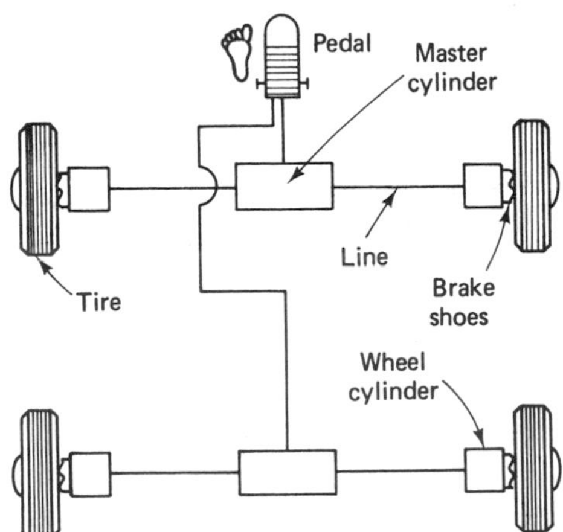

Figure P2.3. A braking system.

2.4. Construct a fault tree for the domestic hot-water system in Problem 1.6 of Chapter 1. Take as a top event the rupture of a water tank. Develop a secondary failure listing.

2.5. The reset switch in the schematic of Fig. P2.5 is closed to latch the circuit and provide current to the light bulb. The system boundary conditions for fault tree construction are:

Top event: No current in circuit 1.
Initial conditions: Switch closed. Reset switch is closed momentarily and then opened.
Not-allowed events: Wiring failures, operator failures, switch failure.
Existing events: Reset switch open.

Draw the fault tree, clarifying how it is terminated. (From Fussell, J. B., "Particularities of Fault Tree Analysis," N00030-74-C-0007, Aerojet Nuclear Co., Idaho Nat'l. Lab., September, 1974.)

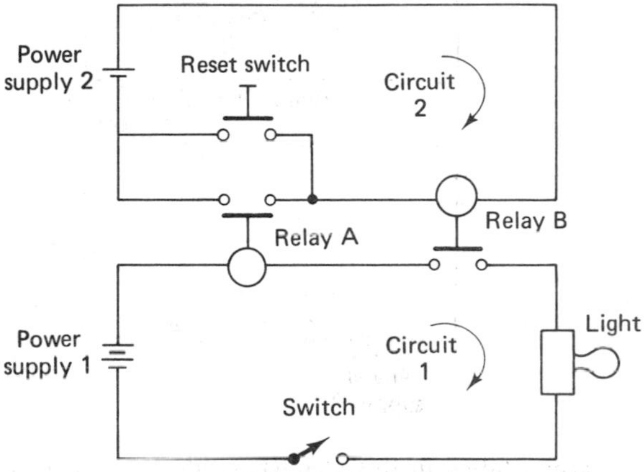

Figure P2.5. A relay circuit.

2.6. A system (Fig. P2.6) has two electric heaters which can fail by short circuiting to ground. Each heater has a switch connecting it to the power supply. If either heater fails with its switch closed, then the resulting short circuit will cause the power supply to short circuit, and the total system fails.

If one switch fails open or is opened in error before its heater fails, then only that side of the system fails, and we operate at half power.

Draw the fault tree, and identify events which are mutually exclusive.

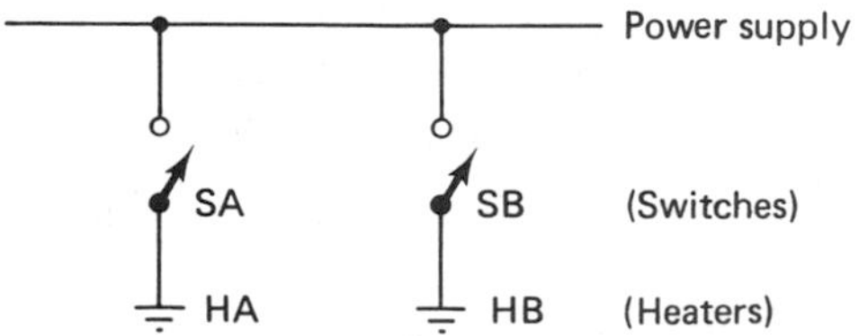

Figure P2.6. Flowsheet for Problem 6.

2.7. The purpose of the system of Fig. P2.7 is to provide light from the bulb. When the switch is closed, the relay contacts close and the contacts of the circuit breaker, defined here as a normally closed relay, open. Should the relay contacts transfer open, the light will go out and the operator will immediately open the switch which, in turn, causes the circuit breaker contacts to close and restore the light. The system boundary conditions are, then:

Top event: No light.
Initial conditions: Switch closed.
Not-allowed events: Operator failures, wiring failures, secondary failures.

Draw the fault tree, and identify dependent basic events.

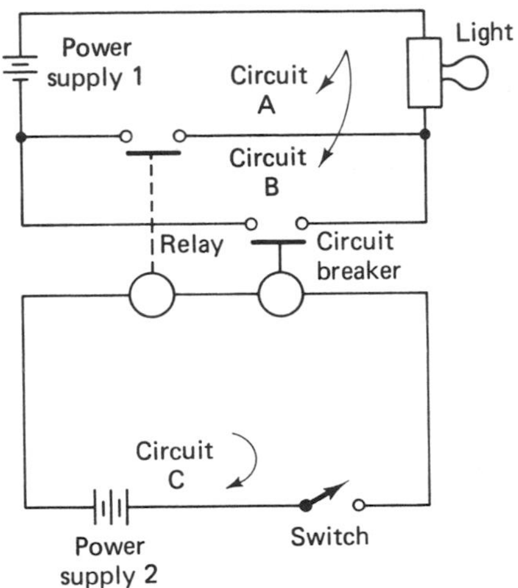

Figure P2.7.

2.8. Using the decision tables in this chapter, synthesize a fault tree for the heat exchange system of Fig. 2.28, with a standby pump switching on when the principal pump fails.

3

QUALITATIVE ASPECTS OF SYSTEM ANALYSIS

System failure can occur in many different ways. Each unique way is a *system failure mode*, and involves single or multiple component failures. To reduce the chance of a system failure, we must first identify the failure modes and then eliminate the most frequently occurring and/or highly probable ones. The fault tree methods discussed in the previous chapter facilitate the discovery of failure modes, and most of the analytical methods of this chapter are predicated on the existence of a system fault tree.

3.1 CUT SETS

For a given fault tree, the system failure modes are clearly defined by the concept of a cut set. A *cut set* is a collection of basic events; if all these basic events occur, the top event is guaranteed to occur. For example, for the fault tree of Fig. 2.24, if events 2 and 4 occur simultaneously, the top event occurs, i.e., if "contacts failure (stuck closed)" and "switch failure (stuck closed)" coexist, the top event, "pressure tank rupture," happens. Thus, set $\{1,2\}$ is a cut set. Also, $\{1\}$ and $\{3,5\}$ are cut sets.

3.2 PATH SETS

A *path set* is a *dual concept* to the cut set. It is a collection of basic events, and if none of the events in the set occur, the top event is guaranteed to not occur. When the system has only one top event, the non-occurrence of the basic failure events in a path set ensures successful system operation. The non-occurrence does not guarantee system success when more than one top event is specified. In such cases, a path set only ensures the non-occurrence of a particular top event.

For the fault tree of Fig. 2.24, if events 1, 2, and 3 do not occur, the top event cannot happen. Hence, if the tank, contacts, and timer are normal, the tank will not rupture. Thus, $\{1,2,3\}$ is a path set. Another path set of the fault tree is $\{1,4,5,6\}$; i.e., the tank will not rupture if these failure events do not happen.

3.3 MINIMAL CUT SETS AND MINIMAL PATH SETS

A large system has an enormous number of failure modes; hundreds of thousands of cut sets are possible for systems having between 40 and 90 components. If there are hundreds of components, billions of cut sets may exist. It is necessary, therefore, to reduce the number of failure modes to simplify the analysis. We require only those failure modes which are *general*, in the sense that one or more of them must happen for a system failure mode to occur. Nothing is lost by this restriction. If it were possible to improve the system in such a way as to eliminate all general failure modes, that would automatically result in the elimination of all system failure modes.

The concept of a minimal cut set clearly defines a general failure mode. A *minimal cut set* is such that if any basic event is removed from the set, the remaining events collectively are no longer a cut set. A cut set including some other sets is not a *minimal cut set*. The minimal cut set concept enables us to reduce the number of cut sets and the number of basic events involved in each cut set. This simplifies the analysis.

The fault tree of Fig. 2.24 has seven minimal cut sets $\{1\},\{2,4\}$, $\{2,5\},\{2,6\},\{3,4\},\{3,5\},\{3,6\}$. Cut set $\{1,2,4\}$ is not minimal because it includes $\{1\}$ or $\{2,4\}$. Both failure modes $\{1\}$, and $\{2,4\}$ must occur for mode $\{1,2,4\}$ to occur. All failure modes would be prevented from occurring if we were able to eliminate the possibility of the modes defined by the minimal cut sets.

The fault tree of Fig. 2.30 has the minimal cut sets, $\{1\}$ and $\{3\}$. Set $\{1\}$ is a cut set because the non-occurrence of event 2 implies the occurrence of event 3. Set $\{1\}$ is a minimal cut set, since it consists of one event. This rather subtle, minimal cut set shows up because the fault tree has *mutually exclusive* basic events 2 and 3. In subsequent sections of this chapter, we classify fault trees on the basis of whether they have mutually exclusive basic events or not. Note that exclusive events are not related to the exclusive OR gate.

A *minimal path set* is a path set such that if any basic event is removed from the set, the remaining events collectively are no longer a path set.

The fault tree of Fig. 2.24 has two minimal path sets, $\{1,2,3\}$ and $\{1,4,5,6\}$. If either $\{1,2,3\}$ or $\{1,4,5,6\}$ do not fail, the tank operates successfully.

3.4 MINIMAL CUT SET GENERATION (NO EXCLUSIVE BASIC EVENTS)

When a fault tree has no mutually exclusive events, the MOCUS computer code can be used to generate minimal cut sets.[1] MOCUS is based on the observation that OR gates *increase the number* of cut sets, whereas AND gates *enlarge the size* of the cut sets. The MOCUS algorithm can be stated as follows:

1) Alphabetize each gate.
2) Number each basic event.
3) Locate the uppermost gate in the first row and column of a matrix.
4) Iterate either of the fundamental permutations a) or b) below in a top-down fashion. (When events appear in rectangles, replace them by equivalent gates or basic events.)
   a) Replace OR gates by a vertical arrangement of the input to the gates, and increase the cut sets.
   b) Replace AND gates by a horizontal arrangement of the input to the gates, and enlarge the size of the cut sets.
5) When all gates are replaced by basic events, obtain the minimal cut sets by removing supersets. A *superset* is a cut set including some other cut sets.

As an example, consider the fault tree of Fig. 2.24. The gates and the basic events have already been labeled. The uppermost gate A is located in the first row and column:

$$A$$

This is an OR gate, and it is replaced by a vertical arrangement of the input to the gate:

$$\begin{matrix} 1 \\ B \end{matrix}$$

---

[1]Fussell, J. B., E. B., Henry, and N. H. Marshall, "MOCUS—A Computer Program to Obtain Minimal Cut Sets from Fault Trees," ANCR-1156, 1974.

Since B is an AND gate, it is permutated by a horizontal arrangement of its input to:

1
- - - - - - - - -
C, D
_____

The OR gate C is transformed into a vertical arrangement of its input:

1
- - - - - - - - -
$2, D$
$3, D$
_____

The OR gate D is replaced by a vertical arrangement of its input:

1
- - - - - - - - -
2,4
$2, E$
- - - - - - - - -
3,4
$3, E$
_____

Finally, the OR gate E is permutated by a vertical arrangement of the input:

1
- - - - - - - - -
2,4
- - - - - - - - -
2,5
2,6
- - - - - - - - -
3,4
- - - - - - - - -
3,5
3,6
_____

We have seven cut sets, $\{1\}, \{2,4\}, \{2,5\}, \{2,6\}, \{3,4\}, \{3,5\}$, and $\{3,6\}$. All seven are minimal, since there are no supersets.

When supersets are involved, we remove them in the process of replacing the gates. Assume the following result at one stage of the replacement.

$$1,2,G$$
$$1,2,3,G$$
$$1,2,K$$

A cut set derived from $\{1,2,3,G\}$ always includes a set from $\{1,2,G\}$. However, the cut set from $\{1,2,3,G\}$ may not include any sets from $\{1,2,K\}$ because the development of K may differ from that of G. We have the following simplified result:

$$1,2,G$$
$$1,3,K$$

When an event appears more than two times in a horizontal arrangement, we should aggregate these events into a single event. For example, the arrangement $\{1,2,3,2,H\}$ should be changed to $\{1,2,3,H\}$.†

3.5 MINIMAL-PATH GENERATION (NO EXCLUSIVE BASIC EVENTS)

The MOCUS algorithm for the generation of minimal path sets makes use of the fact that AND gates *increase* the path sets, whereas OR gates *enlarge* the size of the path sets. The algorithm proceeds in the following way:

1) Alphabetize each gate.
2) Number each basic event.
3) Locate the uppermost gate as an entry in the first row of the first column.
4) Iterate either of the fundamental permutations a) or b) below in a top-down fashion.
   a) Replace OR gates by a horizontal arrangement of the input to the gates, and enlarge the size of the path sets.
   b) Replace AND gates by a vertical arrangement of the input to the gate, and increase the path sets.
5) When all the gates are replaced by basic events, obtain minimal path sets by removing supersets.

†The MOCUS algorithm is an example of a *top-down* search. *Bottom-up* algorithms (P. Chatterjee, *Fault Tree Analysis: Reliability Theory and Systems Safety Analysis*, ORC 74-34, Univ. of California, Berkeley, 1974) are said to be more effective if there are many repeated events in the trees. The F-TAP computer program is said to be capable of handling larger trees than MOCUS (private communication, H. Lambert, TERA Corp., 2150 Shattuck Ave., Berkeley, CA 94704, 1979.).

As an example, consider again the fault tree of Fig. 2.24. The MOCUS algorithm generates the minimal path sets in the following way.

A
replacement of A

$1, B$
replacement of B

$1, C$
$1, D$
replacement of C

$1,2,3$

- - - - - - - - -

$1, D$
replacement of D

$1,2,3$

- - - - - - - - -

$1,4,E$
replacement of E

$1,2,3$

- - - - - - - - -

$1,4,5,6$

- - - - - - - - -

We have two paths sets

$$\{1,2,3\}, \{1,4,5,6\}$$

These two are minimal since there are no supersets.

3.6 MINIMAL-CUT GENERATION (EXCLUSIVE BASIC EVENTS)

A fault tree may have *mutually exclusive basic events*. For example, the fault tree of Fig. 2.30 has two mutually exclusive events, "pump normal" and "pump stops." When constructing a fault tree, this type of mutual exclusivity is avoided if at all possible; however it is not possible to avoid this problem if the system hardware is *multi-state*, i.e., has plural failure modes. For example, a generator may have the mutually exclusive failure events "generator stops" and "generator surge." One of these faults precludes the other.

When a fault tree contains mutually exclusive events, the MOCUS algorithm does not always produce the minimal cut sets. MOCUS when applied to the tree of Fig. 2.30, for example, yields the cut sets {1,2} and {3}. Thus, minimal cut set {1} cannot be obtained by MOCUS, although it would be apparent to an engineer.

We now develop a method of obtaining cut sets which can be applied to the case of exclusive events. The procedure consists of first using MOCUS to obtain path sets from which block diagrams are constructed. The last step involves a *classification method* whereby minimal cut sets are identified.

MOCUS is modified in such a way as to remove *inconsistent path sets* from the outputs, inconsistent path sets being sets with mutually exclusive events covering all possible component states. An example is "generator normal," "pump normal," "pump stops," when the pump has only two modes, "pump normal" and "pump stops." For this path set, at least one of the primary events is always occurring, so we cannot achieve non-occurrence of all basic events in the path set. The inconsistent set does not satisfy the definition of a path set and should be removed.

For the fault tree of Fig. 2.30, MOCUS generates path sets in the following way:

A
B,3
1,3 2,3

Set {2,3} is inconsistent; thus, only path set {1,3} is obtained as an output of the modified MOCUS.

The top event occurs if and only if at least one basic event occurs in each of the path sets obtained by the modified MOCUS. Therefore, the top event can be expressed by the block diagram of Fig. 3.1.

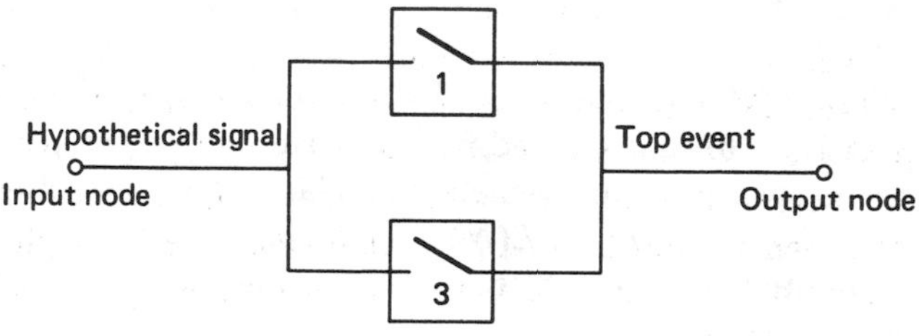

Figure 3.1. Block diagram with path set {1,3}.

The switch in the block diagram means that it is either closed or open, depending on the occurrence of the basic event declared in the block. The occurrence closes the switch, whereas non-occurrence opens it. The top event occurs if and only if the hypothetical signal passes through the block diagram to the output node.

If MOCUS identifies three consistent path sets, $\{1,3,4\}$, $\{1,2,4\}$, and $\{1,2,3\}$, the top event is represented by the block diagram of Fig. 3.2. Note that each path set is a parallel circuit of basic events. The circuit forms the series structure of the path sets.

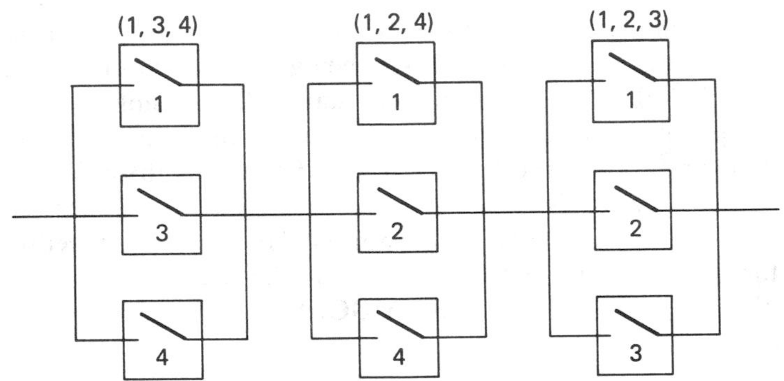

Figure 3.2. Block diagram with path sets $\{1,3,4\}$, $\{1,2,4\}$ and $\{1,2,3\}$.

The *classification method* converts the path sets into the minimal cut sets. It is based on the fact that a necessary and sufficient condition for a set to be a cut set is that every parallel structure passes a hypothetical signal when all the basic events in the set close the corresponding switches. The collection of cut sets is divided into increasingly smaller subclasses, and each minimal cut set is obtained as the smallest set in a subclass by the following procedure:

A basic event L appearing in the block diagram is selected and a *dichotomous classification* is made, yielding two disjoint subclasses $L(Y)$ and $L(N)$. $L(Y)$ is a class of the cut sets containing the basic event L. Any cut set in $L(Y)$ does not have basic events exclusive to L. Class $L(N)$ is a class of cut sets not containing basic event L. The symbols Y and N denote "Yes" and "No," respectively.

For the class $L(Y)$, switch L is closed because the basic event L is accepted in $L(Y)$. The corresponding exclusive events are prohibited in $L(Y)$, and their switches are opened. For class $L(N)$, switch L is open because event L is prohibited in $L(N)$. The corresponding exclusive events are neither accepted nor prohibited in $L(N)$, and their switches remain undetermined.

Each newly generated subclass receives a similar dichotomous classification, and the cut sets are classified into increasingly smaller subclasses. The states of the switches are determined accordingly, and the block diagram eventually passes or blocks the signal when the corresponding subclass becomes sufficiently small. In each block diagram, the switches of basic events are opened if and only if the events are prohibited in the corresponding subclass. The switches are closed if and only if the events are accepted in the subclass.

A subclass *blocking* the signal indicates that there exists a parallel structure in which switches are all opened by events prohibited in the subset. We can actually open all these switches, since all the path sets are consistent. Therefore, we cannot pass the signal whenever we choose basic events not prohibited in the subclass. This means that the subclass under consideration actually has no cut sets. The search for the cut sets is thus terminated for this class.

A subclass *passing* the signal indicates that the basic events accepted in the subclass close at least one switch in each parallel structure. Thus, the set of these basic events is itself a cut set. This set is the smallest set in the subclass. This minimal property suggests that most cut sets obtained in this way are minimal cut sets. Indeed, it can be proven that all minimal cut sets are obtained, along with some supersets.

As an example, consider the block diagram of Fig. 3.1. The tree of Fig. 3.3 shows the classification process. Subclasses (a), (b), (c), and (d), respectively lead to the block diagram shown in Fig. 3.4. The diagram of subclass (c) is obtained from that of subclass (b) by closing switch 3. Similarly, diagram (d) is obtained by opening switch 3 in diagram (b).

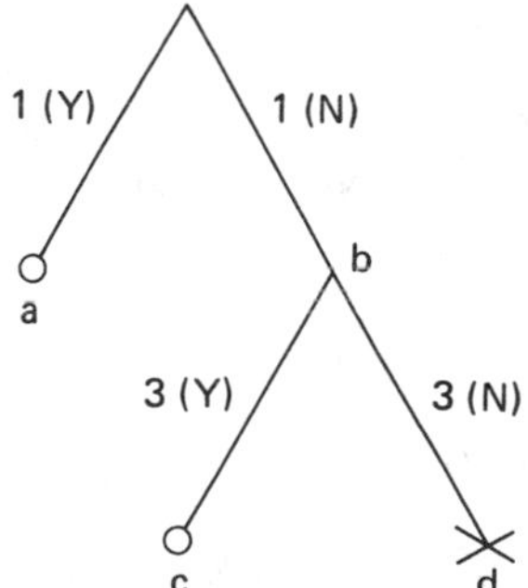

Figure 3.3. Classification tree for the block diagram of Fig. 3.1.

The circle in the classification tree of Fig. 3.3 means that the corresponding diagram passes the signal, whereas the cross shows blockage. We see that subclasses (a) and (c) contain a cut set, while (d) does not. The smallest set in subclass (c) is $\{3\}$. We have the minimal cut sets $\{1\}$ and $\{3\}$, since there is no inclusion relation between these two sets.

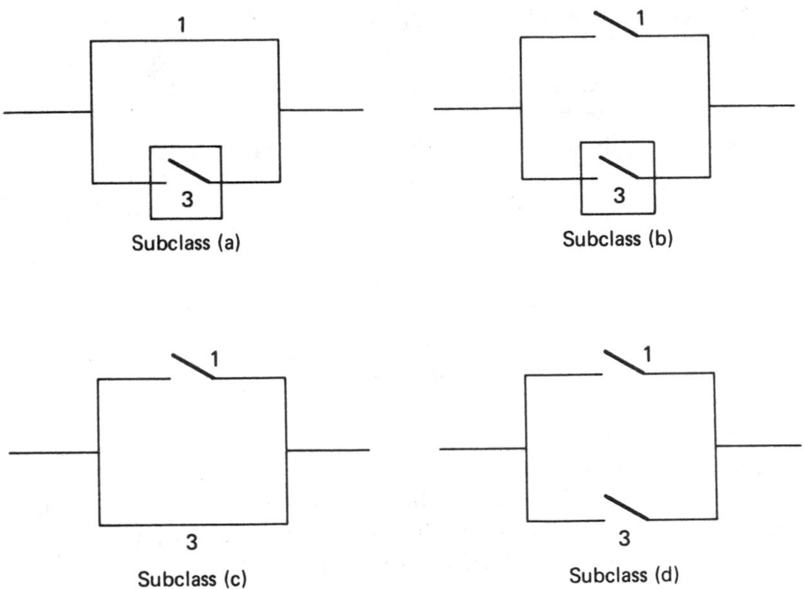

Figure 3.4. Block diagram for each subclass.

As an another example, take the diagram of Fig. 3.2. Assume that events 2, 3, and 4 are mutually exclusive. These events are, for example, "wire shorted," "wire cut", and "wire normal," respectively. The classification tree is shown in Fig. 3.5. We obtain minimal cut set $\{1\}$; there are no other minimal cut sets.

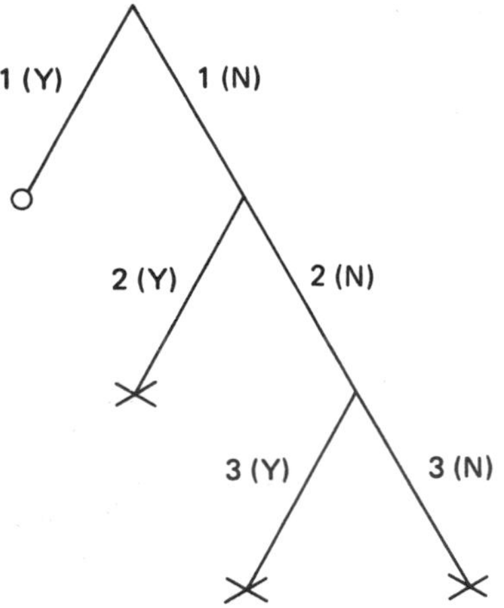

Figure 3.5. Classification tree for the block diagram of Fig. 3.2.

3.7 MINIMAL-PATH GENERATION (EXCLUSIVE BASIC EVENTS)

We obtain a so-called *dual fault tree* when we replace, in the original fault tree, AND and OR gates, respectively, by OR and AND gates. The cut sets of the dual fault tree coincide with the path sets of the original fault tree. Thus, the minimal path sets of the original fault tree are obtained by applying the method of the previous section to the dual fault tree.

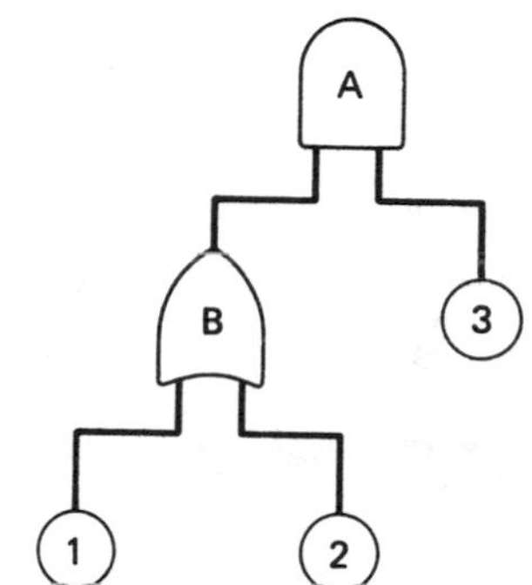

Figure 3.6. Dual fault tree for Fig. 2.30.

As an example, we will obtain the minimal path sets for the tree of Fig. 2.30. The dual fault tree is given by Fig. 3.6. MOCUS gives path sets of the dual fault tree in the following way:

A
B 3
1,2 3

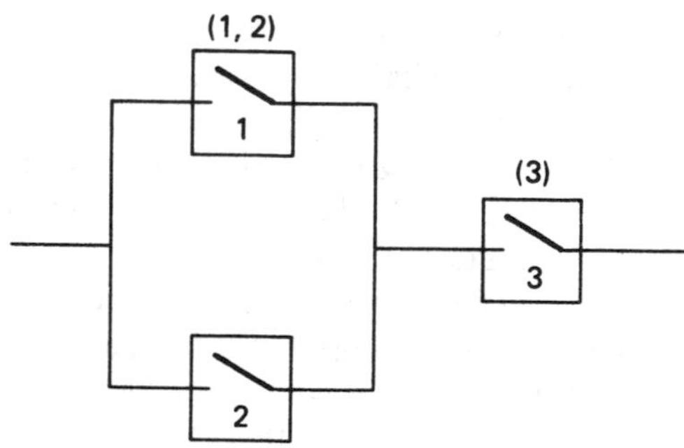

Figure 3.7. Block diagram with path sets {1,2} and {3}.

Since there are no inconsistent path sets, we construct a block diagram, using path sets $\{1,2\}$ and $\{3\}$. The resulting diagram is shown in Fig. 3.7, and the classification tree is shown by Fig. 3.8. We have a minimal cut set $\{1,3\}$ for the dual fault tree, and this cut set is a minimal path set of the original fault tree of Fig. 2.30.

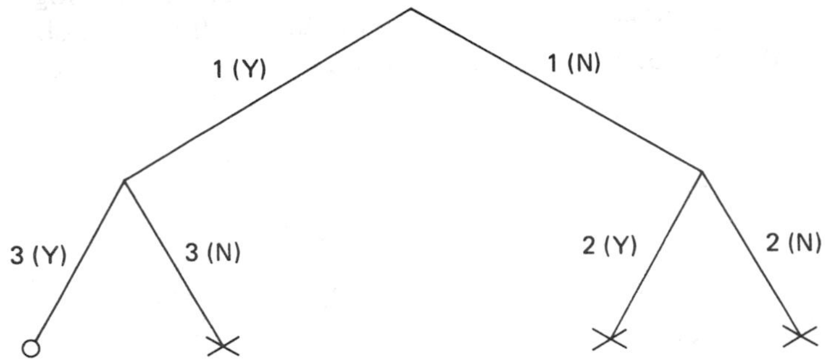

Figure 3.8. Classification tree for the block diagram of Fig. 3.7.

3.8 COMMON-MODE FAILURE ANALYSIS

3.8.1 Common-Mode Cut Sets

Consider a system consisting of valves A and B. Assume that one of the valves is redundant; i.e., either valve has enough capacity to shut down the system and take valve failure as the top failure event. The resulting fault tree has a minimal cut set:

$$\{\text{malfunction of valve } A, \text{ malfunction of valve } B\}$$

This valve system will be far more reliable than a system with a single valve, if one valve fails independently of the other. Coexistence of the two malfunctions is almost a miracle. However, if one valve is liable to fail under the same conditions as the other, the double-valve system is only slightly more reliable than the single-valve system.

Two valves will fail simultaneously, for example, if both contain identical manufacturing faults. Under these conditions, two are only as reliable as one. Therefore, there is no significant difference in reliability between the two systems. A condition or an event which causes multiple basic events is called a *common cause*. An example of a common cause is a flood which causes all supposedly redundant components to fail simultaneously.

The minimal-cut generation methods discussed in the previous sections give minimal cuts of various sizes. A cut set consisting of n basic events is called an *n-event cut set*.

One-event cut sets are significant contributors to the top event unless their probability of occurrence is very small. Generally, hardware failures occur with low frequencies; hence, two-or-more event cut sets can often be neglected if one-event sets are present because co-occurrence of rare events have extremely low probabilities. However, when a common cause is involved, it may cause multiple basic event failures, so we cannot always neglect higher-order cut sets because some two-or-more event cut sets behave like one-event cut sets.

A cut set is called a *common-mode cut set* when a common cause results in the co-occurrence of all events in the cut set. J. R. Taylor reported on the frequency of common causes in the U. S. power reactor industry:[2] "Of 379 component failures or groups of failures arising from independent causes, 78 involved common causes." In system failure mode analysis, it is therefore very important to identify all common-mode cut sets.

3.8.2 Common Causes and Basic Events

As was shown in Fig. 2.15, causes creating component failures come from one or more of the following four sources:

1) Aging
2) Plant personnel
3) System environment
4) System components (or subsystems)

There are a large number of common causes in each source category, and these can also be *further* classified into sub-categories. For example, the causes "water hammer" and "pipe whip" in a piping subsystem can be classified into the category "impact." Some categories and examples are listed in Table 3.1.[3]

For each common cause, we have to identify the basic events affected. In doing so, a *domain* for each common cause, as well as *physical location* of the basic events and components are identified. Some common causes have only limited domains of influence, and the basic events located outside the domain are not affected by the causes. A liquid spill may be confined to one room, so electric components will not be damaged by the spill if they are in another room and no conduit exists between the two rooms. The basic events caused by a common cause are called *common-mode events* of the cause.

[2]Taylor, J. R., RISØ National Laboratory, Rotskild, Denmark, private communication.

[3]Wagner, D. P., C. L. Cate, and J. B. Fussell, "Common Cause Failure Analysis for Complex Systems," *Nuclear Systems Reliability Engineering and Risk Assessment*, J. Fussell and G. Burdick, (eds.), *SIAM*, Philadelphia PA, p. 289, 1977.

TABLE 3.1 CATEGORIES AND EXAMPLES OF COMMON CAUSES

Source	Symbol	Category	Examples
Environment, System Components, or Subsystems	I	Impact	Pipe whip, water hammer, missiles, earthquake, structural failure
	V	Vibration	Machinery in motion, earthquake
	P	Pressure	Explosion, out-of-tolerance system changes (pump overspeed, flow blockage)
	G	Grit	Airborne dust, metal fragments generated by moving parts with inadequate tolerances
	S	Stress	Thermal stress at welds of dissimilar metals, thermal stresses and bending moments caused by high conductivity and density
	T	Temperature	Fire, lightning, welding equipment, cooling system faults, electrical short circuits
	E	Loss of energy source	Common drive shaft, same power supply
	C	Calibration	Misprinted calibration instruction
	F	Manufacturer	Repeated fabrication error, such as neglect to properly coat relay contacts. Poor workmanship. Damage during transportation.
Plant Personnel	IN	Installation contractor	Same subcontractor or crew
	M	Maintenance	Incorrect procedure, inadequately trained personnel
	O	Operator or operation	Operator disabled or overstressed, faulty operating procedures
	TS	Test procedure	Faulty test procedures which may affect all component normally tested together
Aging	A	Aging	Components of same materials

Consider the fault tree of Fig. 3.9.[3] The floor plan is shown in Fig. 3.10. This figure also includes the location of the basic events. We consider 20 common causes. Each common cause has the set of common-mode events shown in Table 3.2. This table also shows the domain of each common cause.

Only two basic events, 6 and 3, are caused by impact I1, whereas basic events 1, 2, 7, 8 are caused by impact I2. This difference arises because each impact has its own *domain of influence*, and each basic event has it own *location of occurrence*. Neither event 4 nor event 12 are caused by impact

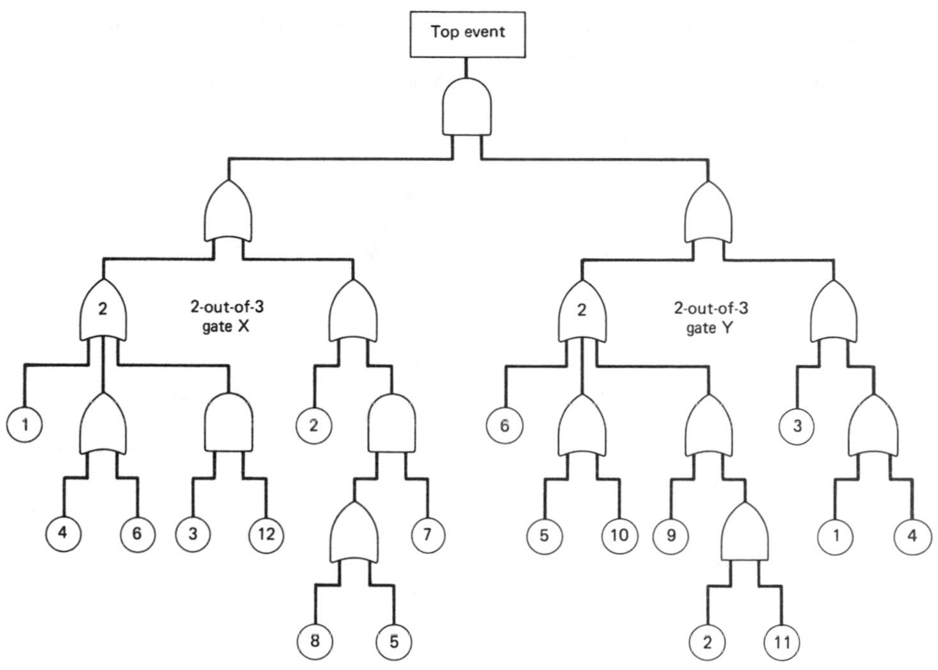

Figure 3.9. Fault tree for the example problem.

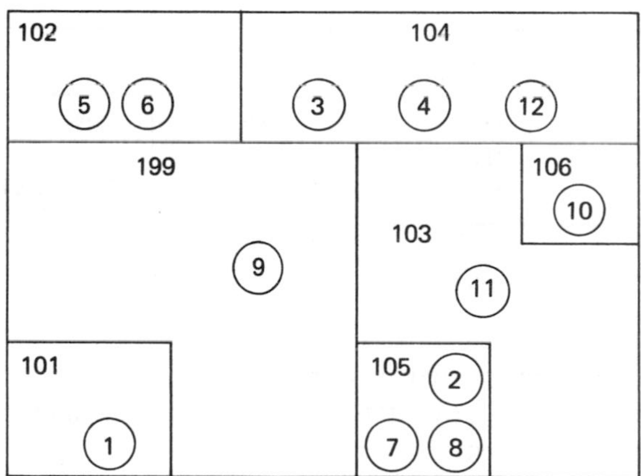

Figure 3.10. Example floor plan and location of basic events.

TABLE 3.2 COMMON CAUSES, DOMAINS, AND COMMON-MODE EVENTS OF EXAMPLE

Category	Common Cause	Domain	Common-Mode Events
Impact	I1	102, 104	6, 3
	I2	101, 103, 105	1, 2, 7, 8
	I3	106	10
Stress	S1	103, 105, 106	11, 2, 7, 10
	S2	199	9
	S3	101, 102, 104	1, 4
Temperature	T1	106	10
	T2	101, 102, 103, 104, 105, 199	5, 11, 8, 12, 3, 4
Vibration	V1	102, 104, 106	5, 6, 10
	V2	101, 103, 105, 199	7, 8
Operator	O1	All	1, 3, 12
	O2	All	5, 7, 10
Energy source	E1	All	2, 9
	E2	All	1, 12
Manufacturer	F1	All	2, 11
Installation Contractor	IN1	All	1, 12
	IN2	All	6, 7, 10
	IN3	All	3, 4, 5, 8, 9, 11
Test procedure	TS1	All	2, 11
	TS2	All	4, 8

I1 although they are located in domain 104 of I1. This is because these events occur independently of the impact, although they share the same physical location as event 3.

3.8.3 Obtaining Common-Mode Cut Sets

Assume a list of common causes, common-mode events, and basic events. We can readily obtain common-mode cut sets if we have all the minimal cut sets of a given fault tree. Large fault trees, however, may have an astronomically large number of minimal cut sets, and it is time-consuming to obtain them. For such a fault tree, the generation methods discussed in the previous sections are frequently truncated to give only two-or-less-event cut sets. However, this truncation should not be used when there is a possibility of common-cause failures because three-or-more-event cut sets may behave like one-event cut sets and hence should not be neglected.

We now develop a new approach to the common-cause problem, using a simplified fault tree. Another approach, due to Fussell and Wagner, based on dissection of fault trees is provided in footnote 3, page 121.

A basic event is called a *neutral event* vis-a-vis a common cause if it is independent of the cause. For a given common cause, a basic event is thus either a neutral event or a common-mode event. The present approach assumes a *probable situation* for each common cause. This situation is defined by the statement: "Assume a common cause. Since most neutral events have far smaller possibilities of occurrence than common-mode events, these neutral events are assumed not to occur in the given fault tree." Other situations violating the above requirement can be neglected because they imply the occurrence of one or more rare events.

The probable-situation classification simplifies the fault tree. It uses the fundamental simplifications of Fig. 2.34 in a bottom-up fashion. For the simplified fault tree we can easily obtain the minimal cut sets. These minimal cut sets automatically become the common-mode cut sets.

As an example, consider the fault tree of Fig. 3.9. Assume that there are no exclusive basic events. Note that the two-out-of-three gates, X and Y, can be rewritten as shown in Fig. 3.11.

Let us first analyze common cause 1. The common-mode events of the cause are 1, 3, and 12. Thus, the neutral events are 2, 4, 5, 6, 7, 8, 9, 10, and 11. Assume these neutral events have far smaller probabilities than the common-mode events when common cause 1 occurs. The fundamental simplification of Fig. 2.34 yields the simplified fault tree of Fig. 3.12. MOCUS is applied to the simplified fault tree of Fig. 3.12 in the following way:

A

B, C

$1, 3, 12, C$

$1, 3, 12, 3$——$1, 3, 12$
$1, 3, 12, 1$——$1, 3, 12$

We have one common-mode cut set $\{1, 3, 12\}$ for the common cause 1.

Next, consider common cause I3 in Table 3.2. The neutral basic events are 1, 2, 3, 4, 5, 6, 7, 8, 9, 11, and 12. The fundamental simplifications yield

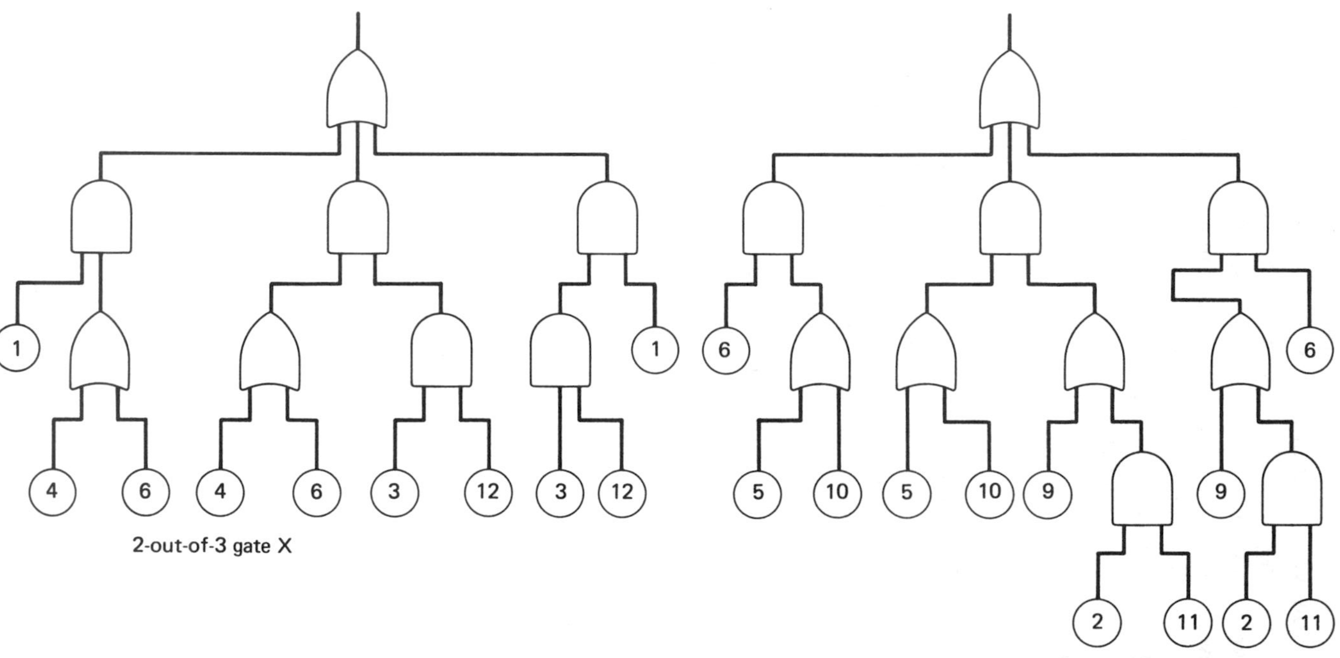

Figure 3.11. Equivalent expressions for 2-out-of-3 gates *X* and *Y*.

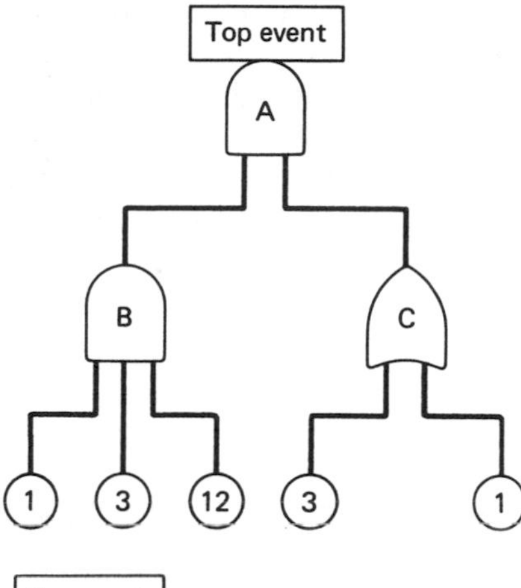

Figure 3.12. Simplified fault tree for common cause 1.

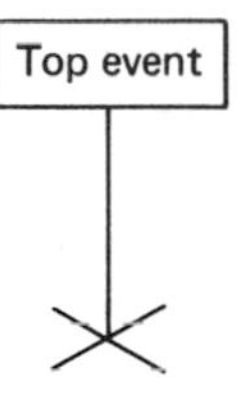

Figure 3.13. Simplified fault tree for common cause I3.

the reduced fault tree of Fig. 3.13. We see there are no common-mode cut sets for common cause I3.

We repeat the procedure for all other common causes to obtain the common-mode cut sets listed in Table 3.3.

TABLE 3.3 COMMON CAUSES AND COMMON-MODE CUT SETS

Common Cause	Common-Mode Cut Set
I2	{1,2}
I2	{1,7,8}
S3	{1,4}
S1	{2,10,11}
T2	{3,4,12}
1	{1,3,12}

3.9 SIMPLIFICATION OF DECISION TABLES: PRIME IMPLICANTS

Decision Tables were introduced in Chapter 2, where they were used to automate the fault tree construction process. System fault trees produced from component decision tables usually contain mutually exclusive events as well as working states, thus it is not feasible to apply MOCUS-like algorithms to obtain unique failure modes. Systems which contain other than simple fault modes connected by AND-OR logic gates, i.e., EOR gates, working states, etc., are termed *non-coherent* and their unique failure modes are called *prime implicants*. More rigorous definitions of non-coherency will be given later in this book; in this section we show how prime implicants are obtained for non-coherent trees using *consensus methods*. Furthermore, we develop a methodology, based on decision tables, for obtaining system prime implicants directly from component decision tables without the intermediate step of constructing fault trees.

We refer to a row consisting of input events of the decision table as a *term*. A term r_1 is said to *subsume* a term r_2 if r_1 and r_2 yield the same output event and every input event in r_2 also occurs in r_1. If this situation exists, the term r_1 is absorbed into the shorter term r_2 by removing it from the table.

Table 3.4 shows the cut sets obtained from a fault tree. Obviously, row 4 is redundant here, since if the situation described by row 3 occurs, the system will have failed. Thus, row 4 can be removed from the table. Similarly, Table 3.5 can be simplified to give Table 3.6. This is equivalent to the absorption law in Boolean algebra:

$$A \text{ or } (A \text{ and } B) = A$$

TABLE 3.4 Table of Cut Sets (Critical Transition Table)

Row	A	B	C	D	Output
1	F	—	F	—	F
2	F	—	—	—	F
3	—	F	F	—	F
4	—	F	F	F	F

TABLE 3.5 A Decision Table

A	B	Output
T	—	T
T	T	T

TABLE 3.6 Simplified Decision Table

A	B	Output
T	—	T

The decision table for the top event, after elimination of all intermediate variables, will henceforth be referred to as the *critical transition table*. A term (row) in the table is a prime implicant of the top event if it subsumes no shorter term in the critical transition table. A similar definition is possible for a decision table which models component I/O relations; a term is a prime implicant for an output event if it subsumes no shorter term having the same output event.

However, the above procedure, absorption, does not guarantee a complete set of prime implicants, particularly if multi-state components and success states exists. A method originally proposed by Quine[4,5] and extended by Tilson[6] to truth tables can be modified to obtain all the prime implicants in systems with multi-state input variables and success states. This method is called the *consensus operation*, since it creates a new term out of terms already in the table by mixing and matching their input events. We demonstrate the method by a series of examples.

Example 1.

Consider the table.

Row	Input				Output
	A	*B*	*C*	*D*	*O*
1	W	N	F	W	2
2	—	F	—	W	2
3	W	R	—	—	2

Assume each input variable has associated with it the following input events.

$$A \in \{W, F\}, \quad B \in \{N, F, R\}, \quad C \in \{W, F\}, \quad D \in \{W, F\}.$$

where W denotes working; F, failed; R, reversed; N, normal.

This table is a critical transition table if output event 2 represents a top event. Alternately, this could be regarded as part of a decision table for a component, if event 2 is a particular output event for the component. Obtain all prime implicants of the table.

[4]Quine, W. V., "The Problem of Simplifying Truth Functions," *American Mathematical Monthly*, Vol. 59, pp. 521–531, 1952.

[5]Quine, W. V., "A Way to Simplify Truth Functions," *American Mathematical Monthly*, Vol. 62, pp. 627–631, 1955.

[6]Tison, P., "Generalization of Consensus Theory and Application to the Minimization of Boolean Functions," *IEEE Trans. on Electronic Computers*, Vol. EC-16, No. 4, pp. 446–456, 1967.

Solution: The consensus procedure is summarized in the following table.

Step Number	Initial Set at the Beginning of the Step	n-Form Variable P in the Initial Set	Residue Respect to P			New Consensi (Products of Residues for Different P_i)	Final Set at the End
			P_1	$\cdots$	P_n		
1	~~$A_W B_N C_F D_W$~~ $B_F D_W$ $A_W B_R$	B	$A_W C_F D_W$	D_W	A_W	$A_W C_F D_W$	$B_F D_W$ $A_W B_R$ $A_W C_F D_W$
2	$B_F D_W$ $A_W B_R$ $A_W C_F D_W$	None	None			None	Procedure is terminated

The initial set at the beginning of step 1 is a set of all terms in the decision table,

$$\left.\begin{matrix} A_W B_N C_F D_W \\ B_F D_W \\ A_W B_R \end{matrix}\right\}$$

These are, respectively, rows 1, 2, and 3 in the decision table. The symbol A_W means that input variable A takes the value of W, etc. Row 2 in the decision table yields term $B_F D_W$ and variables, A or C do not appear.

We begin by searching for an n-event variable P such that each of the n events appears in at least one term in the initial set. Such a variable is called *n-form* in the set. Variable B is three-form because B_N is in the first term, B_F in the second, and B_R in the last term.

The residue with respect to event P_i is the term obtained by removing P_i from a term containing it. Thus, residues $A_W C_F D_W$, D_W, and A_W are obtained for events B_N, B_F, and B_R, respectively. The residues are classified into n groups according to which event is removed from the terms. In the present case, we have three groups $\{A_W C_F D_W\}$, $\{D_W\}$, and $\{A_W\}$, each of them consisting of one residue.

The new consensi are all products of residues from different groups. In the current case, each group has only one residue, and a single consensus $A_W C_F D_W$ is obtained. If a consensus has mutually exclusive events, it is removed from the list of the new consensi. As soon as a consensus is found, it is compared to the other consensi and to the terms in the initial set, and the longer products are removed from the table. We see that the term $A_W B_N C_F D_W$ can be removed from the table because of consensus $A_W C_F D_W$.

The final set of step 1 is the union of the initial set and the set of new consensi. We have the final set

$$\left.\begin{matrix} B_F D_W \\ A_W B_R \\ A_W C_F D_W \end{matrix}\right\}$$

This set becomes the initial set for step 2.

Since there is no *n*-form variable in this initial set, the procedure is terminated. We obtain $B_F D_W$, $A_W B_R$, and $A_W C_F D_W$ as the prime implicants. In other words, the decision table is simplified into

A	*B*	*C*	*D*	*O*
—	F	—	W	2
W	R	—	—	2
W	—	F	W	2

As illustrated by the above example, the new consensi are formed through successive steps. Step i corresponds to the seeking of all consensi existing among the terms of the initial set *S* at step i with respect to a variable *P*, which is *n*-form in the set. As soon as a consensus is found, it is compared to the other terms in set *S* or to the other consensi already generated in step i, and longer terms are eliminated. The non-eliminated terms and consensi are put in the set at the end of the step and the set becomes the initial set of the next step.

Example 2. Merging

Consider the truth table

A	*B*
T	F
T	T

(T, true; F, false)

The output column is not shown; we assume the same output event for each row of input events. Simplify the table.

Solution: According to convention, we denote A_T, B_T, and B_F by A, B, and $\bar{B}$, respectively. We have the following procedure.

Step	Initial Set *S*	Biform Variable	Residues		New Consensi	Final Set
1	~~$A\bar{B}$~~ ~~AB~~	*B*	*A*	*A*	*A*	*A*

There is only one prime implicant *A*, and the simplified table is

A
T

This simplification is called *merging*, which can be expressed in terms of the Boolean equation

$$(A \text{ and } \bar{B}) \text{ or } (A \text{ and } B) = A$$

If two terms in a truth table are the same except for exactly one input entry and the input entries in the two terms have opposite values, the two terms can be merged (see Section 2.6).

Example 3. Extended Version of Merging

$$B \in \{-2, -1, 0, 1, 2\}, \qquad A \in \{W, F\}.$$

Simplify the table

A	B
W	0
W	−1
W	2
W	−2
W	1

assuming identical outputs.

Solution: We have the following procedure.

Step	Initial Set S	n-form P	Residues					New Consensi	Final Set
1	~~$A_W B_0$~~ ~~$A_W B_{-1}$~~ ~~$A_W B_2$~~ ~~$A_W B_{-2}$~~ ~~$A_W B_1$~~	B	A_W	A_W	A_W	A_W	A_W	A_W	A_W

The simplified table is

A
W

This *extended version of merging* applies when a decision table includes multi-state input variable B. If n terms in a decision table are the same except for exactly one n-event input entry, and every one of the n possible states of the input variable exist in these terms, the n terms can be merged.

Example 4. Reduction

Simplify:

A	B	C
T	T	T
T	F	—

assuming identical outputs.

Solution: The procedure:

Step	Initial Set S	Biform Variable	Residues		New Consensi	Final Set
1	~~ABC~~ $A\bar{B}$	B	AC	A	AC	$A\bar{B}$ AC

The simplified table is

A	B	C
T	—	T
T	F	—

The Boolean expression is

$$(A \text{ and } B \text{ and } C) \text{ or } (A \text{ and } \bar{B}) = (A \text{ and } C) \text{ or } (A \text{ and } \bar{B})$$

This relation is called *reduction*; if two terms in a truth table are comparable except for exactly one input entry, the larger of the two terms can be reduced by that input entry if its entries have opposite entries in the two terms.

A slight modification similar to that made for the merging is required if B is an n-event variable. In order to reduce the largest term, every one of the n states of B must be present in one of the comparable terms. We see that the simplification in Example 1 is reduction with respect to three-event variable B.

The simplification operations (absorption, merging, reduction) are applied to the terms in the decision table in cycles, until none of them is applicable. The table is no longer reducible when this occurs.

Example 5. Reactor System

Obtain a fault tree and the minimal cut sets for the reactor system represented by the following I/O relation (block diagram).

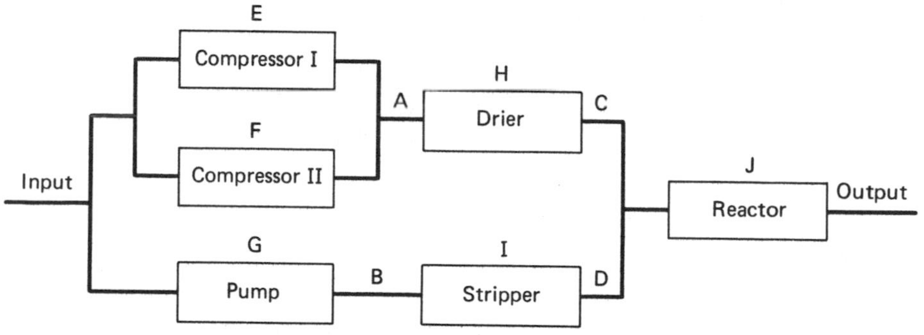

Legend:

A – Feed from compressor to drier
B – Feed from pump to stripper
C – Feed from drier to reactor
D – Feed from stripper to reactor
E – Internal mode of compressor I
F – Internal mode of compressor II
G – Internal mode of pump
H – Internal mode of drier
I – Internal mode of stripper
J – Internal mode of reactor

Decision tables:

1) Reactor.

C	D	J	Output from Reactor
W	W	W	W
W	W	F	F
W	F	W	F
W	F	F	F
F	W	W	F
F	W	F	F
F	F	W	F
F	F	F	F

2) Drier (after simplification).

	H	A	C
e	W	W	W
f	F	—	F
g	—	F	F

3) Subsystem of compressors (after simplification).

	E	F	A
h	—	W	W
i	W	—	W
j	F	F	F

4) Stripper (after simplification).

	B	I	D
k	W	W	W
l	F	—	F
m	—	F	F

5) Pump.

	G	B
n	W	W
o	F	F

Solution: The consensus operation for simplifying the decision table of the reactor is summarized in Table 3.7. The prime implicants are D_F, C_F, and J_F. The

TABLE 3.7 Consensus Operation for Simplifying Reactor Decision Table

Step	Initial Set S	n-form P	Residues		New Consensi
1	~~$C_W D_W J_F$~~ ~~$C_W D_F J_W$~~ ~~$C_W D_F J_F$~~ ~~$C_F D_W J_W$~~ ~~$C_F D_W J_F$~~ ~~$C_F D_F J_W$~~ ~~$C_F D_F J_F$~~	C	$D_W J_F$ $D_F J_W$ $D_F J_F$	$D_W J_W$ $D_W J_F$ $D_F J_W$ $D_F J_F$	$D_W J_F$ $D_F J_W$ $D_F J_F$
2	~~$C_F D_W J_W$~~ ~~$D_F J_W$~~ ~~$D_F J_F$~~	D	$C_F J_W$ J_F	J_W J_F	$C_F J_W$ J_F
3	~~$D_F J_W$~~ ~~$C_F J_W$~~ J_F	J	D_F C_F	1	D_F C_F

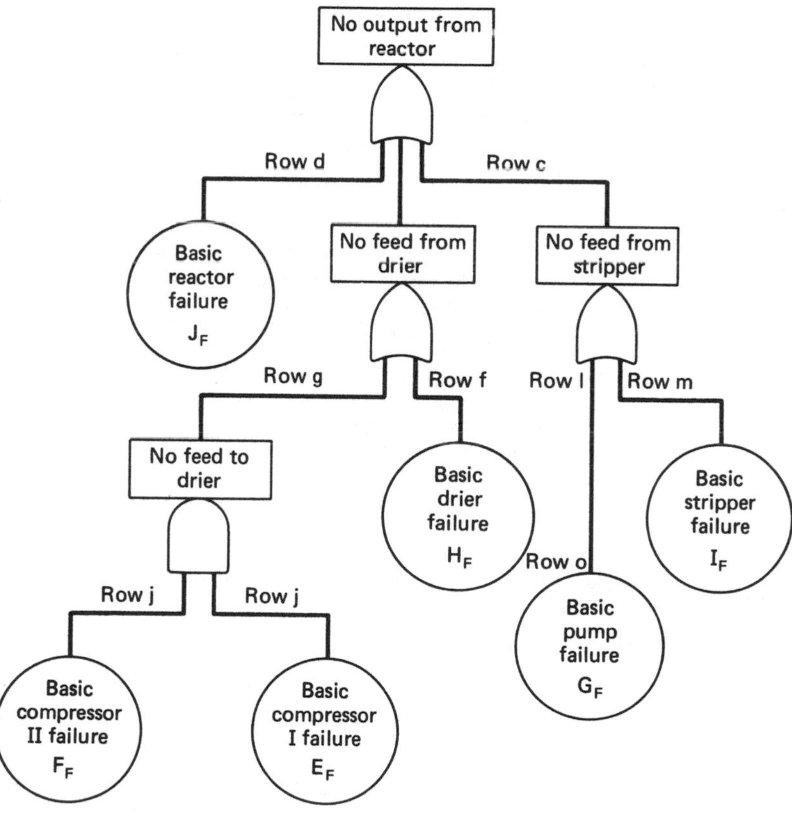

Figure 3.14. Fault tree for reactor system.

simplified table is

	C	D	J	Output from Reactor
a	W	W	W	W
b	F	—	—	F
c	—	F	—	F
d	—	—	F	F

The simplified decision tables above were used to construct the fault tree of Fig. 3.14. The minimal cut sets are $\{J_F\}$, $\{E_F, F_F\}$, $\{H_F\}$, $\{G_F\}$, and $\{I_F\}$.

Example 6. Critical Transition Table

Obtain the critical transition table for the system of Fig. 3.15. Assume the following simplified decision tables.

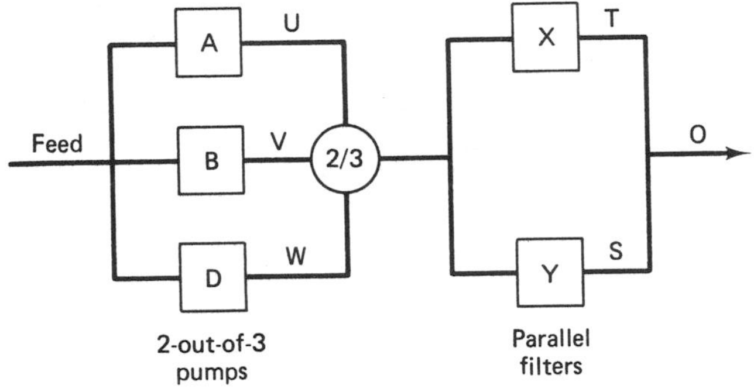

Legend:
O – Output; the top event is no output
S – Output from filter Y
T – Output from filter X
X – Internal state of filter X
Y – Internal state of filter Y
U – Output from pump A
V – Output from pump B
W – Output from pump D
A – Internal state of pump A
B – Internal state of pump B
D – Internal state of pump D

Figure 3.15. I/O relation for system in Example 6.

1) Table for variable O.

	S	T	O
	W	W	W
a	F	—	F
b	—	F	F

2) Table for variable S.

	U	W	V	X	S
	—	W	W	W	W
	W	—	W	W	W
	W	W	—	W	W
c	F	F	—	—	F
d	F	—	F	—	F
e	—	F	F	—	F
f	—	—	—	F	F

3) Table for variable U.

	A	U
	W	W
g	F	F

4) Table for variable W.

	D	W
	W	W
h	F	F

5) Table for variable V.

	B	V
	W	W
i	F	F

6) Table for variable T.

	U	W	V	Y	T
	—	W	W	W	W
	W	—	W	W	W
	W	W	—	W	W
j	F	F	—	—	F
k	F	—	F	—	F
l	—	F	F	—	F
m	—	—	—	F	F

The top event is defined as variable O taking the value of F. In the flow diagram (which is equivalent to Fig. 7.20) two of the three pumps must work for there to be feed to the two filters.

Solution: The critical transition table is a representation of the top event in terms of basic variables A, B, D, X, and Y. The table is obtained by eliminations of intermediate variables S, U, W, V, and T.

1) Initial top event table. From the table for variable O,

	S	T	O
a	F	—	F
b	—	F	F

2) Elimination of variable S. Event S_F in row a can be replaced by rows c, d, e, and f. Also, "don't care" event S_- in row b can be represented as

	U	W	V	X	S
n	—	—	—	—	—

Thus, the top event table becomes

	S					
	U	W	V	X	T	O
c	F	F	—	—	—	F
d	F	—	F	—	—	F
e	—	F	F	—	—	F
f	—	—	—	F	—	F
n	—	—	—	—	F	F

No simplification is possible for this table.

3) Elimination of variable U. Event U_F in rows c and d in the above top event table can be replaced by row g. Event U_- is replaced by A_-.

	U / A	W	V	X	T	O
g	F	F	—	—	—	F
g	F	—	F	—	—	F
e	—	F	F	—	—	F
f	—	—	—	F	—	F
n	—	—	—	—	F	F

No simplification is possible for this table.

4) Elimination of variables W and V. The resulting top event table is

A	W / D	V / B	X	T	O
F	F	—	—	—	F
F	—	F	—	—	F
—	F	F	—	—	F
—	—	—	F	—	F
—	—	—	—	F	F

5) Elimination of variable *T*.

				T				
A	*D*	*B*	*X*	*U*	*W*	*V*	*Y*	*O*
F	F	—	—	—	—	—	—	F
F	—	F	—	—	—	—	—	F
—	F	F	—	—	—	—	—	F
—	—	—	F	—	—	—	—	F
—	—	—	—	F	F	—	—	F
—	—	—	—	F	—	F	—	F
—	—	—	—	—	F	F	—	F
—	—	—	—	—	—	—	F	F

6) Elimination of variable *U*.

A	*D*	*B*	*X*	*U* *A*	*W*	*V*	*Y*	*O*
F	F	—	—	—	—	—	—	F
F	—	F	—	—	—	—	—	F
—	F	F	—	—	—	—	—	F
—	—	—	F	—	—	—	—	F
—	—	—	—	F	F	—	—	F
—	—	—	—	F	—	F	—	F
—	—	—	—	—	F	F	—	F
—	—	—	—	—	—	—	F	F

We note that this top event table has two columns for variable *A*. The two columns should be aggregated into one column. The following rules apply.

(—,—)⇒—

(—,F)⇒F

(F,—)⇒F

(—,W)⇒W

(W,—)⇒W

(F,F)⇒F

(W,W)⇒W

(F,W)⇒Eliminate the row including this pair

(W,F)⇒Eliminate the row including this pair

The aggregation for variable *A* yields

A	*D*	*B*	*X*	*W*	*V*	*Y*	*O*
F	F	—	—	—	—	—	F
F	—	F	—	—	—	—	F
—	F	F	—	—	—	—	F
—	—	—	F	—	—	—	F
F	—	—	—	F	—	—	F
F	—	—	—	—	F	—	F
—	—	—	—	F	F	—	F
—	—	—	—	—	—	F	F

No simplification is possible for this table.

7) Elimination of variable *W*.

A	*D*	*B*	*X*	*W* *D*	*V*	*Y*	*O*
F	F	—	—	—	—	—	F
F	—	F	—	—	—	—	F
—	F	F	—	—	—	—	F
—	—	—	F	—	—	—	F
F	—	—	—	F	—	—	F
F	—	—	—	—	F	—	F
—	—	—	—	F	F	—	F
—	—	—	—	—	—	F	F

The aggregation for variable *D* and the absorption law gives:

A	*D*	*B*	*X*	*V*	*Y*	*O*
F	F	—	—	—	—	F
F	—	F	—	—	—	F
—	F	F	—	—	—	F
—	—	—	F	—	—	F
F	—	—	—	F	—	F
—	F	—	—	F	—	F
—	—	—	—	—	F	F

No simplification is possible for this table.

8) Elimination of variable *V*.

A	*D*	*B*	*X*	*V* *B*	*Y*	*O*
F	F	—	—	—	—	F
F	—	F	—	—	—	F
—	F	F	—	—	—	F
—	—	—	F	—	—	F
F	—	—	—	F	—	F
—	F	—	—	F	—	F
—	—	—	—	—	F	F

The aggregation for variable *B* yields:

	A	*D*	*B*	*X*	*Y*	*O*
	F	F	—	—	—	F
o	F	—	F	—	—	F
p	—	F	F	—	—	F
	—	—	—	F	—	F
q	F	—	F	—	—	F
r	—	F	F	—	—	F
	—	—	—	—	F	F

This table can be simplified, since row q subsumes row o, and row r subsumes row p. Rows q and r are absorbed into rows o and p, respectively.

A	*D*	*B*	*X*	*Y*	*O*
F	F	—	—	—	F
F	—	F	—	—	F
—	F	F	—	—	F
—	—	—	F	—	F
—	—	—	—	F	F

No simplification is possible for this table. Further, all the input variables are basic variables. Thus, the above table is the critical transition table. We observe five minimal cut sets (prime implicants) for the system.

$$\{A_F, D_F\}, \{A_F, B_F\}, \{D_F, B_F\}, \{X_F\}, \{Y_F\}.$$

Note that minimal cut sets can be obtained directly from critical transition tables without constructing system fault trees.

3.10. DECISION TABLE APPROACH TO SYSTEMS WITH CONTROL LOOPS

Control loops make the construction of fault trees extremely difficult because certain logical complexities cannot easily be described in terms of simple Boolean functions (Henley and Kumamoto, 1977).[7] For example,

1) The order of component failures becomes important. A system fails dangerously because a failure in a temperature controller causes overheating. Suppose that the system has a temperature-sensitive cut-off switch on its heater. If the controller fails before the cut-off switch, no accident results. If the switch fails and then the controller fails, the result may be serious. This kind of problem involving a

[7]Henley, E. J., and H. Kumamoto, "Comments on Computer-Aided Synthesis of Fault Trees," *IEEE Trans. Reliability*, **R-26**, No. 5, 316, 1977.

sequence of component failures is discussed further in Chapter 11 as a protective system hazard analysis. A more complete treatment of the problem is developed in the computer code, PROTECT, and its user's manual.

2) The occurrence of the system failure is dependent on the internal system state, since outputs of feedback controllers are functions of the system state. In a cooling system, a cooling-water, control-valve, reversed action is serious when the system state, i.e., the temperature at the exit of the heat exchanger is high. The reversed action causes no failure when the temperature is low because the cooling water increases by the action.

The system state is a complicated function of histories of disturbances and component failures and the initial system state, which makes the situation more difficult. However, if we concentrate on the system's steady-state behavior, the problem becomes less difficult because we can obtain explicit representations of the steady states in terms of the disturbances and the component failures. The analysis of the control loops in Section 2.7 of Chapter 2 is based on this steady-state approach.

The approach developed in this section aims at non-steady-state analysis of the systems with control loops. We extended the concept of the critical transition set to the case involving the system's internal states, which are time (or sequence) dependent.

The approach is based on the methodology of Kumamoto and Henley, which involves the use of the decision table models for system components,[8] and on the simplification method for the tables, which has been developed and computerized by Ogunbiyi and Henley.[9] The computer code is called PITE (Prime Implicant of Top Event). It uses the Quine's consensus theory and generates the prime implicants of a function with multi-state variables when the function is expressed in the form of a decision table.

The PITE code is capable of constructing critical transition tables for a number of systems and obtaining the prime implicants of the top event. A cascade control system is used in this section as an example to demonstrate the usefulness of this methodology in modeling systems with control loops.

3.10.1 System Description

Figure 3.16 is the block diagram for a cascade liquid-level control system. The inner loop controls the fluid flow; the outer loop the fluid flow

[8]Kumamoto, H., and E. J. Henley, "Safety and Reliability Analysis of Systems with Control Loops," *AIChE Journal*, **25**, No. 1, 108, Jan., 1979.

[9]Ogunbiyi, E., Ph.D. Thesis, University of Houston, Dept. of Chemical Engineering, Sept., 1979.

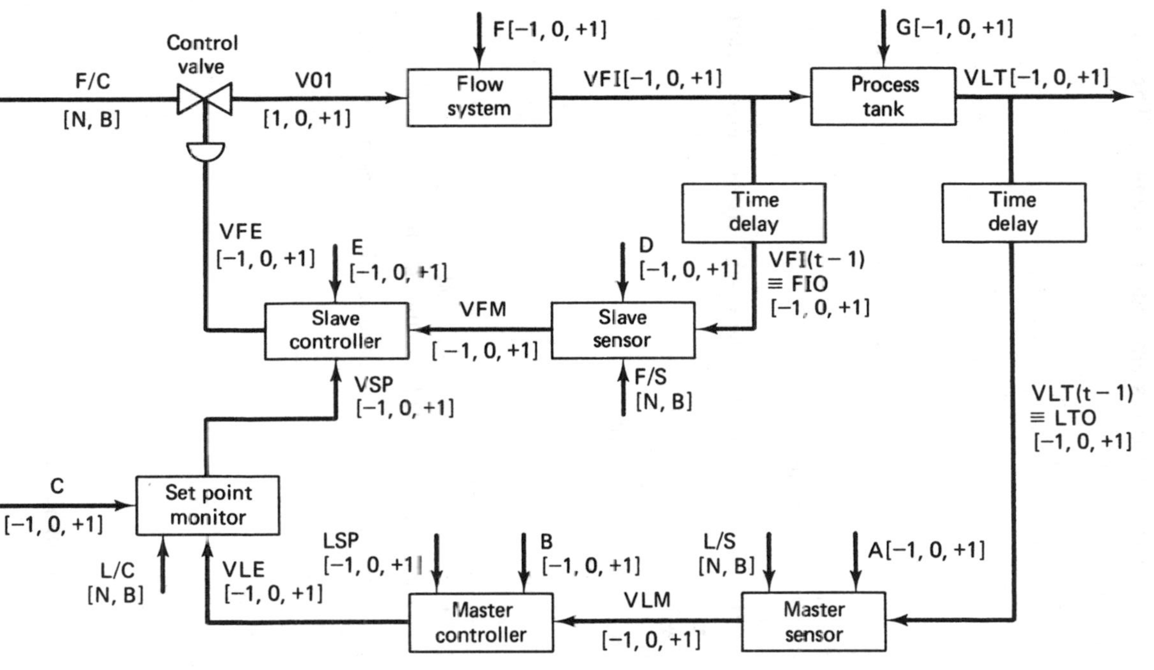

Figure 3.16. Cascade control system diagram.

set point to the inner loop. There are unspecified disturbances entering at each node of the loop. Time delays are introduced in the feedback loops as a representation of the internal system state or memory. The variables in the system are defined as follows:

A: External disturbance on master sensor.
B: External disturbance on master controller.
C: External disturbance on set point monitor.
D: External disturbance on slave sensor.
E: External disturbance of slave controller.
F: External disturbance on flow system.
FIO: Output of flow control loop time delay=VFI(t−1), internal state.
F/C: Internal mode of control valve.
F/S: Internal mode of flow sensor.
G: External disturbance on process tank.
LSP: Level control loop set point.
LTO: Output of level control loop time delay=VLT(t−1), internal state.
L/C: Internal mode of set point monitor.
L/S: Internal mode of level sensor.
t: Discrete time, an integer.
V01: Output of flow control valve.
VFE: Output of flow controller.
VFI: Output of flow system.
VFM: Output of flow monitor.
VLE: Output of master controller.
VLM: Output of level monitor.
VLT: Level of fluid in process tank; controlled variable.
VSP: Flow loop set point; output of set point monitor.

The following representations are used for the states of hardware, flows, and errors.

−1: Low.
0: Normal.
+1: High.
B: Broken.
N: Normal.
−: Don't care.

3.10.2 System Definition and Preliminary Considerations

The top event for this system is the event that the controlled variable (level of fluid in process tank) is higher than the desired level. With reference to Fig. 3.16, this is the event that the variable VLT has the value +1.

By preliminary consideration of the order in which the intermediate variables are encountered in the top event table, the component decision table models are arranged as follows:

	Component	*Typical Qualitative Aspects at Normal Phase*
1)	Process tank	The level is proportional to the flow rate.
2)	Flow system	The flow rate is proportional to V01, the output from the flow control valve.
3)	Control valve	Output V01 is proportional to input command VFE.
4)	Slave controller	Output VFE is inversely proportional to flow measurement VFM and is proportional to flow rate set point VSP.
5)	Set point monitor	Output VSP, the flow rate set point, is proportional to VLE, the command from the master controller.
6)	Master controller	Output VLE is inversely proportional to the level measurement VLM, provided that the level set point is normal.
7)	Master sensor	Level measurement VLM is proportional to level LTO.
8)	Slave sensor	Flow measurement VFM is proportional to flow rate FIO.

The decision tables for the components are shown in Table 3.8. These tables were constructed by considering the typical qualitative characteristics of the normal components and the possible deviations caused by disturbances and failures. The component tables are not unique. It is advisable to experiment by constructing several tables for a component, especially when some disturbances compete with one another.

The consensus operations in the PITE code were applied, and simplified decision tables were obtained as shown in Table 3.9.

TABLE 3.8 Decision Tables for System Components

Process Tank			Flow System		
G	VFI	VLT	V01	F	VFI
−1	−1	−1	0	−1	−1
0	−1	−1	+1	−1	+1
+1	−1	+1	0	0	0
−1	0	−1	+1	0	+1
0	0	0	0	+1	+1
+1	0	+1	+1	+1	+1
−1	+1	+1			
0	+1	+1			
+1	+1	+1			

TABLE 3.8 (continued).

Control Valve

F/C	VFE	V01
N	−1	0
B	−1	0
N	0	0
B	0	0
N	+1	+1
B	+1	0

Master Controller

LSP	B	VLM	VLE
−1	−1	−1	−1
0	−1	−1	−1
+1	−1	−1	+1
−1	0	−1	−1
0	0	−1	+1
+1	0	−1	+1
−1	+1	−1	−1
0	+1	−1	+1
+1	+1	−1	+1
−1	−1	0	−1
0	−1	0	−1
+1	−1	0	+1
−1	0	0	−1
0	0	0	0
+1	0	0	+1
−1	+1	0	−1
0	+1	0	+1
+1	+1	0	+1
−1	−1	+1	−1
0	−1	+1	−1
+1	−1	+1	+1
−1	0	+1	−1
0	0	+1	−1
+1	0	+1	+1
−1	+1	+1	−1
0	+1	+1	+1
+1	+1	+1	+1

Slave Controller

VSP	E	VFM	VFE
−1	−1	−1	−1
0	−1	−1	−1
+1	−1	−1	+1
−1	0	−1	−1
0	0	−1	+1
+1	0	−1	+1
−1	+1	−1	−1
0	+1	−1	+1
+1	+1	−1	+1
−1	−1	0	−1
0	−1	0	−1
+1	−1	0	+1
−1	0	0	−1
0	0	0	0
+1	0	0	+1
−1	+1	0	−1
0	+1	0	+1
+1	+1	0	+1
−1	−1	+1	−1
0	−1	+1	−1
+1	−1	+1	+1
−1	0	+1	−1
0	0	+1	−1
+1	0	+1	+1
−1	+1	+1	−1
0	+1	+1	+1
+1	+1	+1	+1

Set Point Monitor

C	L/C	VLE	VSP
−1	N	−1	−1
0	N	−1	0
+1	N	−1	+1
−1	B	−1	−1
0	B	−1	0
+1	B	−1	+1
−1	N	0	−1
0	N	0	0
+1	N	0	+1
−1	B	0	−1
0	B	0	0
+1	B	0	+1
−1	N	+1	+1
0	N	+1	+1
+1	N	+1	+1
−1	B	+1	−1
0	B	+1	0
+1	B	+1	+1

TABLE 3.8 (continued)

Master Sensor				Slave Sensor			
A	L/S	LTO	VLM	D	F/S	FIO	VFM
−1	N	−1	−1	−1	N	−1	−1
0	N	−1	−1	0	N	−1	−1
+1	N	−1	−1	+1	N	−1	−1
−1	B	−1	0	−1	B	−1	0
0	B	−1	0	0	B	−1	0
+1	B	−1	0	+1	B	−1	0
−1	N	0	−1	−1	N	0	−1
0	N	0	0	0	N	0	0
+1	N	0	+1	+1	N	0	+1
−1	B	0	0	−1	B	0	0
0	B	0	0	0	B	0	0
+1	B	0	0	+1	B	0	0
−1	N	+1	+1	−1	N	+1	+1
0	N	+1	+1	0	N	+1	+1
+1	N	+1	+1	+1	N	+1	+1
−1	B	+1	0	−1	B	+1	0
0	B	+1	0	0	B	+1	0
+1	B	+1	0	+1	B	+1	0

TABLE 3.9 Simplified Decision Tables for System Components

Process Tank

G	VFI	VLT
−1	−1	−1
0	−1	−1
−1	0	−1
+1	—	+1
—	+1	+1
0	0	0

Flow System

V01	F	VFI
0	−1	−1
+1	—	+1
—	+1	+1
0	0	0

Control Valve

F/C	VFE	V01
—	−1	0
—	0	0
B	—	0
N	+1	+1

Slave Controller

VSP	E	VFM	VFE
−1	—	—	−1
0	−1	—	−1
0	0	+1	−1
+1	—	—	+1
0	0	−1	+1
0	+1	—	+1
0	0	0	0

Set Point Monitor

C	L/C	VLE	VSP
−1	—	−1	−1
−1	—	0	−1
−1	B	—	−1
0	—	−1	0
0	—	0	0
0	B	—	0
+1	—	—	+1
—	N	+1	+1

Master Controller

LSP	B	VLM	VLE
−1	—	—	−1
0	−1	—	−1
0	0	+1	−1
+1	—	—	+1
0	0	−1	+1
0	+1	—	+1
0	0	0	0

TABLE 3.9 (continued).

Master Sensor				Slave Sensor			
A	L/S	LTO	VLM	D	F/S	FIO	VFM
—	N	−1	−1	—	N	−1	−1
−1	N	0	−1	−1	N	0	−1
—	B	—	0	—	B	—	0
0	—	0	0	0	—	0	0
+1	N	0	+1	+1	N	0	+1
—	N	+1	+1	—	N	+1	+1

TABLE 3.10 Critical Transition Table in Minimal Form

G	F/C	C	L/0	LSP	B	A	L/S	LTO	E	D	F/S	FIO	F	VLT
+1	—	—	—	—	—	—	—	—	—	—	—	—	—	+1
—	N	+1	—	—	—	—	—	—	—	—	—	—	—	+1
—	N	—	N	+1	—	—	—	—	—	—	—	—	—	+1
—	N	—	N	0	0	—	N	−1	—	—	—	—	—	−1
—	N	—	N	0	0	−1	N	0	—	—	—	—	—	−1
—	N	—	N	0	+1	—	—	—	—	—	—	—	—	+1
—	N	0	—	−1	—	—	—	—	0	−1	N	0	—	+1
—	N	0	—	—	−1	—	—	—	0	−1	N	0	—	+1
—	N	0	—	—	—	—	—	—	0	—	N	−1	—	+1
—	N	0	—	—	0	—	—	—	0	−1	N	0	—	+1
—	N	0	B	—	—	—	—	—	0	−1	N	0	—	+1
—	N	0	—	−1	—	—	—	—	+1	—	—	+1		
—	N	0	—	—	−1	—	—	—	+1	—	—	—	—	+1
—	N	0	—	—	0	—	—	—	+1	—	—	—	—	+1
—	N	0	B	—	—	—	—	—	+1	—	—	—	—	+1
—	—	—	—	—	—	—	—	—	—	—	—	—	1	+1
—	N	0	—	—	—	—	—	1	0	−1	N	0	—	+1
—	N	0	—	—	—	+1	—	—	0	−1	N	0	—	+1
—	N	0	—	—	—	—	—	1	0	−1	N	0	—	+1
—	N	0	—	—	—	—	B	—	0	−1	N	0	—	+1
—	N	0	—	—	—	0	—	—	0	−1	N	0	—	+1
—	N	0	—	+1	—	—	—	—	0	−1	N	0	—	+1
—	N	0	—	—	—	—	—	−1	+1	—	—	—	—	+1
—	N	0	—	—	—	+1	—	—	+1	—	—	—	—	+1
—	N	0	—	—	—	—	—	+1	+1	—	—	—	—	+1
—	N	0	—	—	—	—	B	—	+1	—	—	—	—	+1
—	N	0	—	—	—	0	—	—	+1	—	—	—	—	+1
—	N	0	—	+1	—	—	—	—	+1	—	—	—	—	+1
—	N	0	—	0	+1	—	—	—	0	−1	N	0	—	+1
—	N	0	N	—	+1	—	—	—	0	−1	N	0	—	+1
—	N	0	N	0	—	—	—	—	0	−1	N	0	—	+1
—	N	0	N	0	—	—	—	—	+1	—	—	—	—	+1

3.10.3 Critical Transition Table

The PITE code was used to construct the critical transition table by eliminating intermediate variables. Table 3.10, which was obtained after the elimination, consists of 28 rows, all of which are minimal in the sense that no rows subsume others. Each row consists of component failure modes and system internal states. The total time for the construction of the table on a Honeywell 66/10 computing system was 7.06 seconds.

The table was reduced to an irredundant form using Quine's consensus method. The resulting critical transition table is shown as Table 3.11 which is the list of all (and only) prime implicant of the top event in terms of component failure modes and system internal states.

TABLE 3.11 CRITICAL TRANSITION TABLE IN IRREDUNDANT FORM

G	F/C	C	L/C	LSP	B	A	L/S	LTO	E	D	F/S	FIO	F	VLT	MODE
+1	—	—	—	—	—	—	—	—	—	—	—	—	—	+1	1
—	N	+1	—	—	—	—	—	—	—	—	—	—	—	+1	2
—	N	—	N	+1	—	—	—	—	—	—	—	—	—	+1	3
—	N	—	N	0	0	—	N	−1	—	—	—	—	—	+1	4
—	N	—	N	0	0	−1	N	0	—	—	—	—	—	+1	5
—	N	—	N	0	+1	—	—	—	—	—	—	—	—	+1	6
—	N	0	—	—	—	—	—	—	0	−1	N	0	—	+1	7
—	N	0	—	—	—	—	—	—	0	—	N	−1	—	+1	8
—	N	0	—	—	—	—	—	—	+1	—	—	—	—	+1	9
—	—	—	—	—	—	—	—	—	—	—	—	—	+1	+1	10

Neglecting normal and don't care events in each row of the critical transition table, we obtained 10 system's failure modes by which an increase of the tank level occurs.

Mode 1) G = +1: High external disturbance on process tank.
Mode 2) C = +1: High external disturbance on set point monitor.
Mode 3) LSP = +1: Level control loop set point is high.
Mode 4) LTO = −1: Low output from level control time delay; if the level is low, the negative control loop increases the level and results in an overshoot. The sensitivity of the control loop should be decreased to prevent the excess control.
Mode 5) A = −1: Master sensor is biased low. Because of this failure, the master controller increases the flow-rate set point in spite of the actual level of the fluid being high.
Mode 6) B = +1: Master controller is biased high.
Mode 7) D = −1: Valve sensor is biased low; the slave controller increases the flow rate in spite of the actual flow rate in the flow system being high.
Mode 8) FIO = −1: Low output from flow control time delay; the flow control loop has the possibility of overshooting when it works to adjust the low flow rate.
Mode 9) E = +1: The slave controller is biased high.
Mode 10) F = +1: High external disturbance on flow system.

PROBLEMS

3.1. Figure P3.1 shows a simplified fault tree of a domestic hot-water system in Problem 1.6. (1) Find the minimal cut sets. (2) Find the minimal path sets.

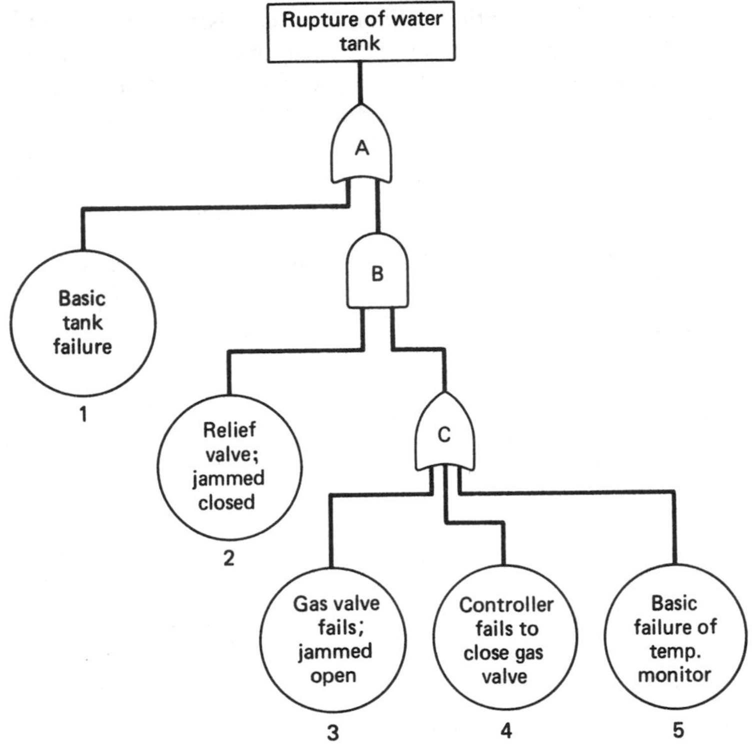

Figure P3.1. A simplified fault tree of a domestic hot water system.

3.2. P. S. Ufford, *Chem. Eng. Prog.*, **68**, No. 3, 47, 1972, shows the simplified block diagram of Fig. P3.2 for a chemical plant. Construct the fault tree, and find the minimal path sets and cut sets.

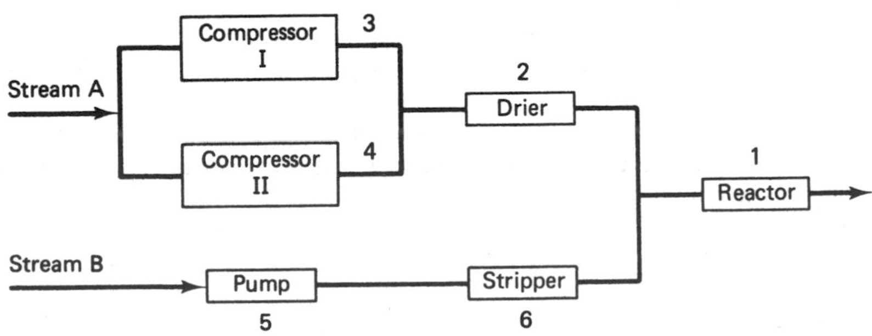

Figure P3.2.

3.3. Fig. P3.3 shows a fault tree for the heater system of Problem 2.6. Obtain the minimal cut sets, noting the exclusive events.

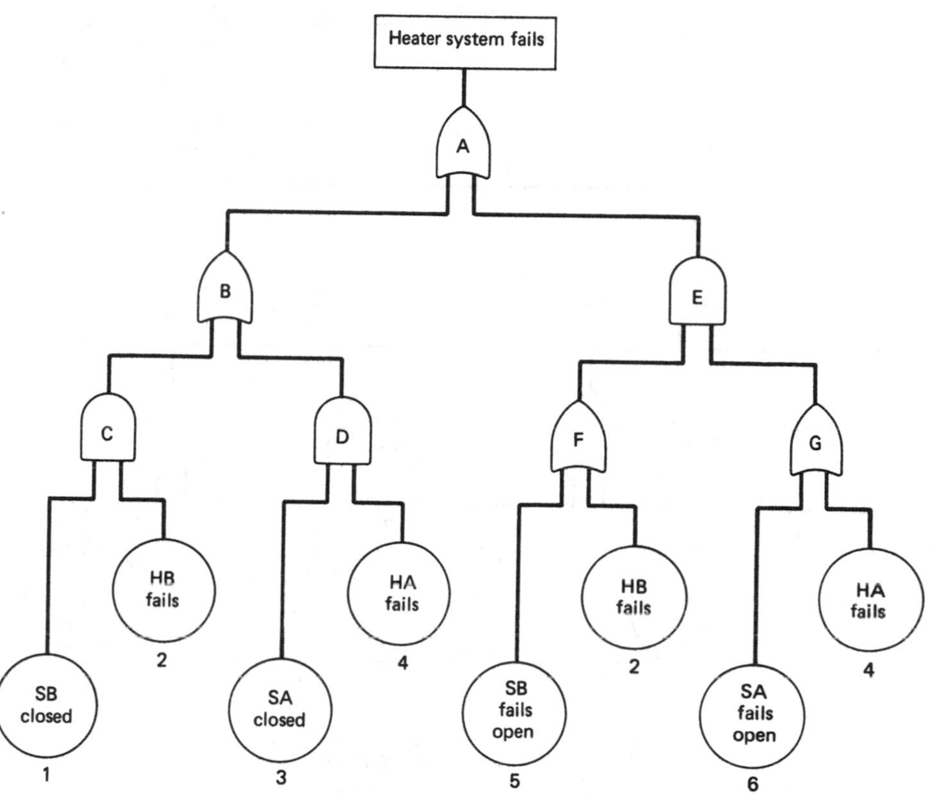

Figure P3.3. A fault tree of a heater system.

3.4. The relay system of Problem 2.7 has the fault tree shown in Fig. P3.4 on page 152. Obtain the minimal cut sets, noting the mutually exclusive events 4 and 9, and 5 and 6.

3.5. Verify the common-mode cut sets in Table 3.3 for causes S3, S1, and T2.

3.6. Simplify Table 2.8, using the consensus procedure.

3.7. Obtain the critical transition table and the prime implicant for the system of Fig. 2.28 for the top event, "high temperature of outflow acid."

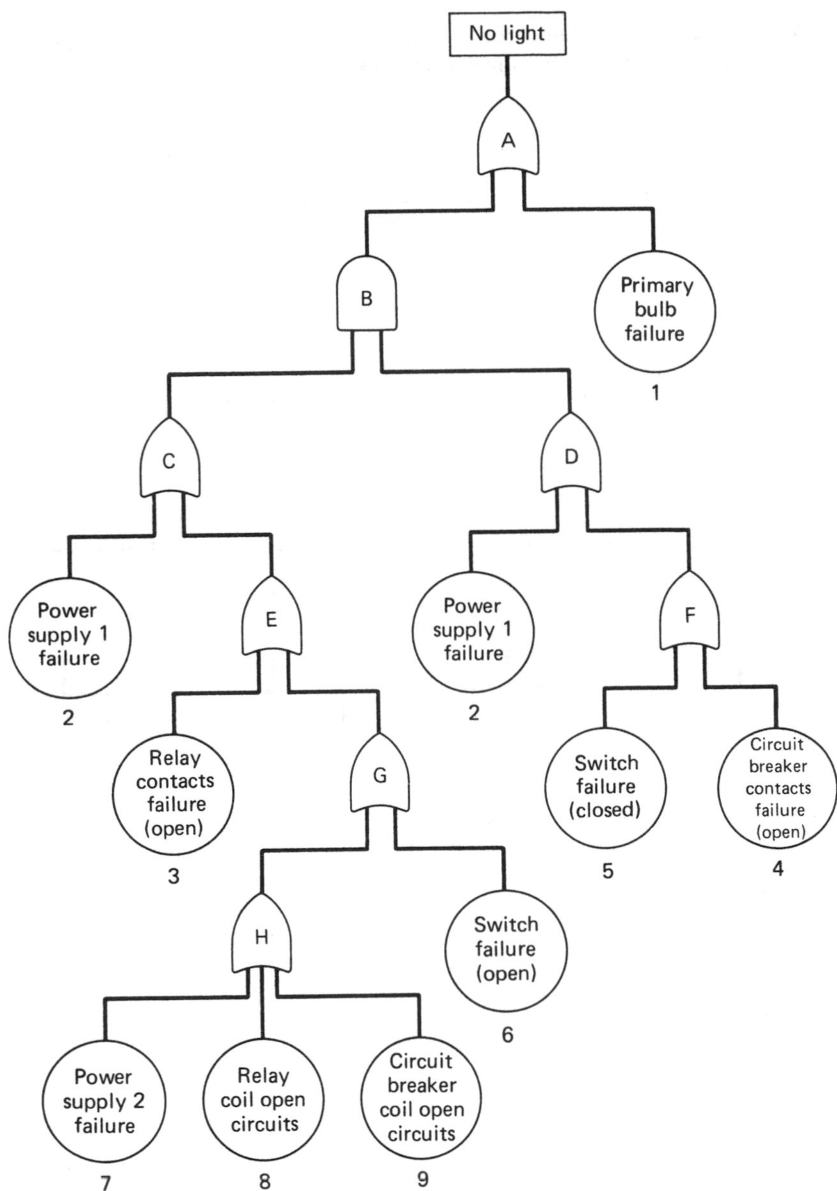

Figure P3.4. A fault tree for a relay system.

4

QUANTIFICATION OF BASIC EVENTS

All systems eventually fail; nothing is perfectly reliable, nothing endures forever. A reliability engineer must assume that a system will fail and, therefore, concentrate on decreasing the frequency of failure to an economically and socially acceptable level. That is a more realistic and tenable approach than are political slogans such as "zero pollution," "no risk," "accident-free," etc.

Probabilistic statements are not unfamiliar to the public. We have become accustomed, for example, to a weather forecaster predicting that "there is a twenty percent risk of thundershowers."[†] Likewise, the likelihood that a person will be drenched if his umbrella malfunctions can be expressed probabilistically. For instance, one might say that there is a ninety percent chance that a one year old umbrella will work as designed. This probability is, of course, time dependent. The reliability of an umbrella would be expected to decrease with time; a two year old umbrella is more likely to fail than a one year old umbrella.

Reliability is by no means the only performance criterion by which a device such as an umbrella can be characterized. If it malfunctions or breaks, it can be repaired. Since the umbrella cannot be used while it is being repaired, one might also measure its performance in terms of *availability*, i.e., the fraction of time it is available for use and functioning

[†]A comedian once asked whether this statement meant that, if you stopped ten people in the street and asked them if it would rain, two of them would say "yes."

properly. Repairs cost money, so we also want to know the *expected number of failures* during any given time interval.

Intuitively, one feels that there are analytical relationships between descriptions such as reliability, availability, and expected number of failures. In this chapter, these relationships are developed. An accurate description of component failures and failure modes is central to the identification of system hazards, since these are caused by combinations of component faults. If there are no system-dependent component faults, then the quantification of basic (component) faults is independent of a particular system, and generalizations can be made. Unfortunately, that is not usually the case.

In this chapter, we first quantify basic events related to system components with *binary states*, i.e., normal state and failed state. Here, by components we mean elementary devices, equipment, subsystems, etc. Then, this quantification is extended to components having plural failure modes. Finally, quantitative aspects of human errors and impacts from the environment are discussed.

We assume that the reader has some knowledge of statistics. Statistical concepts generic to reliability are developed in the chapter and additional material can be found in Appendix 4.3 to this chapter. A useful glossary of definitions appears as Appendix 4.6.

4.1 CONDITIONAL AND UNCONDITIONAL PROBABILITIES

There are a seemingly endless number of sophisticated definitions and equations in this chapter, and the reader may wonder whether this degree of detail and complexity is justified or whether it is a purely academic indulgence.

The first version of this chapter, which was written in 1975, was considerably simpler and contained fewer definitions. When this material was distributed at the NATO Advanced Study Institute on Risk Analysis in Italy in 1978, it became clear during the ensuing discussion that the (historical) absence of very precise and commonly understood definitions for failure parameters had resulted in theories of limited validity and computer programs which purport to calculate the same parameters but don't. In rewriting this chapter, we tried to set things right, and to label all parameters so that their meanings are clear. Much existing confusion centers around the lack of rigor in defining failure parameters as being *conditional* or *unconditional*. Clearly, the probability of a man's dying the day after his 30th birthday party is not the same as the probability of a man's living for 30 years and 1 day. The later probability is unconditional, which the former is conditional on his having survived to age thirty.

The notation and theorems pertaining to conditional probabilities may be unfamiliar to some readers, so we now summarize them.

4.1.1 Definition of Conditional Probabilities

Conditional probability $\Pr(A|C)$ is the *probability of the occurrence* of event A, given that event C occurs. This probability is defined by

$\Pr(A|C)$ = proportion of the things resulting in event A among the set of things yielding event C

The conditional probability is different either from unconditional probabilities $\Pr(A)$ or $\Pr(C)$ or $\Pr(A,C)$:

$\Pr(A)$ = proportion of the things resulting in event A among the set of all things

$\Pr(C)$ = proportion of the things resulting in event C among the set of all things

$\Pr(A,C)$ = proportion of the things resulting in the simultaneous occurrence of events A and C among the set of all things

Example 1A.

There are six balls which are small or medium or large; red or white or blue.

BALL 1	BALL 2	BALL 3	BALL 4	BALL 5	BALL 6
SMALL	SMALL	MEDIUM	LARGE	SMALL	MEDIUM
BLUE	RED	WHITE	WHITE	RED	RED

Obtain the following probabilities.

1) Pr(BLUE BALL)
2) Pr(SMALL BALL)
3) Pr(BLUE BALL, SMALL BALL)
4) Pr(BLUE BALL|SMALL BALL)

Solution: There are six balls. Among them, one is blue, three are small, and one is blue and small. Thus,

$$\Pr(\text{BLUE BALL}) = \tfrac{1}{6}$$

$$\Pr(\text{SMALL BALL}) = \tfrac{3}{6} = \tfrac{1}{2}$$

$$\Pr(\text{BLUE BALL, SMALL BALL}) = \tfrac{1}{6}$$

Among the three small balls, only one is blue. Thus,

$$\Pr(\text{BLUE BALL}|\text{SMALL BALL}) = \tfrac{1}{3}$$

Conditional probability $\Pr(A|B,C)$ is the probability of the occurrence of event A, given that both events B and C occur. This probability is defined by

$\Pr(A|B,C)$ = proportion of the things yielding event A among the set of things resulting in the simultaneous occurrence of events B and C

Example 2A.

Obtain

1) $\Pr(\text{BALL } 2)$
2) $\Pr(\text{SMALL BALL, RED BALL})$
3) $\Pr(\text{BALL } 2, \text{SMALL BALL, RED BALL})$
4) $\Pr(\text{BALL } 2|\text{SMALL BALL, RED BALL})$
5) $\Pr(\text{BALL } 1|\text{SMALL BALL, RED BALL})$

Solution: Among the six balls, two are small and red, and only ball 2 is at the same time small and red. Thus,

$$\Pr(\text{BALL } 2) = \tfrac{1}{6}$$
$$\Pr(\text{SMALL BALL, RED BALL}) = \tfrac{2}{6} = \tfrac{1}{3}$$
$$\Pr(\text{BALL } 2, \text{SMALL BALL, RED BALL}) = \tfrac{1}{6}$$

Ball 2 is one of the two small red balls; therefore,

$$\Pr(\text{BALL } 2|\text{SMALL BALL, RED BALL}) = \tfrac{1}{2}$$

Ball 1 does not belong to the set of the two small red balls. Thus,

$$\Pr(\text{BALL } 1|\text{SMALL BALL, RED BALL}) = \tfrac{0}{2} = 0$$

4.1.2 Chain Rule

The simultaneous *existence* of events A and C is equivalent to the existence of event C plus the existence of event A under the occurrence of event C. Symbolically,

$$(A,C) \Leftrightarrow C \text{ and } (A|C)$$

This equivalence can be extended to probabilities:

$$\Pr(A,C) = \Pr(C)\Pr(A|C) \tag{4.1}$$

More generally,

$$\Pr(A_1, A_2, \ldots, A_n) = \Pr(A_1)\Pr(A_2|A_1)\ldots\Pr(A_n|A_1, A_2, \ldots, A_{n-1}) \quad (4.2)$$

If we think of the world (the entire population) as having a certain property W, then (4.1) becomes:

$$\Pr(A, C|W) = \Pr(C|W)\Pr(A|C, W) \quad (4.3)$$

These equations are the *chain rule* relationships. They are useful for calculating simultaneous (unconditional) probabilities from conditional probabilities. Some conditional probabilities can be calculated more easily than unconditional probabilities, since conditions narrow the world under consideration.

Example 3A.

Confirm the chain rules:

1) Pr(BLUE BALL, SMALL BALL)
 = Pr(SMALL BALL) Pr(BLUE BALL|SMALL BALL)
2) Pr(BALL 2, SMALL BALL|RED BALL)
 = Pr(SMALL BALL|RED BALL) Pr(BALL 2|SMALL BALL, RED BALL)

Solution: From Example 1A

Pr(BLUE BALL, SMALL BALL) = $\frac{1}{6}$
Pr(SMALL BALL) = $\frac{1}{2}$
Pr(BLUE BALL|SMALL BALL) = $\frac{1}{3}$.

The first chain rule is confirmed, since

$$\tfrac{1}{6} = (\tfrac{1}{2})(\tfrac{1}{3})$$

Among the three red balls, two are small, and ball 2 is the only one which is small ball 2. Thus,

Pr(BALL 2, SMALL BALL|RED BALL) = $\frac{1}{3}$
Pr(SMALL BALL|RED BALL) = $\frac{2}{3}$

Only one ball is ball 2 among the two small red balls.

Pr(BALL 2|SMALL BALL, RED BALL) = $\frac{1}{2}$

Thus, the second chain rule is confirmed, since

$$(\tfrac{1}{3}) = (\tfrac{2}{3})(\tfrac{1}{2})$$

4.1.3 Alternative Expression of Conditional Probabilities

From the chain rule of (4.1) and (4.3), we have

$$\Pr(A|C)=\frac{\Pr(A,C)}{\Pr(C)} \tag{4.4}$$

$$\Pr(A|C,W)=\frac{\Pr(A,C|W)}{\Pr(C|W)} \tag{4.5}$$

We see that the conditional probability is the ratio of the unconditional simultaneous probability to the probability of condition C.

Example 4A.

Confirm:

1) $\Pr(\text{BLUE BALL}|\text{SMALL BALL}) = \dfrac{\Pr(\text{BLUE BALL, SMALL BALL})}{\Pr(\text{SMALL BALL})}$

2) $\Pr(\text{BALL 2}|\text{SMALL BALL, RED BALL}) = \dfrac{\Pr(\text{BALL 2, SMALL BALL}|\text{RED BALL})}{\Pr(\text{SMALL BALL}|\text{RED BALL})}$

Solution: From Example 3A

$$\tfrac{1}{3}=\frac{\frac{1}{6}}{\frac{1}{2}}=\tfrac{1}{3}, \text{ for the first equation}$$

$$\tfrac{1}{2}=\frac{\frac{1}{3}}{\frac{2}{3}}=\tfrac{1}{2}, \text{ for the second equation}$$

4.1.4 Independence

Event A is *independent* of event C if and only if

$$\Pr(A|C)=\Pr(A) \tag{4.6}$$

This means that the probability of event A is unchanged by the occurrence of event C. Equations (4.4) and (4.6) give

$$\Pr(A,C)=\Pr(A)\Pr(C)$$

This is another expression for independence. We see that if event A is independent of event C, then event C is also independent of event A.

Example 5A.

Is event "BLUE BALL" independent of "SMALL BALL"?

Solution: It is not independent because

$$\Pr(\text{BLUE BALL}) = \tfrac{1}{6} \text{ (Example 1A)}$$
$$\Pr(\text{BLUE BALL}|\text{SMALL BALL}) = \tfrac{1}{3} \text{ (Example 1A)}$$

Event "BLUE BALL" is more likely to occur when "SMALL BALL" occurs. In other words, the possibility "BLUE BALL" is increased by the observation, "SMALL BALL."

4.1.5 Bridge Rule

It becomes easier to calculate the conditional probability $\Pr(A|C)$ as causal relations between events A and C become clearer. To achieve the clarification, we introduce intermediate events $B_1, \ldots, B_n$, each of which acts as a *bridge* from event C to event A (see Fig. 4.1). We assume that

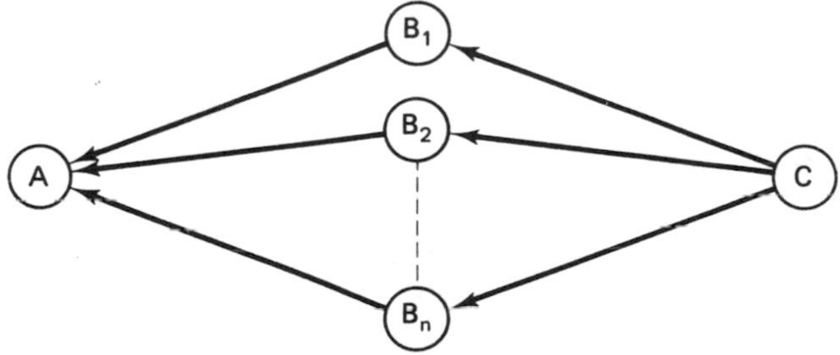

Figure 4.1. Bridges $B_1, \ldots, B_n$.

events $B_1, \ldots, B_n$ are mutually exclusive and cover all cases; i.e.,

$$\Pr(B_i, B_j) = 0, \quad \text{for } i \neq j \tag{4.7}$$

$$\Pr(B_1 \text{ or } B_2 \text{ or} \ldots \text{or } B_n) = 1 \tag{4.8}$$

Then, the conditional probability $\Pr(A|C)$ can be written as

$$\Pr(A|C) = \sum_{i=1}^{n} \Pr(B_i|C)\Pr(A|B_i, C) \tag{4.9}$$

Event A can occur through any one of the n events $B_1, \ldots, B_n$: Intuitively speaking, $\Pr(B_i|C)$ is the probability of the choice of bridge B_i, and $\Pr(A|B_i, C)$ is the probability of the occurrence of event A when we have passed through bridge B_i.

Example 6A.

Calculate $\Pr(\text{BLUE BALL}|\text{SMALL BALL})$ by letting B_i be "BALL i."

Solution: Equation (4.9) becomes

$$\begin{aligned}
&\Pr(\text{BLUE BALL}|\text{SMALL BALL})\\
&=\Pr(\text{BALL 1}|\text{SMALL BALL})\Pr(\text{BLUE BALL}|\text{BALL 1, SMALL BALL})\\
&=\Pr(\text{BALL 2}|\text{SMALL BALL})\Pr(\text{BLUE BALL}|\text{BALL 2, SMALL BALL})\\
&+\dots\\
&+\Pr(\text{BALL 6}|\text{SMALL BALL})\Pr(\text{BLUE BALL}|\text{BALL 6, SMALL BALL})\\
&=(\tfrac{1}{3})(1)+(\tfrac{1}{3})(0)+(0)(0)+(0)(0)+(\tfrac{1}{3})(0)+(0)(0)\\
&=\tfrac{1}{3}
\end{aligned}$$

When there is no ball satisfying the condition, the corresponding conditional probability is zero. Thus,

$$\Pr(\text{BLUE BALL}|\text{BALL 3, SMALL BALL})=0$$

This confirms the result of Example 1A.

4.2 PROBABILISTIC PARAMETERS OF COMPONENTS WITH BINARY STATES

We assume that at any given time a component is either functioning normally or failed, and that the component state changes as time evolves. Possible transitions of state are shown in Fig. 4.2. A new component "jumps" into a normal state, is there for some time; then it fails and experiences a transition to the failed state. The failed state continues forever if the component is non-repairable. A repairable component remains in the failed state for a period, then undergoes a transition to the normal state when the repair is completed. It is assumed that the component changes its state instantaneously when the transition takes place. It is further assumed that, at most, one transition occurs in a sufficiently small time interval and that the possibility of two or more transitions is negligible.

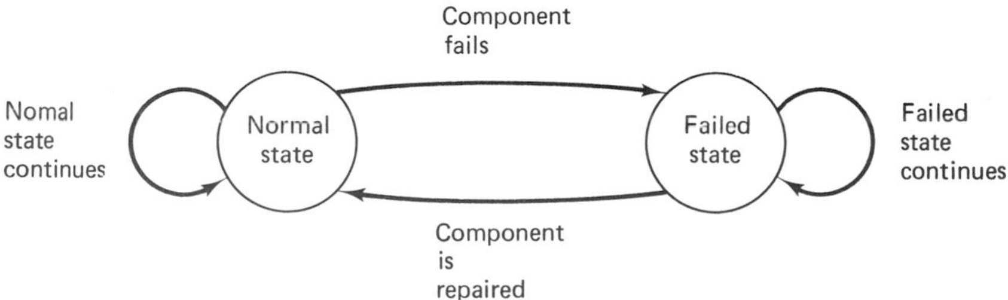

Figure 4.2. Transition diagram of component states.

The transition to the normal state is called *repair*, whereas the transition to the failed state is *failure*. We assume that repairs restore the component to a condition as good as new, so we can regard the production of a factory component as a repair. The entire cycle thus consists of repetitions of the repair-to-failure and the failure-to-repair process. We first discuss the *repair-to-failure process*, then the *failure-to-repair* process, and finally the *whole process*.

4.2.1 A Repair-to-Failure Process

A life cycle is a typical repair-to-failure process. Here, repair means birth and failure corresponds to death.

We cannot predict a man's exact lifetime, since his failure is a random variable whose characteristics must be established by considering him as a sample from a large population of men. His failure can be characterized only by the stochastic properties of the population as a whole.

The reliability $R(t)$, in this example, is the probability of survival to (inclusive) age t and is the number surviving at t divided by the total sample. Similarly, the unreliability $F(t)$ is the probability of death to age t (t is not included) and is obtained by dividing the total number of deaths before age t by the total population. From the mortality data in Table 4.1, which lists lifetimes for a population of 1,023,102, the reliability and the unreliability are calculated in Table 4.2 and plotted in Fig. 4.3.

TABLE 4.1 MORTALITY DATA[a]
t = age in years; $L(t)$ = number of living at age t

t	$L(t)$	t	$L(t)$	t	$L(t)$	t	$L(t)$
0	1,023,102	15	962,270	50	810,900	85	78,221
1	1,000,000	20	951,483	55	754,191	90	21,577
2	994,230	25	939,197	60	677,771	95	3,011
3	990,114	30	924,609	65	577,822	99	125
4	986,767	35	906,554	70	454,548		
5	983,817	40	883,342	75	315,982		
10	971,804	45	852,554	80	181,765		

[a]After Bompas-Smith, J. H., *Mechanical Survival: The Use of Reliability Data*, McGraw-Hill Book Company, New York, 1971.

The curve of $R(t)$ versus t is a *survival distribution*, whereas the curve of $F(t)$ versus t is a *failure distribution*. The survival distribution represents both the probability of survival of an individual to age t and the proportion of the population that is expected to survive to any given age t. The failure distribution $F(t)$ is the probability of death of an individual before age t. It also represents the proportion of the population that is predicted

TABLE 4.2 Human Reliability

t Age in Years	$L(t)$, Number Living at Age t	$R(t)=\dfrac{L(t)}{N}$	$F(t)=1-R(t)$
0	1,023,102	1.	0.
1	1,000,000	0.9774	0.0226
2	994,230	0.9718	0.0282
3	990,114	0.9678	0.0322
4	986,767	0.9645	0.0355
5	983,817	0.9616	0.0384
10	971,804	0.9499	0.0501
15	962,270	0.9405	0.0595
20	951,483	0.9300	0.0700
25	939,197	0.9180	0.0820
30	924,609	0.9037	0.0963
35	906,554	0.8861	0.1139
40	883,342	0.8634	0.1366
45	852,554	0.8333	0.1667
50	810,900	0.7926	0.2074
55	754,191	0.7372	0.2628
60	677,771	0.6625	0.3375
65	577,882	0,5648	0.4352
70	454,548	0.4443	0.5557
75	315,982	0.3088	0.6912
80	181,765	0.1777	0.8223
85	78,221	0.0765	0.9235
90	21,577	0.0211	0.9789
95	3,011	0.0029	0.9971
99	125	0.0001	0.9999
100	0	.0	1.

to die before age t. The difference $F(t_2)-F(t_1), t_2>t_1$ is the proportion of the population expected to die between ages t_1 and t_2 (t_1 is included and t_2 is not).

Since the number of deaths at each age is known, a *histogram* such as the one in Fig. 4.4 can be drawn. The height of each bar in the histogram represents the number of deaths in a particular life band. This is proportional to the difference $F(t+\Delta)-F(t)$, where Δ is the width of the life band.

If the width is reduced, the steps in Fig. 4.4 draw progressively closer, until a continuous curve is formed. This curve, when normalized by the total sample, is the *failure density* $f(t)$. The probability of death during a small life band $[t, t+dt)$ is given by $f(t)dt$ and is equal to $F(t+dt)-F(t)$.

The probability of death between any two ages t_1 and t_2 is the area under the curve and can be obtained by integrating the curve between the

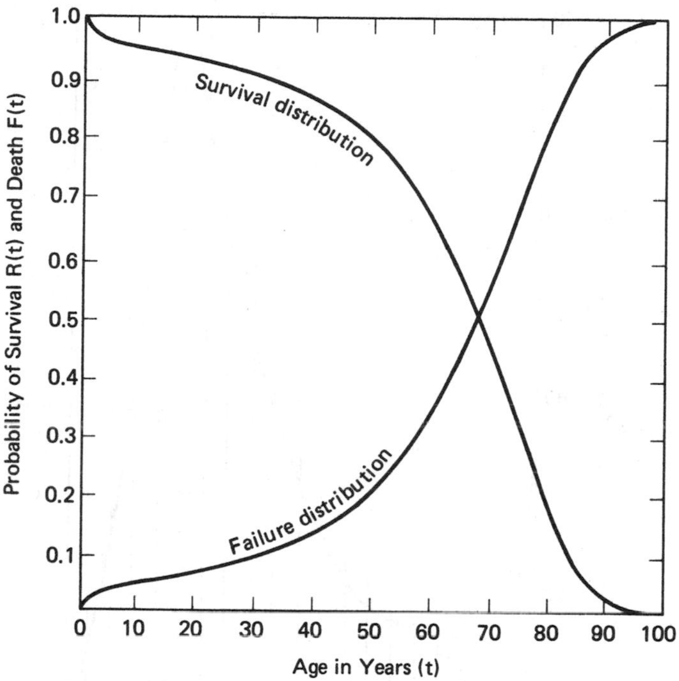

Figure 4.3. Survival and failure distributions.

ages

$$F(t_2) - F(t_1) = \int_{t_1}^{t_2} f(t)\,dt$$

This identity indicates that the failure density $f(t)$ is given by

$$f(t) = \frac{dF(t)}{dt}$$

and can be approximated by the numerical differentiation

$$f(t) \cong \frac{F(t+\Delta) - F(t)}{\Delta}$$

Letting

N = total number of sample = 1,023,102
$n(t)$ = number of deaths before age t
$n(t+\Delta)$ = number of deaths before age $t+\Delta$

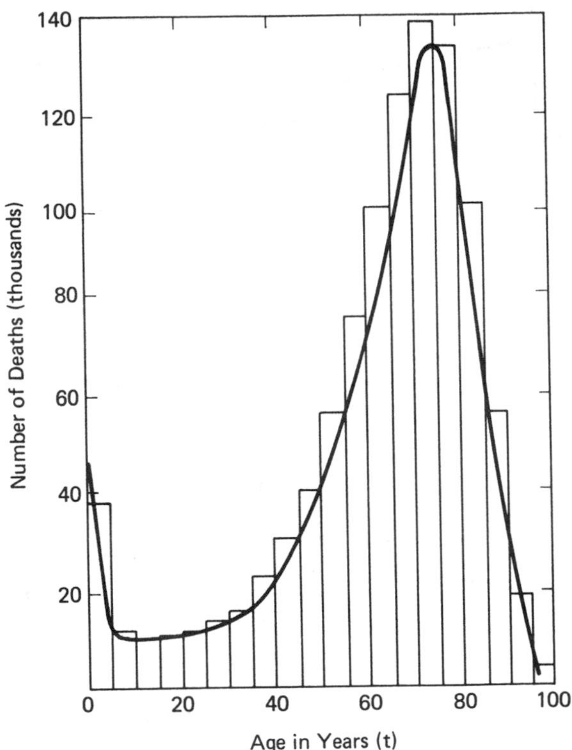

Figure 4.4. Histogram and smooth curve.

the quantity $[n(t+\Delta)-n(t)]/N$ is the proportion of the population expected to die during $[t, t+\Delta)$ and, hence, equals $F(t+\Delta)-F(t)$. Thus,

$$f(t) \cong \frac{n(t+\Delta)-n(t)}{\Delta \cdot N} \tag{4.10}$$

The quantity $[n(t+\Delta)-n(t)]$ is equal to the height of the histogram in a life band $[t, t+\Delta)$. Thus, the numerical differentiation formula (4.10) is equivalent to the normalization of the histogram of Fig. 4.4 divided by the total sample N and the band width Δ.

Calculated values for $f(t)$ are given in Table 4.3 and plotted in Fig. 4.5. Column 4 of Table 4.3 is based on the differentiation of the smooth curve of $F(t)$, and column 3 on the numerical differentiation (i.e., the normalized histogram). A smooth failure distribution can be obtained by a polynomial approximation of discrete values of $F(t)$.

Consider now a new population consisting of *the individuals surviving at age t*. The *failure rate* $r(t)$ is the probability of death per unit time at age t

TABLE 4.3 Failure Density Function $f(t)$

Age in Years	$n(t+\Delta)-n(t)$ No. of Failures (death)	$f_2(t)=\dfrac{n(t+\Delta)-n(t)}{N\cdot\Delta}$	$f_2(t)=\dfrac{dF(t)}{dt}$
0	23,102	0.02260	0.00540
1	5,770	0.00564	0.00454
2	4,116	0.00402	0.00284
3	3,347	0.00327	0.00330
4	2,950	0.00288	0.00287
5	12,013	0.00235	0.00192
10	9,534	0.00186	0.00198
15	10,787	0.00211	0.00224
20	12,286	0.00240	0.00259
25	14,588	0.00285	0.00364
30	18,055	0.00353	0.00393
35	23,212	0.00454	0.00436
40	30,788	0.00602	0.00637
45	41,654	0.00814	0.00962
50	56,709	0.01110	0.01367
55	76,420	0.01500	0.01800
60	99,889	0.01950	0.02200
65	123,334	0.02410	0.02490
70	138,566	0.02710	0.02610
75	134,217	0.02620	0.02460
80	103,554	0.02020	0.01950
85	56,634	0.01110	0.00970
90	18,566	0.00363	0.00210
95	2,886	0.00071	—
99	125	0.00012	—
100	0		—

for the individual in this population. Thus, for sufficiently small Δ, the quantity $r(t)\cdot\Delta$ is estimated by the number of deaths during $[t,t+\Delta)$ divided by the number of individuals surviving at age t:

$$r(t)\Delta \cong \frac{\text{number of deaths during } [t,t+\Delta)}{\text{number of survivals at age } t} = \frac{[n(t+\Delta)-n(t)]}{L(t)}$$

If we divide the numerator and the denominator by the total sample ($N=1,023,102$), then, we have

$$r(t)\Delta = \frac{f(t)\Delta}{R(t)}$$

since $R(t)$ is the number of survivals at age t divided by the population,

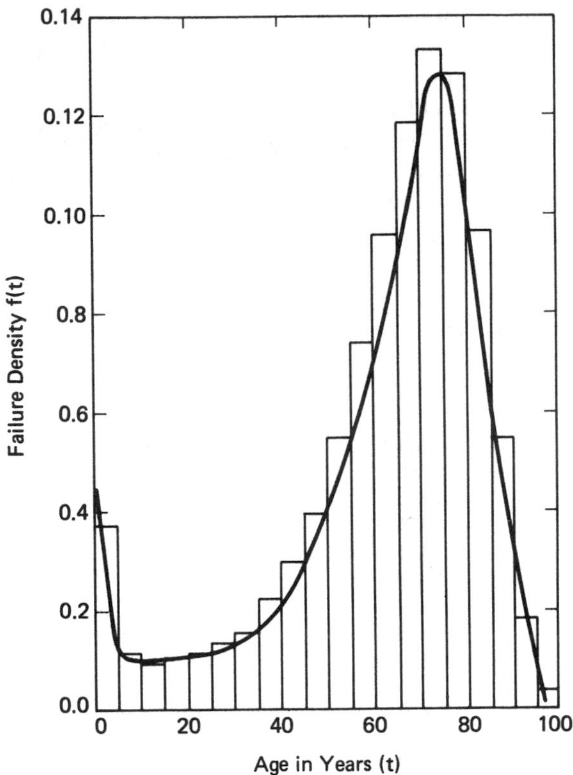

Figure 4.5. Failure density $f(t)$.

TABLE 4.4 CALCULATION OF FAILURE RATE $r(t)$

Age in Years	No. of Failures (death)	$r(t)=\frac{f(t)}{1-F(t)}$	Age in Years	No. of Failures (death)	$r(t)=\frac{f(t)}{1-F(t)}$
0	23,102	0.02260	40	30,788	0.00697
1	5,770	0.00570	45	41,654	0.00977
2	4,116	0.00414	50	56,709	0.01400
3	3,347	0.00338	55	76,420	0.02030
4	2,950	0.00299	60	99,889	0.02950
5	12,013	0.00244	65	123,334	0.04270
10	9,534	0.00196	70	138,566	0.06100
15	10,787	0.00224	75	134,217	0.08500
20	12,286	0.00258	80	103,554	0.11400
25	14,588	0.00311	85	56,634	0.14480
30	18,055	0.00391	90	18,566	0.17200
35	23,212	0.00512	95	2,886	0.24000
			99	125	1.20000

and the numerator is equivalent to (4.10). This can also be written as

$$r(t) = \frac{f(t)}{1 - F(t)}$$

This method of calculating the failure rate $r(t)$ results in the data summarized in Table 4.4 and plotted in Fig. 4.6. The curve of $r(t)$ is known as a *Bathtub curve*. It is characterized by a relatively high early failure rate (the *burn-in* period) followed by a fairly constant, *prime of life* period where failures occur randomly, and then a final *wearout* or *burn-out* phase. Ideally, critical hardware is put into service after a burn-in period and is replaced before it enters the wearout phase.

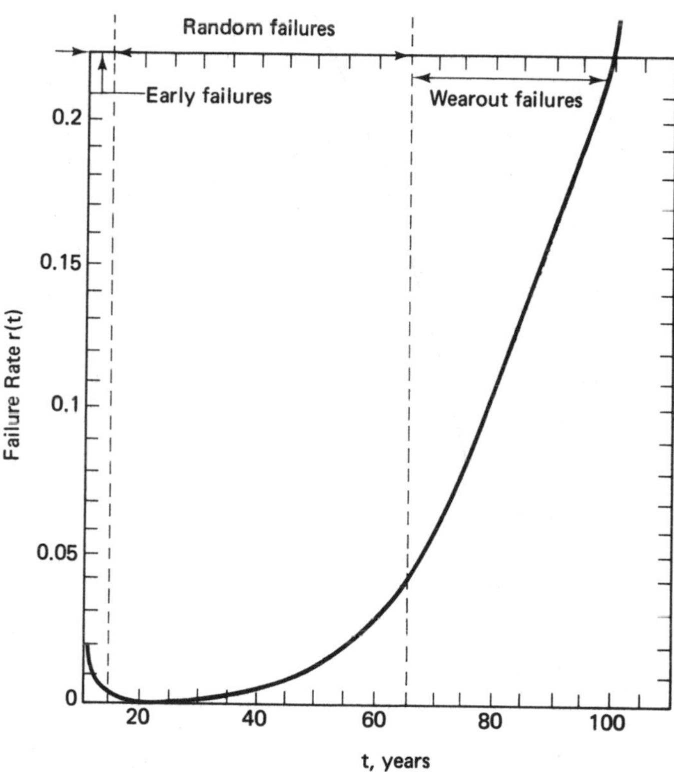

Figure 4.6. Failure rate $r(t)$ versus t.

Example 1.

Calculate, using the mortality data of Table 4.1, the reliability $R(t)$, unreliability $F(t)$, failure density $f(t)$, and failure rate $r(t)$ for:

(a) A man's living to be 75 years old.
(b) A man on the day after his 75th birthday party.

Solution:

(a) At age 75, for a man (neglecting the additional day):

$$R(t)=0.3088, \qquad F(t)=0.6912 \text{ (Table 4.2)}$$

$$f(t)=0.02620 \text{ (Table 4.3)}$$

$$r(t)=0.08500 \text{ (Table 4.4)}$$

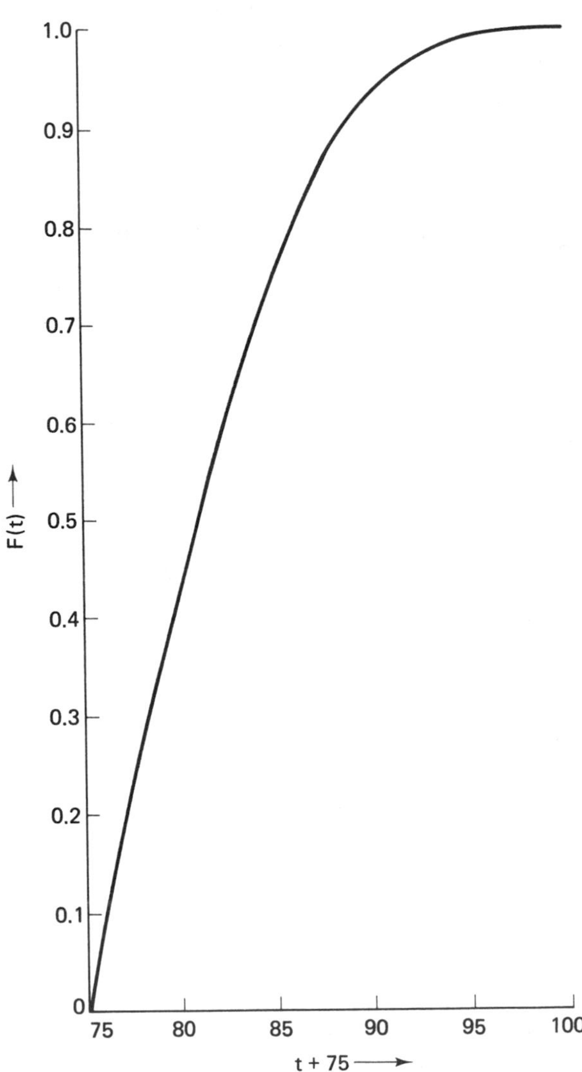

Figure 4.7. Failure $F(t)$ for Example 1.

(b) In effect, we start with a new population of $N=315{,}982$ having the following characteristics, where $t=0$ means 75 years.

t	$L(t)$	$L(t)/N$, $R(t)$	$1-R(t)$, $F(t)$	Table 4.3, $n(t+\Delta)-n(t)$	$\frac{n(t+\Delta)-n(t)}{N\cdot\Delta}$, $f(t)$	$\frac{f(t)}{R(t)}$, $r(t)$
0	315,982	1	0	134,217	0.0850	0.0850
5	181,765	0.575	0.425	103,554	0.0655	0.1139
10	78,221	0.248	0.752	56,634	0.0358	0.1444
15	21,577	0.0683	0.9317	18,566	0.0118	0.1728
20	3,011	0.0095	0.9905	2,886	0.0023	0.2421
24	125	0.0004	0.9996	125	0.0004	1.0000
25	0	0	1	0	0	—

By linear interpolation techniques, at 75 years and 1 day.

$$R(t)=1+\frac{0.575-1}{5\times 365}=0.9998$$

$$F(t)=1-R(t)=0.0002$$

$$f(t)=0.085+\frac{0.0655-0.0850}{5\times 365}=0.0850$$

$$r(t)=0.0850$$

Figure 4.7 shows the failure distribution for this population.

4.2.2 A Whole (Repair-Failure-Repair) Process

A repairable component experiences repetitions of the repair-to-failure and failure-to-repair process. The characteristics of such components can be obtained by considering the component as a sample from a population of similar components undergoing identical repetitions. The time varying history of each sample in a population of 10 is illustrated in Fig. 4.8. All samples are assumed to jump into the normal state at time zero; i.e., each component is as good as new at $t=0$. The following probabilistic parameters describe the population of Fig. 4.8.

Availability $A(t)$ at time t is the probability of the component's being normal at time t. This is the number of the normal components at time t divided by the total sample. For our sample, we have $A(5)=\frac{6}{10}=0.6$. Note that the normal components at time t have different ages, and that these differ from t. For example, component 1 in Fig. 4.8 has age 0.5 at time 5, whereas component 4 has age 1.2.

Unavailability $Q(t)$ is the probability that the component is in failed state at time t and is equal to the number of the failed components at time t divided by the total sample.

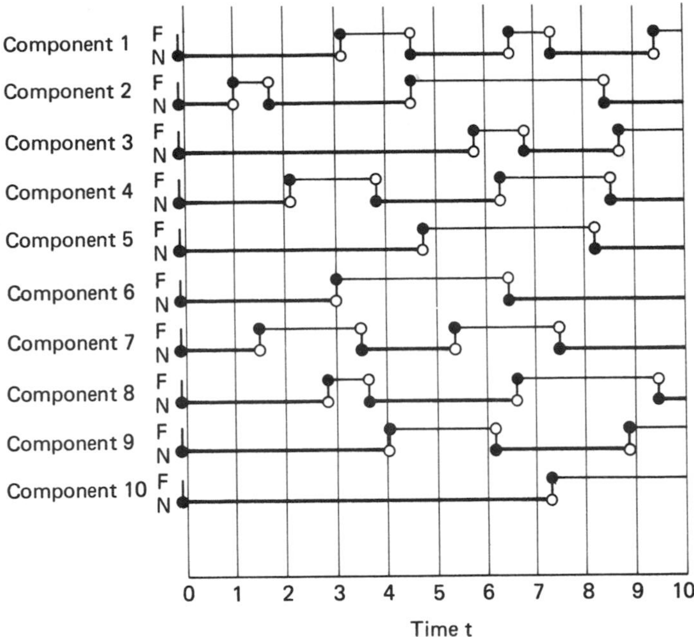

Figure 4.8. Histories of component states: (F, failure; N, normal).

Unconditional failure intensity $w(t)$ is the probability that the component fails per unit time at time t. Figure 4.8 shows that components 3 and 7 fail during time period $[5,6)$ so $w(5)$ is approximated by $\frac{2}{10}=0.2$.

The quantity $w(5)\times 1$ is equal to the *expected number of failures* $W(5,6)$ during the time interval $[5,6)$. The expected number of failures $W(0,6)$ during $[0,6)$ is evaluated by

$$W(0,6)=w(0)\times 1+\ldots+w(5)\times 1$$

The exact value of $W(0,6)$ is given by the integration

$$W(0,6)=\int_0^6 w(t)\,dt$$

Unconditional repair intensity $v(t)$ *and expected number of repairs* $V(t_1,t_2)$ can be defined similarly to $w(t)$ and $W(t_1,t_2)$, respectively. The costs due to failures and repairs during $[t_1,t_2)$ can be related to $W(t_1,t_2)$ and $V(t_1,t_2)$, respectively, if the cost-to-repair and production losses for failure are known.

There is yet another failure parameter to be obtained. Consider another population of components which are normal at time t. When $t=5$, this population consists of the components 1, 3, 4, 7, 8, and 10. A *conditional failure intensity* $\lambda(t)$ is the proportion of the (normal) population that is expected to fail per unit time at time t. For example, $\lambda(5)\times 1$ is estimated as $\frac{2}{6}$, since components 3 and 7 fail during [5,6). A conditional repair intensity $\mu(t)$ is defined in a similar way. Large values of $\lambda(t)$ means that the component is about to fail, whereas large values of $\mu(t)$ state that the component will be repaired in the near future.

Example 2.

Calculate values for $R(t)$, $F(t)$, $f(t)$, $r(t)$, $A(t)$, $Q(t)$, $w(t)$, $W(0,t)$, and $\lambda(t)$ for the 10 components of Fig. 4.8 at 5 hr and 9 hr.

Solution: We need times to failures, i.e., lifetimes, to calculate $R(t)$, $F(t)$, $f(t)$, and $r(t)$, since these are parameters in the repair to failure process.

Component	Repair t	Failure t	TTF
1	0	3.1	3.1
1	4.5	6.6	2.1
1	7.4	9.5	2.1
2	0	1.05	1.05
2	1.7	4.5	2.8
3	0	5.8	5.8
3	6.8	8.8	2.0
4	0	2.1	2.1
4	3.8	6.4	2.6
5	0	4.8	4.8
6	0	3	3
7	0	1.4	1.4
7	3.5	5.4	1.9
8	0	2.85	2.85
8	3.65	6.7	3.05
9	0	4.1	4.1
9	6.2	8.95	2.75
10	0	7.35	7.35

The following mortality data is obtained from these times to failures.

t	$L(t)$	$R(t)$	$F(t)$	$n(t+\Delta)-n(t)$	$f(t)$	$r(t)=\dfrac{f(t)}{R(t)}$
0	18	1	0	0	0	0
1	18	1	0	3	0.1667	0.1667
2	15	0.8333	0.1667	10	0.5556	0.6667
3	5	0.2778	0.7222	1	0.0556	0.2001
4	4	0.2222	0.7778	2	0.1111	0.5000
5	2	0.1111	0.8889	1	0.0556	0.5005
6	1	0.0556	0.9444	0	0	0
7	1	0.0556	0.9444	1	0.0556	1.0000
8	0	0	1	0	0	—
9	0	0	1	0	0	—

Thus, at age 5,

$$R(5)=0.1111, \quad F(5)=0.8889, \quad f(5)=0.0556, \quad r(5)=0.5005$$

And, at age 9,

$$R(9)=0, \quad F(9)=1, \quad f(9)=0, \quad r(9)\text{: undefined}$$

Parameters $A(t)$, $Q(t)$, $w(t)$, $W(0,t)$, and $\lambda(t)$ are obtained from the whole repair-failure-repair process shown in Fig. 4.8. At time 5,

$$A(5)=\tfrac{6}{10}=0.6, \qquad Q(5)=0.4, \qquad w(5)=0.2$$
$$W(0,5)=\frac{[2+2+2+3]}{10}=0.9, \qquad \lambda(5)=\tfrac{2}{6}=\tfrac{1}{3}$$

and, at time 9,

$$A(9)=\tfrac{6}{10}=0.6, \qquad Q(9)=0.4, \qquad w(9)=0.1$$
$$W(0,9)=W(0,5)+\frac{2+3+1+2}{10}=1.7, \qquad \lambda(9)=\tfrac{1}{6}$$

4.2.3 Probabilistic Parameters of Repair-To-Failure Processes

We return now to the problem of characterizing the reliability parameters for repair-to-failure processes. These processes apply to non-repairable components and also to repairable components if we restrict our attention to times to the first failures. We first restate some of the concepts introduced in Section 4.2.1, in a more formal manner, and then deduce some new relations.

Consider a process starting at a repair and ending in its first failure. Shift the time axis appropriately, and take $t=0$ as the time at which the component is repaired, so that the component is then as good as new at time zero. The probabilistic definitions and their notations are summarized as follows:

$R(t)=$ *reliability at time* t

The probability that the component experiences no failure during the time interval $(0,t]$, *given that the component was repaired at time zero.*

The curve $R(t)$ versus t is a survival distribution. The distribution is monotonically decreasing, since the reliability gets smaller as time increases. A typical survival distribution is shown in Fig. 4.3.

The following asymptotic properties hold:

$$\lim_{t\to 0} R(t)=1 \tag{4.11}$$

$$\lim_{t\to\infty} R(t)=0 \tag{4.12}$$

Equation (4.11) shows that almost all the components are surviving around time zero, whereas (4.12) indicates a vanishingly small probability of a component's surviving forever.

$F(t)$ = unreliability at time t

The probability that the component experiences the first failure during $[0,t)$, given that it is repaired at time zero.

The curve $F(t)$ versus t is called a *failure distribution* and is a monotonically increasing function of t. A typical failure distribution is shown in Fig. 4.3.

The following asymptotic properties hold:

$$\lim_{t\to 0} F(t)=0 \tag{4.13}$$

$$\lim_{t\to\infty} F(t)=1 \tag{4.14}$$

Equation (4.13) shows that few components fail just after the repair, whereas (4.14) indicates an asymptotic approach to complete failure.

Since the component either remains normal or experiences its first failure at time t,

$$R(t)+F(t)=1 \tag{4.15}$$

Now, let $t_1 \le t_2$. The difference $F(t_2)-F(t_1)$ is the probability that the component experiences its first failure during the interval $[t_1,t_2)$, given that it was as good as new at time zero. This probability is illustrated in Fig. 4.9.

$f(t)$ = failure density of $F(t)$

The first order derivative of $F(t)$.

This was shown previously to be the first derivative of $F(t)$.

$$f(t)=\frac{dF(t)}{dt} \tag{4.16}$$

or, equivalently,

$$f(t)\,dt=F(t+dt)-F(t). \tag{4.17}$$

Thus, $f(t)\,dt$ is the probability that the first component failure occurs during the small interval $[t,t+dt)$, given that the component was repaired at time zero.

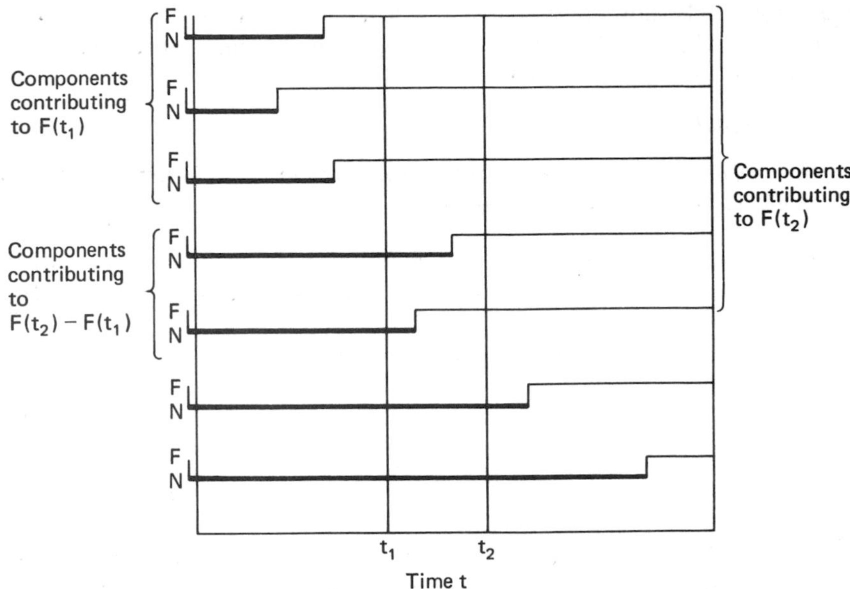

Figure 4.9. Illustration of the probability $F(t_2)-F(t_1)$.

The unreliability $F(t)$ is obtained by integration,

$$F(t)=\int_0^t = f(u)\,du \tag{4.18}$$

Similarly, the difference in the unreliability

$$F(t_2)-F(t_1)=\int_{t_1}^{t_2} f(u)\,du \tag{4.19}$$

is the reliability

$$R(t)=\int_t^{\infty} f(u)\,du \tag{4.20}$$

These relationships are illustrated in Fig. 4.10.

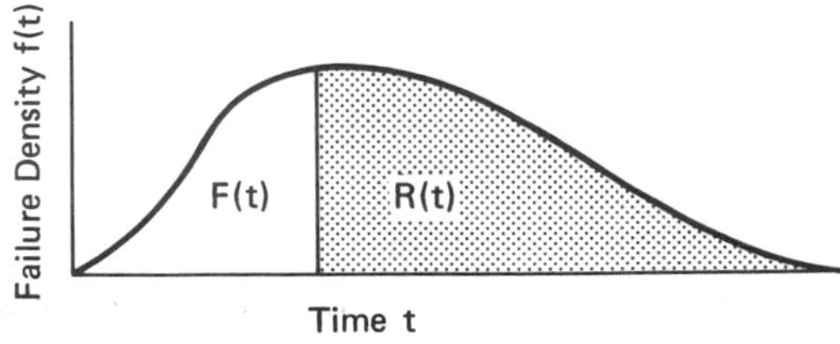

Figure 4.10. Integration of failure density $f(t)$.

$r(t)$ = failure rate

The probability that the component experiences a failure per unit time at time t, given that the component was repaired at time zero and has survived to time t.

The quantity $r(t)\,dt$ is the probability that the component fails during $[t, t+dt)$, given that the component age is t. Here age t means that the component was working at time zero and has survived to time t. The rate is simply designated as r when it is independent of the age t. The component with a constant failure rate r is considered as good as new if it is functioning.

TTF = time to failure:

The span of time from repair to first failure.

The time to failure TTF is a random variable, since we cannot predict the exact time of the first failure.

MTTF = mean time to failure

The expected value of the time to failure.

This is obtained by

$$\text{MTTF} = \int_0^\infty t f(t)\,dt \tag{4.21}$$

The quantity $f(t)\,dt$ is the probability that the TTF is around t, so (4.21) is the average of all possible TTF's.

Example 3.

Table 4.5 shows failure data for 250 germanium transistors, all of which failed during the test period. Calculate the unreliability $F(t)$, the failure rate $r(t)$, the failure density $f(t)$, and the MTTF.

Solution:

TABLE 4.5 FAILURE DATA FOR TRANSISTORS

Time to Failure (min) t	Cumulative Failures, $n(t)$
0	0
20	9
40	23
60	50
90	83
160	113
230	143
400	160
900	220
1200	235
2500	240
→2500	→250

The unreliability $F(t)$ at a given time t is simply the number of transistors failed divided by the total number (250) of samples tested. The results are summarized in Table 4.6, and the failure distribution is plotted in Fig. 4.11.

TABLE 4.6 TRANSISTOR RELIABILITY, UNRELIABILITY, FAILURE DENSITY, AND FAILURE RATE

t	$L(t)$	$R(t)$	$F(t)$	$n(t+\Delta)-n(t)$	Δ	$\frac{n(t+\Delta)-n(t)}{N\cdot\Delta}$, $f(t)$	$\frac{f(t)}{R(t)}$, $r(t)$
0	250	1	0	9	20	0.00180	0.00180
20	241	0.9640	0.0360	14	20	0.00280	0.00290
40	227	0.9080	0.0920	27	20	0.00540	0.00595
60	200	0.8000	0.2000	33	30	0.00440	0.00550
90	167	0.6680	0.3320	30	70	0.00171	0.00256
160	137	0.5480	0.4520	30	70	0.00171	0.00312
230	107	0.4280	0.5720	17	170	0.00040	0.00093
400	90	0.3600	0.6400	60	500	0.00048	0.00133
900	30	0.1200	0.8800	15	300	0.00020	0.00167
1200	15	0.0600	0.9400	5	1300	0.00002	0.00033
2500	10	0.0400	0.9600	—	—	—	—

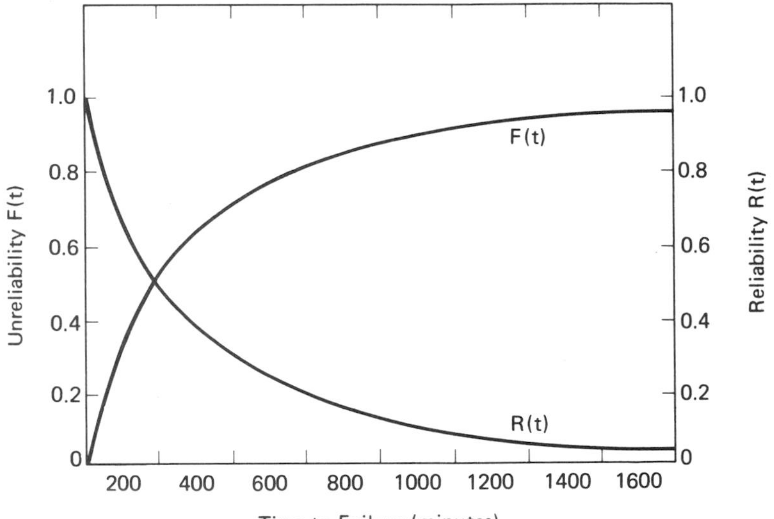

Figure 4.11. Transistor reliability and unreliability.

The failure density $f(t)$ and the failure rate $r(t)$ are calculated in a similar manner to the mortality case (Example 1) and are listed in Table 4.6. The first-order approximation of the rate is a constant failure rate $r(t)=r=0.0026$, the averaged value. In general, the constant failure rate describes solid state components, and systems and equipment which are in their *prime of life*, for example, an automobile having mileage of 3000 to 20,000 miles.

If the failure rate is constant, then, as shown in Section 4.4, $\text{MTTF}=1/r=384.6$. Alternatively, equation (4.21) could be used, giving

$$\text{MTTF}\cong 10\times 0.0018\times 20+30\times 0.0028\times 20+\ldots+1850\times 0.00002\times 1300=501.4$$

4.2.4 Probabilistic Parameters of Failure-to-Repair Processes

Consider a process starting with a failure and ending at the first repair. We shift the time axis appropriately and take $t=0$ as the time at which the component failed. The probabilistic parameters are conditioned by the fact that the component failed at time zero.

$G(t)$ = *repair probability at time t*

The probability that the repair is completed before time t, given that the component failed at time zero.

The curve $G(t)$ versus t is a repair distribution and has properties similar to that of the failure distribution $F(t)$. A non-repairable component has $G(t)$ identically equal to zero. The repair distribution $G(t)$ is a monotonically increasing function for the repairable component, and the following asymptotic property holds:

$$\lim_{t\to 0} G(t)=0 \tag{4.22}$$

$$\lim_{t\to\infty} G(t)=1 \tag{4.23}$$

$g(t)$ = *repair density of* $G(t)$

The first order derivative of $G(t)$.

This can be written as

$$g(t)=\frac{dG(t)}{dt} \tag{4.24}$$

or

$$g(t)\,dt=G(t+dt)-G(t) \tag{4.25}$$

Thus, the quantity $g(t)dt$ is the probability that component repair is completed during $[t,t+dt)$, given that the component failed at time zero.

The repair density is related to the repair distribution and its difference in the following way:

$$G(t)=\int_0^t g(u)\,du \tag{4.26}$$

$$G(t_2)-G(t_1)=\int_{t_1}^{t_2} g(u)\,du \tag{4.27}$$

Note that the difference $G(t_2)-G(t_1)$ is the probability that the first repair is completed during $[t_1,t_2)$, given that the component failed at time zero.

m(t) = repair rate

The probability that the component is repaired per unit time at time t given that the component failed at time zero and has been failed to time t.

The quantity $m(t)\,dt$ is the probability that the component is repaired during $[t,t+dt)$, given that the component's failure age is t. Failure age t means that the component failed at time zero and has been failed to time t. The rate is simply designated as m when it is independent of the failure age t. A component with a constant repair rate has the same chance of being repaired whenever it is failed, and a non-repairable component has a repair rate of zero.

TTR = time to repair

The length of time from the failure to the succeeding first repair.

The time to repair is a random variable because the first repair occurs at random.

MTTR = mean time to repair

Expected value of the time to repair, TTR.

The mean time to repair is given by

$$\text{MTTR}=\int_0^{\infty} t\cdot g(t)\,dt \tag{4.28}$$

Example 4.

The following repair times (i.e., TTR's) for the repair of electric motors have been logged in:

Repair No.	Time (hr)	Repair No.	Time (hr)
1	3.3	10	0.8
2	1.4	11	0.7
3	0.8	12	0.6
4	0.9	13	1.8
5	0.8	14	1.3
6	1.6	15	0.8
7	0.7	16	4.2
8	1.2	17	1.1
9	1.1		

Using these data obtain the values for $G(t)$, $g(t)$, $m(t)$, and MTTR.

Solution: $N=17=$ total number of repairs.

t(TTR)	Number of Completed Repairs $M(t)$	$G(t)=\dfrac{M(t)}{N}$	$g(t)=\dfrac{G(t+\Delta)-G(t)}{\Delta}$	$m(t)=\dfrac{g(t)}{1-G(t)}$
0.0	0	0	0	0
0.5	0	0	0.9412	0.9412
1.0	8	0.4706	0.5882	1.110
1.5	13	0.7647	0.2354	1.0004
2.0	15	0.8824	0	0
2.5	15	0.8824	0	0
3.0	15	0.8824	0.1176	1
3.5	16	0.9412	0	0
4.0	16	0.9412	0.1176	2.0000
4.5	17	1	—	—

Equation (4.28) gives

$$\begin{aligned}\text{MTTR} &= (0.25\times 0+0.75\times 0.9412+\ldots+4.25\times 0.1176)\times 0.5\\ &= 1.3676\end{aligned}$$

The average of repair times also gives MTTR:

$$\begin{aligned}\text{MTTR} &= \frac{3.3+1.4+\ldots+1.1}{17}\\ &= 1.3588\end{aligned}$$

4.2.5 Probabilistic Parameters of Whole Processes

Consider a process consisting of repetitions of the repair-to-failure process and the failure-to-repair process. Assume that the component jumped into the normal state at time zero so that it is good as new at $t=0$. A number of failures and repairs may occur to time $t>0$. Figure 4.8 shows that time t for the whole process differs from the time t for the repair-to-failure process because the latter time is measured from the latest repair before time t of the whole process. Both time scales coincide with each other if and only if the component has been normal to time t. In this case, the time scale of the repair-to-failure is measured from time zero of the whole process because the component is assumed to jump into the normal state at time zero. Similarly, time t of the whole process differs from the time t of the failure-to-repair process. The probabilistic concepts for the whole process are summarized as follows:

$A(t)$ = *availability at time t*

> *The probability that the component is normal at time t, given that it was as good as new at time zero.*

Reliability generally differs from availability because the reliability requires the continuation of the normal state over the whole interval $(0,t]$. A component contributes to the availability $A(t)$ but not to the reliability $R(t)$ if the component failed before time t, is then repaired, and is normal at time t. Thus, the availability $A(t)$ is larger, or equal to the reliability $R(t)$:

$$A(t) \geq R(t) \tag{4.29}$$

The equality in (4.29) holds for a non-repairable component because the component is normal at time t if and only if it has been normal to time t: Thus,

$$A(t) = R(t), \quad \text{for non-repairable components} \tag{4.30}$$

The availability of a non-repairable component decreases to zero as t becomes larger, whereas the availability of the repairable component converges to a non-zero positive number. Typical curves of $A(t)$ are shown in Fig. 4.12.

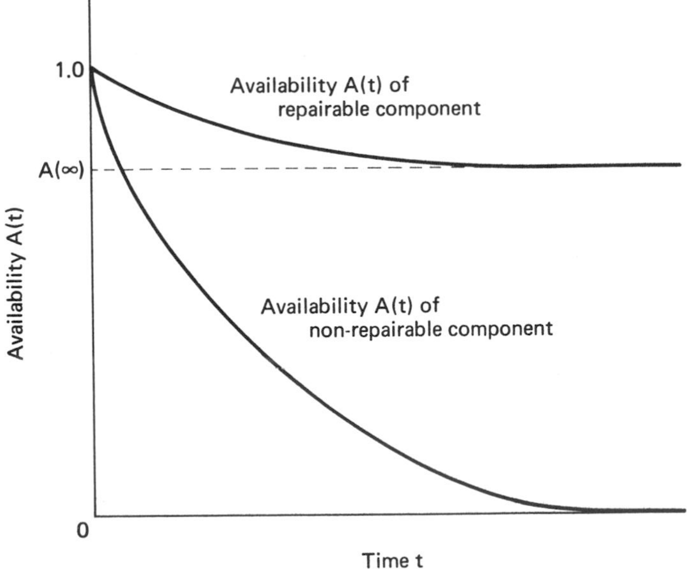

Figure 4.12. Schematic curves of availability $A(t)$.

$Q(t)$ = *unavailability at time t*

The probability that a component is in the failed state at time t, given that it jumped into the normal state at time zero.

Since a component is either in the normal state or in the failed state at time t, the unavailability $Q(t)$ is obtained from the availability and vice versa:

$$A(t)+Q(t)=1 \tag{4.31}$$

From (4.15), (4.29), and (4.31), we have the inequality

$$Q(t) \le F(t) \tag{4.32}$$

In other words, the unavailability $Q(t)$ is less than or equal to the unreliability $F(t)$. The equality holds for non-repairable components:

$$Q(t)=F(t), \quad \text{for non-repairable components} \tag{4.33}$$

The unavailability of a non-repairable component approaches unity as t gets larger, whereas the unavailability of a repairable component remains smaller than unity.

$\lambda(t)$ = *conditional failure intensity*

The probability that the component fails per unit time at time t, given that it is in the normal state at time zero and is normal at time t.

The quantity $\lambda(t)\,dt$ is the probability that a component fails during the small interval $[t, t+dt)$, given that the component was as good as new at time zero and normal at time t. Note that the quantity $r(t)\,dt$ represents the probability that the component fails during $[t, t+dt)$, given that the component was repaired (or as good as new) at time zero and has been normal to time t. $\lambda(t)\,dt$ differs from $r(t)\,dt$ because the latter quantity assumes the continuation of the normal state to time t, i.e., no failure in the interval $[0, t]$.

$$\lambda(t) \neq r(t), \quad \text{for the general case}$$

The failure intensity $\lambda(t)$ coincides with the failure rate $r(t)$ if the component is non-repairable because the component is normal at time t if and only if it has been normal to time t:

$$\lambda(t)=r(t), \quad \text{for non-repairable component} \tag{4.34}$$

Also, it is proven in Appendix 4.1 at the end of this chapter that the failure intensity $\lambda(t)$ becomes the failure rate if the rate is a constant r:

$$\lambda(t)=r, \quad \text{for constant failure rate } r \tag{4.35}$$

$w(t)$ = *unconditional failure intensity at time t*

> *The probability that a component fails per unit time at time t, given that it jumped into the normal state at time zero.*

In other words, the quantity $w(t)dt$ is the probability that the component fails during $[t, t+dt)$, given that the component was as good as new at time zero. For a non-repairable component, the unconditional failure intensity $w(t)$ coincides with the failure density $f(t)$.

Both the quantities $\lambda(t)$ and $w(t)$ refer to the failure per unit time at time t. These quantities, however, assume different populations. The conditional failure intensity $\lambda(t)$ presumes a set of components as good as new at time zero and normal at time t, whereas the unconditional failure intensity $w(t)$ assumes components simply as good as new at time zero. Thus, they are different quantities. For example, Fig. 4.13 gives

$$\begin{aligned} \lambda(t)\,dt &= \frac{0.7\,dt}{70} = 0.01\,dt \\ w(t)\,dt &= \frac{0.7\,dt}{100} = 0.007\,dt \end{aligned} \tag{4.36}$$

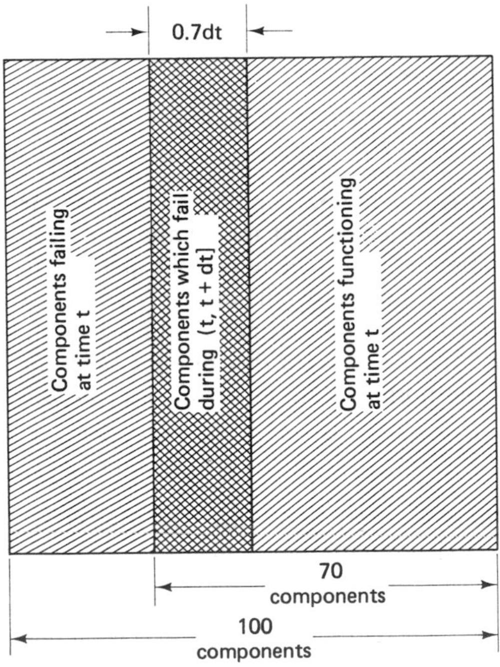

Figure 4.13. Definition of conditional failure intensity $\lambda(t)$ and unconditional failure intensity $w(t)$.

$W(t,t+dt)$ = expected number of failures (ENF)

Expected number of failures during $[t,t+dt)$, given that the component jumped into the normal state at time zero.

From the definition of the expected values, we have

$$W(t,t+dt)=\sum_{i=1}^{\infty} i\cdot \Pr\{i \text{ failures during } (t,t+dt]|C\} \tag{4.37}$$

where C = component jumped into the normal state at time zero. At most one failure occurs during $[t,t+dt)$ and we obtain

$$W(t,t+dt]=\Pr\{\text{one failure during } (t,t+dt]|C\} \tag{4.38}$$

or, equivalently,

$$W(t,t+dt)=w(t)\,dt \tag{4.39}$$

The expected number of failures is calculated from the unconditional failure intensity $w(t)$ by integration.

$W(t_1,t_2)$ = ENF over an interval

Expected number of failures during (t_1,t_2), given that the component jumped into the normal state at time zero.

$W(t_1,t_2)$ is given by the integration of $W(t,t+dt)$ over the interval $[t_1,t_2)$. Thus, we have

$$W(t_1,t_2)=\int_{t_1}^{t_2} w(t)\,dt \tag{4.40}$$

The $W(0,t)$ of a non-repairable component is equal to $F(t)$ and approaches unity as t gets larger. The $W(0,t)$ of a repairable component diverges to infinity as t becomes larger. Typical curves of $W(0,t)$ are shown in Fig. 4.14. The asymptotic behavior of W and other parameters are summarized in Table 4.9.

$\mu(t)$ = conditional repair intensity

The probability that a component is repaired per unit time at time t, given that it jumped into the normal state at time zero and is failed at time t.

The repair intensity generally differs from the repair rate $m(t)$. Similarly, to the relationship between $\lambda(t)$ and $r(t)$ we have the following special

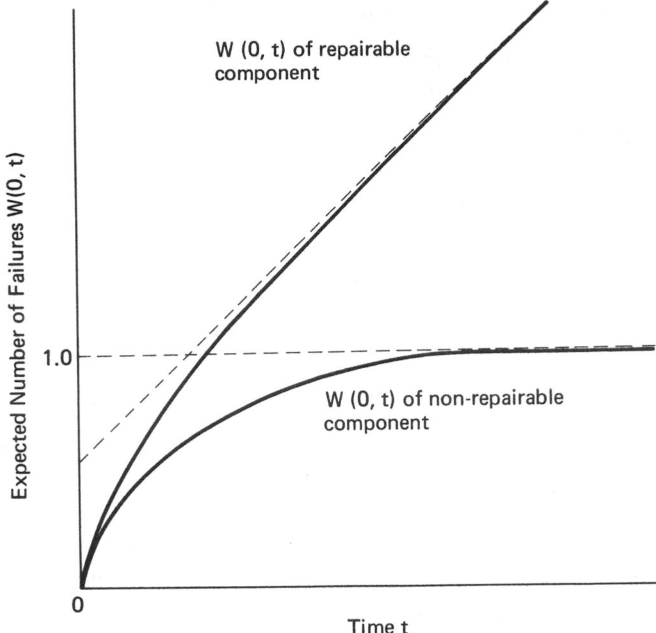

Figure 4.14. Schematic curves of expected number of failures $W(0,t)$.

cases:

$$\mu(t)=m(t)=0, \quad \text{for a non-repairable component} \tag{4.41}$$

$$\mu(t)=m, \quad \text{for constant repair rate } m \tag{4.42}$$

$v(t)$ = *unconditional repair intensity at time* t

The probability that the component is repaired per unit time at time t, *given that it jumped into the normal state at time zero.*

The intensities $v(t)$ and $\mu(t)$ are different quantities, since they involve different populations.

$V(t,t+dt)$ = *Expected number of repairs*

Expected number of repairs during $[t,t+dt)$, *given that the component jumped into the normal state at time zero.*

Similarly to (4.39) the following relation holds:

$$V(t,t+dt)=v(t)\,dt \tag{4.43}$$

$V(t_1,t_2)$ = *expected number of repairs in an interval*

Expected number of repairs during $[t_1, t_2)$, given that the component jumped into the normal state at time zero.

Analogously to (4.40), we have

$$V(t_1, t_2) = \int_{t_1}^{t_2} v(t)\,dt \tag{4.44}$$

The expected number of repairs $V(0,t)$ is zero for a non-repairable component. A repairable component has $V(0,t)$ approaching infinity as t gets larger. It is proven in the next section that the difference $W(0,t) - V(0,t)$ equals the unavailability $Q(t)$.

MTBF = mean time between failures

Expected length of time between two consecutive failures.

The mean time between failure is equal to the sum of MTTF and MTTR:

$$\text{MTBF} = \text{MTTF} + \text{MTTR} \tag{4.45}$$

MTBR = mean time between repairs

Expected length of time between two consecutive repairs.

The MTBR equals the sum of MTTF and MTTR and hence MTBF:

$$\text{MTBR} = \text{MTBF} = \text{MTTF} + \text{MTTR} \tag{4.46}$$

Example 5.

For the data of Fig. 4.8, calculate $\mu(7)$, $v(7)$, and $V(0,5)$.

Solution: Six components are failed at $t = 7$. Among them, only two components are repaired during unit interval $[7, 8)$. Thus,

$$\mu(7) = \tfrac{2}{6} = \tfrac{1}{3}$$

$$v(7) = \tfrac{2}{10} = 0.2$$

$$V(0,5) = \tfrac{1}{10} \sum_{i=0}^{4} \text{[total number of repairs in } (i, i+1)]$$

$$= \tfrac{1}{10} \times (0 + 1 + 0 + 3 + 1) = 0.5$$

4.3 FUNDAMENTAL RELATIONS AMONG PROBABILISTIC PARAMETERS

In the previous section, we defined various probabilistic parameters and their interelationships. These relations and the characteristics of the probabilistic parameters are summarized in Tables 4.7, 4.8, and 4.9. Table 4.7 refers to the repair-to-failure process, Table 4.8 to the failure-to-repair process, and Table 4.9 to the whole process. These tables include some new and important relations which are deduced in this section.

TABLE 4.7 Relations among Probabilistic Parameters for Repair-to-Failure Process (the First Failure)

Failure Rate $r(t)$	General $r(t)$	(1) $R(t)+F(t)=1$	(7) $R(t)=\int_t^{\infty} F(u)\,du$
		(2) $R(0)=1, R(\infty)=0$	(8) $\text{MTTF}=\int_0^{\infty} tf(t)\,dt$
		(3) $F(0)=0, F(\infty)=1$	(9) $r(t)=\dfrac{f(t)}{[1-F(t)]}$
		(4) $f(t)=\dfrac{dF(t)}{dt}$	(10) $R(t)=\exp\left[-\int_0^t r(u)\,du\right]$
		(5) $f(t)\,dt=F(t+dt)-F(t)$	(11) $F(t)=1-\exp\left[-\int_0^t r(u)\,du\right]$
		(6) $F(t)=\int_0^t f(u)\,du$	(12) $f(t)=r(t)\exp\left[-\int_0^t r(u)\,du\right]$
	Const. $r(t)=\lambda$	(13) $R(t)=e^{-\lambda t}$	(14) $F(t)=1-e^{-\lambda t}$
		(15) $f(t)=\lambda e^{-\lambda t}$	(16) $\text{MTTF}=\dfrac{1}{\lambda}$

TABLE 4.8 Relations among Probabilistic Parameters for Failure-to-Repair Process

Repair Rate $m(t)$	General $m(t)$	(1) $G(t)=g(t)=m(t)=0$ for non-repairable component	(6) $G(t_2)-G(t_1)=\int_{t_1}^{t_2} g(u)\,du$
		(2) $G(0)=0, G(\infty)=1$	(7) $\text{MTTR}=\int_0^{\infty} tg(t)\,dt$
		(3) $g(t)=\dfrac{dG(t)}{dt}$	(8) $m(t)=\dfrac{g(t)}{1-G(t)}$
		(4) $g(t)\,dt=G(t+dt)-G(t)$	(9) $G(t)=1-\exp\left[-\int_0^t m(u)\,du\right]$
		(5) $G(t)=\int_0^t g(u)\,du$	(10) $g(t)=m(t)\exp\left[-\int_0^t m(u)\,du\right]$
	Const. $m(t)=\mu$	(11) $G(t)=1-e^{-\mu t}$	(12) $g(t)=\mu e^{-\mu t}$
		(13) $\text{MTTR}=\dfrac{1}{\mu}$	

TABLE 4.9 RELATIONS AMONG PROBABILISTIC PARAMETERS FOR THE WHOLE PROCESS

	Repairable	Non-repairable
Fundamental Relations	(1) $A(t)+Q(t)=1$	$A(t)+Q(t)=1$
	(2) $A(t)>R(t)$	$A(t)=R(t)$
	(3) $Q(t)<F(t)$	$Q(t)=F(t)$
	(4) $w(t)=f(t)+\int_0^t f(t-u)v(u)\,du$	$w(t)=f(t)$
	(5) $v(t)=\int_0^t g(t-u)w(u)\,du$	$v(t)=0$
	(6) $W(t,t+dt)=w(t)dt$	$W(t,t+dt)=w(t)dt$
	(7) $V(t,t+dt)=v(t)dt$	$V(t,t+dt)=0$
	(8) $W(t_1,t_2)=\int_{t_1}^{t_2} w(u)\,du$	$W(t_1,t_2)=\int_{t_1}^{t_2} w(u)\,du$ $=F(t_2)-F(t_1)$
	(9) $V(t_1,t_2)=\int_{t_1}^{t_2} v(u)\,du$	$V(t_1,t_2)=0$
	(10) $Q(t)=W(0,t)-V(0,t)$	$Q(t)=W(0,t)=F(t)$
	(11) $\lambda(t)=\dfrac{w(t)}{1-Q(t)}$	$\lambda(t)=\dfrac{w(t)}{1-Q(t)}$
	(12) $\mu(t)=\dfrac{v(t)}{Q(t)}$	$\mu(t)=0$
Stationary Values	(13) MTBF=MTBR=MTTF+MTTR	MTBF=MTBR=∞
	(14) $0<A(\infty)<1, 0<Q(\infty)<1$	$A(\infty)=0, Q(\infty)=1$
	(15) $0<w(\infty)<\infty, 0<v(\infty)<\infty$	$w(\infty)=0, v(\infty)=0$
	(16) $w(\infty)=v(\infty)$	$w(\infty)=v(\infty)=0$
	(17) $W(0,\infty)=\infty, V(0,\infty)=\infty$	$W(0,\infty)=1, V(0,\infty)=0$
Remark	(18) $w(t)\neq\lambda(t), v(t)\neq\mu(t)$	$w(t)\neq\lambda(t), v(t)=\mu(t)=0$
	(19) $\lambda(t)\neq r(t), \mu(t)\neq m(t)$	$\lambda(t)=r(t), \mu(t)=m(t)=0$
	(20) $w(t)\neq f(t), v(t)\neq g(t)$	$w(t)=f(t), v(t)=g(t)=0$

4.3.1 Relations among Repair-to-Failure Processes

We shall derive the following relations:

$$r(t)=\frac{f(t)}{1-F(t)} \tag{4.47}$$

$$F(t)=1-\exp\left[-\int_0^t r(u)\,du\right] \tag{4.48}$$

$$R(t)=\exp\left[-\int_0^t r(u)\,du\right] \tag{4.49}$$

$$f(t)=r(t)\exp\left[-\int_0^t r(u)\,du\right] \tag{4.50}$$

The first identity is used to obtain the failure rate $r(t)$ when the unreliability $F(t)$ and the failure density $f(t)$ are given. The second through the fourth identities can be used to calculate $F(t)$, $R(t)$, and $f(t)$ when the failure rate $r(t)$ is given.

The flow chart of Fig. 4.15 shows general procedures for calculating the probabilistic parameters for the repair-to-failure process. The number adjacent to each arrow corresponds to the relation identified in Table 4.7. Note that the first step in processing failure data (such as the data in Tables 4.1 and 4.5) is to plot it as a histogram (Fig. 4.4) or to fit it, by parameter estimation techniques, to a standard distribution (exponential, normal, etc.). Parameter estimation techniques and failure distributions are discussed later in this chapter.

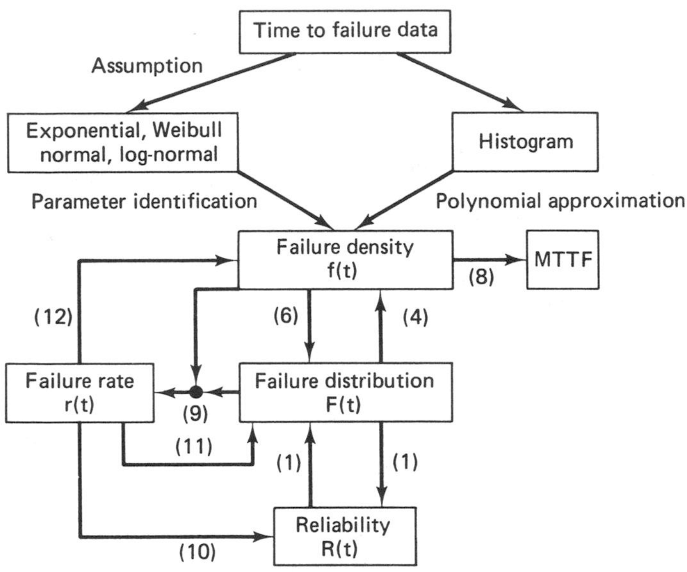

Figure 4.15. Flow chart for calculating probabilistic parameters for repair-to-failure process.

We now begin the derivation of identities (4.47) through (4.50) with a statement of the definition of a conditional probability [see equation (4.5)].

$$\Pr\{A|C,W\} = \frac{\Pr\{A,C|W\}}{\Pr\{C|W\}} \tag{4.51}$$

The quantity $r(t)dt$ coincides with the conditional probability $\Pr\{A|C,W\}$ where

A = the component fails during $[t, t+dt)$,

C = the component has been normal to time t,

and

W = the component was repaired at time zero.

The probability $\Pr\{C|W\}$ is the reliability $R(t)=1-F(t)$. Further, $\Pr\{A,C|W\}$ is given by $f(t)dt$. Thus, from (4.51), we have

$$r(t)\,dt=\frac{f(t)\,dt}{1-F(t)} \tag{4.52}$$

yielding (4.47). Note that $f(t)=dF/dt$, so that we obtain

$$r(t)=\frac{dF/dt}{1-F(t)} \tag{4.53}$$

We can rewrite (4.53) as

$$r(t)=-\frac{d}{dt}\log[1-F(t)] \tag{4.54}$$

Integrating both sides of (4.54), we obtain

$$\int_0^t r(u)\,du=\log[1-F(0)]-\log[1-F(t)] \tag{4.55}$$

Substituting $F(0)=0$ into (4.55), we have

$$\int_0^t r(u)\,du=-\log[1-F(t)] \tag{4.56}$$

yielding (4.48). The remaining two identities are obtained from (4.15) and (4.16).

The flow chart of Fig. 4.15 indicates that $R(t)$, $F(t)$, $f(t)$, and $r(t)$ can be obtained if any one of the parameters is known.

4.3.2 Relations among Failure-to-Repair Processes

Similarly to the case of the repair-to-failure process, we obtain the following relations for the failure-to-repair process:

$$m(t)=\frac{g(t)}{1-G(t)} \tag{4.57}$$

$$G(t)=1-\exp\left[-\int_0^t m(u)\,du\right] \tag{4.58}$$

$$g(t)=m(t)\exp\left[-\int_0^t m(u)\,du\right] \tag{4.59}$$

The first identity is used to obtain the repair rate $m(t)$ when the repair probability $G(t)$ and the repair density $g(t)$ are given. The second and the third identities calculate $G(t)$ and $g(t)$ when the repair rate $m(u)$ is given.

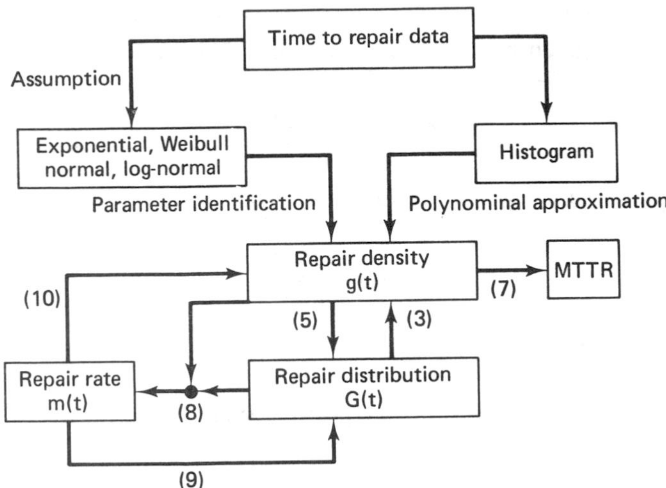

Figure 4.16. Flow chart for calculating probabilistic parameters for failure-to-repair process.

The flow chart, Fig. 4.16, shows the general procedures for calculating the probabilistic parameters related to the failure-to-repair process. The number adjacent to each arrow corresponds to the relation identification number of Table 4.8. We can calculate $G(t)$, $g(t)$, and $m(t)$ if any one of them is known.

4.3.3 Relations among the Whole Process

General procedures for calculating whole process probabilistic parameters are shown in Fig. 4.17. The identification numbers in the flow chart are listed in Table 4.9. The chart includes some new and important relations which we now derive.

The Unconditional Intensities $w(t)$ and $v(t)$. As shown in Fig. 4.18, the components which fail during $(t, t+dt]$ are classified into two types:

Type 1. A component which was repaired during $[u, u+du)$, has been normal to time t, and fails during $[t, t+dt)$, given that the component jumped into the normal state at time zero.

Type 2. A component which has been normal to time t and fails during $[t, t+dt)$, given that it jumped into the normal state at time zero.

The probability for the first type of component is $v(u)du \cdot f(t-u)dt$, since [see equation (4.2)]

$$v(u)\,du = \textit{the probability that the component is repaired during } [u, u+du), \textit{ given that it is as good as new at time zero,}$$

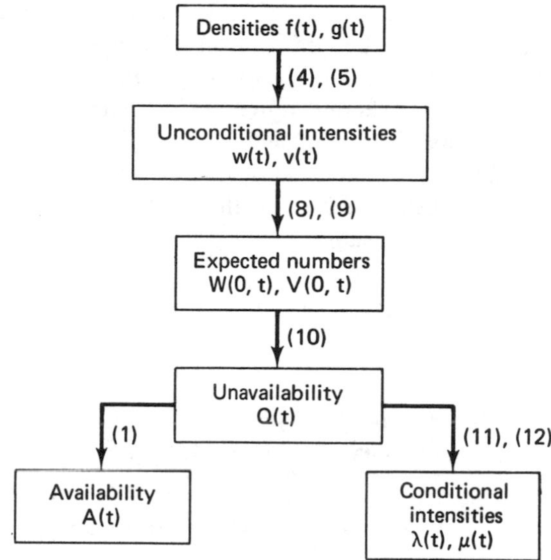

Figure 4.17. Flow chart for calculating probabilistic parameters for the whole process.

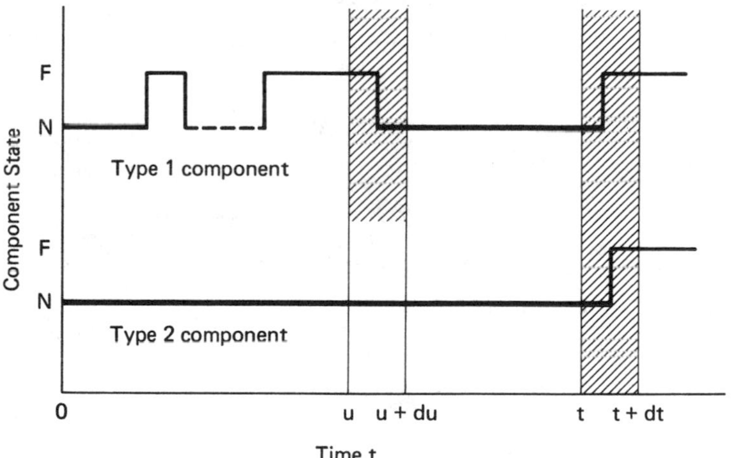

Figure 4.18. Component which fails during $[t, t+dt)$.

and

$f(t-u)\,dt =$ *the probability that the component has been normal to time t and failed during* $(t, t+dt]$, *given that it was as good as new at time zero and was repaired at time u.*

Note that we add the condition to the definition of $f(t-u)dt$ because the component failure characteristics depend only on the survival age $t-u$ at time t and are independent of the history before u.

The probability for the second type of the components is $f(t)dt$ as shown by (4.17). The quantity $w(t)dt$ is the probability that the component fails during $[t,t+dt)$, given that it jumped into the normal state at time zero. Since this probability is a sum of the probabilities for the first and second type components, we have

$$w(t)\,dt = f(t)\,dt + dt\int_0^t f(t-u)v(u)\,du \tag{4.60}$$

or, equivalently,

$$w(t) = f(t) + \int_0^t f(t-u)v(u)\,du \tag{4.61}$$

On the other hand, the components which are repaired during $[t,t+dt)$ consist of components of the following type.

Type 3. A component which was failed during $[u,u+du)$, has been failed to time t, and is repaired during $[t,t+dt)$, given that the component jumped into the normal state at time zero.

The behavior for this type of component is illustrated in Fig. 4.19.

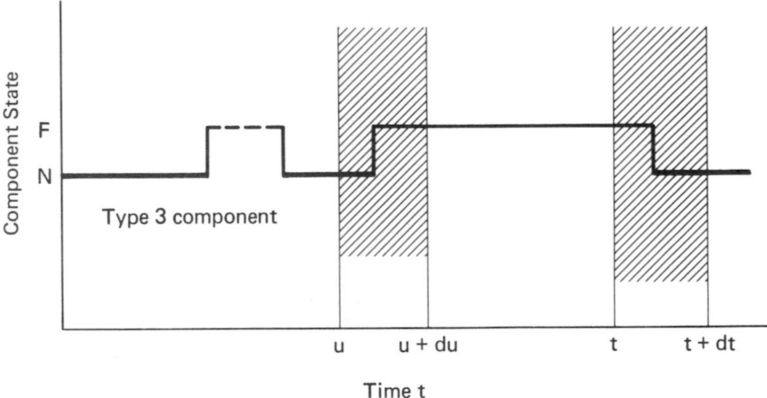

Figure 4.19. Component which is repaired during $[t,t+dt)$.

The probability for the third type of the components is $w(u)du\cdot g(t-u)dt$. Thus, we have

$$v(t)\,dt = dt\int_0^t g(t-u)w(u)\,du \tag{4.62}$$

or, equivalently,

$$v(t)=\int_0^t g(t-u)w(u)\,du \tag{4.63}$$

From (4.61) and (4.63), we have the following simultaneous identity:

$$\left.\begin{aligned} w(t)&=f(t)+\int_0^t f(t-u)v(u)\,du \\ v(t)&=\int_0^t g(t-u)w(u)\,du \end{aligned}\right\} \tag{4.64}$$

The unconditional failure intensity $w(t)$ and the repair intensity $v(t)$ are calculated by an iterative numerical integration of (4.64) when densities $f(t)$ and $g(t)$ are given. If a rigorous, analytical solution is required, Laplace transforms can be used.

Relations for Calculating the Unavailability, $Q(t)$. Let $x(t)$ be an indicator variable defined by

$$x(t)=1, \quad \text{if the component is in a failed state} \tag{4.65}$$

and

$$x(t)=0, \quad \text{if the component is in a normal state} \tag{4.66}$$

Represent by $x_{0,1}(t)$ and $x_{1,0}(t)$ the numbers of failures and repairs to time t, respectively. Then, we have

$$x(t)=x_{0,1}(t)-x_{1,0}(t) \tag{4.67}$$

For example, if the component has experienced three failures and two repairs to time t, the component state $x(t)$ at time t is given by

$$x(t)=3-2=1 \tag{4.68}$$

Taking an expectation of (4.67) then, as shown in Appendix 4.2 of this chapter, we have

$$Q(t)=W(0,t)-V(0,t) \tag{4.69}$$

In other words, the unavailability $Q(t)$ is given by the difference between the expected number of failures $W(0,t)$ and repairs $V(0,t)$ to time t. The expected numbers are obtained from the unconditional failure intensity $w(u)$ and the repair intensity $v(u)$, according to (4.40) and (4.44). We can rewrite (4.69) as

$$Q(t)=\int_0^t [w(u)-v(u)]\,du \tag{4.70}$$

Relations for Calculating Conditional Failure Intensity $\lambda(t)$. As was described by the "chain rule" (Section 4.1.2) the simultaneous occurrence of events A and C is equivalent to the occurrence of event C followed by event A [see (4.3)]:

$$\Pr(A,C|W)=\Pr(A|C,W)\Pr(C|W) \tag{4.71}$$

Substitute the following events into (4.71):

$C=$ the component is normal at time t,
$A=$ the component fails during $(t,t+dt]$,
$W=$ the component jumped into the normal state at time zero.

At most, one failure occurs during a small interval, and the event A implies event C. Thus, the simultaneous occurrence of A and C reduces to the occurrence of A, and (4.71) can be written as

$$\Pr(A|W)=\Pr(A|C,W)\Pr(C|W) \tag{4.72}$$

According to the definition of availability $A(t)$, conditional failure intensity $\lambda(t)$, and unconditional failure intensity $w(t)$, we have

$$\Pr(A|W)=w(t)\,dt \tag{4.73}$$

$$\Pr(A|C,W)=\lambda(t)\,dt \tag{4.74}$$

$$\Pr(C|W)=A(t) \tag{4.75}$$

Thus, from (4.72),

$$w(t)=\lambda(t)A(t) \tag{4.76}$$

or, equivalently,

$$w(t)=\lambda(t)[1-Q(t)] \tag{4.77}$$

and

$$\lambda(t)=\frac{w(t)}{1-Q(t)} \tag{4.78}$$

Identity (4.78) is used to calculate the conditional failure intensity $\lambda(t)$ when the unconditional failure intensity $w(t)$ and the unavailability $Q(t)$ are given. $w(t)$ and $Q(t)$ can be obtained by (4.64) and (4.69), respectively.

In the case of a constant failure rate, the conditional failure intensity coincides with the failure rate r as was shown by (4.35). Thus, $\lambda(t)$ is known and (4.77) is used to obtain $w(t)$ from $\lambda(t)=r$ and $Q(t)$.

Relations for Calculating $\mu(t)$. As in the case of $\lambda(t)$, we have the following identities for the conditional repair intensity $\mu(t)$:

$$\mu(t)=\frac{v(t)}{Q(t)} \tag{4.79}$$

$$v(t)=\mu(t)Q(t) \tag{4.80}$$

$\mu(t)$ can be calculated by (4.79) when the unconditional repair intensity $v(t)$ and the unavailability are known. $v(t)$ and $Q(t)$ can be obtained by (4.64) and (4.69), respectively.

When the component has a constant repair rate $m(t)=m$, the conditional repair intensity becomes m and is known. In this case, (4.80) is used to calculate the unconditional repair intensity $v(t)$, given $\mu(t)$ and $Q(t)$.

If the component has a time-varying failure rate $r(t)$, the conditional failure intensity $\lambda(t)$ does not coincide with $r(t)$. Similarly, a time-varying repair rate $m(t)$ is not equal to the conditional repair intensity $\mu(t)$. Thus, in general,

$$w(t)\neq r(t)[1-Q(t)] \tag{4.81}$$

$$v(t)\neq m(t)Q(t) \tag{4.82}$$

Example 6.

Use the results of Examples 2 and 5 to confirm, in Table 4.9, relations (2), (3), (4), (5), (10), (11), and (12). Obtain the TTF's, TTR's, TBF's, and TBR's for component 1.

Solution:

1) Inequality (2): From Example 2,

$$A(5)=0.6>R(5)=0.1111$$

2) Inequality (3):

$$Q(5)=0.4<F(5)=0.8889 \text{ (Example 2)}$$

3) Equality (4): We shall show that

$$w(5)=f(5)+\int_0^5 f(5-u)v(u)\,du$$

From Example 2,

$$w(5)=0.2$$

The probability $f(5)\times 1$ refers to the component which has been normal to time 5 and failed during $[5,6)$, given that it was as good as new at time zero. Component

3 is identified, and we have

$$f(5)=\tfrac{1}{10}$$

The integral on the right-hand side refers to the components shown below:

Repaired	Normal	Failed	Components
[0, 1)	[1, 5)	[5, 6)	None
[1, 2)	[2, 5)	[5, 6)	None
[2, 3)	[3, 5)	[5, 6)	None
[3, 4)	[4, 5)	[5, 6)	Component 7

Therefore,

$$\int_0^5 f(5-u)v(u)\,du=\tfrac{1}{10}$$

Equation (4) is confirmed since

$$0.2=\tfrac{1}{10}+\tfrac{1}{10}$$

4) Equality (5): We shall show that

$$v(7)=\int_0^7 g(t-u)w(u)\,du$$

From Example 5,

$$v(7)=0.2$$

The integral on the right-hand side refers to the components listed below:

Fails	Failed	Repaired	Components
[0, 1)	[1, 7)	[7, 8)	None
[1, 2)	[2, 7)	[7, 8)	None
[2, 3)	[3, 7)	[7, 8)	None
[3, 4)	[4, 7)	[7, 8)	None
[4, 5)	[5, 7)	[7, 8)	None
[5, 6)	[6, 7)	[7, 8)	Component 7
[6, 7)	—	[7, 8)	Component 1

Thus, the integral is $\tfrac{2}{10}=0.2$, and we confirm the equality.

5) Equality (10): We shall show that

$$Q(5)=W(0,5)-V(0,5)$$

From Example 2,

$$Q(5)=0.4, \qquad W(0,5)=0.9$$

From Example 5,

$$V(0,5)=0.5$$

Thus,

$$0.4=0.9-0.5$$

6) Equality (11): From Example 2,

$$Q(5)=0.4, \qquad w(5)=0.2, \qquad \lambda(5)=\tfrac{1}{3}$$

Thus,

$$\tfrac{1}{3}=\frac{0.2}{1-0.4}$$

and

$$\lambda(5)=\frac{w(5)}{1-Q(5)}$$

is confirmed.

7) Equality (12): We shall show that

$$\mu(7)=\frac{v(7)}{Q(7)}$$

From Example 5,

$$\mu(7)=\tfrac{1}{3}, \qquad v(7)=0.2$$

From Fig. 4.8,

$$Q(7)=\tfrac{6}{10}=0.6$$

This is now confirmed, since

$$\tfrac{1}{3}=\frac{0.2}{0.6}$$

8) TTF's, TTR's, TBF's, and TBR's are shown in Fig. 4.20.

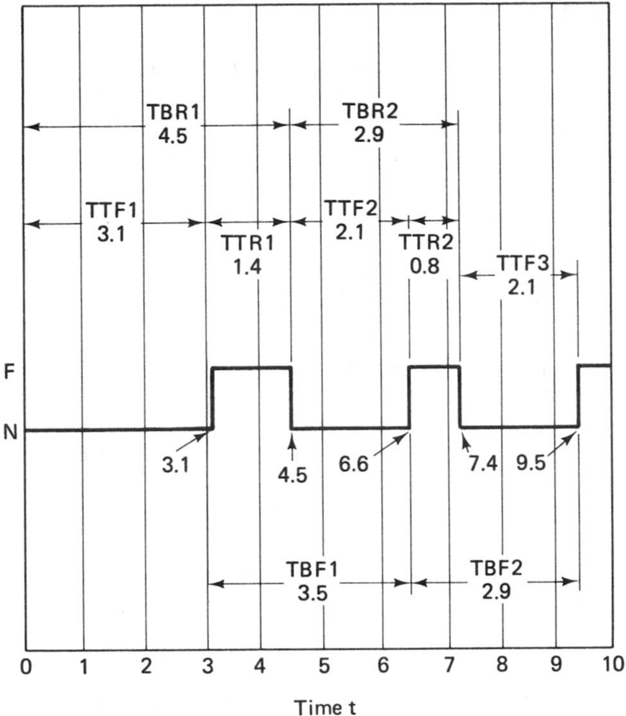

Figure 4.20. Time history of component 1.

4.4 ANALYSIS OF THE CONSTANT FAILURE RATE AND REPAIR RATE MODEL

An example of a pseudo-constant failure rate process was given in Example 3, Section 4.2.3. We now extend and generalize the treatment of these processes.

4.4.1 Repair-to-Failure Process with Constant Failure Rate

Constant failure rates greatly simplify systems analysis and are, accordingly, very popular with mathematicians, systems analysts, and optimization specialists.

The assumption of the constant rate is viable if

1) the component is in its "prime of life,"
2) the component is a solid state electronic device,
3) the component in question is a large one with many sub-components which have different failure rates or ages,

4) the data are so limited that elaborate mathematical treatments are unjustified.

Identity (4.35) shows that there is no difference between the failure rate $r(t)$ and the conditional failure intensity $\lambda(t)$ when the rate r is constant. Therefore, we denote by λ both the constant failure rate and the constant conditional failure intensity.

Substitute λ into (4.48), (4.49), and (4.50), and we obtain

$$F(t)=1-e^{-\lambda t} \tag{4.83}$$

$$R(t)=e^{-\lambda t} \tag{4.84}$$

$$f(t)=\lambda e^{-\lambda t} \tag{4.85}$$

The distribution (4.83) is called an *exponential distribution*, and its characteristics are given in Table 4.10.

The MTTF is defined by (4.21), and is

$$\text{MTBF}=\int_0^{\infty} t\lambda e^{-\lambda t}\,dt=\frac{1}{\lambda} \tag{4.86}$$

The MTTF can be obtained from an arithmetical mean of the time to failure data. The conditional failure intensity λ is the reciprocal of the MTTF.

On a plot of $F(t)$ versus t, the value of $F(t)$ at $t=\text{MTTF}$ is $(1-e^{-1})=$ 0.63 (Fig. 4.21). When the failure distribution is known, we can obtain

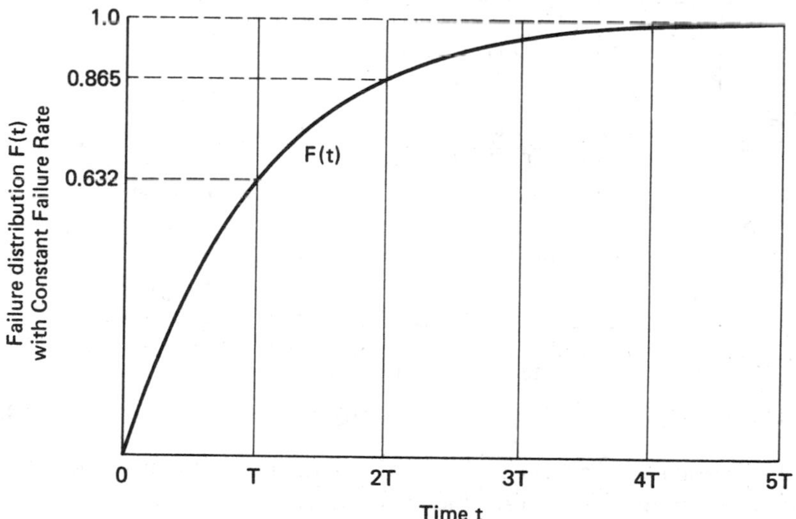

Figure 4.21. Determination of mean time to failure T.

TABLE 4.10 SUMMARY OF CONSTANT RATE MODEL

	Repairable	Non-repairable
Repair to Failure	$r(t)=\lambda$ $R(t)=e^{-\lambda t}$ $F(t)=1-e^{-\lambda t}$ $f(t)=\lambda e^{-\lambda t}$ $\text{MTTF}=1/\lambda$	$r(t)=\lambda$ $R(t)=e^{-\lambda t}$ $F(t)=1-e^{-\lambda t}$ $f(t)=\lambda e^{-\lambda t}$ $\text{MTTF}=1/\lambda$
Failure to Repair	$m(t)=\mu$ $G(t)=1-e^{-\mu t}$ $g(t)=\mu e^{-\mu t}$ $\text{MTTR}=1/\mu$	$m(t)=\mu=0$ $G(t)=0$ $g(t)=0$ $\text{MTTR}=\infty$
Dynamic Behavior of Whole Process	$Q(t)=\frac{\lambda}{\lambda+\mu}[1-e^{-(\lambda+\mu)t}]$ $A(t)=\frac{\mu}{\lambda+\mu}+\frac{\lambda}{\lambda+\mu}e^{-(\lambda+\mu)t}$ $w(t)=\frac{\lambda\mu}{\lambda+\mu}+\frac{\lambda^2}{\lambda+\mu}e^{-(\lambda+\mu)t}$ $v(t)=\frac{\lambda\mu}{\lambda+\mu}[1-e^{-(\lambda+\mu)t}]$ $W(0,t)=\frac{\lambda\mu}{\lambda+\mu}t+\frac{\lambda^2}{(\lambda+\mu)^2}[1-e^{-(\lambda+\mu)t}]$ $V(0,t)=\frac{\lambda\mu}{\lambda+\mu}t-\frac{\lambda\mu}{(\lambda+\mu)^2}[1-e^{-(\lambda+\mu)t}]$ $\frac{dQ(t)}{dt}=-(\lambda+\mu)Q(t)+\lambda,\ Q(0)=0$	$Q(t)=1-e^{-\lambda t}=F(t)$ $A(t)=e^{-\lambda t}=R(t)$ $w(t)=\lambda e^{-\lambda t}$ $v(t)=0$ $W(0,t)=1-e^{-\lambda t}=F(t)$ $V(0,t)=0$ $\frac{dQ(t)}{dt}=-\lambda Q(t)+\lambda,\ Q(0)=0$
Stationary Value of Whole Process	$Q(\infty)=\frac{\lambda}{\lambda+\mu}=\frac{\text{MTTR}}{\text{MTTF}+\text{MTTR}}$ $A(\infty)=\frac{\mu}{\lambda+\mu}=\frac{\text{MTTF}}{\text{MTTF}+\text{MTTR}}$ $w(\infty)=\frac{\lambda\mu}{\lambda+\mu}=\frac{1}{\text{MTTF}+\text{MTTR}}$ $v(\infty)=\frac{\lambda\mu}{\lambda+\mu}=w(\infty)$ $\frac{Q(t)}{Q(\infty)}=0.632$ for $t=\frac{1}{\lambda+\mu}$ $0=-(\lambda+\mu)Q(\infty)+\lambda$	$Q(\infty)=1$ $A(\infty)=0$ $w(\infty)=0$ $v(\infty)=0=w(\infty)$ $\frac{Q(t)}{Q(\infty)}=0.632$ for $t=\frac{1}{\lambda}$ $0=-\lambda Q(\infty)+\lambda$

MTTF by finding the time t which satisfies the equality

$$F(t)=0.63 \tag{4.87}$$

The presence or absence of a constant failure rate can be detected by suitable plotting procedures, which is discussed in the parameter identification section later in this chapter.

4.4.2 Failure-to-Repair Process with Constant Repair Rate

When the repair rate is constant, it coincides with the conditional repair intensity and is designated as μ.

Substituting μ into (4.58) and (4.59), we obtain:

$$G(t)=1-e^{-\mu t} \tag{4.88}$$

$$g(t)=\mu e^{-\mu t} \tag{4.89}$$

The distribution of (4.88) is an exponential repair distribution. The MTTR is given by

$$\text{MTTR}=\int_0^\infty t\mu e^{-\mu t}\,dt=\frac{1}{\mu} \tag{4.90}$$

The MTTR can be estimated by an arithmetical mean of the time to repair data, and the constant repair rate μ is the reciprocal of the MTTR.

When the repair distribution $G(t)$ is known, the MTTR can also be evaluated by noting the time t satisfying

$$G(t)=0.63 \tag{4.91}$$

The assumption of a constant repair rate can be verified by suitable plotting procedures. A numerical example which demonstrates this technique will be presented shortly.

4.4.3 Laplace Transform Analysis for the Whole Process

When constant failure rate and constant repair rate apply we can simplify the analysis of the whole process to such an extent that analytical solutions become possible. The solutions, which are summarized in Table 4.10, are now derived. First, we make a few comments regarding Laplace transforms.

A Laplace transform $L[h(t)]$ of $h(t)$ of t is a function of a variable s and is defined by

$$L[h(t)]=\int_0^\infty e^{-st}h(t)\,dt \tag{4.92}$$

For example, the transformation of e^{-at} is given by

$$L[e^{-at}]=\int_0^\infty e^{-st}e^{-at}=\frac{1}{s+a} \tag{4.93}$$

An inverse Laplace transform $L^{-1}[H(s)]$ is a function of t having the Laplace transform $H(s)$. Thus, the inverse transformation of $1/(s+a)$ becomes e^{-at}:

$$L^{-1}\left(\frac{1}{s+a}\right)=e^{-at} \tag{4.94}$$

One of the significant characteristics of the Laplace transform is the following identity:

$$L\left[\int_0^t h_1(t-u)h_2(u)\,du\right]=L[h_1(t)]\cdot[h_2(t)]$$

In other words, the transformation of the convolution can be represented by the product of the two Laplace transforms $L[h_1(t)]$ and $L[h_2(t)]$. The convolution integral can be treated as an algebraic product in the Laplace-transformed domain.

Now we take the Laplace transform of (4.64):

$$\left.\begin{aligned} L[w(t)]&=L[f(t)]+L[f(t)]\cdot L[v(t)] \\ L[v(t)]&=L[g(t)]\cdot L[w(t)] \end{aligned}\right\} \tag{4.95}$$

The constant failure rate λ and the repair rate μ give

$$L[f(t)]=L[\lambda e^{-\lambda t}]=\lambda\cdot L[e^{-\lambda t}]=\frac{\lambda}{s+\lambda} \tag{4.96}$$

$$L[g(t)]=\frac{\mu}{s+\mu} \tag{4.97}$$

Thus, (4.95) becomes

$$\left.\begin{aligned} L[w(t)]&=\frac{\lambda}{s+\lambda}+\frac{\lambda}{s+\lambda}L[v(t)] \\ L[v(t)]&=\frac{\mu}{s+\mu}L[w(t)] \end{aligned}\right\} \tag{4.98}$$

The identity (4.98) is a simultaneous algebraic equation for $L[w(t)]$ and $L[v(t)]$ and can be solved:

$$L[w(t)]=\frac{\lambda\mu}{v+\mu}\left(\frac{1}{s}\right)+\frac{\lambda^2}{\lambda+\mu}\left(\frac{1}{s+\lambda+\mu}\right) \tag{4.99}$$

$$L[v(t)]=\frac{\lambda\mu}{\lambda+\mu}\left(\frac{1}{s}\right)-\frac{\lambda\mu}{v+\mu}\left(\frac{1}{s+\lambda+\mu}\right) \tag{4.100}$$

Taking the inverse Laplace transform of (4.99) and (4.100) we have:

$$w(t)=\frac{\lambda\mu}{\lambda+\mu}L^{-1}\left(\frac{1}{s}\right)+\frac{\lambda^2}{\lambda+\mu}L^{-1}\left(\frac{1}{s+\lambda+\mu}\right) \qquad (4.101)$$

$$v(t)=\frac{\lambda\mu}{\lambda+\mu}L^{-1}\left(\frac{1}{s}\right)-\frac{\lambda\mu}{\lambda+\mu}L^{-1}\left(\frac{1}{s+\lambda+\mu}\right) \qquad (4.102)$$

From (4.94),

$$w(t)=\frac{\lambda\mu}{\lambda+\mu}+\frac{\lambda^2}{\lambda+\mu}e^{-(\lambda+\mu)t} \qquad (4.103)$$

$$v(t)=\frac{\lambda\mu}{\lambda+\mu}-\frac{\lambda\mu}{\lambda+\mu}e^{-(\lambda+\mu)t} \qquad (4.104)$$

The expected number of failures $W(0,t)$ and the expected number of repairs $V(0,t)$ are given by the integration of (4.40) and (4.44) with $t_1=0$ and $t_2=t$:

$$W(0,t)=\frac{\lambda\mu}{\lambda+\mu}t+\frac{\lambda^2}{(\lambda+\mu)^2}\left(1-e^{-(\lambda+\mu)t}\right) \qquad (4.105)$$

$$V(0,t)=\frac{\lambda\mu}{\lambda+\mu}t-\frac{\lambda\mu}{(\lambda+\mu)^2}\left(1-e^{-(\lambda+\mu)t}\right) \qquad (4.106)$$

The unavailability $Q(t)$ is obtained by (4.69):

$$Q(t)=W(0,t)-V(0,t)=\frac{\lambda}{\lambda+\mu}\left(1-e^{-(\lambda+\mu)t}\right) \qquad (4.107)$$

The availability is given by (4.31):

$$A(t)=1-Q(t)=\frac{\mu}{\lambda+\mu}+\frac{\lambda}{\lambda+\mu}e^{-(\lambda+\mu)t} \qquad (4.108)$$

The stationary unavailability $Q(\infty)$ and the stationary availability $A(\infty)$ are

$$Q(\infty)=\frac{\lambda}{\lambda+\mu}=\frac{1/\mu}{1/\lambda+1/\mu} \qquad (4.109)$$

$$A(\infty)=\frac{\mu}{\lambda+\mu}=\frac{1/\lambda}{1/\lambda+1/\mu} \qquad (4.110)$$

Equivalently, the steady-state unavailability and availability can be expressed as

$$Q(\infty) = \frac{\text{MTTR}}{\text{MTTF} + \text{MTTR}} \tag{4.111}$$

$$A(\infty) = \frac{\text{MTTF}}{\text{MTTF} + \text{MTTR}} \tag{4.112}$$

We also have

$$\frac{Q(t)}{Q(\infty)} = 1 - e^{-(\lambda+\mu)t} \tag{4.113}$$

Thus, 63.2% and 86.5% of the stationary steady-state unavailability is attained at time T and $2T$, respectively, where

$$T = \frac{1}{\lambda+\mu} = \frac{\text{MTTF}\cdot\text{MTTR}}{\text{MTTF}+\text{MTTR}} \tag{4.114}$$

$$\cong \text{MTTR if MTTR} \ll \text{MTTF} \tag{4.115}$$

Example 7.

Assume constant failure and repair rates for the components shown in Fig. 4.8. Obtain $Q(t)$ and $w(t)$ at $t=5$ and $t=\infty$ (stationary values).

Solution: TTF's in Example 2 give

$$\text{MTTF} = \frac{54.85}{18} = 3.05$$

Further, we have the following TTR data.

Component	Fails At	Repaired At	TTR
1	3.1	4.5	1.4
1	6.6	7.4	0.8
2	1.05	1.7	0.65
2	4.5	8.5	4.0
3	5.8	6.8	1.0
4	2.1	3.8	1.7
4	6.4	8.6	2.2
5	4.8	8.3	3.5
6	3.0	6.5	3.5
7	1.4	3.5	2.1
7	5.4	7.6	2.2
8	2.85	3.65	0.8
8	6.7	9.5	2.8
9	4.1	6.2	2.1

Thus,

$$\text{MTTR}=\frac{28.75}{14}=2.05$$

$$\lambda=\frac{1}{\text{MTTF}}=0.328$$

$$\mu=\frac{1}{\text{MTTR}}=0.488$$

$$Q(t)=\frac{0.328}{0.328+0.488}(1-e^{-(0.328+0.488)t})$$

$$=0.402(1-e^{-0.816t})$$

$$w(t)=\frac{0.328\times 0.488}{0.328+0.488}+\frac{0.328^2}{0.328+0.488}e^{-(0.328+0.488)t}$$

$$=0.196+0.131e^{-0.816t}$$

and, finally,

$$Q(5)=0.395, \qquad Q(\infty)=0.402$$

$$w(5)=0.198, \qquad w(\infty)=0.196$$

yielding a good agreement with the results in Example 2.

4.4.4 Markov Analysis of the Whole Process with Constant Rates

We now present another approach for analyzing the whole process for the case of constant failure and repair rates. This approach is called a *Markov analysis*.

Let $x(t)$ be the indicator variable defined by (4.65) and (4.66). The definition of the conditional failure intensity λ can be used to give

$$\begin{aligned}
P(1|0)&\equiv\Pr[x(t+dt)=1|x(t)=0]=\lambda\,dt\\
P(0|0)&\equiv\Pr[x(t+dt)=0|x(t)=0]=1-\lambda\,dt\\
P(1|1)&\equiv\Pr[x(t+dt)=1|x(t)=1]=1-\mu\,dt\\
P(0|1)&\equiv\Pr[x(t+dt)=0|x(t)=1]=\mu\,dt
\end{aligned}\tag{4.116}$$

$\Pr[x(t+dt)=1|x(t)=0]$ is the probability of failure at $t+dt$, given that the component is working at time t, etc. The quantities $P(1|0)$, $P(0|0)$, $P(1|1)$, and $P(0|1)$ are called *transition probabilities*. The state transitions are summarized by the Markov diagram of Fig. 4.22.

The conditional intensities λ and μ are the known constants r and m, respectively. A Markov analysis cannot handle the time-varying rates $r(t)$ and $m(t)$, since the conditional intensities are time-varying unknowns.

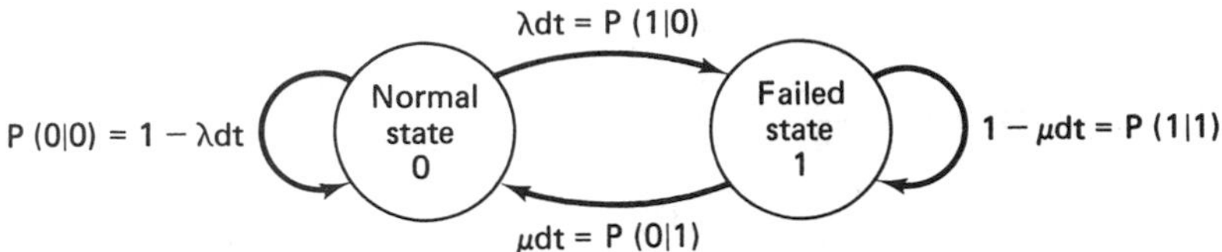

Figure 4.22. Markov transition diagram.

The unavailability $Q(t+dt)$ is the probability of $x(t+dt)=1$, which is, in turn, expressed in terms of the two possible states of $x(t)$ and the corresponding transitions to $x(t+dt)=1$:

$$\begin{aligned} Q(t+dt) &= \Pr[x(t+dt)=1] \\ &= P(1|0)\Pr[x(t)=0]+P(1|1)\Pr[x(t)=1] \\ &= \lambda dt[1-Q(t)]+(1-\mu dt)Q(t) \end{aligned} \tag{4.117}$$

This identity can be rewritten as

$$Q(t+dt)-Q(t)=dt(-\lambda-\mu)Q(t)+\lambda dt$$

yielding

$$\frac{dQ(t)}{dt}=-(\lambda+\mu)Q(t)+\lambda \tag{4.118}$$

with the initial condition at $t=0$ of

$$Q(0)=0 \tag{4.119}$$

The solution of this linear differential equation is

$$Q(t)=\frac{\lambda}{\lambda+\mu}(1-e^{-(\lambda+\mu)t}) \tag{4.120}$$

Thus, we reach the result given by (4.107).

The unconditional intensities $w(t)$ and $v(t)$ are obtained from (4.77) and (4.80) since $Q(t)$, λ, and μ are known. We have the results previously obtained: (4.103) and (4.104).

The expected number of failures $(0,t)$ and repairs $V(0,t)$ can be calculated by (4.40) and (4.44), yielding (4.105) and (4.106), respectively.

4.5 STATISTICAL DISTRIBUTIONS FOR REPAIR-TO-FAILURE AND FAILURE-TO-REPAIR PARAMETERS

The commonly used distributions are listed in Table 4.11. For components which have an increasing failure rate with time, the normal or log-normal, or the Weibull distribution with the shape parameter β larger than unity apply. The normal distributions arise by pure chance, resulting from a sum of a large number of small disturbances. Repair times are frequently best fitted by the log-normal distribution, since some repair times can be much greater than the mean because a few repairs take a long time due to a lack of spare parts or local expertise. A detailed description of these distributions is given in Appendix 4.3 of this chapter and in most textbooks on statistics or reliability.

When enough data are available, a histogram like Fig. 4.4 can be constructed. The density can be obtained analytically through a piecewise polynomial approximation of the normalized histogram.

4.6 ANALYSIS OF THE WHOLE PROCESS WITH GENERAL FAILURE AND REPAIR RATES

Consider a histogram such as Fig. 4.5. This histogram was constructed from the mortality data shown in Fig. 4.4 after dividing by the total number of individuals, 1,023,102. A piecewise polynomial interpolation of the histogram yields the following failure density:

$$f(t)=\begin{cases} 0.00638-0.001096t+0.951\times10^{-4}t^2-0.349\times10^{-5}t^3+0.478\times10^{-7}t^4, & \text{for } t\le 30 \\ 0.0566-0.279\times10^{-2}t+0.259\times10^{-4}t^2+0.508\times10^{-6}t^3-0.573\times10^{-8}t^4, & \text{for } 30<t\le 90 \\ -0.003608+0.777\times10^{-3}t-0.755\times10^{-5}t^2, & \text{for } t>90 \end{cases} \tag{4.121}$$

The failure density is plotted in Fig. 4.5. Assume now that repair data are also available and have been fitted to a log-normal distribution.

$$g(t)=\frac{1}{\sqrt{2\pi}\,\sigma t}\exp\left[-\frac{1}{2}\left(\frac{\log t-\mu}{\sigma}\right)^2\right] \tag{4.122}$$

with parameter values of

$$\mu=1.0, \qquad \sigma=0.5 \tag{4.123}$$

TABLE 4.11 SUMMARY OF TYPICAL DISTRIBUTIONS

Distributions → / Parameter ↓	Exponential	Normal	Log-Normal	Weibull	Poisson
pdf, $f(t)$	$\lambda \exp(-\lambda t)$	$\frac{1}{\sigma\sqrt{2\pi}} \exp\left[-\frac{1}{2}\left(\frac{t-\mu}{\sigma}\right)^2\right]$	$\frac{1}{\sigma t\sqrt{2\pi}} \exp\left[-\frac{1}{2}(\log t-\mu)^2/\sigma^2\right]$	$\frac{\beta(t-\gamma)^{\beta-1}}{\sigma^\beta} \exp\left[-\left(\frac{t-\gamma}{\sigma}\right)^\beta\right]$	$\frac{e^{-\lambda t}(\lambda t)^n}{n!}$
Unreliability, $F(t)$	$1-e^{-\lambda t}$	$\frac{1}{\sigma\sqrt{2\pi}} \int_o^t \exp\left[-(t-\mu)^2/2\sigma^2\right] dt$	$\frac{1}{\sigma\sqrt{2\pi}} \int_o^t \frac{1}{t} \exp\left[-\frac{1}{2}(\log t-\mu)^2/\sigma^2\right] dt$	$1-\exp\left[-\left(\frac{t-\gamma}{\sigma}\right)^\beta\right]$	$\sum_{j=0}^{n} \frac{(\lambda t)^j}{j!} \exp^{(-\lambda t)}$ (n = no. of failures)
Failure rate, $r(t)$	λ	$\frac{f(t)}{1-F(t)}$	$\frac{f(t)}{1-F(t)}$	$\frac{\beta(t-\gamma)^{\beta-1}}{\sigma^\beta}$	—
Mean time to failure	$1/\lambda$	μ	$\exp[(\mu+\frac{1}{2}\sigma^2)]$	$\gamma+\sigma\Gamma\left(\frac{1+\beta}{\beta}\right)$	λt
$f(t)$	$f(t)$; λ; $\lambda e^{-\lambda t}$; 0; t	$f(t)$; 0.4; 0.399; 0.242; 0.2; 0; $\mu-\sigma$ μ $\mu+\sigma$; t	$f(t)$; $\sigma = 0.3$; $\sigma = 1.5$; $\sigma = 1.0$; 0; e^μ; t	$f(t)$; $\beta = 1/2$; $\beta = 4$; $\beta = 1$; 0; σ; $t-\gamma$	$f(n)$; 0; n
$F(t)$	$F(t)$; 1; 0; t	$F(t)$; 1.0; 0.811; 0.5; 0.159; 0; $\mu-\sigma$ μ $\mu+\sigma$; t	$F(t)$; 1.0; 0.5; $\sigma = 0.3$; $\sigma = 1.0$; $\sigma = 1.5$; 0; e^μ; t	$F(t)$; 1.0; 0.5; $\beta = 4$; $\beta = 1$; $\beta = 1/2$; 0; σ; $t-\gamma$	
$r(t)$	$r(t)$; λ; 0; t	$r(t)$; $\frac{\sqrt{2}}{\sigma\sqrt{\pi}}$; 0; t	$r(t)$; $\sigma = 0.3$; $\sigma = 1.0$; $\sigma = 1.5$; 0; e^μ; t	$r(t)$; $\beta = 4$; $\beta = 1$; $\beta = 1/2$; 0; σ; $t-\gamma$	

We now differentiate the fundamental identity (4.64):

$$\left.\begin{aligned} \frac{w(t)}{dt} &= f'(t) + f(0)v(t) + \int_0^t f'(t-u)v(u)\,du \\ \frac{v(t)}{dt} &= g(0)w(t) + \int_0^t g'(t-u)w(u)\,du \end{aligned}\right\} \tag{4.124}$$

where $f'(t)$ and $g'(t)$ are defined by

$$f'(t) = \frac{f(t)}{dt}, \qquad g'(t) = \frac{g(t)}{dt} \tag{4.125}$$

The differential equation (4.124) is now integrated, yielding the results shown in Fig. 4.23.

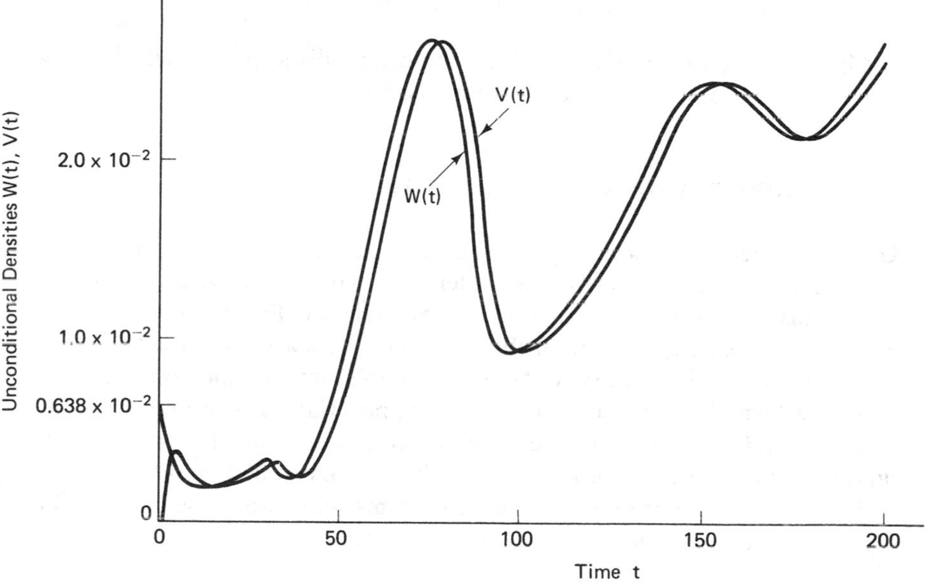

Figure 4.23. Result of integration of (4.124).

The expected number of failures $W(0,t)$ and repairs $V(0,t)$ can be calculated by integration of (4.40) and (4.44).

The unreliability $Q(t)$ is given by (4.69), and the transient behavior is shown in Fig. 4.24. The conditional failure intensity $\lambda(t)$ can be calculated by (4.78). It is found that $\lambda(t)$ considerably differs from the failure rate $r(t)$ of Fig. 4.6. This difference arises because $r(t)$ or $m(t)$ is not constant.

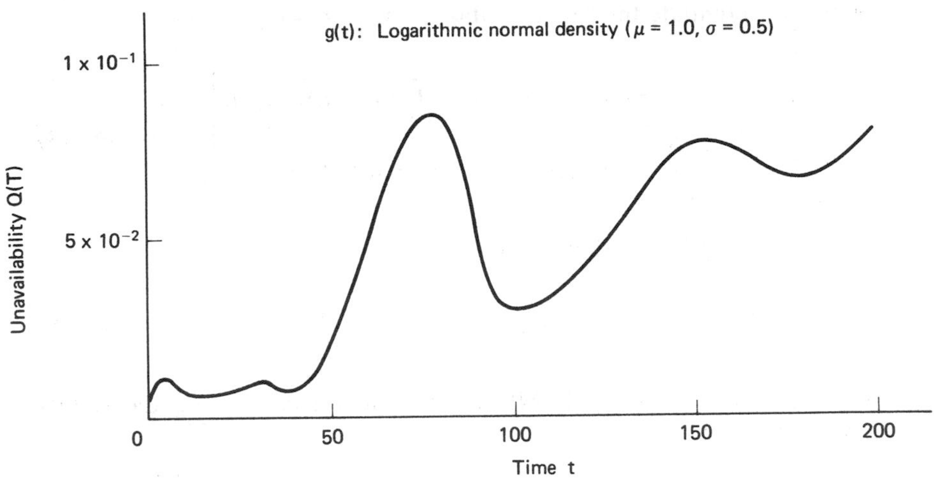

Figure 4.24. Curve of unavailability $Q(t)$.

Given failure and repair densities, the probabilistic parameters for any process can be obtained in the manner shown.

4.7 IDENTIFYING DISTRIBUTION PARAMETERS

Given sufficient data, a histogram such as Fig. 4.5 can be constructed and the failure or repair distribution determined by a piecewise polynomial approximation, as was demonstrated in Section 4.6. The procedure of Fig. 4.17 is then applied, and the probabilistic concepts are quantified.

When only fragmentary data are available we cannot construct the complete histogram. In such a case, an appropriate distribution must be assumed and its parameters evaluated from the data. The component quantification can then be made using the flow chart of Fig. 4.17.

In this section, parameter identification methods are presented for the repair-to-failure and the failure-to-repair process.

4.7.1 Parameter Identification for Repair-to-Failure Process

In parameter identification based on test data, three cases arise:

1) All components concerned proceed to failure and no component is taken out of use before failure. (All samples fail.)
2) Not all components being tested proceed to failure because they have been taken out of service before failure. (Incomplete failure data.)

3) Only a small portion of the sample is tested to failure. (Early failure data.)

Case 1: All Samples Fail. Consider the failure data for the 250 germanium transistors in Table 4.5. Assume a constant failure rate λ. The existence of the constant λ can be checked as follows:

The survival distribution is given by

$$R(t) = e^{-\lambda t}$$

This can be written as

$$\log\left[\frac{1}{R(t)}\right] = \lambda t \tag{4.126}$$

So, if the natural log of $1/R(t)$ is plotted against t, we should have a straight line with slope λ.

Values of log $[1/R(t)]$ versus t from Table 4.5 are plotted in Fig. 4.25. The best straight line is passed through the points and the slope is readily calculated from the graph:

$$\lambda = \frac{y_2 - y_1}{x_2 - x_1} = \frac{1.08 - 0.27}{400 - 100} = 0.0027 \tag{4.127}$$

Note that this λ is consistent with constant rate r in Example 3.

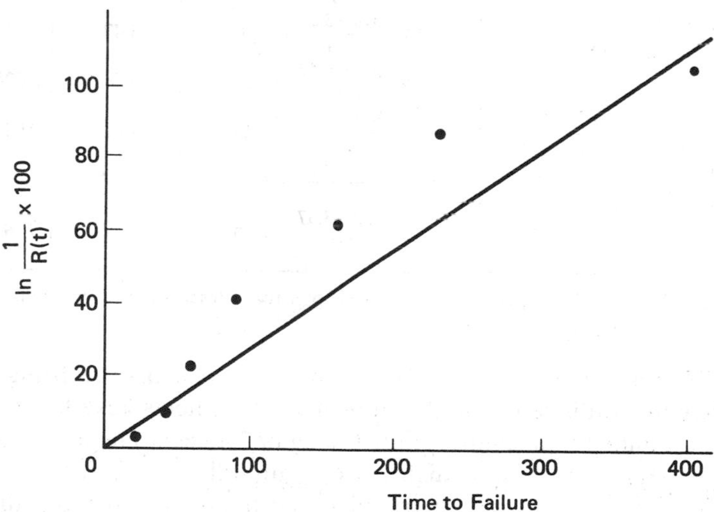

Figure 4.25. Test for constant λ.

Case 2: Incomplete Failure Data. In some tests, components are taken out of service for reasons other than failures. This will affect the number of components exposed to failure at any given time and a correction factor must be used in calculating the reliability. As an example, consider the lifetime to failure for bearings given in Table 4.12.[1] The original number of bearings exposed to failure is 202; however, between each failure some of the bearings are taken out of service before failure has occurred.

The unreliability $F(t)$ is calculated by dividing the cumulative number of failures expected if all original components had been allowed to proceed to failure by the original number components exposed to failure (202). The failure distribution $F(t)$ for data of Table 4.12 is plotted in Fig. 4.26. That curve represents only the portion of the mortality curve which corresponds to early wearout failures.

TABLE 4.12 Bearing Test Data

(1)	(2)	(3)	(4)[a]	(5)	(6)	(7)
Lifetime to Failures in hours	Number of Failures	Number Exposed to Failure	Number of Failures Expected if All Original Population Had Been Allowed to Proceed to Failure	Cumulative Number of Failures Expected	$F(t)$	$R(t)$
141	1	202		1.0	0.0049	0.9951
337	1	177	$1\times\frac{202-1}{177}=1.135$	2.135	0.0106	0.9894
364	1	176	$1\times\frac{202-2.135}{176}=1.135$	3.27	0.0162	0.9838
542	1	165	$1\times\frac{202-3.27}{165}=1.20$	4.47	0.0221	0.9779
716	1	156	$1\times\frac{202-4.47}{156}=1.27$	5.74	0.0284	0.9716
765	1	153	$1\times\frac{202-5.74}{153}=1.28$	7.02	0.0347	0.9613
940	1	144	$1\times\frac{202-7.02}{144}=1.35$	8.37	0.0414	0.9586
986	1	143	$1\times\frac{202-8.37}{143}=1.35$	9.72	0.0481	0.9519

[a]See the discussion in Appendix 4.4 of this chapter for a description of the computational procedure.

Case 3: Early Failure Data. Generally, when n items are being tested for failure the test is terminated before all of the n items have failed, either because of limited time available for testing or for economical reasons. For such a situation the failure distribution can still be estimated from the available data by assuming a particular distribution and plotting the data

[1]Bompas-Smith, J. H., *Mechanical Survival: The Use Of Reliability Data*, McGraw-Hill Book Company, New York, 1971.

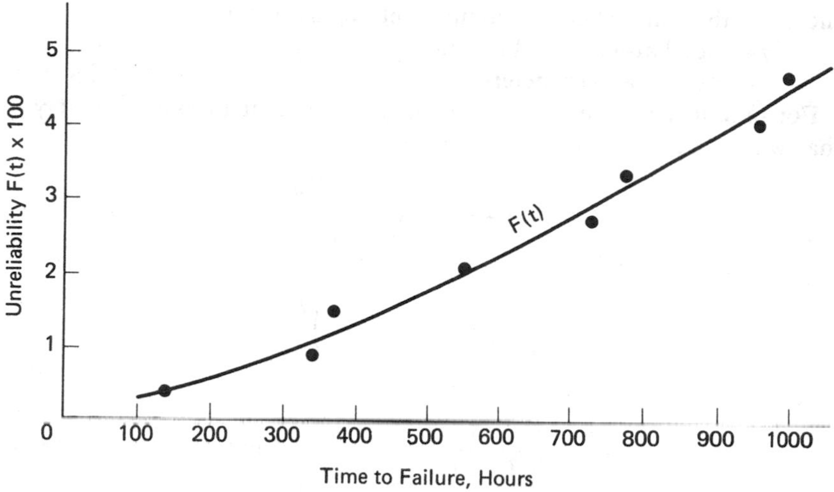

Figure 4.26. Bearing failure distribution.

for the assumed distribution. The closeness of the plotted data to a straight line indicates whether the model represents the data reasonably. As an example, consider the time to failure for the first seven failures of 20 guidance systems ($n=20$) given in Table 4.13.[2]

TABLE 4.13 FAILURE DATA OF GUIDANCE SYSTEMS

Failure No.	Time to Failure (hr)
1	1
2	4
3	5
4	6
5	15
6	20
7	40

Suppose it is necessary to estimate the number of failures to $t=100$ hr and $t=300$ hr. First, let us assume that the data can be described by a three-parameter Weibull distribution for which the equation is (see Table 4.11):

$$F(t)=\begin{cases} 1-\exp\left[-\left(\dfrac{t-\gamma}{\sigma}\right)^{\beta}\right], & \text{for } t\geq\gamma \\ 0, \quad \text{for } 0\leq t<\gamma \end{cases} \tag{4.128}$$

[2]Hahn, G. J., and S. S. Shapiro, *Statistical Models in Engineering*, John Wiley & Sons, Inc., New York, 1967.

where $\gamma=$ the time when the component begins to fail,
$\sigma=$ the characteristic life, and
$\beta=$ the shape parameter.

For practical reasons, it is frequently convenient to assume that $\gamma=0$. That will reduce the above equation to:

$$F(t)=1-\exp\left[-\left(\frac{t}{\sigma}\right)^{\beta}\right]$$

or

$$\frac{1}{1-F(t)}=\exp\left[\left(\frac{t}{\sigma}\right)^{\beta}\right]$$

and,

$$\log\log\left[\frac{1}{1-F(t)}\right]=\beta\log t-\beta\log\sigma \tag{4.129}$$

This is the basis for the Weibull probability plots, where $\log\log\{1/[1-F(t)]\}$ plots as a straight line against $\log t$ with slope β and y-intercept of $-\beta\log\sigma$:

$$\text{slope}=\beta \tag{4.130}$$

$$y\text{-intercept}=-\log\sigma \tag{4.131}$$

$$\sigma=\exp\left(\frac{-y\,\text{int.}}{\beta}\right) \tag{4.132}$$

To use this equation to extrapolate failure probabilities it is necessary to estimate the two parameters σ and β from the time to failure data. This is done by plotting the data of Table 4.14 in Fig. 4.27. The median rank plotting position is obtained by the method described in Appendix 4.5 of this chapter.

TABLE 4.14 Plotting Points

Failure No. i	Time to Failure	Plotting Points (% Failure) $F(t)=\dfrac{(i-\frac{1}{2})100}{n}$
1	1	2.5
2	4	7.5
3	5	12.5
4	6	17.5
5	15	22.5
6	20	27.5
7	40	32.5

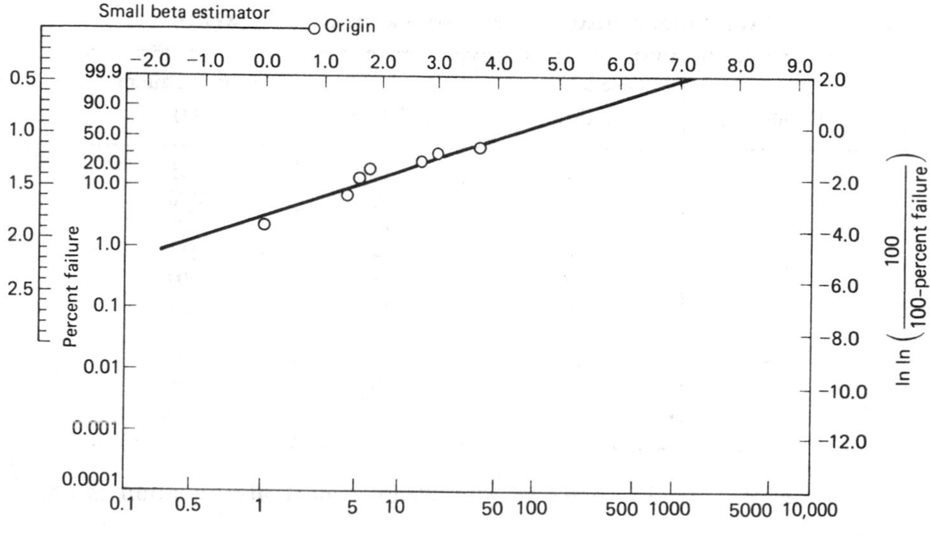

Figure 4.27. Test data plot.

From the graph, the parameters β and σ are identified:

$$\beta = \text{slope} = \frac{y_2 - y_1}{x_2 - x_1} = \frac{2.0-(-3.0)}{7.25-0.06} = 0.695 \tag{4.133}$$

$$\sigma = e^{-(y\,\text{int.})/\beta} = e^{-(-3.4/0.695)} = 132.85 \tag{4.134}$$

Thus,

$$F(100) = 1 - \exp\left[-\left(\frac{100}{132.85}\right)^{0.695}\right] = 0.56$$

The number of failures at $t = 100$ is $0.56 \times 20 = 11.2$. Also,

$$F(300) = 1 - \exp\left[-\left(\frac{300}{132.85}\right)^{0.695}\right] = 0.828$$

or, the number of failures to time $t = 300$ is $0.828 \times 20 = 16.6$.

Table 4.15 gives the actual time to failure for all 20 components. The comparison of the above results with the actual number of failures to 100 and 300 hr demonstrates our serendipity in choosing the Weibull distribution.

Once the functional form of the failure distribution has been established and the constants determined, other reliability factors of the repair-to-failure process can be readily obtained. For example, to calculate the

TABLE 4.15 ACTUAL FAILURE DATA FOR THE GUIDANCE SYSTEM

Failure No.	Time to Failure (hr)	Failure No.	Time to Failure (hr)
1	1	11	95
2	4	12	106
3	5	13	125
4	6	14	151
5	15	15	200
6	20	16	268
7	40	17	459
8	41	18	827
9	60	19	840
10	93	20	1089

failure density the derivative of the Weibull mortality equation is employed.

$$F(t)=1-\exp\left[-\left(\frac{t}{\sigma}\right)^{\beta}\right] \tag{4.135}$$

Then,

$$f(t)=\frac{dF(t)}{dt}=\frac{\beta\cdot t^{\beta-1}}{\sigma^{\beta}}\exp\left[-\left(\frac{t}{\sigma}\right)^{\beta}\right] \tag{4.136}$$

Substituting the values for σ and β gives

$$f(t)=\frac{0.02324}{t^{0.305}}\exp\left[-\left(\frac{t}{132.85}\right)^{0.695}\right] \tag{4.137}$$

The calculated values of $f(t)$ are given in Table 4.16 and plotted in Fig. 4.28. These values represent the probability that the component fails per unit time at time t, given that the component was operating at time zero and has been normal to time t.

The expected number of times the failures occur in the interval t to $t+dt$ is $w(t)\,dt$, and its integral over an interval is the expected number of failures. Once the failure density and the repair density are known, the unconditional failure intensity $w(t)$ may be obtained from the identity (4.64).

Assume that the component is as good as new at time zero. Assume further that once the component fails at time $t>0$ it cannot be repaired (non-repairable component). Then, the repair density is identically equal to

TABLE 4.16 Failure Density for Guidance System

Time to Failure (hr)	$f(t)$	Time to Failure (hr)	$f(t)$
1	0.0225	95	0.0026
4	0.0139	106	0.0024
5	0.0128	125	0.0020
6	0.0120	151	0.0017
15	0.0082	200	0.0012
20	0.0071	268	0.0008
40	0.0049	459	0.0003
40	0.0049	827	0.0001
60	0.0037	840	—
93	0.0027	1089	—

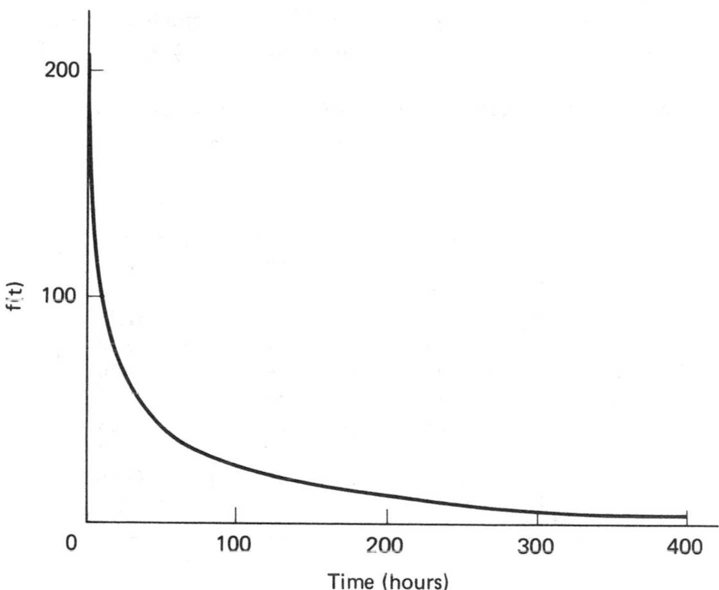

Figure 4.28. Failure density for guidance system.

zero, and the unconditional repair intensity $v(t)$ of (4.64) becomes zero. Thus, (4.64) reduces to

$$w(t)=f(t)=\frac{0.02324}{t^{0.305}}\exp\left[-\left(\frac{t}{132.85}\right)^{0.695}\right] \tag{4.138}$$

Thus, the unconditional failure intensity is also the failure density for the non-repairable component. The values of $f(t)$ in Table 4.16 represent $w(t)$ as well.

The expected number of failures $W(t_1, t_2)$ can be obtained by integrating the above equation over the t_1 to t_2 time interval and is equal to $F(t_2) - F(t_1)$:

$$\begin{aligned} W(t_1, t_2) &= \int_{t_1}^{t_2} w(t)\, dt \\ &= \int_{t_1}^{t_2} f(t)\, dt = F(t_2) - F(t_1) \\ &= \left\{ 1 - \exp\left[-\left(\frac{t}{\sigma}\right)^{\beta} \right] \right\}_{t_1}^{t_2} \end{aligned} \tag{4.139}$$

The ENF (expected number of failures) values $W(0, t)$ for the data of Table 4.16 are given in Table 4.17. In this case, since no repairs can be made, $W(0, t) = F(t)$, and (4.139) is equivalent to (4.135).

TABLE 4.17 EXPECTED NUMBER OF FAILURES OF GUIDANCE SYSTEM

Δt, $t=0$	ENF×20	Δt, $t=0$	ENF×20
1	.66	95	10.94
4	1.68	106	11.49
5	1.95	125	12.33
6	2.19	151	13.30
15	2.94	200	14.70
20	4.71	268	16.08
40	7.04	459	18.13
40	7.04	827	19.43
60	8.75	840	19.46
93	10.84	1089	19.73

Parameter Identification in a Wearout Situation. This example concerns a retrospective Weibull analysis carried out on an Imperial Chemicals Industries Ltd. (ICI) furnace. The furnace was commissioned in 1962 and had 176 tubes. Early in 1963, tubes began to fail after 475 days on-line, the first four failures being logged at the times listed in Table 4.18.

As far as can be ascertained, operation up to the time of these failures was perfectly normal; there had been no unusual excursions of temperature or pressure. Hence, it appears that tubes were beginning to wear out, and if normal operation were continued it should be possible to predict the

TABLE 4.18 TIMES TO FAILURE OF FIRST FOUR REFORMER TUBES

Failure	On-Line (days)
1	475
2	482
3	541
4	556

likely number of failures in a future period on the basis of the pattern of failures which these early failures were building up. In order to make this statement, however, it is necessary to make one further assumption. It may well be that the failures occurred at a weak weld in the tubes; one would expect the number of tubes with weak welds to be limited.

If, for example, six tubes had poor welds, then two further failures would clear this failure mode out of the system, and no further failures would take place until a different failure mode such as corrosion became significant. If we assume that all 176 tubes can fail for the same reason, then we are liable to make a pessimistic prediction of the number of failures in a future period. However, without being able to shut the furnace down to determine the failure mode, this is the most useful assumption which can be made.

The problem, therefore, is to predict future failures based on the assumption of a wearout failure mechanism.

The median rank plotting positions for the first four failures are listed in Table 4.19. The corresponding points are then plotted and the best straight line is drawn through the four points: i.e., line (a) of Fig. 4.29.

TABLE 4.19 MEDIAN RANK PLOTTING POSITIONS FOR THE FIRST FOUR FAILURES

Failure	On-Line (days)	Median Rank (%)
1	475	0.40
2	482	0.96
3	541	1.53
1	556	2.10

The line intersects the time axis at around 400 days and is extremely steep, corresponding to an apparent Weibull shape factor β of around 10. Both of these observations suggest that if we were able to plot the complete failure distribution, it would curve over towards the time axis as failures accumulated, indicating a three-parameter Weibull model rather than the

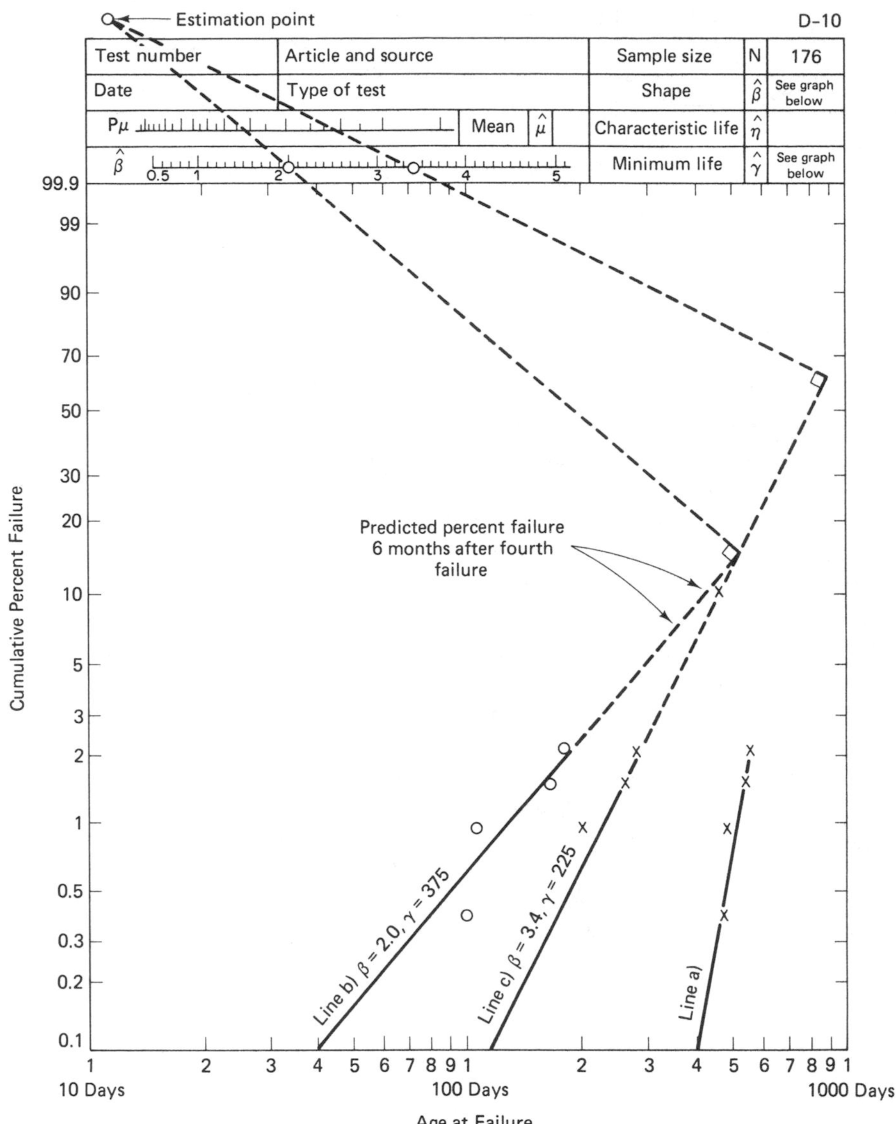

Figure 4.29. Weibull plots of reformer tube failures.

simplest two-parameter model which can be represented by a straight line on the plotting paper.

From the point of view of making predictions about the future number of failures, a straight line is clearly easier to deal with than a small part of a line of unknown curvature. Physically, the three-parameter Weibull model

$$F(t)=\begin{cases} 1-\exp\left[-\left(\frac{t-\gamma}{\alpha}\right)^{\beta}\right], & t\geq\gamma \\ 0, & 0<t<\gamma \end{cases} \tag{4.140}$$

implies that no failure occurs during the initial period $(0,\gamma)$. Similarly to (4.129), we have for $t\geq\gamma$,

$$\log\log\left[\frac{1}{1-F(t)}\right]=\beta\log(t-\gamma)-\beta\log\sigma \tag{4.141}$$

Thus, mathematically, the Weibull model can be reduced to the two-parameter model and is represented by a straight line by making the transformation

$$t'=t-\gamma$$

Graphically, this is equivalent to plotting the failure plots with a fixed time subtracted from the times to failure.

The correct time has been selected when the transformed plot

$$\left\{\log(t-\gamma),\log\log\left[\frac{1}{1-F(t)}\right]\right\}$$

becomes the asymptote of the final part of the original curved plot

$$\left\{\log t,\log\log\left[\frac{1}{1-F(t)}\right]\right\}$$

because $\log t\cong\log(t-\gamma)$ for large values of t.

In this case it is impossible to decide empirically what the transformation should be, since only the initial part of the curved plot is available. However, we know that we are dealing with a wearout phenomenon, and from experience we know that when these phenomena are represented by a two-parameter Weibull model, the Weibull shape parameter generally takes a value $2\leq\beta\leq3.4$. Hence, fixed times are subtracted from the four times to failures until by trial and error the straight lines drawn through the plotted points have apparent values of β of 2 and 3.4. These are, respectively, lines (b) and (c) of Fig. 4.29. The transformation formed by trial and error is shown in Table 4.20.

In Fig. 4.29, the two lines have been projected forward to predict the likely number of failures in the 6 months after the fourth failure (i.e., to

TABLE 4.20 TRANSFORMATION TO YIELD APPARENT VALUES OF β OF 2 AND 3.4

1 Failure	2 On-Line (days)	3 Col. 2 ($\gamma=375$ days)	4 Col. 2 ($\gamma=275$ days)	5 Median Rank (%)
1	475	100	200	0.40
2	482	107	207	0.96
3	541	166	266	1.53
4	556	181	281	2.10

182 days after the fourth failure). The respective predictions are of 9 and 14 further failures.

The furnace was, in fact, operated for more than 6 months after the fourth failure and, in the 6 month period referred to, 11 further failures took place.

4.7.2 Parameter Identification for Failure-to-Repair Process

Time to repair (TTR), or downtime, consists not only of the time it takes to repair a failure but also of waiting time for spare parts, time for replacement, etc. The availability $A(t)$ is the proportion of population of the components expected to function at time t. This availability is related to the population ensemble. We can consider another availability based on an average over a time ensemble. It is defined by

$$\bar{A}=\frac{\sum_{i=1}^{N}\mathrm{TTF}_i}{\sum_{i=1}^{N}\mathrm{TTF}_i+\mathrm{TTR}_i} \tag{4.142}$$

where $(\mathrm{TTF}_i, \mathrm{TTR}_i)$, $i=1,\ldots,N$ are consecutive pairs of times to failure and times to repair of a particular component. The number N of the cycles $\mathrm{TTF}_i+\mathrm{TTR}_i$ is assumed sufficiently large. The time ensemble availability represents percentiles of the component functioning in one cycle. The so-called ergodic theorem states that the time ensemble availability $\bar{A}$ coincides with the stationary values of the population ensemble availability $A(\infty)$.

As an example, consider the consecutive 20 sets of TTF and TTR given in Table 4.21.[3] The time ensemble availability is

$$\bar{A}=\frac{1102}{1151.8}=0.957 \tag{4.143}$$

[3]Locks, M. O., *Reliability, Maintainability, and Availability Assessment*, Hayden Book Co., Inc., New York, 1973.

TABLE 4.21 TIME TO FAILURE AND TIME TO REPAIR DATA

TTF (hr)	TTR (hr)	TTF (hr)	TTR (hr)
125	1.0	58	1.0
44	1.0	53	0.8
27	9.8	36	0.5
53	1.0	25	1.7
8	1.2	106	3.6
46	0.2	200	6.0
5	3.0	159	1.5
20	0.3	4	2.5
15	3.1	79	0.3
12	1.5	27	3.8
	Subtotal:	1102	49.8
	Total:	1151.8	

The mean time to failure and the mean time to repair are

$$\text{MTTF} = \frac{1102}{20} = 55.10 \tag{4.144}$$

$$\text{MTTR} = \frac{49.8}{20} = 2.49 \tag{4.145}$$

Like the failure parameters, the TTR data of Table 4.21 form a distribution for which parameters can be estimated. Table 4.22 is an ordered listing of the repair times in Table 4.21 (see Appendix 4.5 for the method used for plotting points in Table 4.22).

TABLE 4.22 ORDERED LISTING OF REPAIR TIMES

Repair No. i	TTR	Plotting Points $(i-\frac{1}{2})100/n$	Repair No. i	TTR	Plotting Points $(i-\frac{1}{2})100/n$
1	0.2	2.5	11	1.5	52.5
2	0.3	7.5	12	1.5	57.5
3	0.3	12.5	13	1.7	62.5
4	0.5	17.5	14	2.5	67.5
5	0.8	22.5	15	3.0	72.5
6	1.0	27.5	16	3.1	77.5
7	1.0	32.5	17	3.6	82.5
8	1.0	37.5	18	6.0	87.5
9	1.0	42.5	19	9.8	92.5
10	1.2	47.5	20	9.8	97.5

Let us assume that these data can be described by the log normal distribution where the natural log of times to repair are *distributed according to normal distribution with mean* μ *and* variance σ^2. The mean μ then may best be estimated by plotting the TTR data on a log-normal probability paper against the plotting points (Fig. 4.30) and finding the natural logarithm of the plotted 50th percentile. The 50th percentile is 1.43, so the parameter $\mu = \log 1.43 = 0.358$.

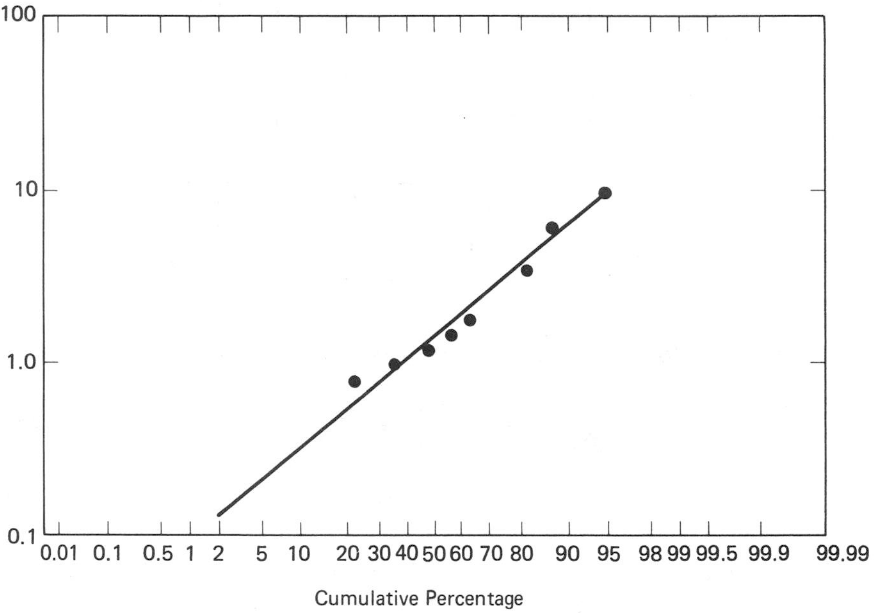

Figure 4.30. Plot of TTR data.

The μ is not only the median (50th percentile) of the normal distribution of log (TTR) but also a mean value of log (TTR). Thus, the parameter μ may also be estimated by the arithmetical mean of the natural log of TTR's in Table 4.21. This yields $\bar{\mu} = 0.368$, almost the same result as 0.358 obtained from the log-normal probability paper.

Note that the $T = 1.43$ satisfying $\log T = \mu = 0.358$ is not the expected value of the time to repair, although it is a 50th percentile of the log-normal distribution. The expected value, or the mean time to repair, can be estimated by averaging observed times to repair data in Table 4.21 and was given as 2.49 by (4.145). This is considerably larger than the 50 percentile ($T = 1.43$) because, in practice, there are usually some unexpected breakdowns which take a long time to repair. A time to repair distribution with this property frequently follows the log-normal distribution which decreases gently for large values of TTR.

The parameter σ^2 is a variance of log (TTR) and can be estimated by

$$\sigma^2 = \frac{\sum_{i=1}^{N} (\log \mathrm{TTR}_i - \bar{\mu})^2}{N-1} \tag{4.146}$$

where N = total number of sample.

Table 4.21 gives $\sigma = 1.081$.

Assume that the TTF is distributed with the constant failure rate $\lambda = 1/\mathrm{MTTF} = 1/55.1 = 0.01815$. Since both of the distributions for repair-to-failure process and failure-to-repair process are known, the general procedure of Fig. 4.17 can be used. The results are shown in Fig. 4.31. Note that the stationary unavailability $Q(\infty) = 0.04295$ agrees to the time ensemble availability $\bar{A} = 0.957$ of (4.143).

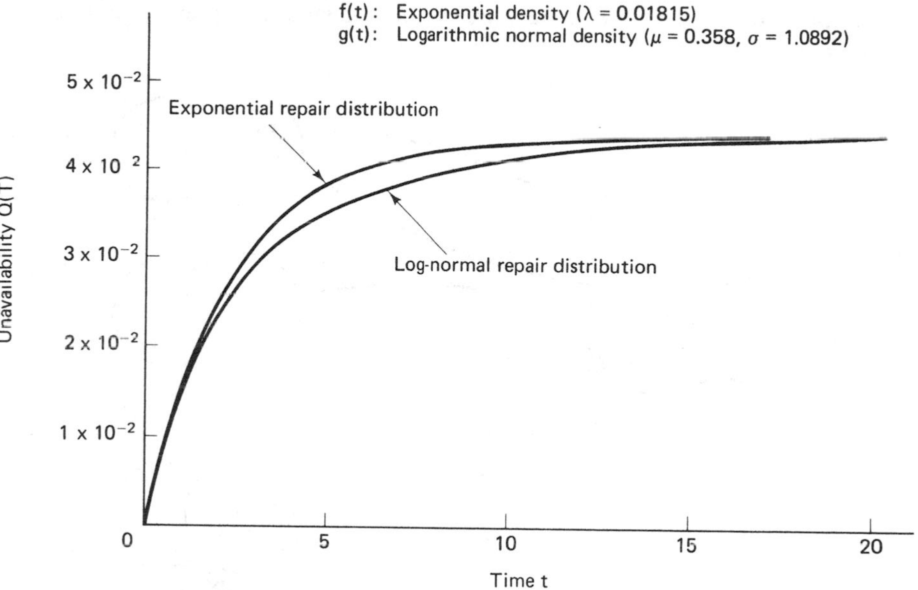

Figure 4.31. Unavailability $Q(t)$.

Consider now the case where the repair distribution is approximated by a constant repair rate model. The constant m is given by $m = 1/\mathrm{MTTR} = 1/2.49 = 0.40161$. The unavailabilities $Q(t)$ as calculated by (4.107) are plotted in Fig. 4.31. This $Q(t)$ is a good approximation to the unavailability obtained by the log-normal assumption. This is not an unusual situation. The constant rate model frequently gives a first-order approximation and should be tried prior to more complicated distributions. We can ascertain trends by using the constant rate model and recognize system improvements. Usually, the constant rate model itself gives sufficiently accurate results.

4.8 QUANTIFICATION OF COMPONENTS WITH MULTIPLE FAILURE MODES

As was pointed out in Chapter 1, many components have more than one failure mode. In any practical application of fault trees, if a basic event is a component failure, then the exact failure modes must be stated. When the basic event refers to more than one failure mode, it can be developed through "OR gates" to more basic events, each of which refers to a single failure mode. Thus, we can assume that every basic event has associated with it only one failure mode, although a component itself may suffer from multiple failure modes. The state transition for such components is represented by Fig. 4.32.

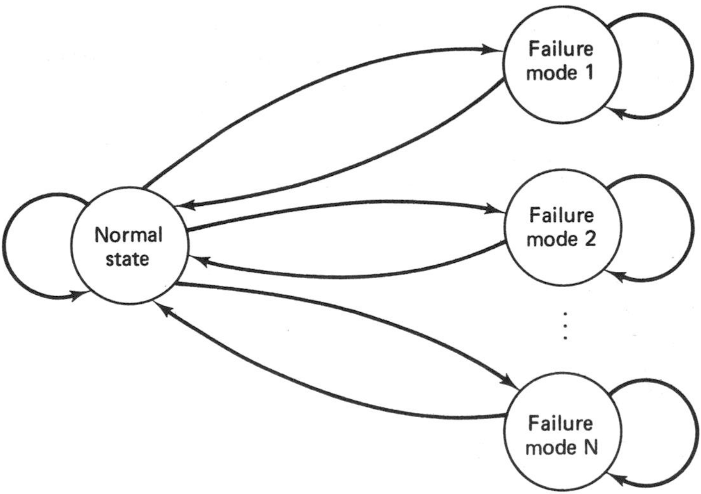

Figure 4.32. Transition diagram for components with multiple failure modes.

Suppose that a basic event is a single failure mode, say mode 1 in Fig. 4.32. Then, the normal state, and modes 2 to N result in non-existence of the basic event, and this can be expressed by Fig. 4.33. This diagram is analogous to Fig. 4.2, and quantification techniques developed in the previous sections can apply without major modifications: the reliability $R(t)$ becomes the probability of non-occurrence of mode 1 failures to time

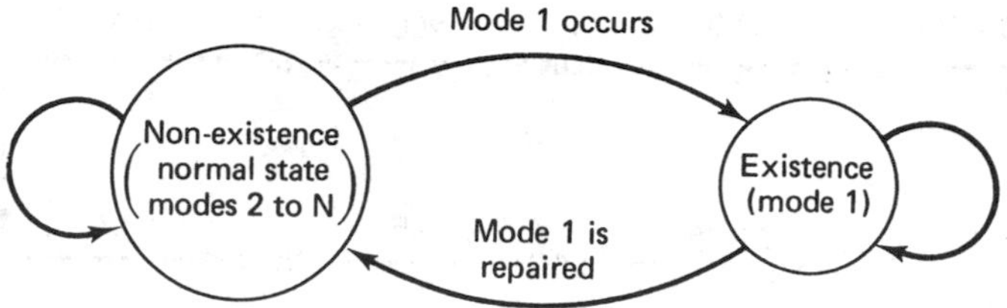

Figure 4.33. Transition diagram for a basic event.

t, the unavailability $Q(t)$ is the existence probability of mode 1 failure at time t, etc.

Example 8.

Consider a time history of the valve shown in Fig. 4.34. The valve has two failure modes, "stuck open" and "stuck closed." Assume a basic event with the failure mode "stuck closed." Calculate MTTF, MTTR, $R(t)$, $F(t)$, $A(t)$, $Q(t)$, $w(t)$, and $W(0,t)$ by assuming constant failure and repair rates.

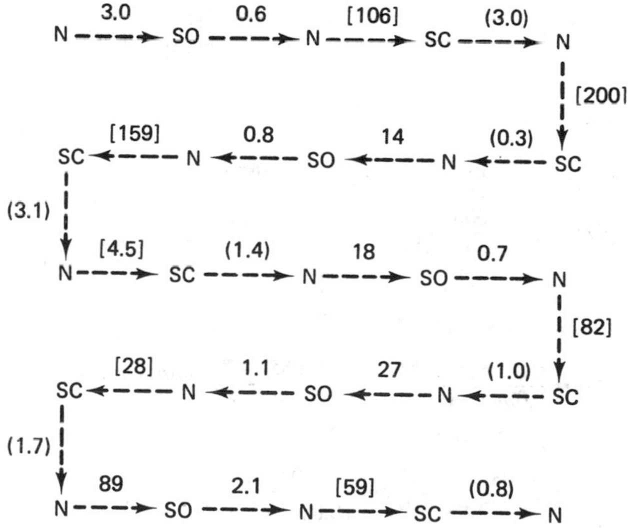

Figure 4.34. A time history of a valve.

Solution: The "valve normal" and "valve stuck open" denotes non-existence of the basic event. Thus, Fig. 4.34 can be rewritten as Fig. 4.35, where the symbol "NON" denotes the non-existence.

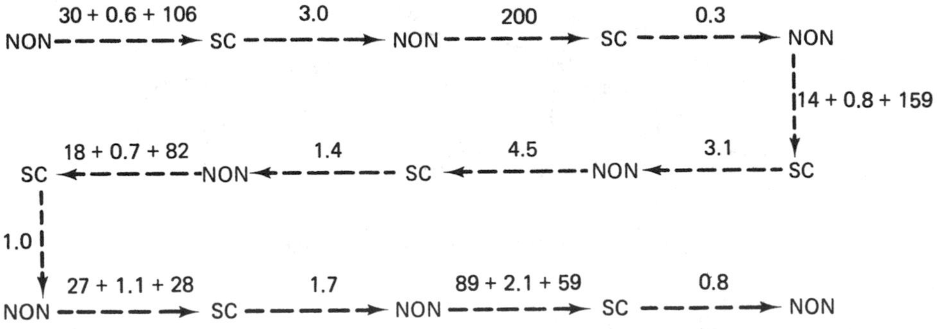

Figure 4.35. TTF's and TTR's of "stuck closed" event for the valve.

$$\text{MTTF} = \frac{136.6+200+173.8+4.5+100.7+56.1+150.1}{7}$$

$$= 117.4$$

$$\text{MTTR} = \frac{3.0+0.3+3.1+1.4+1.0+1.7+0.8}{7}$$

$$= 1.61$$

$$\lambda = \frac{1}{\text{MTTF}} = 0.0085, \qquad \mu = \frac{1}{\text{MTTR}} = 0.619$$

Table 4.10 yields

$$F(t) = 1 - e^{-0.0085t}, \qquad R(t) = e^{-0.0085t}$$

$$Q(t) = \frac{0.0085}{0.0085+0.619}[1 - e^{-(0.0085+0.619)t}]$$

$$= 0.0135 \times [1 - e^{-0.6275t}]$$

$$A(t) = 0.9865 + 0.0135e^{-0.6275t}$$

$$w(t) = \frac{0.0085 \times 0.619}{0.0085+0.619} + \frac{0.0085^2}{0.085+0.619}e^{-(0.085+0.619)t}$$

$$= 0.0084 + 0.0001e^{-0.6275t}$$

$$W(0,t) = \frac{0.0085 \times 0.619}{0.0085+0.619}t + \frac{0.0085^2}{(0.0085+0.619)^2}[1 - e^{-(0.0085+0.619)t}]$$

$$= 0.0002 + 0.0084t - 0.0002e^{-0.6275t}$$

These calculations hold only approximately, since the three-state valve is modeled by the two-state diagram of Fig. 4.33. However, MTTR for "stuck open" is usually small, and the approximation error is negligible. If rigorous analysis is required, we can start with the Markov transition diagram of Fig. 4.36 and apply the differential equations that are described in Chapters 7, 8, and 9 for the calculation of $R(t)$, $Q(t)$, $w(t)$, etc.

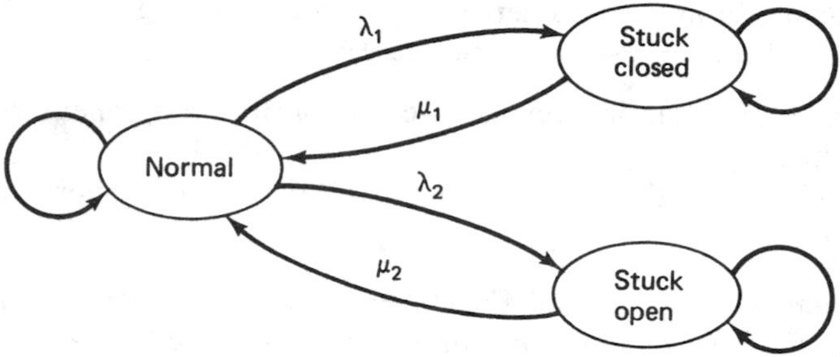

Figure 4.36. Markov transition diagram for the valve.

As is described in Chapter 6 (Data Bases), some data on component failure modes are available in the form of "frequency = failures/period." The frequency can be converted into the constant failure intensity λ in the following way.

From Table 4.10, the stationary value of the frequency is

$$W(\infty) = \frac{\lambda\mu}{\lambda + \mu} \tag{4.147}$$

Usually, MTTF is much greater than MTTR; i.e.,

$$\lambda \ll \mu \tag{4.148}$$

Thus,

$$w(\infty) = \lambda \tag{4.149}$$

The frequency itself can be used as the conditional failure intensity λ, provided that MTTR is sufficiently small. When this is not true, (4.147) is used to calculate λ for given MTTR and frequency data. The MTTR can be estimated, since a failure mode has been specified.

Example 9.

The frequency $w(t)$ in Example 8 yields $w(\infty) = 0.0084$ ("stuck closed" failures/time unit). Recalculate the unconditional failure intensity λ.

Solution: $\lambda = w(\infty) = 0.0084$ by (4.149). This gives good agreement with $\lambda = 0.0085$ in Example 8.

4.9 QUANTIFICATION OF ENVIRONMENTAL IMPACTS

As explained in Section 2.4.1, system hazards are caused by one or a set of system components generating failure events. The environment, plant personnel, and aging can affect the system only through the system components.

Component failure characteristics are shown in Fig. 2.15. As to the environmental impacts, we have two cases:

(I). Environmental impact as a cause of the component secondary failures.
(II). Environmental impact as a cause of the component command faults.

Environmental Impacts and Command Faults. The command fault was defined as the component being in a non-working state due to improper control signals or noise and, frequently, *repair action is not required* to return the component to the working state. Thus, a command fault exists if and only if improper commands exists. Thus, the quantification of the command fault reduces to that of improper commands. As a matter of fact, improper commands such as "area power failure" and "inadvertent environmental noise" appear as basic events in fault trees (see "area power failure" in Fig. 2.26). The environmental impacts can be quantified in the same way as components.

Example 10.

Assume MTTF=0.5 year and MTTR=30 min for an area power failure. Calculate $R(t)$ and $Q(t)$ at $t=1$ year.

Solution:

$$\lambda=\frac{1}{0.5}=2 \text{ year}^{-1}$$

$$\text{MTTR}=\frac{30}{365\times 24\times 60}=5.71\times 10^{-5} \text{ year}$$

$$\mu=\frac{1}{\text{MTTR}}=1.75\times 10^{4} \text{ year}^{-1}$$

$$R(1)=e^{-2\times 1}=0.135$$

$$Q(1)=\frac{2}{2+17{,}500}[1-e^{-(2+17{,}500)\times 1}]=1.14\times 10^{-4}$$

Environmental Impacts and Secondary Failures. A secondary failure is the same as a primary failure except that the component is not considered as accountable for the failure. Past or present excessive stresses placed on the component are responsible for the secondary failure. Note that repair action on the component is required to return the component to the working state; i.e., the removal of the excessive stresses do not guarantee a working state because the stresses have damaged the component and it must be repaired.

In qualitative fault tree analysis, a primary failure and the corresponding secondary failure are aggregated into a single basic event. Thus, it is natural to quantify the resulting basic event. The event occurs if the primary failure or secondary failure occurs. If we assume constant failure and repair rates for the two failures, we have the transition diagram of Fig. 4.37. Here, $\lambda_{(P)}$ and $\lambda_{(s)}$ are conditional failure intensities for primary and secondary failures, respectively, and μ is the repair intensity which is assumed to be the same for primary and secondary failures. The diagram can be used to quantify basic events, including secondary component failures resulting from environmental impacts.

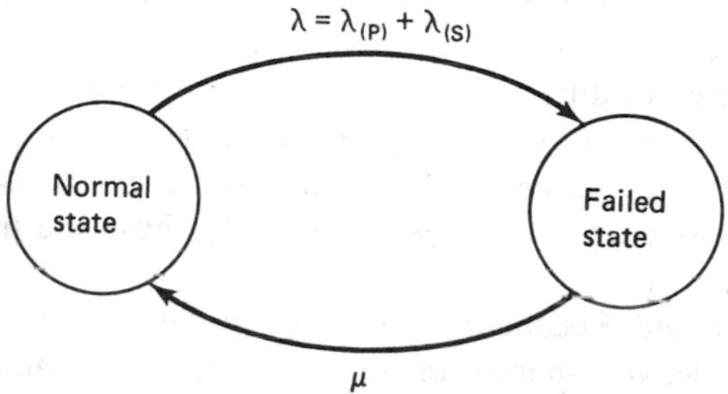

Figure 4.37. Transition diagram for primary and secondary failures.

Example 11.

Assume that an earthquake occurs once in 60 years. When it occurs, there is a 50% chance of a tank being destroyed. Assume that MTTF = 30 (years) for the tank under normal environment. Assume further that it takes 0.1 year to repair the tank. Calculate $R(10)$ and $Q(10)$ for the basic event, which is obtained by the aggregation of the primary and secondary tank failure.

Solution: The tank is destroyed by the earthquakes once in 120 years. Thus,

$$\lambda_{(s)} = \frac{1}{120} = 8.33 \times 10^{-3} \text{ (year}^{-1}\text{)}$$

Further,

$$\lambda_{(P)} = \frac{1}{30} = 3.33 \times 10^{-2} \text{ (year}^{-1}\text{)}$$

$$\lambda = \lambda_{(P)} + \lambda_{(s)} = 4.163 \times 10^{-2} \text{ (year}^{-1}\text{)}$$

$$\mu = \frac{1}{0.1} = 10 \text{ (year}^{-1}\text{)}$$

Thus, at 10 years

$$R(10) = e^{-4.163 \times 10^{-2} \times 10} = 0.659$$

$$Q(10) = \frac{4.163 \times 10^{-2}}{4.163 \times 10^{-2} + 10}\left(1 - e^{-(4.163 \times 10^{-2} + 10) \times 10}\right)$$

$$= 4.15 \times 10^{-3}$$

In most cases, environmental impacts act as common causes. The quantification of basic events involved in common causes is developed in Chapter 8.

4.10 QUANTIFICATION OF HUMAN ERROR

Similarly to the environmental impacts, human errors are causes of a component command fault or secondary failure.

Human Error and Command Faults. Plant operators act in response to demands. A typical fault tree representation is shown in Fig. 2.3. The operator error is also included in Figs. 2.24 and 2.26. As explained in Section 2.5.2, various conditions are introduced by OR and AND gates. We may use these conditions to quantify operator error, since the operator may be 99.99% perfect at a routine job, but relatively useless if he panics in an emergency. Probabilities of operator error are usually time invariant and can be expressed as "*error per demand.*" These are listed in Fig. 6.9.

Human Error and Secondary Failures. When a plant operator breaks components or leaves memories in them, he is a cause of secondary failures. For example, if he inadvertently closes a valve, then that valve suffers a secondary failure because the valve cannot open by itself.

This type of secondary failure can be quantified by using transition diagrams similar to Fig. 4.37. The only difference is that the parameter $\lambda_{(s)}$ must be estimated based on frequencies of demand and the frequency of the operator error for given demand.

Example 12.

An operator receives the demand "open valve 1" once a month. There is 1% chance of the operator's closing valve 1 or leaving it closed. Within the design envelope, the valve "fails closed" once in 10 years. It takes 10 hr on the average to repair the inadvertent valve closure. Calculate $R(1)$ and $Q(1)$ for the basic event "inadvertent valve closure."

Solution:

$$\text{Mean time between demands} = \frac{1}{12} \text{ (year)}$$

$$\text{Mean time to operator error} = \frac{1}{12} \times 100 \text{ (year)}$$

$$\lambda_{(s)} = \frac{1}{\frac{1}{12} \times 100} = \frac{12}{100} = 0.12 \text{ (year}^{-1}\text{)}$$

$$\lambda_{(P)} = \frac{1}{10} = 0.1$$

$$\lambda = 0.12 + 0.1 = 0.22$$

$$\text{MTTR} = \frac{10}{365 \times 24}$$

$$\mu = 365 \times \frac{24}{10} = 876 \text{ (year}^{-1}\text{)}$$

Thus, at 1 year,

$$R(1) = e^{-0.22} = 0.803$$

$$Q(1) = \frac{0.22}{0.22+876}[1-e^{-(0.22+876)}] = 2.51 \times 10^{-4}$$

Finally, we come to so-called *system dependent basic events*, typified by the "secondary fuse failure" of Fig. 2.19. This failure can also be analyzed by a diagram similar to Fig. 4.37. The parameter $\lambda_{(s)}$ is given by the sum of conditional failure intensities for "wire shorted" and "generator surge," since "excessive current to fuse" is expressed by Fig. 2.21.

Example 13.

Assume the following conditional failure intensities:

Wire shorted: $\frac{1}{10{,}000}$ (hour^{-1})

Generator surge: $\frac{1}{50{,}000}$ (hour^{-1})

Primary fuse failure: $\frac{1}{25{,}000}$ (hour^{-1})

Conditional repair intensity μ is estimated as

$$\mu = \tfrac{1}{2} \text{ (hour}^{-1}\text{)}$$

To obtain conservative results, the mean repair time, $1/\mu$, should be that to repair "broken fuse," "shorted wire," and "generator surge" because, without repairing all of them, we cannot return the fuse to the system. Calculate $R(1000)$ and $Q(1000)$.

Solution:

$$\lambda = \frac{1}{10{,}000} + \frac{1}{50{,}000} + \frac{1}{25{,}000} = 0.00016 \text{ (hour}^{-1}\text{)}$$

$$R(1000) = e^{-0.00016 \times 1000} = 0.852$$

$$Q(1000) = \frac{0.00016}{0.00016+0.5}[1-e^{-(0.00016+0.5)\times 1000}] = 3.20 \times 10^{-4}$$

APPENDIX 4.1 DERIVATION OF (4.35)

The failure during $[t, t+dt)$ occurs in a repair-to-failure process. Let s be a survival age of the component which is normal at time t. In other words, assume that the component has been normal since time $t-s$ and is normal at time t. The bridge rule of (4.9) can be written in integral form as

$$\Pr(A|c) = \int \Pr(A|s,C)P(s|C)\,ds \tag{1}$$

where $P(s|C)$ is the conditional probability density of B, given that event C occurs. The term $P(s|C)\,ds$ is the probability of "bridge $[s,s+ds)$]," and the term $\Pr(A|s,C)$ is the probability of the occurrence of event A when we have passed through the bridge. The integration in (1) is the representation of $\Pr(A|C)$ by the sum of all possible bridges. Define the following events and parameter s.

- A: Failure during $(t,t+dt)$.
- s: The normal component has the survival age s at time t.
- C: The component was as good as new at time zero and is normal at time t.

Since the component failure characteristics at time t are assumed to depend only on the survival age s at time t, we have

$$\Pr(A|s,C)=\Pr(A|s)=r(s)\,dt \tag{2}$$

Further, from the definition of $\lambda(t)$, we obtain

$$\Pr(A|C)=\lambda(t)\,dt \tag{3}$$

Substituting (2) and (3) into (1), we have

$$\lambda(t)\,dt=dt\cdot\int r(s)P(s|C)\,ds \tag{4}$$

For the constant failure rate r, we have

$$\lambda(t)\,dt=dt\cdot r\cdot\int P(s|C)\,ds=dt\cdot r\cdot 1 \tag{5}$$

yielding (4.35).

APPENDIX 4.2 DERIVATION OF (4.69)

Denote by $E(\cdot)$ the operation of taking the expected value. In general,

$$E(\chi(t))=E(\chi_{0,1}(t))-E(\chi_{1,0}(t)) \tag{1}$$

holds. The expected value $E(\chi(t))$ of $\chi(t)$ is

$$E(\chi(t))=1\times\Pr(\chi(t)=1)+0\times\Pr(\chi(t)=0) \tag{2}$$

$$=\Pr(\chi(t)=1) \tag{3}$$

yielding

$$E(\chi(t))=Q(t) \tag{4}$$

Since $\chi_{0,1}(t)$ is the number of failures to time t, $E(\chi_{0,1}(t))$ is the expected number of failures to that time.

$$E(\chi_{0,1}(t)) = W(0,t) \tag{5}$$

Similarly,

$$E(\chi_{1,0}(t)) = V(0,t) \tag{6}$$

Equations (1), (4), (5), and (6) yield (4.69).

APPENDIX 4.3 DISTRIBUTIONS

For a random variable X, the distribution $F(\chi)$ is defined by

$$F(\chi) = \text{probability of } X \text{ being less than } \chi$$

If X is a continuous variable, the probability density is defined as the first derivative of $F(\chi)$.

$$f(\chi) = \frac{dF(\chi)}{d\chi} \tag{1}$$

The small quantity $f(\chi)\,d\chi$ is the probability that a random variable takes the value in the interval $[\chi, \chi + d\chi)$. For a discrete random variable, the probability $\Pr(X = \chi_i)$ is denoted by $\Pr(\chi_i)$ and is given by

$$\Pr(\chi_i) = F(\chi_{i+1}) - F(\chi_i)$$

provided that

$$\chi_1 < \chi_2 < \chi_3 \cdots$$

Different families of distribution may be described by their particular parameters. However, as an alternative one may use the values of certain related measures such as the mean, median, mode, etc.

Mean. The *mean*, sometimes called the *expected value* $E(X)$, is the average of all the values that make up the distribution. Mathematically, it may be defined as $\int_{-\infty}^{\infty} xf(x)\,dx$ if X is a continuous random variable with probability density function $f(x)$, and as $\sum_i x_i \Pr(x_i)$ if x is a discrete random variable with probability function $\Pr(x_i)$.

Median. The *median* is the midpoint of the distribution. For a continuous *Pdf*, $f(x)$, with midpoint Z, this is

$$\int_{-\infty}^{Z} f(x)\,dx = 0.5$$

and for a discrete random variable it is

$$\sum_{i=1}^{Z} P(x_i) = 0.5$$

Mode. The *mode* for a continuous variate is the value associated with the maximum of the probability density function, and for a discrete random variable it is that value of the random variable that has the highest probability.

The approximate relationship among mean, median, and mode is shown graphically for three different probability densities in Fig. 4.38.

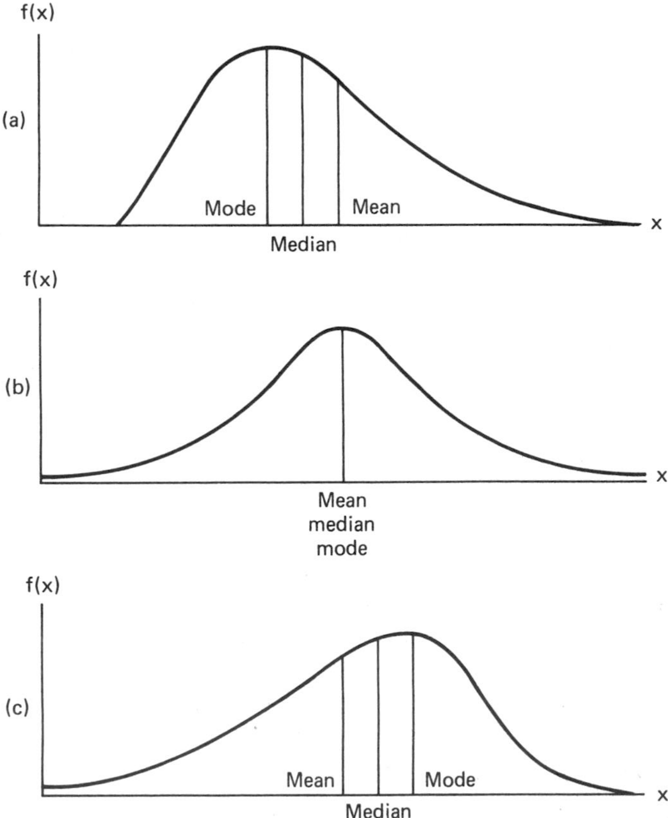

Figure 4.38.

Variance and Standard Deviation. In addition to the measures of tendency discussed above, it is often necessary to describe the *distribution spread*, *symmetry*, and *peakedness*. One such measure is the moment which is defined for the Kth moment about the mean as

$$\mu_k = E[X - E(X)]^k$$

where μ_k is the Kth moment and $E(X)$ is the mean or expected value of X. The second moment about the mean and its square root are measures of dispersion and are called *variance* and *standard deviation*, respectively. Hence, the variance is given by

$$E[X - E(X)]^2$$

which may be proved to be

$$E(X^2) - E(X^2)$$

The standard deviation is the square root of the above expression and is denoted by σ.

Exponential Distribution. *Exponential distributions* are used frequently for the analysis of time-dependent data when the rate at which events occur does not vary. The defining equation for $f(t)$, $F(t)$, $r(t)$ and their graphs for the exponential distribution and other distributions discussed here are shown in Table 4.11.

Normal and Log-Normal Distributions. The *normal* (*or Gaussian*) distribution is the best known two-parameter distribution. All normal distributions are symmetric, and the two parameters of the distribution, μ and σ, are its mean and standard deviation.

Normal distributions are frequently used to describe equipment which has increasing failure rates with time. The equation for $f(t)$, $F(t)$, and $r(t)$ for a normal distribution is shown in Table 4.11. The mean time to failure, μ, is obtained by simple averaging and is frequently called the *first moment*. The standard deviation, σ, is calculated as follows:

$$\sigma = \sqrt{\frac{(t_1 - \mu)^2 + (t_2 - \mu)^2 + \ldots + (t_n - \mu)^2}{N - 1}}$$

where t_i is the time to failure for sample n_i, N is the total population, and μ is the arithmetic mean of t_i's.

$F(t)$ is difficult to evaluate; however, there are tabulations of integrals in statistics and/or reliability texts. Special graph paper, which can be used to transform an S-shaped $F(t)$ curve to a straight line function, is available.

A *log-normal distribution* is similar to a normal distribution with the exception that the logarithms of the values of the random variables, rather than the values themselves, are assumed to be normally distributed. Thus, all values are positive, the distribution is skewed to the right, and the skewness is a function of σ. The availability of log-normal probability paper makes it relatively easy to test experimental data to see whether they are distributed log-normally.

Log-normal distributions are encountered frequently in metal fatigue testing, maintainability data (time to repair), and chemical process equipment failures and repairs.

Weibull Distribution. Among all the distributions available for reliability calculation, the *Weibull distribution* is regarded as unique to the field. In his original paper, "A Distribution of Wide Applicability," Professor Weibull argued that normal distributions require that initial metallurgical strengths be normally distributed and that what was needed was a function that could embrace a great variety of distributions (including the normal).

The Weibull distribution is a three parameter (γ, σ, β) distribution (unlike the normal which has only two), where:

γ = the time at which $F(t)=0$ and is a datum parameter; i.e., failures start occurring at time t,
σ = the characteristic life and is a scale parameter,
β = a shape parameter.

As can be seen from Table 4.11, the Weibull distribution assumes a great variety of shapes. If $\gamma=0$, i.e., $F(t)=0$ at time zero, or if the time axis is shifted to conform to this requirement, then we see that

1) For $\beta<1$, we have a decreasing failure rate (such as may exist at the beginning of a bathtub curve).
2) For $\beta=1, r(t)=\lambda$, and we have an exponential reliability curve.
3) For $1<\beta<2$ (not shown), we have a skewed (low failure rates at low times) normal distribution.
4) For $\beta>2$, the curve approaches a normal distribution.

By suitable arrangement of the variables, σ can be obtained by reading the value of $(t-\gamma)$ at $F(t)=63\%$. Methods for obtaining γ, which must frequently be guessed, have been published.[4]

[4]Hahn and Shapiro (op cit.).

Poisson Distribution. The *Poisson distribution* is an approximation to the binomial distribution function for a large number of samples. In reality, it is a form of a multi-state, discrete function and is used to calculate the probability of a certain number of events occurring in a large system. One would use the $F(t)$ in Table 4.11 to calculate, for example, the probability of a certain number of failed pieces, given a constant failure rate λ and a time t.

For this situation, the $F(t)$ is really an $F(n)$, i.e., the probability of having n or less failures in time t. In an expansion of $F(n)$, i.e.,

$$F(n)=e^{-\lambda t}+\lambda t e^{-\lambda t}+\frac{(\lambda t)^2}{2}e^{-\lambda t}\ldots\frac{+(\lambda t)^n e^{-\lambda t}}{n!}$$

the first term defines the probability of no component failures, the second term defines the probability of one component failure, etc.

APPENDIX 4.4 COMPUTATIONAL PROCEDURE FOR INCOMPLETE TEST DATA

When N items which fail, in turn, at lives $t_1, t_2, \ldots, t_n$ are being considered, then the probability of failure at lifetime t_1 can be approximated by $F(t_1)=1/N$, at lifetime t_2 by $F(t_2)=2/N$, and, in general by $F(t_r)=r/N$.

The above approximation is applicable when all the items concerned continue to failure. However, in many cases some items are taken out of use for reasons other than failures, hence affecting the numbers exposed to failures at different lifetimes. Therefore, a correction to take this into account must be included in the calculation.

Suppose that N items have been put into use and failure occurs at lives $t_1, t_2, t_3, \ldots,$ the number of failures occurring at each lifetime being $r_1, r_2, r_3, \ldots,$ and the number of items actually exposed to failure at each lifetime being $N_1, N_2, N_3, \ldots.$

At t_1 the proportion expected to fail is r_1/N_1, and this gives us our first point on the curve. We now calculate the number that would have failed at t_2, provided the original number (N_1) had been allowed to proceed to failure.

Since r_1 failed at t_1, the original number has been reduced to $N_1 - r_1$. The proportion actually failing at t_2 is r_2/N_2, so the number that would have failed, had N_1 proceeded to failure, is

$$(N_1 - r_1)\frac{r_2}{N_2}$$

and the proportion of N_1 expected to fail at t_2 is

$$(N_1 - r_1)\frac{r_2}{N_2 N_1}$$

We now proceed in the same manner to estimate the proportion of N_1 that would fail at t_3.

If the original number had been allowed to proceed to failure, the number exposed to failure at t_3 would be

$$N_1 - \left[r_1 + (N_1 - r_1)\frac{r_2}{N_2} \right]$$

and the proportion of N_1 expected to fail at t_3 is

$$\left\{ N_1 - \left[r_1 + (N_1 - r_1)\frac{r_2}{N_2} \right] \right\} \frac{r_3}{N_1 N_3}$$

The same process can be repeated for subsequent values.

APPENDIX 4.5 MEDIAN RANK PLOTTING POSITION[5]

In order to approximate the distribution function, plotting points are used, each corresponding to a cumulative probability F and associated with one of the order statistics, the values arranged in increasing order. For n observations in the sample, let

$$X_i, \quad i = 1, \ldots, n, \qquad X_1 < \ldots < X_n$$

denote the order statistics. Then corresponding to each X_i, we have the plotting point

$$\hat{F}_i = \frac{i - \frac{1}{2}}{n}$$

which gives the approximate probability that a value is below X_i. The use of the caret symbol (ˆ) denotes that the value is estimated.

[5] Locks, M. (op cit.).

APPENDIX 4.6 BASIC DEFINITIONS OF FAILURE AND REPAIR

Failure			Repair		
$R(t)$	Reliability	Probability that a component does not fail during an interval $(0,t]$, given that it was as good as new at time $t=0$.			
$F(t)$	Unreliability	Probability that a component experiences its first failure during the interval $(0,t]$, given that it was as good as new at time $t=0$.	$G(t)$	Repair distribution	Probability that a component is repaired to time t, given that the component failed at time $t=0$.
$f(t)$	Failure density	Probability the first component failure occurs during the small interval $[t,t+dt)$.	$g(t)$	Repair density	Probability that the component's first repair is completed during $[t,t+dt)$, given the component failed at $t=0$. $g(t)=dG(t)/dt$.
$r(t)$	Failure rate	Probability that a component experiences a failure per unit time, given that the component was as good as new at $t=0$ and has survived to time t.	$m(t)$	Repair rate	Probability that the component is repaired per unit time at time t, given the component failed at time zero and has been failed to time t.
$A(t)$	Availability	Probability of a component working at time t, given it was as good as new at time $t=0$.			

Failure			Repair		
$Q(t)$	Unavailability	Probability that a component is in a failed state at time t, given it was as good as new at $t=0$.			
$w(t)$	Unconditional failure intensity	Probability that a component fails per unit time at time t, given it is working at time t.	$v(t)$	Unconditional repair intensity	Probability that a component is repaired per unit time at time t, given it was as good as new at time $t=0$.
$W(t_1,t_2)$	Expected number of failures	Number of failures over a time interval t_1 to t_2. Equals $\int_{t_1}^{t_2} w(t)\,dt$.	$V(t_1,t_2)$	Expected number of repairs	Integrated value of $v(t)$ in a time interval, given a component was as good as new at time $t=0$.
$\lambda(t)$	Conditional failure intensity	The proportion of the components that is expected to fail per unit time at time t, given it was as good as new at time $t=0$ and is normal at time t.	$\mu(t)$	Conditional repair intensity	Probability that a component is repaired per unit time at time t, given that it was as good as new at time $t=0$ and is failed at time t.
TTF	Time to fail	Span of time from repair to first failure.	TTR	Time to repair	Length of time from failure to succeeding first repair.
MTTF	Mean time to failure	Expected value of the time to failure MTTF $= \int_0^\infty tf(t)\,dt$.	MTTR	Mean time to repair	Expected value of the time to repair $\int_0^\infty tg(t)\,dt$.
MTBF	Mean time between failures	Expected length of time between two consecutive failures.	MTBR	Mean time between repairs	Expected length of time between two consecutive failures.

PROBLEMS

4.1. (a) Suppose we are presented with two indistinguishable urns. Urn I contains 30 red balls and 70 green ones, and Urn II contains 50 red balls and 50 green ones. One urn is selected at random and a ball withdrawn. What is the probability that the ball is red?
(b) Suppose the ball drawn was red. What is the probability of its being from Urn I? (Kapur and Lamberson, op cit., p. 371.)

4.2. Calculate, using the mortality data of Table 4.1, the reliability $R(t)$, failure density $f(t)$, and failure rate $r(t)$ for:
(a) A man living to be 60 years old.
(b) A man living to be 15 years and 1 day after his 60th birthday.

4.3. Calculate values for $R(t)$, $F(t)$, $r(t)$, $A(t)$, $Q(t)$, $w(t)$, $W(0,t)$, and $\lambda(t)$ for the 10 components of Fig. 4.8 at 3 hr and 8 hr.

4.4. Prove that

$$\text{MTTF}=\int_0^\infty R(t)\,dt, \quad \text{assuming } \lim_{t\to\infty} tR(t)=0$$

4.5. Using the values shown in Fig. 4.8, calculate $G(t)$, $g(t)$, $m(t)$, and MTTR.

4.6. Use the data of Fig. 4.8 to obtain $\mu(t)$ and $v(t)$ at $t=3$ and also $V(0,t)$.

4.7. Obtain $f(t)$, $r(t)$, $g(t)$, and $m(t)$, assuming

$$F(t)=1-\tfrac{8}{7}e^{-t}+\tfrac{1}{7}e^{-8t}$$

$$G(t)=1-e^{-6t}$$

4.8. Suppose that

$$f(t)=\tfrac{1}{2}(e^{-t}+3e^{-3t})$$

$$g(t)=1.5e^{-1.5t}$$

(a) Show that the following $w(t)$ and $v(t)$ satisfy the simultaneous equations of (4.64).

$$w(t)=\tfrac{1}{4}(3+5e^{-4t})$$

$$v(t)=\tfrac{3}{4}(1-e^{-4t})$$

(b) Obtain $W(0,t)$, $V(0,t)$, $Q(t)$, $\lambda(t)$, and $\mu(t)$.
(c) Confirm (4.81).

4.9. A device has a constant failure rate of $\lambda=10^{-5}$ failures per hour.
(a) What is its reliability for an operating period of 1000 hr?
(b) If there are 1000 such devices, how many will fail in 1000 hr?

(c) What is the reliability for an operating time equal to the MTTF?
(d) What is the probability of its surviving for an additional 1000 hr, given it has survived for 1000 hr?

4.10. Suppose that

$$f(t)=\tfrac{1}{2}(e^{-t}+3e^{-3t})$$

$$g(t)=1.5e^{-1.5t}$$

Obtain $w(t)$ and $v(t)$, using the inverse Laplace transforms

$$L^{-1}\left[\frac{1}{(s+a)(s+b)}\right]=\frac{1}{b-a}(e^{-at}-e^{-bt})$$

$$L^{-1}\left[\frac{s+Z}{(s+a)(s+b)}\right]=\frac{1}{b-a}[(Z-a)e^{-at}-(Z-b)e^{-bt}]$$

4.11. Given a component for which the failure rate is 0.001 hr^{-1} and the mean time to repair is 20 hr, calculate the parameters of Table 4.10 at 10 hr and 1000 hr.

4.12. (a) Using the failure data for 1000 B-52 aircraft given below, obtain $R(t)$. (Shooman, op cit., p. 168.)

Time to Failure (hr)	Number of Failures
0–2	222
2–4	45
4–6	32
6–8	27
8–10	21
10–12	15
12–14	17
14–16	7
16–18	14
18–20	9
20–22	8
22–24	3

(b) Determine if the above data can be approximated by an exponential distribution, plotting log $[1/R(t)]$ against t.

4.13. (a) Determine a Weibull distribution for the data in Problem 4.12, assuming that $\gamma=0$.
(b) Estimate the number of failures to $t=0.5$ and $t=30$, assuming that the aircraft were non-repairable.

4.14. A thermocouple fails 0.35 times per year. Obtain the failure rate λ, assuming that (a) $\mu=0$ and (b) $\mu=1\ \text{day}^{-1}$, respectively.

5

CONFIDENCE LIMITS FOR RELIABILITY PARAMETERS

5.1 INTRODUCTION

When the statistical distribution of a failure or repair characteristic of a population is known, the probability of obtaining samples having a particular characteristic can be calculated. However, as mentioned in the preceeding chapter, measurement of the characteristic of every item in a population is seldom possible because such a determination would be too time-consuming and expensive, particularly if the measurement destroys the sample. Thus, methods for estimating the characteristics of a population for a given set of sample data are required.

It is difficult to generalize about a given population when we measure only the characteristic of a sample because that sample may not be representative of the population. As the sample size increases, the values of the sample and those of the population will, of course, agree more closely.

Although we cannot be certain that a sample is representative of a population, it is usually possible to associate a *degree of assurance* with a sample characteristic. That degree of assurance is called *confidence* and can be defined as the level of certainty associated with conclusions based on the results of sampling.

To illustrate the above statements, suppose a set of 10 identical components are life-tested for a specified length of time. At the end of the test, we observe five survivors. Based on these experiments, we would expect that the components have an average reliability of 0.5. However, that is far from certain. We would not be surprised if the true reliability was 0.4, but we would deem it unlikely that the reliability was 0.01 or 0.99.

We can associate a *confidence limit* to probabilistic parameters like reliability. That is, we can say that we are $(1-\alpha)$ confident that the true reliability is at least (or at most) a certain value. Thus α, *the level of significance*, must lie between zero and one, and $(1-\alpha)$ is a limit that refers to one end of the reliability; i.e. a *one-sided* or *single-sided* confidence limit, where the *range* is known as the *confidence interval.*

Figure 5.1 illustrates one-sided and two-sided confidence limits (note that for single-sidedness the confidence is $1-\alpha$ and for double-sidedness it is $1-2\alpha$). We see that 19 out of 20 upper confidence limits include the true reliability, whereas 18 out of 20 are contained in the double-sided interval. Note that the confidence interval varies according to the results of life-tests. For example, if we have no test survivors, the confidence interval would be located around zero; if there were only survivors, the value would be around unity.

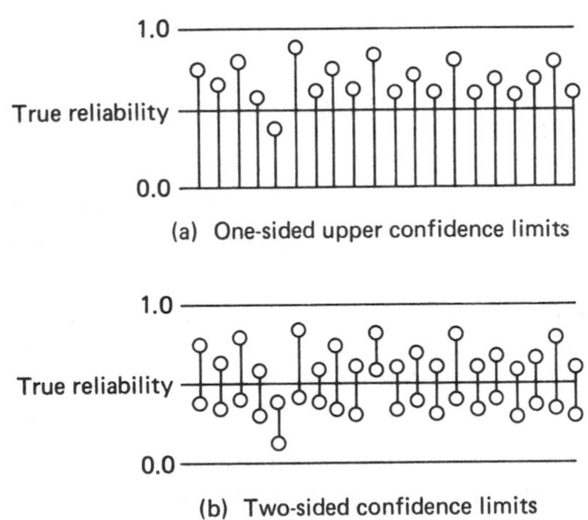

Figure 5.1. Illustration of confidence limits.

5.2 CLASSICAL CONFIDENCE LIMITS

5.2.1 General Principles

Suppose that N random samples $X_1, X_2, \ldots, X_N$ are taken from a population with unknown statistical characteristics (for example, mean and standard deviation). Let the characteristic of the population be represented by an unknown constant parameter θ. The *measured characteristic* $S = g(X_1, \ldots, X_n)$ of the samples may vary considerably from the *true character-*

istic θ. However, S has a probability distribution $F(s;\theta)$ or density $f(s;\theta)$ which depends on θ, so we can say something about θ on the basis of the dependence. Probability distribution $F(s;\theta)$ is the *sampling distribution* of S.

The classical approach uses the sampling distribution to determine two numbers, $s_1(\theta)$ and $s_2(\theta)$, as a function of θ, such that

$$\int_{s_2(\theta)}^{\infty} f(s;\theta)\,ds = \alpha \tag{5.1}$$

$$\int_{-\infty}^{s_1(\theta)} f(s;\theta)\,ds = \alpha \tag{5.2}$$

Figure 5.2 illustrates this definition of $s_1(\theta)$ and $s_2(\theta)$ for a particular θ. Note that (5.1) and (5.2) are equivalent, respectively, to

$$\Pr(S \le s_2(\theta)) = 1-\alpha \tag{5.3}$$

and

$$\Pr(s_1(\theta) \le S) = 1-\alpha \tag{5.4}$$

Since constant α is generally less than 0.5,

$$s_1(\theta) < s_2(\theta) \tag{5.5}$$

Equations (5.3) and (5.4) yield another probability expression,

$$\Pr(s_1(\theta) \le S \le s_2(\theta)) = 1-2\alpha \tag{5.6}$$

Although equations (5.3), (5.4), and (5.6) do not include explicit inequalities for θ, they can be rewritten to express confidence limits for θ.

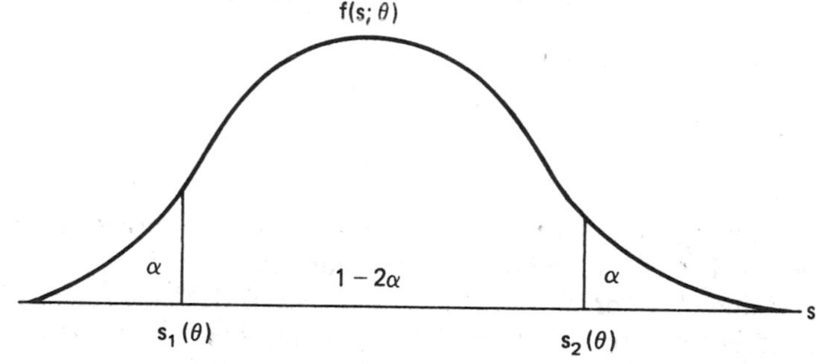

Figure 5.2. Quantities $s_1(\theta)$ and $s_2(\theta)$ for given θ.

Example 1.

Table 5.1 lists 20 samples, $X_1,\ldots,X_{20}$, from a normal population with unknown mean θ and known standard deviation $\sigma=1.5$. Let $S=g(X_1,\ldots,X_N)$ be the arithmetical mean $\bar{X}$ of $N=20$ samples $X_1,\ldots,X_N$ from the population:

$$S=\bar{X}=\frac{1}{N}\sum_{i=1}^{N} X_i=0.6473 \tag{5.7}$$

Obtain $s_1(\theta)$ and $s_2(\theta)$ corresponding to the 95% single-sided confidence limit, or 90% confidence interval, $\alpha=0.05$.

TABLE 5.1 TWENTY SAMPLES FROM A NORMAL POPULATION WITH UNKNOWN MEAN θ AND KNOWN STANDARD DEVIATION $\sigma=1.5$

X_1	0.090	X_{11}	0.049
X_2	−0.105	X_{12}	0.588
X_3	2.280	X_{13}	−0.693
X_4	−0.051	X_{14}	5.310
X_5	0.182	X_{15}	1.280
X_6	−1.610	X_{16}	1.790
X_7	1.100	X_{17}	0.405
X_8	−1.200	X_{18}	0.916
X_9	1.130	X_{19}	−1.200
X_{10}	0.405	X_{20}	2.280

Solution: Sample mean $\bar{X}$ is a normal random variable with mean θ and standard deviation $\sigma/\sqrt{N}=1.5/\sqrt{20}=0.3354$. Normal distribution tables indicate that it is 95% certain that the sample mean is not more than $(\theta+2.58\sigma/\sqrt{N})$:

$$\Pr(\bar{X}\le\theta+0.8651)=0.95 \tag{5.8}$$

Similarly, we are also 95% confident that $\bar{X}$ is not less than $(\theta-2.58\sigma/\sqrt{N})$:

$$\Pr(\theta-0.8651\le\bar{X})=0.95 \tag{5.9}$$

Thus, $s_1(\theta)$ and $s_2(\theta)$ are given by

$$s_1(\theta)=\theta-0.8651 \tag{5.10}$$

$$s_2(\theta)=\theta+0.8651 \tag{5.11}$$

Assume that $s_1(\theta)$ and $s_2(\theta)$ are the monotonically increasing functions of θ shown in Fig. 5.3 (similar representations are possible for monotonically decreasing cases or more general cases). Consider now rewriting (5.3), (5.4), and (5.6) in a form suitable for expressing confidence limits. Equation (5.3) shows that the random variable $S=g(X_1,\ldots,X_N)$ is less than $s_2(\theta)$ with probability $(1-\alpha)$ when we repeat a large number of experiments, each of which yields possibly different sets of N observations

$X_1,\ldots,X_N$. We now define a new random variable θ_2, related to S, such that

$$s_2(\Theta_2) = S \tag{5.12}$$

where S is the observed characteristic and $s_2(\theta)$ the known function of θ. Or equivalently,

$$\Theta_2 = s_2^{-1}(S) \tag{5.13}$$

Variable Θ_2 is illustrated in Fig. 5.3. The inequality $S \le s_2(\theta)$ describes the fact that variable Θ_2, thus defined, falls on the left-hand side of constant θ:

$$S \le s_2(\theta) \Leftrightarrow \Theta_2 \le \theta$$

Hence, from (5.3),

$$\Pr(\Theta_2 \le \theta) = 1 - \alpha \tag{5.14}$$

This shows that random variable Θ_2 determined by S and curve $s_2(\theta)$ is a $(1-\alpha)$ lower confidence limit; variable Θ_2 becomes a lower bound for unknown constant θ, with probability $(1-\alpha)$.

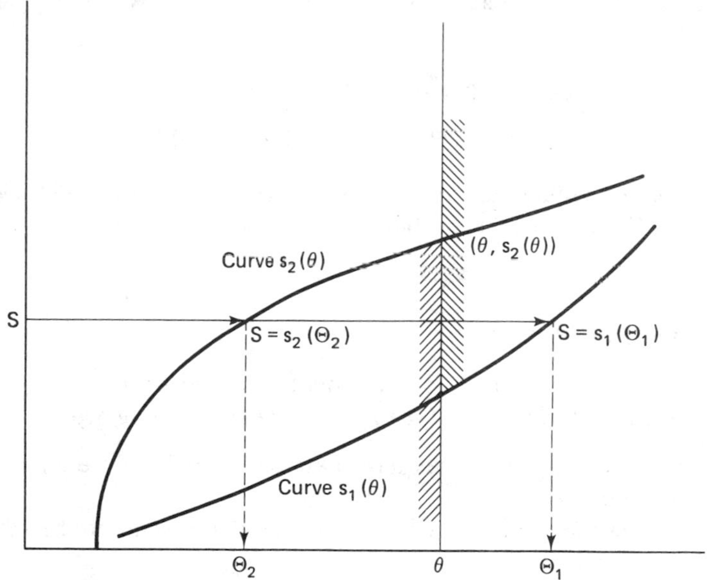

Figure 5.3. Illustration of random variable Θ_2 determined from static S and curve $s_2(\theta)$.

Similarly, we define another random variable Θ_1 by

$$s_1(\Theta_1)=S \tag{5.15}$$

where S is the observed characteristic and $s_1(\theta)$ the known function of θ. Or, equivalently,

$$\Theta_1=s_1^{-1}(S) \tag{5.16}$$

Random variable Θ_1 is illustrated in Fig. 5.3. Equation (5.4) yields

$$\Pr(\theta \le \Theta_1)=1-\alpha \tag{5.17}$$

Thus, variable Θ_1 gives an upper, single-sided confidence limit for constant θ.

Combining (5.14) with (5.17), we have

$$\Pr(\Theta_2 \le \theta \le \Theta_1)=1-2\alpha \tag{5.18}$$

Random interval $[\Theta_2, \Theta_1]$ becomes the $100(1-2\alpha)\%$ confidence interval. In other words, the interval includes true parameter θ with probability $1-2\alpha$.

Example 2. Confidence Limits for the Mean of a Normal Population

Obtain the 95% single-sided limit and the 90% double-sided interval for the population mean θ in Example 1.

Solution: Equations (5.10) and (5.11) and the definition of Θ_1 and Θ_2 [see (5.12) and (5.15)] yield

$$\Theta_1=\bar{X}+0.8651=0.6473+0.8651=1.512 \tag{5.19}$$

$$\Theta_2=\bar{X}-0.8651=0.6473-0.8651=-0.218 \tag{5.20}$$

Variables Θ_1 and Θ_2 are the 95% upper and lower single-sided confidence limits, respectively. The double-sided confidence interval is

$$[\Theta_2, \Theta_1]=[-0.218, 1.512]$$

5.2.2 Types of Life Tests

Suppose that N identical components are placed on life tests. The two test options, which are called *Type I and Type II censoring* are:[1]

Type I censoring. Life test is terminated at time T before all n components have failed.

Type II censoring. Life test is terminated at the time of the rth, $r \le N$ failure.

[1]Mann, N. R., R. E. Schafer, and N. D. Singpurwalla, *Methods for Statistical Analysis of Reliability and Life Data*, John Wiley & Sons, Inc., New York, 1974.

In Type I censoring T is fixed, and the number of failures r and all the failure times $t_1 \le t_2 \le \ldots \le t_r \le T$ are random variables. In Type II censoring the number of failures r is fixed, and the r failure times and $T = t_r$ are random variables.

5.2.3 Lower and Upper Confidence Limits for Mean Time to Failure

Assume Type II censoring for N components, with an exponential time to failure distribution for each component. A point estimate $\hat{\theta}$ for the true mean time to failure θ, is

$$\hat{\theta} = \frac{(N-r)t_r + \sum_{i=1}^{r} t_i}{r} \quad (= S, \text{ the observed characteristic}) \qquad (5.21)^{\dagger}$$

It can be shown that $2rS/\theta$ follows a *chi-square distribution* with $2r$ degrees of freedom[1] (see Appendix 5.1 at the end of this chapter). Let $\chi^2_\alpha(2r)$ and $\chi^2_{1-\alpha}(2r)$ be the 100α and $100(1-\alpha)$ percentiles of the chi-square distribution obtained from standard chi-square tables.[1,2,3] From the definition of percentiles,

$$\Pr\left(\chi^2_\alpha(2r) \le \frac{2rS}{\theta}\right) = \alpha$$

$$\Pr\left(\chi^2_{1-\alpha}(2r) \le \frac{2rS}{\theta}\right) = 1-\alpha$$

or, equivalently,

$$\Pr\left(S \le \chi^2_\alpha(2r)\frac{\theta}{2r}\right) = 1-\alpha \qquad (5.22)$$

$$\Pr\left(\chi^2_{1-\alpha}(2r)\frac{\theta}{2r} \le S\right) = 1-\alpha \qquad (5.23)$$

The last two equations correspond to (5.3) and (5.4), respectively, and we

†This is called the *maximum likelihood estimator* for MTTF.

[2]Catherine, M. T., "Tables of the Percentage Points of the χ^2 Distribution," *Biometrika*, Vol. 32, pp. 188–189, 1941.

[3]Beyer, W. H. (ed.), *Handbook of Tables for Probability and Statistics*, 2nd ed., The Chemical Rubber Company, Cleveland, Ohio. 1968.

see that

$$s_2(\theta)=\chi^2_\alpha(2r)\frac{\theta}{2r}$$
$$s_1(\theta)=\chi^2_{1-\alpha}(2r)\frac{\theta}{2r} \tag{5.24}$$

Equations (5.12) and (5.15) show that Θ_2 and Θ_1 are obtained from

$$\chi^2_\alpha(2r)\frac{\Theta_2}{2r}=S, \qquad \chi^2_{1-\alpha}(2r)\frac{\Theta_1}{2r}=S$$

yielding

$$\Theta_2=\frac{2rS}{\chi^2_\alpha(2r)}, \qquad \Theta_1=\frac{2rS}{\chi^2_{1-\alpha}(2r)} \tag{5.25}$$

Quantities Θ_2 and Θ_1 give $100(1-\alpha)\%$ as the lower and upper confidence limits, whereas the range $[\Theta_2, \Theta_1]$ becomes the 100 $(1-2\alpha)\%$ confidence interval.

We have followed rather laborious routes to the confidence limits in order to show the classical approach. Another way for obtaining the limits is to simply rewrite (5.22) as

$$\Pr\left(\frac{2rS}{\chi^2_\alpha(2r)}\leq\theta\right)=1-\alpha, \qquad \Pr\left(\theta\leq\frac{2rS}{\chi^2_{1-\alpha}(2r)}\right)=1-\alpha \tag{5.26}$$

Example 3. Confidence Limits for MTTF

Assume 30 identical components placed on Type II censoring with $r=20$. The 20th failure has a time to failure of 39.89 min, i.e., $T=39.89$, and the other 19 times to failure are listed in Table 5.2 along with times to failure which would occur if the test were to continue after the 20th failure. Find the 95% two-sided confidence limits for the MTTF.

Solution: $N=30$, $r=20$, $T=39.89$, $\alpha=0.025$.

$$\hat{\theta}=S=\frac{(30-20)\times 39.89+291.09}{20}=34.50$$

From the chi-square table in reference 1

$$\chi^2_\alpha(2r)=\chi^2_{0.025}(40)=59.3417$$
$$\chi^2_{1-\alpha}(2r)=\chi^2_{0.975}(40)=24.4331$$

Equation (5.25) yields

$$\Theta_2 = 2\times 20\times \frac{34.50}{59.3417} = 23.26$$

$$\Theta_1 = 2\times 20\times \frac{34.50}{24.4331} = 56.48$$

Then,

$$23.26 \leq \theta \leq 56.48$$

i.e., we are 95% confident that the mean time to failure (θ) is in the interval 23.26↔56.48. As a matter of fact, TTF's in Table 5.2 were generated from an exponential distribution with the MTTF=26.64. The confidence interval includes this true MTTF.

TABLE 5.2 TTF DATA FOR EXAMPLE 3

TTF's Up to 20th Failure				TTF's After 20th Failure	
t_1	0.26	t_{11}	11.04	t_{21}	(40.84)
t_2	1.49	t_{12}	12.07	t_{22}	(47.02)
t_3	3.65	t_{13}	13.61	t_{23}	(54.75)
t_4	4.25	t_{14}	15.07	t_{24}	(61.08)
t_5	5.43	t_{15}	19.28	t_{25}	(64.36)
t_6	6.97	t_{16}	24.04	t_{26}	(64.45)
t_7	8.09	t_{17}	26.16	t_{27}	(65.92)
t_8	9.47	t_{18}	31.15	t_{28}	(70.82)
t_9	10.18	t_{19}	38.70	t_{29}	(97.32)
t_{10}	10.29	t_{20}	39.89	t_{30}	(164.26)

The reliability of components with exponential distribution was shown to be

$$R(t) = e^{-\lambda t} = e^{-t/\theta} \tag{5.27}$$

Confidence intervals can be obtained by substituting Θ_1 and Θ_2 for θ; hence,

$$e^{-t/\Theta_2} \leq R(t) \leq e^{-t/\Theta_1} \tag{5.28}$$

Thus, for the data in Example 3,

$$e^{-t/23.26} \leq R(t) \leq e^{-t/56.48}$$

Similarly, the confidence interval for failure rate λ is given by

$$\frac{1}{\Theta_1} \leq \lambda \leq \frac{1}{\Theta_2} \tag{5.29}$$

5.2.4 Confidence Limits for Binomial Distributions

Assume N identical components placed in Type I censoring with r failures in test period T. We wish to obtain confidence limits for the component reliability $R(T)$ at time T. We begin by replacing static S by discrete random variable r, where

$$S = r \tag{5.30}$$

The sampling distribution of S is given by the binomial distribution

$$\Pr(s; R) = \frac{N!}{(N-s)!s!} R^{N-s}[1-R]^{s} \tag{5.31}$$

with $R = R(T)$ corresponding to unknown parameter θ in Section 5.2.1.

Equation (5.3) thus becomes

$$\Pr(S \leq s_2(R)) = \sum_{s=0}^{s_2(R)} \frac{N!}{(N-s)!s!} R^{N-s}[1-R]^{s} \geq 1-\alpha \tag{5.32}$$

Here, inequality "$\geq 1-\alpha$" is necessary, since S is discrete. The parameter $s_2(R)$ is defined as the smallest one satisfying (5.32).

A schematic graph of $s_2(R)$ is shown in Fig. 5.4. Note that the graph is a monotonically decreasing step function in R. We can define R_2 for any observed characteristic S as shown in Fig. 5.4.† This R_2 corresponds to the

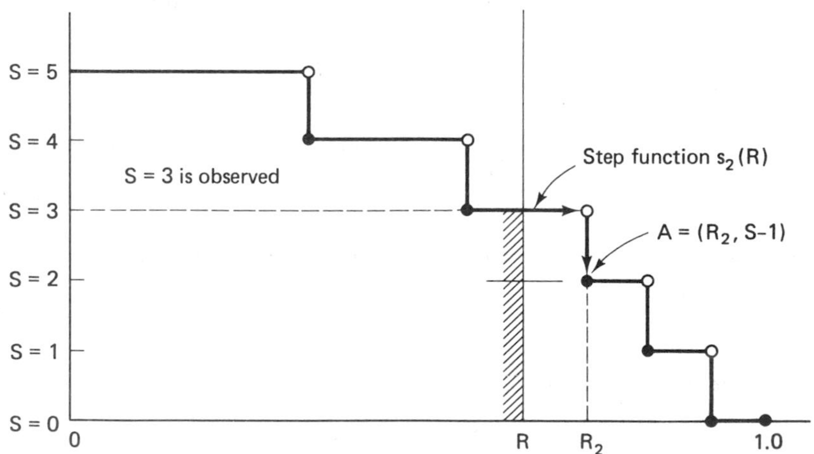

Figure 5.4. Quantity R_2 determined by S and step function $s_2(R)$.

†With the exception that R_2 is defined as unity when $S=0$ is observed.

Θ_2 in Fig. 5.3, where function $s_2(\theta)$ is monotonically increasing. The event $S \le s_2(R)$ occurs if and only if R_2 falls on the right-hand side of R.

$$S \le s_2(R) \leftrightarrow R \le R_2 \tag{5.33}$$

Thus,

$$\Pr(R \le R_2) \ge 1-\alpha \tag{5.34}$$

and R_2 gives $1-\alpha$, the upper confidence limit for reliability R.

Pair $(R_2, S-1)$ is represented by point A in Fig. 5.4. We note that at point A inequality (5.32) reduces to an equality for $S \ne 0$, since the value $s_2(R)$ decreases by one. Thus, the value of R_2 can be obtained for any given S by solving the following equation for R.

$$\sum_{s=S}^{N} \frac{N!}{(N-s)!s!} R^{N-s}[1-R]^s = \alpha, \quad \text{for } S \ne 0 \tag{5.35}$$

$$R_2 = 1, \quad \text{for } S=0 \tag{5.36}$$

The above equation can be solved for R by iterative methods, although tables have been compiled.[4]

Similarly to (5.35), lower confidence limit R_1 for R is given by the solution of the equation

$$\sum_{s=0}^{S} \frac{N!}{(N-s)!s!} R^{N-s}[1-R]^s = \alpha, \quad \text{for } S \ne N \ (R_1 = 0 \text{ for } S=N) \tag{5.37}$$

Example 4. Confidence Limit for Reliability

Assume a test situation which is go-no-go in nature with only two possible outcomes, success or failure. Suppose no failures have occurred during the life-test of N components to a specified time T. Estimate the lower confidence limit for component reliability at time T. (This situation would apply, for example, to the calculation of the probability of having a major nuclear disaster, given that none had ever occurred.)

Solution: Since $S=0$ in (5.37)

$$R^N = \alpha$$

Thus, the lower confidence limit is $R_1 = \alpha^{1/N}$. If $\alpha = 0.05$ and $N = 1000$, then $R_1 = 0.997$. That is, we are 95% confident that the reliability is not less than 0.997.

[4]Burington and May, *Handbook of Probability and Statistics, with Tables*, McGraw-Hill Book Company, New York, 1953.

Example 5. Confidence Limits for Reliability

Assume that $r=1$, $N=20$, and $\alpha=0.1$ in Example 4. Obtain upper and lower confidence limits.

Solution: Since $S=1$, equations (5.35) and (5.37) yield

$$R_2^N = 1-\alpha, \qquad R_2^{20}=0.9$$

$$R_1^N + NR_1^{N-1}[1-R_1]=\alpha, \qquad R_1^{20}+20R_1^{19}[1-R_1]=0.1$$

Thus,

$$R_2 = 0.9^{0.05} = 0.995$$

$$R_1 = 0.819 \quad \text{(from footnote 4)}$$

Thus, we are 80% sure that the true reliability is between 0.819 and 0.995.

5.3 BAYESIAN RELIABILITY AND CONFIDENCE LIMITS

In the previous sections, classical statistics were applied to test data to demonstrate the reliability of a system or component to a calculated degree of confidence. However, in many design situations the designer uses test data, combined with past experience, to meet or exceed a reliability specification. Since the application of classical statistics for predicting reliability parameters does not make use of past experience, an alternate approach is desirable. An example of where that new approach would be required is the case where a designer is redesigning a component similar to one designed previously, but with an improved reliability. Here, if we use the classical approach to predict a failure rate (with a given level of confidence) which is higher than the failure rate for the previous component, then the designer has obtained no really useful information; indeed he may simply reject the premise and its result. So, a method is needed which takes into consideration the designer's past experience.

One such method is based on *Bayesian statistics*, which combines *a priori* experience with hard *posterior* data to provide estimates similar to those obtained using the classical approach, which might simply combine the two.

5.3.1 Basic Concept

Bayes' theorem, in a modified and useful form, may be stated as:

Posterior probabilities $\propto$ *prior probabilities* $\times$ *likelihoods*,

where the symbol $\propto$ means "are proportional to." This relation may be formulated in a general form as follows: If

a) The A_i's are a set of mutually exclusive and exhaustive events, for $i=1,\ldots,n$;
b) $\Pr(A_i)$ is the prior probability of A_i before testing,
c) B is the observation; and
d) $\Pr(B|A_i)$ is the probability of the observation, given that A_i is true, then

$$\Pr(A_i|B)=\frac{\Pr(A_i)\Pr(B|A_i)}{\sum_i \Pr(A_i)\Pr(B|A_i)} \tag{5.38}$$

where $\Pr(A_i|B)$ is the *posterior* probability, meaning the probability of A_i now that B is known. Note that the denominator of (5.38) is simply a normalizing constant for $\Pr(A_i|B)$, ensuring $\Sigma\Pr(A_i|B)=1$.

The transformation from $\Pr(A_i)$ to $\Pr(A_i|B)$ is called the *Bayes' transform*. It utilizes the fact that the likelihood of $\Pr(B|A_i)$ is more easily calculated than $\Pr(A_i|B)$ itself. If we think of probability as a degree of belief, then our prior belief is changed, by the test evidence, to a posterior degree of belief.

To illustrate the application of Bayes' theorem let us consider a hypothetical example. Suppose that we are concerned about the reliability level of a new untested system. Based on past experience we believe there is an 80% chance that the system's reliability is $R_1=0.95$ and a 20% chance it is $R_2=0.75$. Now suppose that we test one system and find that it operates successfully. We would like to know the probability that the reliability level is R_1.

If we define S_i as the event that the system test results in a success then, for the first success S_i, we want $\Pr(R_1|S_1)$, using Bayes' equations:

$$\Pr(R_1|S_1)=\frac{\Pr(R_1)\Pr(S_1|R_1)}{\Pr(R_1)\Pr(S_1|R_1)+\Pr(R_2)\Pr(S_1|R_2)}$$

Substituting numerical values we find that

$$\Pr(R_1|S_1)=\frac{(0.80)(0.95)}{(0.80)(0.95)+(0.20)(0.75)}=0.835$$

Let us assume that a second system was tested and it also was successful. We now want

$$\Pr(R_1|S_1,S_2)=\frac{\Pr(R_1)\Pr(S_1,S_2|R_1)}{\Pr(R_1)\Pr(S_1,S_2|R_1)+\Pr(R_2)\Pr(S_1,S_2|R_2)}$$

which gives

$$\Pr(R_1|S_1,S_2)=\frac{(0.80)(0.95\times0.95)}{(0.80)(0.95\times0.95)+(0.20)(0.75\times0.75)}=0.865$$

Here the probability of event R_1 was updated by applying Bayes' theorem as new information became available.

5.3.2 Bayes' Theorem for Continuous Variables

Let

$x=$ the continuous valued parameter to be estimated,
$\mathbf{y}=(y_1,\ldots,y_N)$: N observations of x,
$p(x)=$ the *a priori* probability density of x,
$p(\mathbf{y}|x)=$ the probability or probability density of the observations given that x is true,
$p(x|\mathbf{y})=$ the *a posteriori* probability density of x.

From the definition of conditional probabilities,

$$p(x|\mathbf{y})=\frac{p(x,\mathbf{y})}{p(\mathbf{y})}=\frac{p(x,\mathbf{y})}{\int[\text{numerator}]\,dx} \tag{5.39}$$

The numerator can be rewritten as

$$p(x,\mathbf{y})=p(x)p(\mathbf{y}|x)$$

yielding Bayes' theorem for the continuous valued parameter x.

$$p(x|\mathbf{y})=\frac{p(x)p(\mathbf{y}|x)}{\int[\text{numerator}]\,dx} \tag{5.40}$$

Example 6. Uniform a Priori Distribution for Reliability

Suppose N components are placed in Type I censoring, where $r(\leq N)$ components failed before specified time T. Define y_i by

$$y_i=\begin{cases}1, & \text{if component } i \text{ failed}\\ 0, & \text{if component } i \text{ survived}\end{cases}$$

Obviously, $\Sigma y_i=r$. Obtain *a posteriori* density $p(R|\mathbf{y})$ for component reliability R at time T, assuming uniform *a priori* distribution in interval [0, 1].

Solution:

$$
\begin{aligned}
p(R) &= \begin{cases} 1, & \text{for } 0 \le R \le 1 \\ 0, & \text{otherwise} \end{cases} \\
p(\mathbf{y}|R) &= R^{N-r}[1-R]^{r}
\end{aligned}
\tag{5.41}
$$

Note that the binomial coefficient $\begin{bmatrix} N \\ r \end{bmatrix}$ is not necessary in the above equation, since the sequence $y_1, \ldots, y_N$ along with total failures r are given. In other words, observation (1,0,1) differs from (1,1,0) or (0,1,1).

$$
p(R|\mathbf{y}) = \begin{cases} \dfrac{R^{N-r}[1-R]^{r}}{\int [\text{numerator}]\, dR}, & \text{for } 0 \le R \le 1 \\ 0, & \text{otherwise} \end{cases}
\tag{5.42}
$$

This *a posteriori* density is a *Beta probability distribution*[1] (see Appendix 5.1 of this chapter). Note that the denominator of (5.42) is a constant when y is given.

Example 7.

Assume that three components are placed in a 10 hr test, and that two components, 1 and 3, failed. Calculate the *a posteriori* probability density for the component reliability at 10 hr, assuming a uniform *a priori* distribution.

Solution: Since components 1 and 3 failed,

$$
y = (1,0,1) \qquad N = 3, \qquad r = 2
$$

Equation (5.42) gives

$$
P(R|\mathbf{y}) = \begin{cases} \dfrac{R^{3-2}[1-R]^2}{\text{const.}} = \dfrac{R[1-R]^2}{\text{const.}}, & \text{for } 0 \le R \le 1 \\ 0, & \text{otherwise} \end{cases}
$$

The normalizing constant in the above equation can be found by

$$
\int_0^1 \frac{R[1-R]^2}{\text{const.}}\, dR = \frac{1}{\text{const.}} \times \frac{1}{12} = 1
$$

or

$$
\text{const.} = \tfrac{1}{12}
$$

Thus,

$$
P(R|\mathbf{y}) = \begin{cases} 12R[1-R]^2, & 0 \le R \le 1 \\ 0, & \text{otherwise} \end{cases}
$$

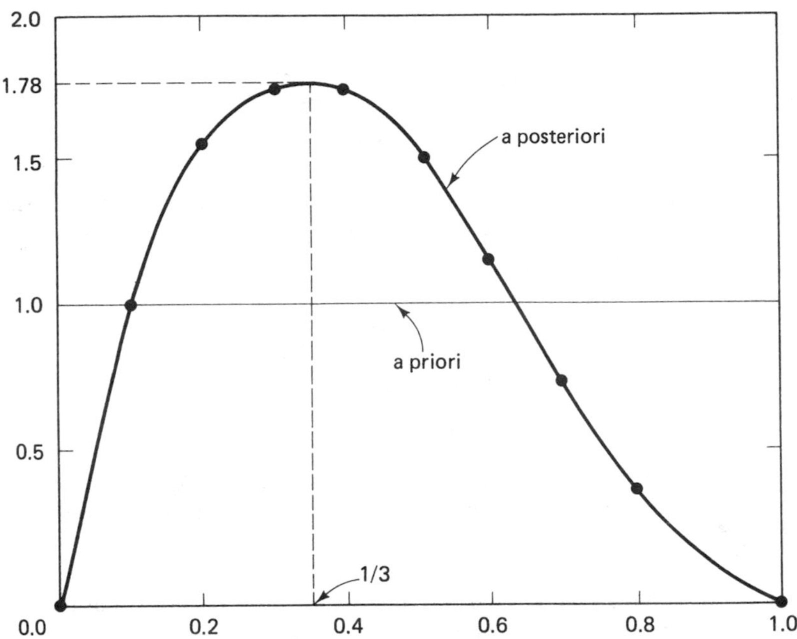

Figure 5.5. A priori and a posteriori densities.

The *a posteriori* and *a priori* density are plotted in Fig. 5.5. We see that the *a posteriori* density is biased toward zero reliability since two out of three components failed.

5.3.3 Confidence Limits

In general, the Bayesian one-sided confidence limit for parameter x based on hard evidence y may be defined as

$$\int_{-\infty}^{L(\mathbf{y})} p(x|\mathbf{y})\,dx = \alpha \tag{5.43}$$

$$\int_{U(\mathbf{y})}^{\infty} p(x|\mathbf{y})\,dx = \alpha \tag{5.44}$$

Quantities $L(\mathbf{y})$ and $U(\mathbf{y})$ are illustrated in Fig. 5.6 and are the constants when hard evidence $\mathbf{y}$ is given. Obviously, $L(\mathbf{y})$ is the Bayesian $(1-\alpha)$ lower confidence limit for x, and $U(\mathbf{y})$ is the Bayesian $(1-\alpha)$ upper confidence limit for x.

An interesting application of the Bayesian approach is in binomial testing situations, where a number of components are placed on test and the results are successes or failures (as described in Section 5.3.2). The Bayesian approach to the problem is to find the smallest $R_1 = L(\mathbf{y})$ in a

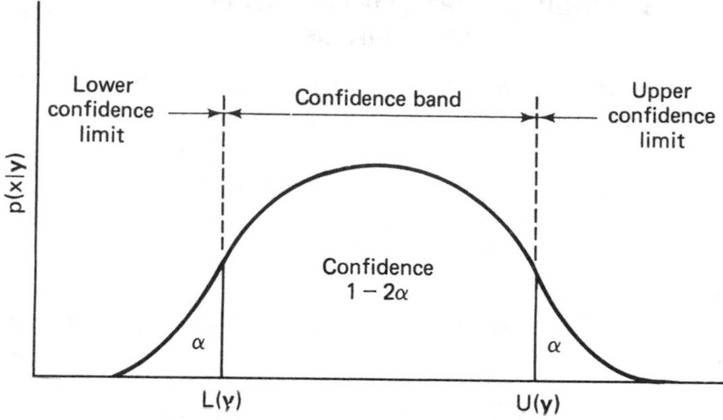

Figure 5.6. Bayesian confidence limits.

table of beta probabilities for $N-r$ successes and r failures, such that the Bayesian can say, "the probability that the true reliability is greater than R_1 is $(1-\alpha)\%$." Similar procedures yield upper bound $R_2 = U(\mathbf{y})$ for the reliability.

Example 8. Bayesian Confidence Limit for Reliability

To illustrate the Bayesian confidence limits, consider Example 5 in Section 5.2.4. Assume a uniform *a priori* distribution for reliability R. Obtain the 90% Bayesian lower confidence limit.

Solution: From (5.42) we see that (5.43) can be written as

$$\int_0^{R_1} \frac{R^{19}[1-R]}{\text{const.}}\,dR = 0.1$$

The beta probability table in footnote reference 5 gives $R_1 = 0.827$. That is, the probability that the true reliability is greater than 0.827 is 90%. Note that the reliability obtained in Example 5 by applying the binomial distribution was 0.819 with 90% confidence. We achieved an improved lower confidence limit by applying Bayesian techniques.

The Bayesian approach applies to confidence limits for reliability parameters such as reliability, failure rate, mean time to failure, etc. The reader is referred to footnote references 1, 6, and 7 for further details.

[5]Harter, H. L., "New Tables of the Incomplete Gamma-Function Ratio and of Percentage Points of the Chi-Square and Beta Distribution," U. S. Government Printing Office, Washington, D. C., 1964.

[6]Walter, R. A., H. F. Martz, Jr., "Bayesian Reliability Estimation: State of the Art for the Time-Dependent Case," Los Alamos Scientific Laboratory, LA-6003, 1975. (Available from National Technical Information Service.)

[7]Martz, H. F., Jr., and R. A. Walter, "The Basics of Bayesian Reliability Estimation from Attribute Test Data," Los Alamos Scientific Laboratory, LA-6126, 1975. (Available from NTIS.)

APPENDIX 5.1 THE χ^2 DISTRIBUTION AND BETA DISTRIBUTION

1. The χ^2 distribution with m degrees of freedom:
   (a) Probability density

$$P(\chi)=\frac{1}{P(m/2)\sqrt{2}\ m}\chi^{(m-2)/2}e^{-\chi/2},\quad \chi\geq 0$$

   (b) Mean $= m$.
   (c) 100α and $100(1-\alpha)$ percentiles, $\chi^2_\alpha(m)$ and $\chi^2_{1-\alpha(m)}$, are defined by

$$\int_0^{\chi^2_\alpha(m)} P(\chi)\,d\chi=\alpha,\qquad \int_0^{\chi^2_{1-\alpha}(m)} P(\chi)\,d\chi=1-\alpha$$

2. The beta distribution:
   (a) Probability density

$$P(\chi)=\frac{\Gamma(m+k)}{\Gamma(m)\Gamma(k)}\chi^{m-1}(1-\chi)^{k-1}\,d\chi,\qquad 0\leq\chi\leq 1$$

   (b) Mean $=\dfrac{m}{m+k}$.

PROBLEMS

5.1. Assume 30 samples, X_i, $i=1,\ldots,30$, from a normal population with unknown mean θ and unknown standard deviation σ:

−0.112,	−0.265,	−0.937,	0.064,	1.236
−1.317,	−1.239,	−0.061,	1.508,	−1.165
−0.082,	0.254,	1.742,	1.706	−1.659
1.211,	−1.532,	1.127,	−0.741,	−0.097
−1.736,	0.252,	−0.379,	−0.875,	−0.598
1.600,	0.694,	0.401,	−1.098,	−0.430

The parameters θ and σ can be estimated respectively by

$$\bar{X}=\frac{1}{30}\sum_{i=1}^{30}X_i,\qquad \bar{\sigma}=\sqrt{\frac{\left[\sum X_i^2\right]-N\bar{X}^2}{N-1}}=\sqrt{\frac{\left[\sum X_i^2\right]-30\bar{X}^2}{29}}$$

(a) Obtain $\bar{X}$ and $\bar{\sigma}$.
(b) Obtain $s_1(\theta)$ and $s_2(\theta)$ for $\alpha=0.05$, using $\bar{\sigma}$ as the true standard deviation σ.
(c) Determine the 90% double-sided confidence interval for the mean θ.

5.2. A test of 15 identical components produced the following times to failures (hours):

118.2,	128.4,	17.0,	161.6,	33.8
55.1,	68.5,	74.7,	15.0,	0.7
25.5,	158.5,	335.5,	306.8,	15.2

(a) Obtain the times to failures for Type I censoring with $T=70$.
(b) Obtain the times to failures for Type II censoring with $r=10$.
(c) Find the 90% two-sided confidence interval of MTTF for Type II censoring, assuming an exponential failure distribution and the chi-square table:

$x_{\gamma}^{2}(\nu)$

ν/γ	0.025	0.05	0.950	0.975
10	3.247	3.940	18.307	20.483
15	6.262	7.261	24.996	27.488
20	9.591	10.851	31.410	34.170

(d) Obtain 90% confidence intervals for the component failure rate λ and the component reliability at $t=100$, assuming Type II censoring.

5.3. A total of 10 identical components were tested using Type I censoring which was terminated at 40 hr. Four components failed during the test. Obtain algebraic equations for 95% upper and lower confidence limits of the component reliability at $t=40$ hr.

5.4. Assume that we are concerned about the reliability of a new system. Past experience gives the following *a priori* information:

Reliability	Probability
$R_1=0.98$	0.6
$R_2=0.78$	0.3
$R_3=0.63$	0.1

Now suppose that we test the two systems and find that the first system operates successfully and the second one fails. Determine the probability that the reliability level is R_i, based on these two test results.

5.5. An *a priori* probability density of reliability R is given by

$$P(R)=\frac{R^{M-m}[1-R]^{m}}{\text{const.}}, \quad M \geq m$$

Prove that the constant is

$$\frac{m!}{(M+1)\times M\times(M-1)\times\ldots\times(M-m+1)}$$

5.6. A Type I censoring of 100 components resulted in a component failure within 200 hr.

(a) Obtain an *a posteriori* distribution $P(R|y)$ of the component reliability at 200 hr, assuming *a priori* information

$$P(R)=\frac{R^{28}(1-R)^2}{\text{const.}}$$

(b) Obtain the mean values of the reliability which distributes according to $P(R)$ and $P(R|y)$, respectively.

(c) Obtain reliabilities which maximize $P(R)$ and $P(R|y)$, respectively.

(d) Graph $P(R)$ and $P(R|y)$.

5.7. Consider an *a posteriori* probability density

$$P(R|y)=\frac{R^{N-r}(1-R)^r}{\text{const.}}$$

Prove that:

(a) The mean value $\bar{R}|y$ of R is

$$\bar{R}|y=\frac{N-r+1}{N+2}$$

(b) The value $\hat{R}|y$ which maximizes $P(R|y)$ is

$$\hat{R}|y=\frac{N-r}{N}$$

5.8. (a) Prove the identities

$$1-\sum_{s=S}^{N}\binom{N}{s}R^{N-s}[1-R]^s=\frac{N!}{(N-S)!(S-1)!}\int_0^R u^{N-S}(1-v)^{S-1}\,du,\quad (S\neq 0) \tag{1}$$

$$1-\sum_{s=0}^{S}\binom{N}{s}R^{N-s}[1-R]^s=\frac{N!}{S!(N-S-1)!}\int_0^{1-R} u^{S}(1-u)^{N-S-1}\,du,\quad (S\neq N) \tag{2}$$

(b) The beta probability density with parameters α,β is defined by

$$P(u)=\frac{(\alpha+\beta-1)!}{(\alpha-1)!(\beta-1)!}u^{\alpha-1}(1-u)^{\beta-1}$$

Prove that:

(1) The upper confidence limit R_2 satisfying (5.35) is

$R_2=100(1-\alpha)$ percentile of the beta distribution with parameters $[N-S+1,S]$

(2) The lower confidence limit R_1 satisfying (5.37) is

$$R_1 = 1 - [100(1-\alpha) \text{ percentile of the beta distribution with parameters } S+1, N-S]$$

(c) Prove that the upper and lower bounds for Problem 5.3 are

$$R_2 = [95 \text{ percentile of the beta distribution with parameters } 7,4]$$
$$R_1 = 1 - [95 \text{ percentile of the beta distribution with parameters } 5,6]$$

5.9. The F distribution with $2k$ and $2l$ degrees of freedom has the probability density

$$P(u) = \frac{(k+l-1)!}{(k-1)!(l-1)!}\left(\frac{k}{l}\right)^k u^{k-1}\left(1+\frac{k}{l}u\right)^{-k-l}, \quad u>0$$

Show that when V is distributed with the beta distribution with parameters k and l, then the distribution of the new random variable

$$U = \frac{l}{k} \times \frac{V}{1-V}$$

is an F distribution with $2k$ and $2l$ degrees of freedom.

5.10. Obtain the upper and lower bounds of the component reliability in Problem 5.3, using the results of Problems 5.8 and 5.9. Assume an F distribution table of:

95 PERCENTILE $F_{0.95}(a,b)$ OF F DISTRIBUTION[a]

b \ a	9	10	11	12	13	14
8	3.3881	3.3472	3.2840	3.2184	3.1503	3.1152
9	3.1789	3.1373	3.0729	3.0061	2.9365	2.9005
10	3.0204	2.9782	2.9130	2.8450	2.7440	2.7372
11	2.8962	2.8536	2.7876	2.7186	2.6464	2.6090
12	2.7144	2.7534	2.6866	2.6169	2.5436	2.5055

$\Pr(x \leqslant F_{0.95}(a,b)) = 0.95$

[a]Note that $F_{0.95}(a,b)$ is denoted by $F_{0.05}(a,b)$ in some standard tables.

5.11. A component reliability has the *a posteriori* distribution

$$P(R|y) = \frac{R^5(1-R)^4}{\text{const.}}$$

Obtain the 90% confidence interval for R.

6

DATA BASES

6.1 DATA BANKS

Quantification of risk, safety, and reliability implies the existence of validated data not only for hardware failures, but also for the rate of occurrence of initiating events, i.e., the anticipated frequency of deviations from normal design environments, and, additionally, data relating to the consequence of an abnormal event.

An initiating event for an accident might be a simultaneous power and water failure due to lightning, rather than to an equipment failure. It might also be a fire started by a workman's careless disposal of a cigarette or, as in the Brown's Ferry reactor incident, a maintenance man's use of a candle to check vaccuum seals. Hardware failure data is typified by the numbers required to obtain top event probabilities for fault trees, i.e., the unreliability of components in a reactor shut-down system, a fire protection unit, or a chemical reaction system. Consequence data needed to obtain human or monetary loss include such factors as structural integrity, meteorological conditions, toxicity, population densities, and environmental impacts.

This chapter acquaints the reader with some of the current data sources and data validation techniques. The critical need for data is now recognized on national and international levels throughout the world, and the past few years have seen the establishment of scores of government and privately sponsored data banks. It is simply too early to predict either the direction or the ultimate viability of many of these newly initiated and very costly activities.

6.2 OCCURRENCE DATA

We deal here with data relating to the *frequency of occurrence* of possible initiating events for an accident. Generally, we focus not on hardware failures, but on occurrences as events that take place at a system level and are depicted in event trees.

Event statistics represent abnormal operating conditions; if nothing else, tabulations of events are useful in alerting the safety analyst in singling out those occurrences pertinent to his operation. To facilitate data retrieval from banks of this type, descriptions have to be stored in keyword formats and carefully documented. Generally, no probability data is stored. An example of a data bank of this type is the U. S. Nuclear Regulatory's License Event Reports, which are mandatory for every plant (Fig. 6.1). Table 6.1 lists two occurrence data sources as well as sources of reliability data. The generic reliability data banks are simply published sources of reliability data, and the event data banks provide data, services, and scenarios.

Unraveling initiating event data requires patience and investigative skills; nearly always a human error in task, omission of task, or an extra act is involved. Two brief initiation event analyses given by Jens Rasmussen, RISØ National Laboratory, Roskilde, Denmark, are in many respects typical:

Case 1. Melt down of fuel element in nuclear reactor. *Nuclear Safety*, September, 1962.

Investigation: Certain tests required several hundred process coolant tubes to be blocked by neoprene disks. Seven disks were left in the system after the test, but were located by a test of the gauge system that monitors water pressure on each individual process tube. For some reason the gauge on one tube was overlooked, and it did not appear in a list of abnormal gauge readings prepared during the test. There was an additional opportunity to spot the blocked tube when a later test was performed on the system. This time the pressure for the tube definitely indicated a blocked tube. The shift supervisor failed, however, to recognize this indication of trouble. The gauge was adjusted at that time by an instrument mechanic to give a midscale reading which for that particular tube was false. This adjustment made it virtually certain that the no-flow condition would exist until serious damage resulted.

Case 2. Docket 50219-167: Two diesel generators set out of service simultaneously.

Event sequence: 8:10 permission to perform surveillance test on containment spray system No. 1 including electrical and mechanical inspection of diesel generator No. 1.

Form NPRD. 4

Nuclear Plant Reliability Data System

REPORT OF FAILURE

This data entry is for:

X New failure report (place a '7' in column two below)

O Correction to previously submitted failure report (place a '5' in column two below)

Input control

	Action		Utility design			Plant			Unit	S/C Code	NSSS	NPRD code for failed system or component						Utility component identification number											Report I.D.	Date of failure Yr.		Mo.		Dy.		Fail No.
1	2	3	4	5	6	7	8	9	10	11	12	13	14	15	16	17	18	19	20	21	22	23	24	25	26	27	28	29	30	31	32	33	34	35	36	37
C	7	E	A	B	C	X	Y	Z	1	3	C	C	H	A	0	0	1												8	7	5	1	0	1	3	1

*Status at time of failure

Event data

Code I.D.	Failure event start Yr.		Mo.		Dy.		Hr.		Min.		Failure event end Yr.		Mo.		Dy.		Hr.		Min.		*	NPRD code for related system or component						Report date Yr.		Mo.		Dy.										
38	39	40	41	42	43	44	45	46	47	48	49	50	51	52	53	54	55	56	57	58	59	60	61	62	63	64	65	66	67	68	69	70	71	72	73	74	75	76	77	78	79	80
G	7	5	1	0	1	3	1	3	2	3	7	5	1	0	2	5	1	4	3	0	A	V	A	L	V	E	X	7	5	1	0	2	8									

Failure description

H DURING ØRDERLY SHUTDØWN, FEEDWATER REG VAL

J VES CAUSED ØSCILLATIØNS, VIBRATIØN, DRAIN

K AND BYPASS LINE BREAKS. VALVES FAILED ØPEN

L . HI RX LEVEL CAUSED TURB TRIP RX SCRAM.

Cause of failure

M MALFUNCTIØN ØF VALVES RESULTED IN UNCØNTRØ

N LLED FLØW TØ RX.

P

Q

Corrective action

R FØLLØWING, DYNAMIC ANALYSIS, VALVE TRIM AND

S ØPERATØRS MØDIFIED TØ PRØVIDE WIDER RANGE

T CAPABILITIES AND PREVENT ØSCILLATIØNS.

U

Failure analysis data

Card I.D.	Type of failure		Mode of failure		Cause of failure A		B						Effect of failure A		B			Failure detection		Action taken A						B		Licensee event report submitted (date) Yr.		Mo.		Dy.									
38	39	40	41	42	43	44	45	46	47	48	49	50	51	52	53	54	55	56	57	58	59	60	61	62	63	64	65	66	67	68	69	70	71	72	73	74	75	76	77	78	79	80
V	A		A	S	A		A	K	B	A					B	C	F	A	F	A	E	A	F			D	H	7	5	1	0	2	0									

Card I.D.	Columns From	To	Remarks
			SYSTEM EXAMPLE

Date prepared: 10/28

Prepared by: J.W. Brown

Reviewed/Approved by: G. B. Jones

Phone No: 714/686-3253

Figure 6.1. NRC Report of Failure.

TABLE 6.1 Data Banks[a]

Occurrences data bank:

1. *NRC Licensee Event Reports* (*LERs*), Nuclear Regulatory Commission, Washington, D. C. 20555, U. S. A.
2. *Failure Data Handbook for Nuclear Facilities, LNEC-Memo*-69-7, available from NTIS, Springfield, VA., 22151, U. S. A.

Generic reliability data bank:

1. *WASH* 1400 "*Reactor Safety Study*," *Appendix III*, available from NTIS, Springfield, VA., 22161, U. S. A.
2. *IEEE Std.* 500–1977, "Guide to the Collection and Presentation of Electrical, Electronic and Sensing Component Reliability Data for Nuclear Power Generating Stations," IEEE Standards, 345 E 47 St., New York, N. Y. 10017, U. S. A.
3. *MIL HANDBOOK* 217 *B*, Rome Air Development Center, RBRS, Griffis Air Force Base, N. Y. 13441, U. S. A.
4. European Space Agency, *Electronic Components Data Bank*, Via G. Galilei, 00044, Frascati, Italy.
5. IEN "Galileo Ferraris," *Reliability Data Bank*, Corso Massimo D'Azeglio, 10125, Torino, Italy.
6. CNET, *Reliability Data Bank on Electronic Equipment*, 22300, Lanniou, France.
7. *Military Electronics Laboratory* (*FTL/FOA*) *Data Bank*, Fack 10450, Stockholm 80, Sweden.

Event data banks:

1. *Nuclear Plant Reliability Data Systems* (*NPRDS*), operated by Southwest Research Institute, 8500 Culebra Road, Building 88, San Antonio, Texas, 78284, U. S. A.
2. *Edison Electric Institute Data Bank*, Edison Electric Institute, 90 Park Avenue, New York, N. Y. 10016, U. S. A.
3. *SYREL—System Reliability Service Data Bank*, UKAEA, Culcheth, Warrington, WA 3 4 NE, U. K.
4. *ENEL Data System for Power Stations*, ENEL CRTN, Bastioni di Porta Volta 10, 20121, Milano, Italy.
5. *ASEA-ATOM Data System for Nuclear Power Plants*, Box 53, 72104, Wasteras 1, Sweden.
6. Gesellschaft für Reaktorsicherheit (GRS) mbH, *Data System for Nuclear Power Plants*, Glockengasse 2, Postfach 101650, D-5000, Koln 1, Germany.
7. *EDF*, *Reliability Data System for Nuclear Power Plants*, SEPTEN, EDF-GDF, Cedex 8, F-92080, Paris-la-Defense, France.
8. *Government/Industry Data Exchange Program* (*GIDEP*), Fleet Missile Systems Analysis and Evaluation Group, Corona, California, 91720, U. S. A.

8:20 permission to take diesel No. 2 out of service for oil addition.

Both systems out of service for 45 min. Foreman overlooked test of No. 1 system when permitting diesel No. 2 operation.

Comments: Coincident unavailability of redundant systems caused by improper timing of routine tasks is difficult to predict due to dependence on station "software," which is vulnerable to changes and oversight due to absence of cues.

[a]After Mancini, G., "Data and Validation," CEC Joint Research Center, ISPRA, Italy, RSA 12/78, June 6, 1978.

In this text, as in real life, component level failure data are required for quantitative assessment of system logic models. This category of failure information has received, by far, the major share of attention throughout the world.

One can stratify reliability data on levels—from "soft" to "hard." At one end of the spectrum we have data originating with "expert judgment," or generic experience. Then, there is data originating with plant experience, usually from a single site and, finally, there are systematic collections and analyses of multiplant event data component failure, repair, and replacement.

First-level data originating with expert judgment was critiqued by Eric Green in the early 1970's; he found that valve failure characteristics could be obtained more accurately from an analysis of valve components than by a survey of professional reliability engineers.[1] However, the use of Bayesian and delphi techniques in conjunction with expert estimation is judged a sound approach by many.[2,3]

The second level of data collection could be termed "ask the man who owns one." Data gathering of this sort is quite popular with engineers, who are accustomed to getting over 70% of the information they need by asking a person whom they think has the answer. For statisticians, this is an unspeakably poor procedure because, unless one knows the sample size, the number of observations, etc., there is nothing one can do with the data (other than use it, of course). To the bureaucrats, this approach is totally unsatisfactory because they want to have, in their filing cabinets and computer memories, in addition to complex failure event data, faceplate descriptions, manufacturing pedigree, and specifications for every item in every plant under their agencies' jurisdiction.

Event data (which is not to be confused with event occurrence scenarios) is composed of four parts:

—Identification of the item (pedigree)
—Identification of service and interfaces
—State of the item before the event
—The event description

Figure 6.2 is a report typical of those required for identifying an item, its interfaces, and service. The *event description* forms have the same

[1]Green, E., "A Review of Systems Reliability Assessment," *Generic Techniques in Reliability Assessment*, Henley, E., and J. Lynn (eds.), Noordhoff Publishing Co., Leyden, Holland, p. 183, 1976.

[2]IEEE Std. 500–1977, "IEEE Guide to the Collection and Presentation of Electrical Electronic and Sensing Component Reliability Data for Nuclear Power Generating Stations," published by IEEE, 1977.

[3]Apostolakis, G., and Ali Mosleh, "A Study on the Quantification of Judgement," ANS Topical Meeting on Probabilistic Analysis, Los Angeles, May, 1978.

From NPRD. 2

Nuclear Plant Reliabilability Data System

REPORT OF ENGINEERING DATA

This data entry is for:

X Engineering data report for original or replacement component or original system (place a '7' in column two below)

O Correction to previous Engineering Data Report (place a '5' in column two below)

Input control

Field	Columns	Entry
(1)	1	C
Action	2	7
(3)	3	E
Utility design	4–6	CWE
Plant	7–9	QAD
Unit	10	1
s/c Code	11	5
NSSS	12	C
NPRD comp. or system code	13–18	PUMPXX
Utility component identification number (* / Report No.)	19–29	1-1302
Report I.D.	30	A
Data start date Yr.	31–32	73
Data start date Mo.	33–34	07
Data start date Dy.	35–36	01
(37)	37	

System or component engineering data

Card I.D.	Field	Columns	Entry
A	NPRD system code	39–44	CEA
A	Utility system code	45–51	1300
A	Safe CLS.	52–53	2
A	Mode	54–56	SBC
A	Environment Int.	57–59	ØFX
A	Environment Ext.	60–61	AF
A	In-service date Yr. / Mo. / Dy.	62–67	71 / 10 / 07
A	Applicable MFGR code or standard	68–80	ASME SEC, 3-68
B	MFGR ref. no.	39–42	B260
B	Manufacturer model number	43–62	4X6X9 BMSD
B	Manufacturer serial number	63–80	270606
C	Supp/vend. ref. no.	39–42	S080
C	Supplier/vendor system or component identification number	43–60	1-1302
C	Source drawing or document number	61–76	RC1C-M-050
D	A	39–40	B
D	B	41–42	C
D	C	43–44	G
D	D	45–46	B
D	E	47–48	E
D	F	49–50	
D	G	53–56	2800
D	Units	57–60	FTHØ
D	H	63–66	1416
D	Units	67–70	GPM
D	J	73–76	4500
D	Units	77–80	RPM

Operation and testing data

Card I.D.	Field	Columns	Entry
E	Estimate of component/system operation as percent of: 1-Reactor critical hours	39–43	00100
E	2-Standby condition hours	44–48	00100
E	3-Reactor shutdown hours	49–53	00100
E	Testing – Check: Freq.	54	1
E	Testing – Check: Interval	55–56	MØ
E	Testing – Check: Out of serv. (hrs.)	57–59	00
E	Testing – Functional: Freq.	60	1
E	Testing – Functional: Interval	61–62	QT
E	Testing – Functional: Out of serv. (hrs.)	63–65	20
E	Testing – Calibration: Freq.	66	0
E	Testing – Calibration: Interval	67–68	00
E	Testing – Calibration: Out of serv. (hrs.)	69–71	000

Remarks

Figure 6.2. NRC Report of Data.

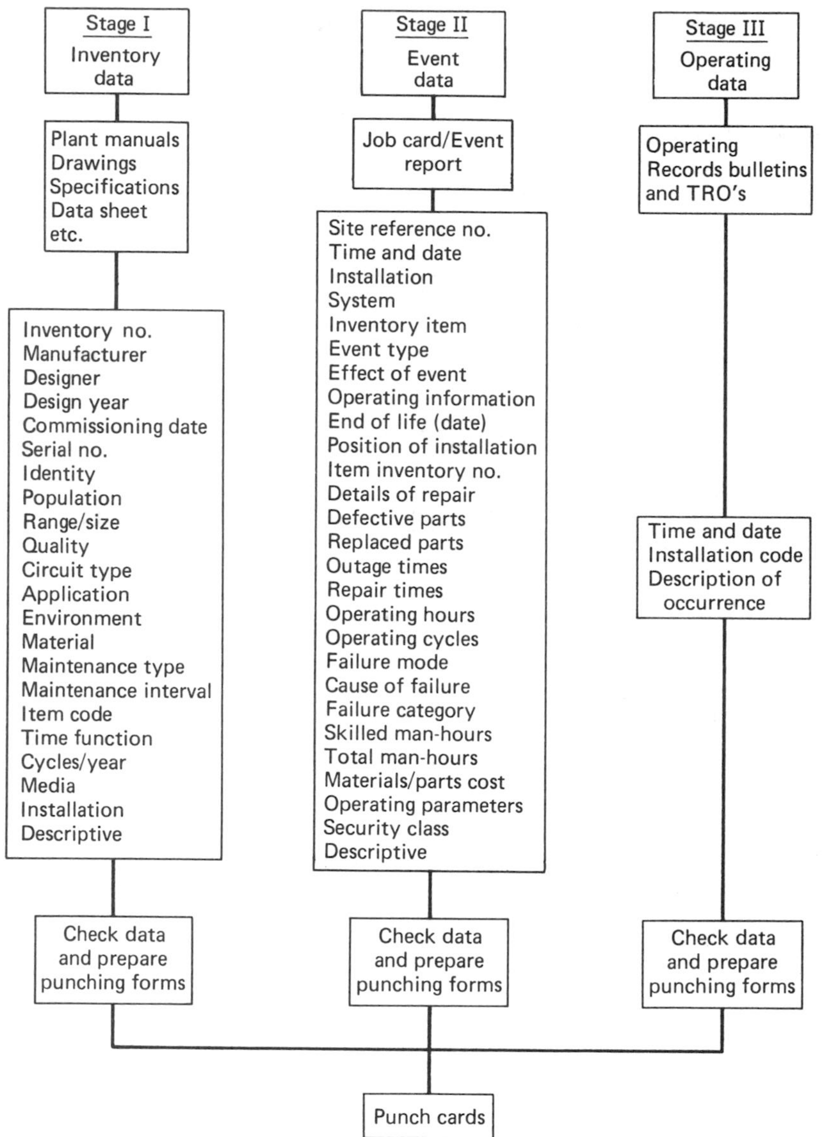

Figure 6.3. The three stages of data collection.

characteristics as the event consequence form, Fig. 6.1 (which is actually an event form). Figure 6.3 is a listing of the data being collected by the British System Reliability Service in their archives.

The calculation of probabilistic component parameters, such as the failure rate $r(t)$ and repair rate $m(t)$ requires considerately more than the raw data found on a single event report form. One needs to know also the total number of items in the sample and the statistical independence of the item with respect to other items in the plant. Perhaps there were common-cause failures, or perhaps the item was in cold standby, which means that the failure reported was a failure to start, not a failure to function.

Figure 6.4 shows some of the ways that event information gathered at various levels can flow in and through a data bank. Ideally, component data should, in addition to being statistically validated, be validated in system models to see if they are *coherent*; i.e., do the component failures predict a (known) system failure?

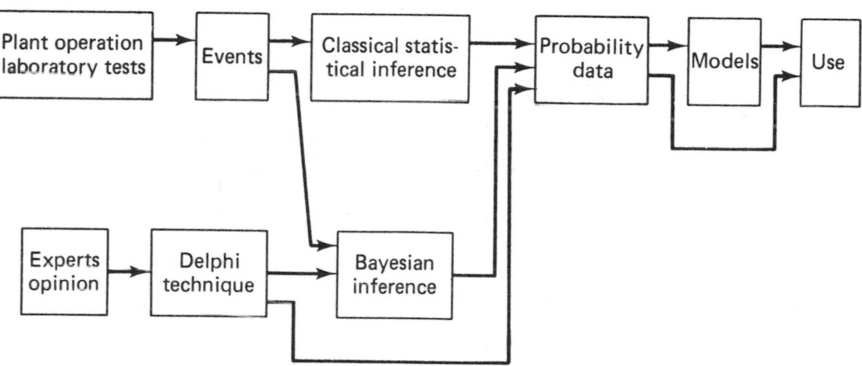

Figure 6.4. Information flow sheet in a reliability data system, (Mancini [1978] op cit.).

The characteristics of the principal data bank systems in the United States and Europe are summarized in Fig. 6.5. All of these activities involve large staffs, large computers, complex information systems, and large expenditures of public money. Indeed, the collection and filing of vast amounts of information of this type by multitudinous government agencies was deemed critical enough by President Jimmy Carter to include in his Emergency Energy Message to Congress in April of 1977 the statement:

> ...that the Commission (NRC) make mandatory the current voluntary reporting of minor mishaps and component failures at operating reactors, in order to develop the reliable data base needed to improve reactor design and operating practice.

Some additional references regarding data banks are given in footnote references 4 through 6.

[4]*Proceedings of the Second Seminar on Reliability Data Banks*, Stockholm, March, 1977. Available from FTL/MEL, FOA 3, 10450 Stockholm 80, Sweden.

[5]Hecht, Linda O., and Joseph R. Fragola, "Reliability Data Bases, A Review," *IEEE Standards*, New York.

[6]Al, A. B. J., S. Capobianchi, and T. Luisi, "Perspectives for a European Reliability Data System for Light Water Reactors," ANS Topical Meeting on "Probabilistic Analysis of Nuclear Reactor Safety," Los Angeles, August, 1978.

Principal data bank system in Europe and U.S.A.	CHARACTERISTICS OF DATA BANK SYSTEMS					
		CODIFICATION REFERENCE SYSTEM FOR THE DATA		COLLECTION OF DATA		
	Pilot experiment or normal operation.	Codification of plants, systems, aggregates components	Codification of events, maintenance, failures, repairs	Raw or reprocessed data	Teleprocessing or batch mode (cards, magnetic and paper tapes)	Format or consistency checks for teleprocessing
Germany (GRS-RWE)	Model case in development	Actually: AKZ and DES Future: KKS	Codification of works order forms, codes of failures, modes and causes in development	Raw data collected in collaboration with RWE and GRS	Batch mode Teleprocessing foreseen	
France (EdF-SRD)	Pilot experiment in development	Sapten and AMN (old codification)	Very sophisticated internal codification	At local level: raw data collected in the field (Bugay and Fessenhelm) At national level: preprocessing data	At local level: teleprocessing mode At national level: Batch mode	Format check
Great Britain (UKAEA-SRS)	Normal operation	Internal decimal modification system	Hierarchical codification in development	Raw data occasionally reprocessed data	Batch mode Punched cards and magnetic tapes	
Italy (ENEL)	Start of pilot operation General system in preparation	At present: codification A.B.C. (alphanumerical) and manufacture code MPL Finally: internal hierarchical codification	In development	Raw data collected in the field	Teleprocessing with paper tapes or real time transmission of data	Format check
U.S.A. (NRPOS)	Normal operation	Internal alphanumeric hierarchical codification system	Very sophisticated internal codification	Raw data collected by Associated Utilities	Batch mode Punched cards and magnetic tapes	

Figure 6.5. Characteristics of data bank systems at the end of 1977.

6.4 CONSEQUENCE DATA BANKS

The third type of data collection for risk evaluation are data dealing with *accident consequences*. Since no such bank now exists one can only speculate on its requirements, which might be:

1) *Material properties*: Fracture toughness, ultimate yield stresses, flammabilities, explosive limits, detonation parameters, etc.
2) *Toxicity data*: This includes both latent and immediate damage to humans and animals. A potential categorization for this type of data

CHARACTERISTICS OF DATA BANK SYSTEMS							
COLLECTION OF DATA		COMPUTING FACILITIES		STORAGE CHARACTERISTICS		COMPUTING	
Input reports prepared by:	Sources of the data	Hardware	DBMS or other software facilities	Storage structure	Principle storage medium	Main calculations	Confidence level assumed to evaluate confidence interval
Nuclear reactor staff in collaboration with reliability experts of GRS.	GRS-RWE, manufacturers at plant components	Siemens 4004-55 in Garching (LRA) AMDAL computer	SESAM System 2000	Indexed sequential	Disks	Constant hazard rates Confidence intervals for constant failure rates	95%
Nuclear reactor staff in collaboration with reliability experts	At present only E d F	IBM 370-155 at Bugay IBM 370-168 at national level in Clamart	CICS-DL1 under DOS IMS/VS under OS	Hierarchical structure	Disks	Totals Mean values Smallest values Largest values Median values Constant hazard rate	
Nuclear reactor staff assisted by reliability experts	Associated members of SRS	ICL 472 by January 1977: ICL 2980	Homemade system RIOS, modified ICR	Sequential	Magnetic tapes and disks	Totals Mean values Smallest values Largest values Constant hazard rates Confidence intervals for constant failure rates	95%
Nuclear reactor staff	At present: ENEL and manufacturers	Honeywell 6600 and terminal 300	Actually homemade system probably IDS	Sequential, random and ring	Magnetic tapes and disks	Totals Mean values Constant hazard rates Confidence intervals for constant failure rates	90% 95% 99%
Nuclear reactor staff	Associated Utilities	IBM 370 series	Actually homemade system	Sequential	Magnetic tapes and disks	Totals	

Figure 6.5. (Continued)

is found in Fig. 6.6, which is a schematic time of exposure versus gas concentration curve for toxic materials. Hamner (op cit.) offers a useful summary of toxicity parameters.

3) *Total risk*: To establish the overall consequences of an accident it is necessary to know the total number of people affected and the total physical damage. These are location and environment specific.

One of the uses of consequence data might be for extrapolation from a small to a large accident.

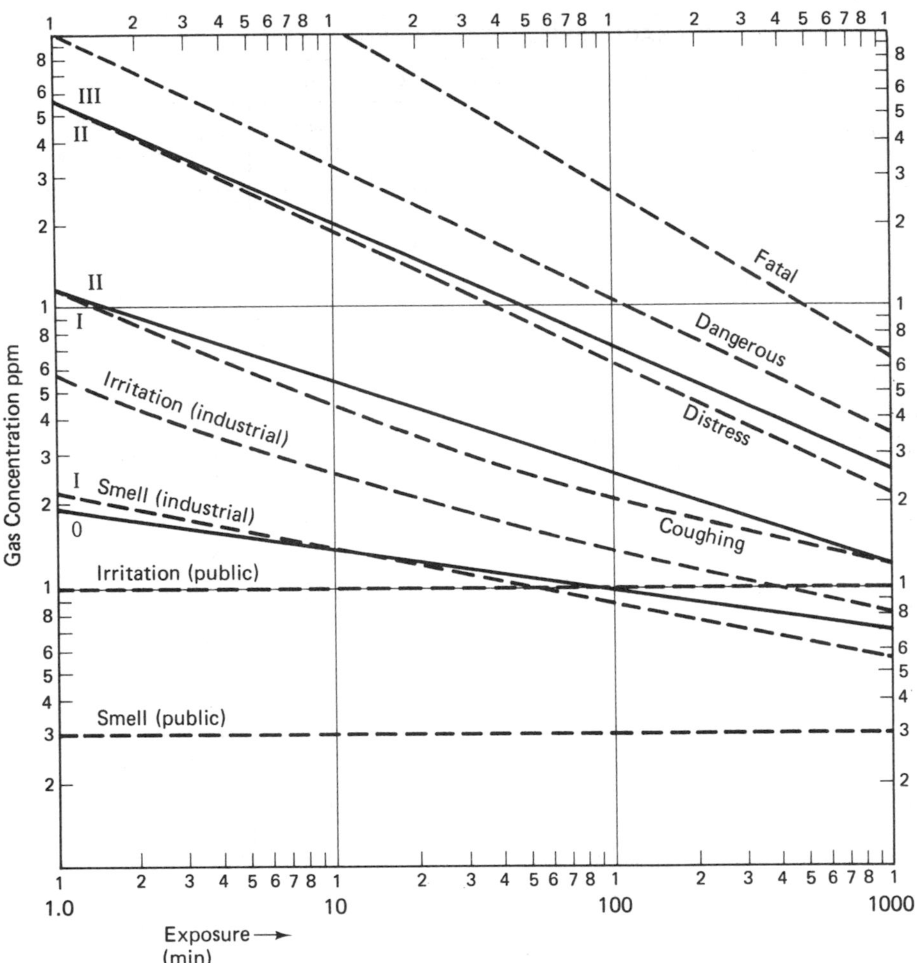

Figure 6.6. Hazard categories for exposure to gas. After Robinson, B. W., "Risk Assessment in the Chemical Industry," CEC, Joint Research Center, ISPRA, Report RSA5/78 (June 30, 1978).

6.5 DATA VALIDATION

A data acquisition program should be interfaced with a *validation effort*. It is not sufficient, for example, to use questionnaires to obtain mean-time-to-failure or repair-time data. The replies must be validated by checking personnel time sheets and records. If a questionnaire has categories such as "Cause of Failure Unknown," that is the one which will invariably be checkmarked. Unless the reliability engineer approaches maintenance personnel properly, there is no hope of obtaining valid information. "How often do these damn pumps break down?" will elicit an entirely different response than "Do you have any trouble keeping those pumps running?" Another critical factor is whether the interrogator is wearing a necktie or coveralls.

Uncertainties in data processing arise because one is dealing in small-sample statistics; to these uncertainties one must add those due to common cause failures, operation errors, and test and maintenance induced failures. Workmen are at least as bad as surgeons when it comes to leaving tools and debris at the working site. In the start-up of chemical plants, it is common practice to put strainers in the lines to catch tools and litter before they cause equipment malfunction. "Small wrench lodged in impeller" is not a frequently reported pump failure mode, but it probably occurs much more frequently than many modes which do appear in reports.

The problem of data validation, the specificity of the information required for any non-academic safety analyses, and the sensitivity of failure data to plant environment and operating and maintenance policies have raised considerable skepticism regarding the integrity and usefulness of large and expensive data banks. They are, undoubtedly, very useful to inexperienced people who are new to the field, have not seen much failure data, and need a frame of reference. For an experienced person it may be a different matter. A good car mechanic, for example, knows that water pumps tend to fail at about 50,000 miles. He would, in general, be correct in predicting that the water pumps on next year's model will also fail at about 50,000 miles. If a component-by-component fault tree analysis using data bank sources, sworn statements by automotive company officials, or bench-scale tests give answers other than 50,000 miles and no major design, component, or material changes have been made, then the mechanic, the "Bayesian Prior," is your best source of data.

6.6 FAILURE DATA TABULATIONS

Published reliability data runs the gamut from estimates to exhaustively tested and verified failure-to-repair distributions. In general, reliability engineers are accustomed to working with failure data which predicts behavior within a factor of ten, and only the most naive people would expect less than a factor-of-two variation. At the lowest level of accuracy we can make use of data such as those given in Fig. 6.7, which is due to Bourne and Green.[7] System reliability estimates based on data like these may be a bit gross, but they are far from useless. Figure 6.8 (Mancini, op

[7]Green, A. E., and A. J. Bourne, *Reliability Technology*, John Wiley & Sons, Inc., New York, p. 538, 1972.

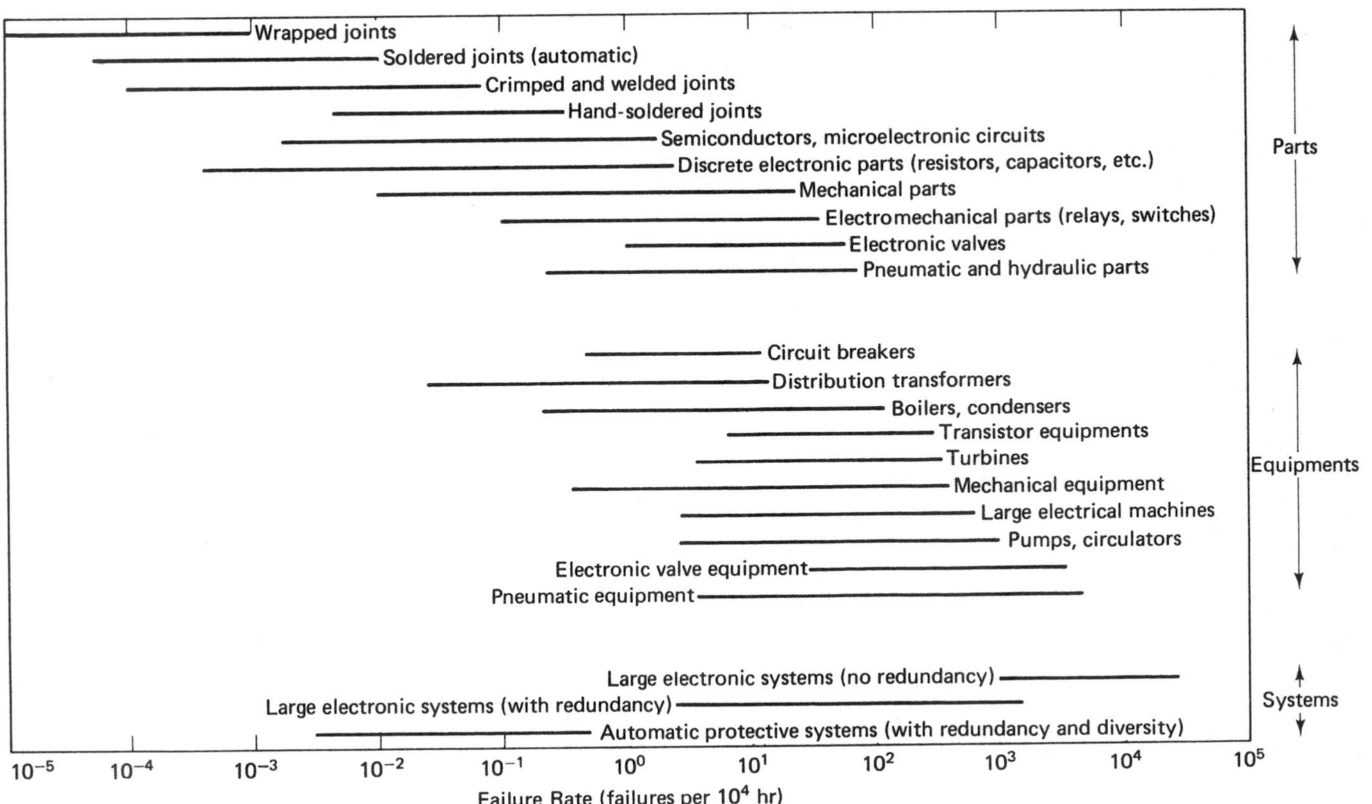

Figure 6.7. Typical ranges of failure rates for parts, equipment, and systems.

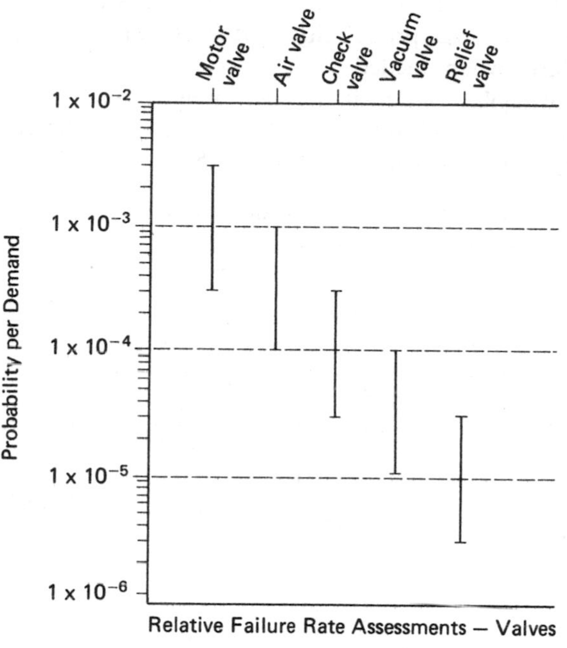

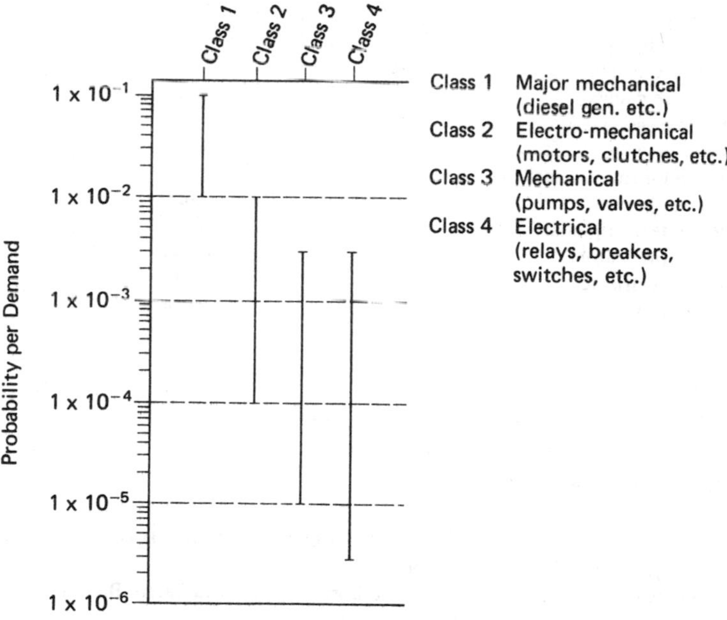

Figure 6.8. Demand probabilities of classes of hardware.

cit.) is another useful generalized tabulation. It describes the probability of failure on demand, i.e., the probability that a redundant or protective system is not available when needed.

The next level of sophistication is the instrument failure data collected by Prof. F. P. Lees, of the University of Loughborough. His very useful tabulation is given as Tables 6.2, 6.3, 6.4, and 6.5.

TABLE 6.2 Instrument Failure Rates

Instrument	Failure Rate		References[a]
	Observed Faults/Year	Assumed/predicted Faults/Year	
Control valve (p)[b]	0.25		1,2,3,4,
		0.26	5,6
Solenoid valve		0.26	5,6
Pressure relief valve		0.022	5,6
Hand valve		0.13	5,6
Differential pressure transmitter (p)	0.76		1,2,3,4,7,8
Variable area flowmeter transmitter (p)	0.68		1,2,3,4,8
Thermocouple		0.088	5,6
Temperature trip amplifier			1,2,7,9
Type A	2.6		7
B	1.7		7
Pressure switch	0.14		1,2,3,4,8
Pressure gauge		0.088	5,6
O_2 analyser	2.5		1,2,4,8
Controller (p)	0.38		7
Indicator (moving coil meter)		0.026	5,6,8
Recorder (strip chart)		0.22	5,6,8
Lamp (indicator)		0.044	5,6
Photoelectric cell		0.13	5,6
Tachometer		0.044	5,6
Stepper motor		0.044	5,6
Relay (p)	0.17		7
Relay (post office)		0.018	5,6

[a]U. K. Atomic Energy Authority: references 1–9.

[b](p) indicates pneumatic.

REFERENCES FOR TABLES 6.2 THROUGH 6.5

1. Hensley, G., *Measurement Control.*, **1**, T72, 1968.
2. Hensley, G., U. K. Atomic Energy Authority, Health and Safety Branch, Rep. AHSB (S) R136, Risley, Lancashire, 1967.
3. Hensley, G., U. K. Atomic Energy Authority, Health and Safety Branch, Rep. AHSB (S) R178, Risley, Lancashire, 1967.

TABLE 6.3 INSTRUMENT FAILURE RATES[a]

Instrument	No. at Risk	Instrument Years	Environment Factor	No. of Faults	Failure Rate Faults/Year
Control value	1531	747	2	447	0.60
Power cylinder	98	39.9	2	31	0.78
Valve positioner	334	158	1	69	0.44
Solenoid valve	252	113	1	48	0.42
Current/pressure transducer	200	87.3	1	43	0.49
Pressure measurement	233	87.9	3	124	1.41
Flow measurement (fluids):	1942	943	3	1069	1.14
Differential pressure transducer	636	324	3	559	1.73
Transmitting variable area flowmeter	100	47.7	3	48	1.01
Indicating variable area flowmeter	857	409	3	137	0.34
Magnetic flowmeter	15	5.98	4	13	2.18
Flow measurement (solids)					
Load cell	45	17.9	—	67	3.75
Belt speed measurement & control	19	7.58	—	116	15.3
Level measurement (liquids):	421	193	4	327	1.70
Differential pressure transducer	130	62	4	106	1.71
Float-type level transducer	158	75.3	4	124	1.64
Capacitance-type level transducer	28	13.4	4	3	0.22
Electrical conductivity probes	100	39.8	4	94	2.36
Level measurement (solids)	11	4.38	—	30	6.86
Temperature measurement (excluding pyrometers):	2579	1225	3	425	0.35
Thermocouple	772	369	3	191	0.52
Resistance thermometer	479	227	3	92	0.41
Mercury-in-steel thermometer	1001	477	2	13	0.027
Vapor pressure bulb	27	10.7	4	4	0.37
Temperature transducer	300	142	3	124	0.88

TABLE 6.3 (Continued)

Instrument	No. at Risk	Instrument Years	Environment Factor	No. of Faults	Failure Rate Faults/Year
Radiation pyrometer	43	30.9	4	67	2.17
Optical pyrometer	4	3.4	4	33	9.70
Controller	1192	575	1	164	0.29
Pressure switch	549	259	2	87	0.34
Flow switch	9	3.59	—	4	1.12
Speed switch	6	2.39	—	0	—
Monitor switch	16	6.38	—	0	—
Flame failure detector	45	21.3	3	36	1.69
Millivolt current transducer	12	4.78	—	8	1.67
Analyzer:	86	39.0	—	331	8.49
pH meter	34	15.8	—	93	5.88
Gas-liquid chromatograph	8	3.43	—	105	30.6
O_2 analyzer	12	5.67	—	32	5.65
CO_2 analyzer	4	1.90	—	20	10.5
H_2 analyzer	11	5.04	—	5	0.99
H_2O analyzer (in gases)	3	1.38	—	11	8.00
Infra red liquid analyzer	3	1.43	—	2	1.40
Electrical conductivity meter (for liquids)	5	1.99	—	33	16.70
Electrical conductivity meter (for water in solids)	3	1.20	—	17	14.2
Water hardness meter	3	1.20	—	13	10.9
Impulse lines	1099	539	3	416	0.77
Controller settings	1231	609	—	84	0.14

[a]Anakora, Engel, and Lees: reference 10.

4. Hensley, G., U. K. Atomic Energy Authority, Systems Reliability Service, Rep. SRS/G1/1, Culcheth, Lancashire.
5. Green, A. E., and A. J., Bourne, U. K. Atomic Energy Authority, Health and Safety Branch, Rep. AHSB (S) R117, Parts 1–3, Risley, Lancashire, 1966.
6. Green, A. E., and A. J., Bourne, *Reliability Technology*, John Wiley & Sons, Inc., New York, 1972.
7. Eames, A. R., *Nuclear Engng.*, **11** (118) March, 189, 1966.
8. Green, A. E., U. K. Atomic Energy Authority, Health and Safety Branch, Rep. AHSB (S) R113, Risley, Lancashire, 1966.
9. Green, A. E., and A. J., Bourne U. K. Atomic Energy Authority, Health and Safety Branch, Rep. AHSB (S) R91, Risley, Lancashire, 1965.
10. Anyakora, S. N., G. F. M., Engel, and F. P., Lees, *Chem. Engr., Lond.*, **225**, 396, 1971.
11. Skala, V., *Instrum. Technol.*, **21** (10), 27, 1974.
12. U. S. Atomic Energy Commission, "*Reactor Safety Study. An Assessment of Accident Risks in U. S. Commercial Nuclear Power Plants. Appendix III, Failure Data*," WASH-1400, Washington, D. C., 1974.

TABLE 6.4 INSTRUMENT FAILURE RATES[a]

Pressure transmitter:	Faults/Year	Controller:	Faults/Year
Type A	0.60	Type A	0.37
B	0.48	B	0.38
C	0.54	C	0.52
Differential pressure transmitter:		D	0.44
Type A	0.74	Control valve:	
B	0.78	Type A	0.50
C	0.58	B	0.49
		C	0.50
		D	0.53
		E	0.47

[a]Skala: reference 11.

TABLE 6.5 INSTRUMENT FAILURE RATES[a]

SECTION 1

Instrument	Failure	Failure rate, Faults/Demand
Valves:		
Motor operated	Failure to operate	$1\times10^{-3}(2\times10^{-4}-7\times10^{-2})$
	Plug	$3\times10^{-5}(6\times10^{-5}-3\times10^{-4})$
Solenoid operated	Failure to operate	$1\times10^{-3}(2\times10^{-5}-6.5\times10^{-3})$
	Plug	3×10^{-5}
Air operated	Failure to operate	$1\times10^{-4}(1\times10^{-6}-2\times10^{-2})$
	Plug	3×10^{-5}
Check	Failure to open	$1\times10^{-4}(2\times10^{-5}-3\times10^{-4})$
Relief	Failure to open	$1\times10^{-5}(1.4\times10^{-5}-3.6\times10^{-5})$
Manual	Plug	$3\times10^{-5}(3\times10^{-4}-3\times10^{-6})$
Switches:		
Pressure switch	Failure to operate	$1\times10^{-4}(5\times10^{-5}-1\times10^{-3})$
Limit switch	Failure to operate	$1\times10^{-4}(1\times10^{-5}-7\times10^{-4})$
		faults/year
Instruments	Failure to operate	0.009 (0.0026 – 0.53)
Battery power supply	Failure to provide proper output (in standby mode)	0.0026 (0.0009 – 0.053)

SECTION 2

Instrument	Number of failures	Failure Rate	
		Faults/Demand	Faults/Year
Valves	102	1×10^{-3}	0.026
Instruments	50	3×10^{-3}	0.009

TABLE 6.6 Failure Data, WASH 1400

	Failure Modes	Assessment -Median-	Upper and Lower Bound
CLUTCH ELEC	Failure to Operate	3×10^{-4}/D	$1\times10^{-4}-1\times10^{-3}$
	Premature Open	1×10^{-6}/HR	$1\times10^{-7}-1\times10^{-5}$
CLUTCH MECH	Failure to Open	3×10^{-7}/HR	$3\times10^{-6}-3\times10^{-8}$
	Failure to Operate	3×10^{-4}/D	$1\times10^{-4}-1\times10^{-3}$
SCRAM RODS	Failure to Insert (Single Rod)	1×10^{-4}/D	$3\times10^{-5}-3\times10^{-4}$
ELECTRIC MOTORS	Failure to Start	3×10^{-4}/D	$1\times10^{-4}-1\times10^{-3}$
	Failure to Run	1×10^{-5}/HR	$3\times10^{-6}-3\times10^{-5}$
	Failure to Run (Extreme ENVIR)	1×10^{-3}/HR	$1\times10^{-4}-1\times10^{-2}$
RELAYS	Failure to Energize	1×10^{-4}/D	$3\times10^{-5}-3\times10^{-4}$
	Failure NO Contact to Close	3×10^{-7}/HR	$1\times10^{-7}-1\times10^{-6}$
	Short Across NO/NC Contact	1×10^{-8}/HR	$1\times10^{-9}-1\times10^{-7}$
	Open NC Contact	1×10^{-7}/HR	$3\times10^{-8}-3\times10^{-7}$
SWITCHS	Limit: Failure to Operate	3×10^{-4}/D	$1\times10^{-4}-1\times10^{-3}$
	Torque: Fail to OPER	1×10^{-4}/D	$3\times10^{-5}-3\times10^{-4}$
	Pressure Fail to OPER	1×10^{-4}/D	$3\times10^{-5}-3\times10^{-4}$
	Manual, Fail to TRANS	1×10^{-5}/D	$3\times10^{-6}-3\times10^{-5}$
	Contacts Short	1×10^{-7}/HR	$1\times10^{-8}-1\times10^{-6}$

	Failure Modes	Assessment -Median-	Upper and Lower Bound
Pumps	Failure To Start	1×10^{-3}/D	$3\times10^{-4}-3\times10^{-3}$
	Failure To Run —Normal	3×10^{-5}/HR	$3\times10^{-6}-3\times10^{-4}$
	Failure To Run —Extreme ENV	1×10^{-3}/HR	$1\times10^{-4}-1\times10^{-2}$
Valves: MOV.	Fails To Operate	1×10^{-3}/D	$3\times10^{-4}-3\times10^{-3}$
	(Plug) Failure To Remain Open	1×10^{-4}/D	$3\times10^{-5}-3\times10^{-4}$
	External Leak Or Rupture	1×10^{-8}/HR	$1\times10^{-9}-1\times10^{-7}$
Valves (SOV)	Fails To Operate	1×10^{-3}/D	$3\times10^{-4}-3\times10^{-3}$
Valves (AOV)	Fails To Operate	3×10^{-4}/D	$1\times10^{-4}-1\times10^{-3}$
	(Plug) Failure To Remain Open	1×10^{-4}/D	$3\times10^{-5}-3\times10^{-4}$
	External Leak—Rupture	1×10^{-8}/HR	$1\times10^{-9}-1\times10^{-7}$
Valves (Check)	Failure To Open	1×10^{-4}/D	$3\times10^{-5}-3\times10^{-4}$
	Reverse Leak	3×10^{-7}/HR	$1\times10^{-7}-1\times10^{-6}$
	External Leak—Rupture	1×10^{-8}/HR	$1\times10^{-9}-1\times10^{-7}$
Valves (Vacuum)	Failure to Operate	3×10^{-5}/D	$1\times10^{-5}-1\times10^{-4}$
	Rupture	1×10^{-8}/HR	$1\times10^{-9}-1\times10^{-7}$
Valves: Orifices, Flow Meters, (Test)	Rupture	1×10^{-8}/HR	$1\times10^{-9}-1\times10^{-7}$
Valves (Manual)	Failure To Remain Open (Plug)	1×10^{-4}/D	$3\times10^{-5}-3\times10^{-4}$

TABLE 6.6 (Continued)

FAILURE MODES		
CIRCUIT BREAKERS — Failure to Operate	1×10^{-3}/D	$3\times10^{-4}-3\times10^{-3}$
CIRCUIT BREAKERS — Premature Transfer	1×10^{-6}/HR	$3\times10^{-7}-3\times10^{-6}$
FUSES — Premature, Open	1×10^{-6}/HR	$3\times10^{-7}-3\times10^{-6}$
FUSES — Failure to Open	1×10^{-5}/D	$3\times10^{-6}-3\times10^{-5}$
WIRES — Open	3×10^{-6}/HR	$1\times10^{-6}-1\times10^{-5}$
WIRES — Short to GND	3×10^{-7}/HR	$3\times10^{-8}-3\times10^{-6}$
WIRES — Short to PWR	1×10^{-8}/HR	$1\times10^{-9}-1\times10^{-7}$
TRANSFORMERS — Open CKT	1×10^{-6}/HR	$3\times10^{-7}-3\times10^{-6}$
TRANSFORMERS — Short	1×10^{-6}/HR	$3\times10^{-7}-3\times10^{-6}$
SOLID STATE DEVICES, HI PWR Application — Fails to Function	3×10^{-6}/HR	$3\times10^{-7}-3\times10^{-5}$
SOLID STATE DEVICES, HI PWR Application — Shorts	1×10^{-6}/HR	$1\times10^{-7}-1\times10^{-5}$
SOLID STATE DEVICES, Low PWR Application — Fails To Function	1×10^{-6}/HR	$1\times10^{-7}-1\times10^{-5}$
SOLID STATE DEVICES, Low PWR Application — Shorts	1×10^{-7}/HR	$1\times10^{-8}-1\times10^{-6}$

FAILURE MODES		
Valves (Relief) — Fail To Open/D	1×10^{-5}/D	$3\times10^{-6}-3\times10^{-5}$
Valves (Relief) — Premature Open/HR	1×10^{-5}/HR	$3\times10^{-6}-3\times10^{-5}$
Pipes > 3" HI Quality — Rupture (Section)	1×10^{-10}/HR	$3\times10^{-12}-3\times10^{-9}$
Pipes < 3" — Rupture	1×10^{-9}/HR	$3\times10^{-11}-3\times10^{-8}$
Gaskets — Leak	3×10^{-6}/HR	$1\times10^{-7}-1\times10^{-4}$
Flanges, Closures, Elbows: —Leak/Rupture	3×10^{-7}/HR	$1\times10^{-8}-1\times10^{-5}$
Welds — Leak	3×10^{-9}/HR	$1\times10^{-10}-1\times10^{-7}$
Diesel (Complete Plant) — Failure to Start Od:	3×10^{-2}/D	$1\times10^{-2}-1\times10^{-1}$
Diesel (Complete Plant) — (Emergency Failure to) (Loads) Run λ_0	3×10^{-3}/HR	$3\times10^{-4}-3\times10^{-2}$
Diesel (Engine Only) — Failure to Run λ_0	3×10^{-4}/HR	$3\times10^{-5}-3\times10^{-3}$
Batteries Power Supplies — NO/Output λ_s	3×10^{-6}/HR	$1\times10^{-6}-1\times10^{-5}$
Instrumentation (Amplification,) Annuciators, Transducers, (Combination) — Failure to Operate λ_s	1×10^{-6}/HR	$1\times10^{-7}-1\times10^{-5}$
Instrumentation (Amplification,) Annuciators, Transducers, (Combination) — Shift Calibration, λ_0	3×10^{-5}/HR	$3\times10^{-6}-3\times10^{-4}$

Estimated rates	Activity
10^{-4}	Selection of a key-operated switch rather than a non-key switch (this value does not include the error of decision where the operator misinterprets situation and believes key switch is correct choice).
10^{-3}	Selection of a switch (or pair of switches) dissimilar in shape or location to the desired switch (or pair of switches), assuming no decision error. For example, operator actuates large-handled switch rather than small switch.
3×10^{-3}	General human error of commission, e.g., misreading label and therefore selecting wrong switch.
10^{-2}	General human error of omission, where there is no display in the control room of the status of the item omitted, e.g., failure to return manually operated test valve to proper configuration after maintenance.
3×10^{-3}	Errors of omission, where the items being omitted are embedded in a procedure rather than at the end as above.
3×10^{-2}	Simple arithmetic errors with self-checking but without repeating the calculation by redoing it on another piece of paper.
$1/x$	Given that an operator is reaching for an incorrect switch (or pair of switches), he selects a particular similar-appearing switch (or pair of switches), where x = the number of incorrect switches (or pair of switches) adjacent to the desired switch (or pair of switches). The 1/x applies up to 5 or 6 items. After that point the error rate would be lower because the operator would take more time to search. With up to 5 or 6 items he doesn't expect to be wrong and therefore is more likely to do less deliberate searching.
10^{-1}	Given that an operator is reaching for a wrong motor-operated valve (MOV) switch (or pair of switches), he fails to note from the indicator lamps that the MOV(s) is (are) already in the desired state and merely changes the status of the MOV(s) without recognizing he had selected the wrong switch(es).
−1.0	Same as above, except that the state(s) of the incorrect switch(es) is (are) *not* the desired state.
−1.0	If an operator fails to operate correctly one of two closely coupled valves or switches in a procedural step, he also fails to correctly operate the other valve.
10^{-1}	Monitor or inspector fails to recognize initial error by operator. Note: With continuing feedback of the error on the annunciator panel, this high error rate would not apply.
10^{-1}	Personnel on different work shift fail to check condition of hardware unless required by check list or written directive.
5×10^{-1}	Monitor fails to detect undesired position of valves, etc., during general walk-around inspections, assuming no check list is used.
0.2 − 0.3	General error rate, given very high stress levels where dangerous activities are occurring rapidly.

Figure 6.9. Operator error estimates (after WASH 1400).

Table 6.2 shows both predicted and observed failure rates taken from work done at the United Kingdom Atomic Energy Establishment.

Table 6.3 is based on a study of instrument maintenance repair slips at large plants, ranging from an acid plant to a water treatment plant. Environmental factors 1 to 4 are directly proportional to the severity of the working fluids and conditions.

Table 6.4 is from a major survey involving 18,400 components in 4800 control loops in a refinery, as reported by Skala.

The ultimate in data refinement is to be found in WASH 1400, Appendix III: 30 different data banks were combed and massaged to provide upper and lower bounds as well as an assessment median for the items listed in Table 6.6. Note that an exhaustive study of this type empowers the tabulators to assign different failure rates to different failure modes.

Lastly, we come to what is probably the most important but the most difficult data to quantify: *human performance*. Its importance lies in the fact that, for many types of plants, hardware faults have been beaten into the ground and, if human factors and common mode failures are not taken into account, non-sensible system safety numbers such as unreliabilities of 10^{-39}, year^{-1}, may result. Lest the reader think that this could never happen, we wish to assure him that that a top event probability smaller than this was predicted by a contractor who did an LNG safety study for the U. S. Coast Guard. Figure 6.9 is a general operator *error rate estimate*. We see the operator is 99.99% perfect at a routine job, but totally useless in an emergency. This, however, is only part of the story: A maintenance man using wrong calibration standards can, in theory, defeat every protective device in a plant, unless precautions are built into the systems. Second-order logic error, i.e., wrong actions taken to correct real (or imagined) malfunctions, are very common and very difficult to predict.

PROBLEMS

6.1. Compare, where possible, the failure data given in Figs. 6.7 and 6.8 and Tables 6.2, 6.3, 6.4, 6.5, and 6.6.

6.2. Rank the following components in the order of their expected MTTF: humans, batteries, analysis sensing elements, gas analyzers, valves, pumps, thermocouples, transducers.

6.3. Use the data in this chapter to obtain failure parameters for the components of the hot-water heater of Problem 1.6.

6.4. Verify, using data in the tables, the failure data given in any of the example problems of Chapter 13.

7

QUANTITATIVE ASPECTS OF SYSTEM ANALYSIS

7.1. INTRODUCTION

The preceding chapters deal with the quantification of basic events. This chapter extends these methods to systems.

System success or failure can be described by a combination of *top events* defined by an OR combination of all system hazards into a composite fault tree (Fig. 7.1). The non-occurrence of all defined top events implies system success. However, occurrence of the top event does not always signify occurrence of a particular system hazard, although it implies a system failure. In general, we can analyze either a particular system hazard or a system success by an appropriate top event and its corresponding fault tree.

The following probabilistic parameters can be defined for the total system. Their interpretation depends on whether the top event refers to a particular system hazard or an OR combination of all system hazards.

1) *System availability* $A_s(t)$ = probability that the top event does not exist at time t. This is the probability of the system's operating successfully when the top event refers to an OR combination of all system hazards. It is the probability of the non-occurrence of a particular hazard when the top event is a single system hazard.
2) *System unavailability* $Q_s(t)$ = probability that the top event exists at time t. This is either the probability of system failure or the probability of a particular system hazard at time t, depending on the defini-

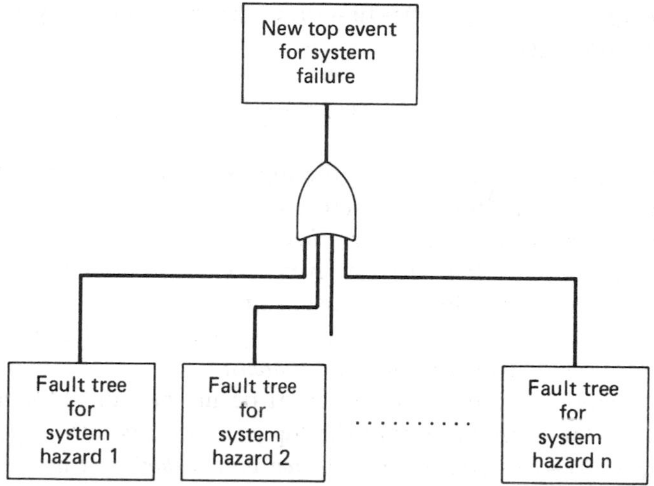

Figure 7.1. Defining a new fault tree by an OR configuration of fault trees for system hazards.

tion of the top event. The system unavailability is complementary to the availability, and the following identity holds:

$$A_s(t) + Q_s(t) = 1 \tag{7.1}$$

3) *System reliability* $R_s(t)$ = probability that the top event does not occur over the time interval $(0, t]$. The system reliability $R_s(t)$ requires continuation of the non-existence of the top event and differs from the system availability $A_s(t)$. The inequality (7.2) holds. Reliability is often used to characterize catastrophic or unrepairable system failures.

$$R_s(t) \le A_s(t) \tag{7.2}$$

4) *System unreliability* $F_s(t)$ = probability that the top event occurs before time t. This is the complement of the system reliability, and the identity

$$R_s(t) + F_s(t) = 1 \tag{7.3}$$

holds. The system unreliability $F_s(t)$ is larger than or equal to the system unavailability:

$$F_s(t) \ge Q_s(t) \tag{7.4}$$

5) *System failure density* $f_s(t)$ = first-order derivative of the system failure distribution $F_s(t)$:

$$f_s(t) = \frac{dF_s(t)}{dt} \tag{7.5}$$

$f_s(t)\,dt$ is the probability that the top event occurs during $[t, t+dt)$, given that it does not occur before time t.

6) $\lambda_s(t)$ = *system conditional failure intensity*
= probability that the top event occurs per unit time at time t, given that it does not exist at time t.

A large value of $\lambda_s(t)$ means that the system is about to fail.

7) $w_s(t)$ = *system unconditional failure intensity*
= probability that the top event occurs per unit time at time t.

$w(t)\,dt$ is the probability that the top event occurs during $[t, t+dt)$.

8) $W_s(t, t+dt)$ = *expected number of top events during* $(t, d+dt]$. Similar to (4.39), the relation

$$W_s(t, t+dt) = w_s(t)\,dt \tag{7.6}$$

holds.

9) $W_s(t_1, t_2)$ = *expected number of top events during* $[t_1, t_2)$. This is given by the integration of the unconditional failure intensity $w_s(t)$:

$$W_s(t_1, t_2) = \int_{t_1}^{t_2} w_s(t)\,dt \tag{7.7}$$

10) MTTF_s = *mean time to first failure*
= expected length of time to the first occurrence of the top event.

The MTTF_s corresponds to average human lifetime and is a suitable parameter for catastrophic system hazards. It is given by

$$\text{MTFF}_s = \int_0^{\infty} t f_s(t)\,dt \tag{7.8}$$

In this chapter we discuss mainly the system availability and unavailability: System reliability and unreliability is quantified in Chapter 9. Unless otherwise stated, all primary events are assumed to be mutually independent. We first demonstrate availability $A_s(t)$ and unavailability $Q_s(t)$ calculations, given relatively simple fault trees. Next, we discuss methods for calculating *lower and upper bounds* for the system unavailability $Q_s(t)$. Then we develop the so-called *kinetic tree theory*[1] which is used to quantify various system parameters for large and complex fault trees.

[1]Vesely, W. E., "A Time-Dependent Methodology for Fault Tree Evaluation," *Nuclear Engr. and Design*, **13**, 337, 1970.

To simplify the nomenclature, we use capital letters B, B_1, C, etc., to represent both the basic events and their *existence* at time t. For a basic event B, the probability $\Pr(B)$ is given by the component unavailability $Q(t)$ when event B is a component failure. The *probability* $\Pr(B)$ becomes an existence probability if the basic event B describes an environmental impact or human error.

As shown in Figs. 2.14 and 2.15 of Chapter 2, each basic event is either:

1) A component primary failure or
2) A component secondary failure or
3) An environmental impact or human error which gives an inadvertent signal to the component and causes a component command fault.

Failure modes should be defined for the component failures and faults. The primary failures are caused by natural aging within the design envelope. Environmental impacts, human error, or system dependent stresses should be identified as possible causes of the secondary failures which create transitions to the failed state. These failure modes and possible causes clarify the basic events and are necessary for successful reliability quantifications.

7.2 AVAILABILITY AND UNAVAILABILITY FOR SIMPLE SYSTEMS WITH INDEPENDENT BASIC EVENTS

7.2.1 Independent Basic Events

The usual assumption regarding basic events $B_1, \ldots, B_n$ is that they are independent, which means that the occurrence of a given basic event is in no way affected by the occurrence of any other basic event. For independent basic events, the simultaneous existence *probability* $\Pr(B_1 \cap B_2 \cap \ldots \cap B_n)$ reduces to

$$\Pr(B_1 \cap B_2 \cap \ldots \cap B_n) = \Pr(B_1)\Pr(B_2)\ldots\Pr(B_n) \tag{7.9}$$

where the symbol $\cap$ represents the intersection of events $B_1 \ldots B_n$. (Appendix 7.1, provides a review of Boolean operations and Venn diagrams.)

7.2.2 System with One AND Gate

Consider the fault tree of Fig. 7.2. Simultaneous existence of basic events $B_1, \ldots, B_n$ results in the top event. Thus, the system unavailability $Q_s(t)$ is given by the probability that all basic events exist at time t:

$$Q_s(t) = \Pr(B_1 \cap B_2 \cap \ldots \cap B_n) \tag{7.10}$$

$$= \Pr(B_1)\Pr(B_2)\ldots\Pr(B_n) \tag{7.11}$$

For an AND gate with two input events, this reduces to

$$Q(t) = \Pr(B_1)\Pr(B_2) \tag{7.12}$$

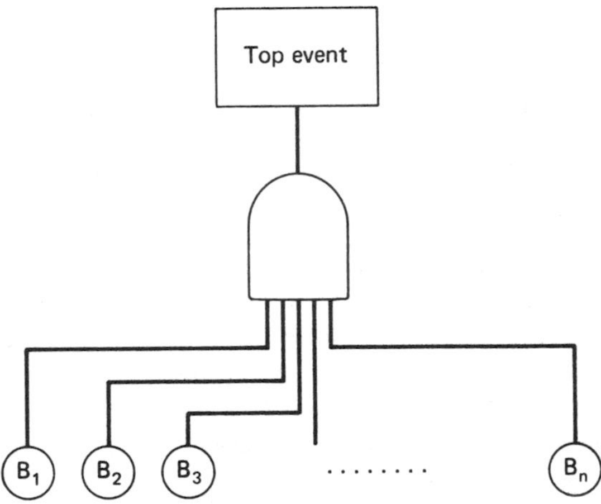

Figure 7.2. Gated AND fault tree.

7.2.3 System with One OR Gate

With reference to Fig. 7.3, the top event exists at time t if and only if at least one of the n basic events occurs at time t. Thus, the system availability $A_s(t)$ and the system unavailability $Q_s(t)$ are given by

$$A_s(t) = \Pr\left(\bar{B}_1 \cap \bar{B}_2 \cap \ldots \cap \bar{B}_n\right) \tag{7.13}$$

$$Q_s(t) = \Pr(B_1 \cup B_2 \cup \ldots \cup B_n) \tag{7.14}$$

where the symbol $\cup$ denotes a union of the events, and $\bar{B}_i$ represents the *complement* of the event B_i; i.e., the event $\bar{B}_i$ means non-occurrence of the event B_i at time t. The independence of the basic events $B_1, \ldots, B_n$ implies the independence of the complementary events $\bar{B}_1, \ldots, \bar{B}_n$. Thus, $A_s(t)$ in (7.13) can be rewritten as

$$\begin{aligned} A_s(t) &= \Pr\left(\bar{B}_1\right)\Pr\left(\bar{B}_2\right)\ldots\Pr\left(\bar{B}_n\right) \\ &= [1-\Pr(B_1)][1-\Pr(B_2)]\ldots[1-\Pr(B_n)] \end{aligned} \tag{7.15}$$

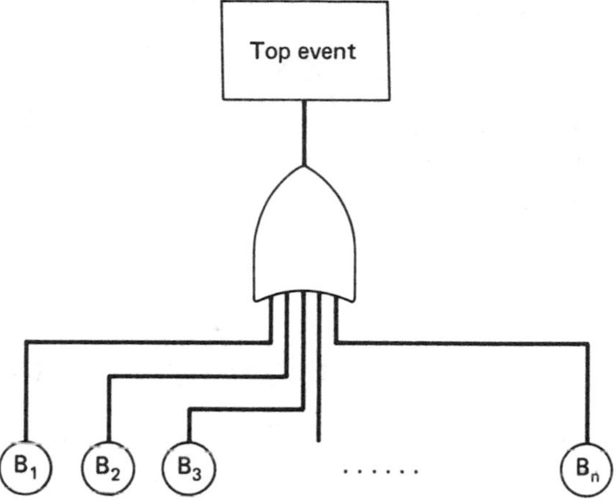

Figure 7.3. Gated OR fault tree.

The unavailability $Q_s(t)$ can be calculated by using (7.1):

$$\begin{aligned} Q_s(t) &= \Pr(B_1 \cup B_2 \cup \ldots \cup B_n) \\ &= 1 - A_s(t) \\ &= 1 - [1 - \Pr(B_1)][1 - \Pr(B_2)] \ldots [1 - \Pr(B_n)] \end{aligned} \tag{7.16}$$

For $n = 2$, we have

$$Q_s(t) = \Pr(B_1 \cup B_2) \tag{7.17}$$

$$= \Pr(B_1) + \Pr(B_2) - \Pr(B_1)\Pr(B_2) \tag{7.18}$$

In other words, the probability $Q_s(t)$ that at least one of the events B_1 and B_2 exists is equal to the sum of the probabilities of each event minus the probability of both events occurring simultaneously. This is shown by the Venn Diagram of Fig. 7.4. For $n = 3$, we obtain

$$\begin{aligned} Q_s(t) &= \Pr(B_1 \cup B_2 \cup B_3) \\ &= \Pr(B_1) + \Pr(B_2) + \Pr(B_3) \\ &\quad - \Pr(B_1)\Pr(B_2) - \Pr(B_2)\Pr(B_3) - \Pr(B_3)\Pr(B_1) \\ &\quad + \Pr(B_1)\Pr(B_2)\Pr(B_3) \end{aligned} \tag{7.19}$$

This unavailability is depicted in Fig. 7.5.

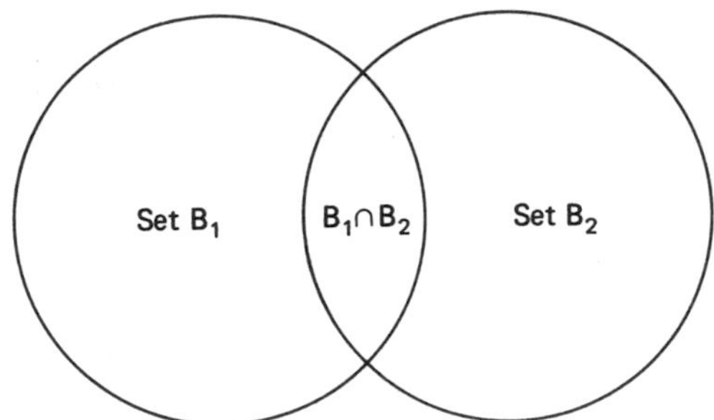

Figure 7.4. $\Pr(B_1 \cup B_2) = \Pr(B_1) + \Pr(B_2) - \Pr(B_1)\Pr(B_2)$.

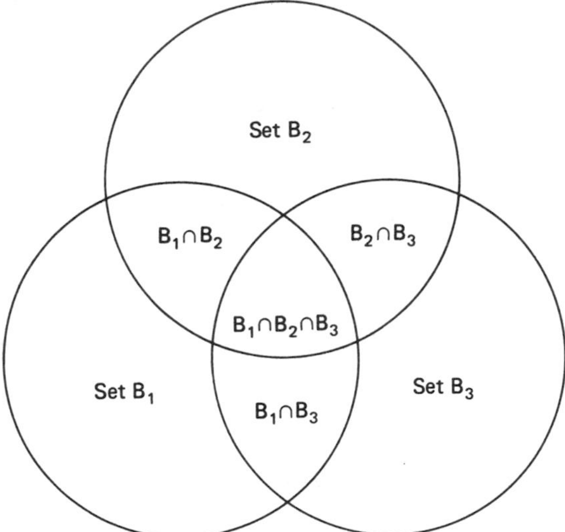

Figure 7.5. Illustration of equation (7.19) for $\Pr(B_1 \cup B_2 \cup B_3)$.

7.2.4 Systems with *m* Out of *n* Voting Gates

The fault tree of Fig. 7.6 appears in a *voting system* which produces an output if more than m components generate an inadvertent command signal. A common application of the m-out-of-n system is in safety systems where it is desirable to avoid expensive shut-downs occasioned by a spurious signal from a single safety monitor.

As an example, consider the two-out-of-three, shut-down device of Fig. 7.7. System shut-down occurs when two out of three safety monitors generate shut-down signals. Consider a case where the system is normal and requires no shut-down. An unnecessary shut-down occurs if more than two safety monitors produce spurious signals. The resulting fault tree is shown in Fig. 7.8, which is a special case of Fig. 7.6.

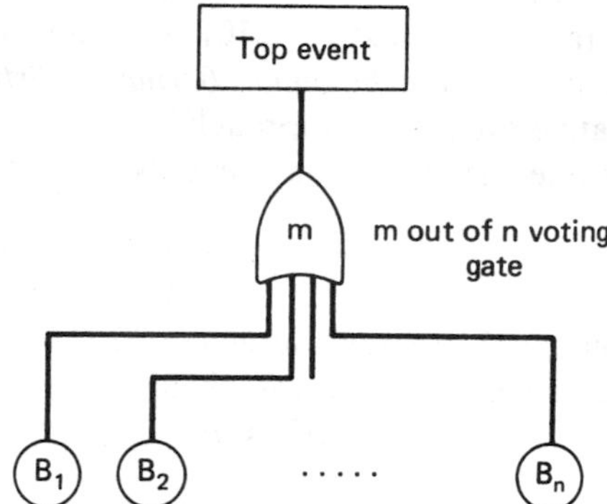

Figure 7.6. *m* out of *n* voting system.

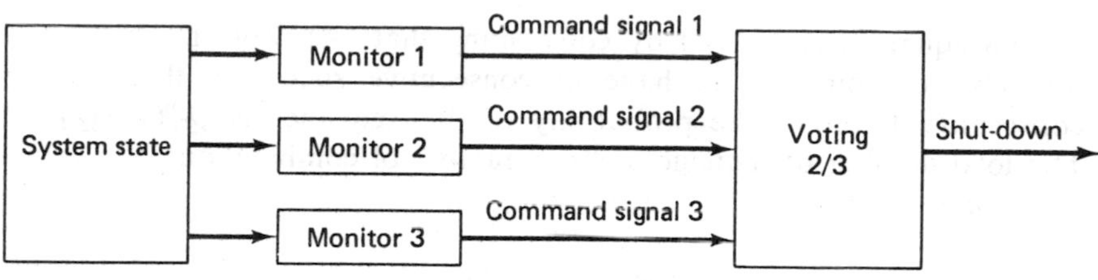

Figure 7.7. Two-out-of-three shut-down system.

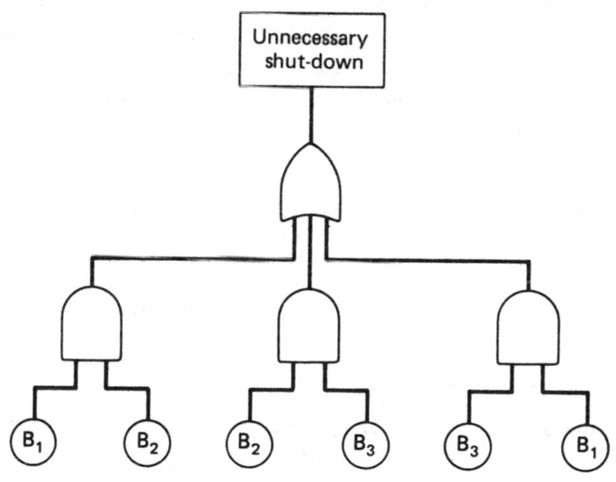

Figure 7.8. Fault tree for two-out-of-three shut-down system (B_i: false signal from monitor i).

Although an m-out-of-n gate such as Fig. 7.8 can always be decomposed (as shown in Fig. 2.10) into equivalent AND and OR gates, direct application of the *binomial Bernoulli distribution* equations represents an alternative analytical approach.

Assume that all primary events have the probability Q.

$$\Pr(B_1) = \Pr(B_2) = \ldots = \Pr(B_n) = Q$$

The binomial distribution gives the probability that a total of m favorable outcomes will occur or not, given the probability of success of any one trial Q and the number of trials n:

$$\Pr(m:n,Q) = \left\{ \begin{matrix} n \\ m \end{matrix} \right\} Q^m(1-Q)^{n-m} \tag{7.20}$$

This equation is derived by considering that one way of achieving m favorable outcomes is to have m consecutive successes, then $(n-m)$ consecutive failures. The probability of this sequence is $Q^m(1-Q)^{n-m}$. The total number of sequences is the number of combinations of n things taken m at a time:

$$\left\{ \begin{matrix} n \\ m \end{matrix} \right\} \equiv \frac{n!}{m!(n-m)!}$$

Therefore, $\Pr(m:n,Q)$ is the sum of all these probabilities and equation (7.20) is proven. In applying it to reliability problems, it is necessary to recognize that the top event will exist if more than m out of the n basic events occur. Thus, it is necessary to sum equation (7.20) over all k successes,

$$Q_s(t) = \Pr(m \le k \le n) = \sum_{k=m}^{n} \left\{ \begin{matrix} n \\ k \end{matrix} \right\} Q^k(1-Q)^{n-k} \tag{7.21}$$

Simple examples which demonstrate the application of the methodology developed in the preceding subsections follow.

Example 1. Two-Out-of-Three System

Calculate the unavailability $Q_s(t)$ for the two-out-of-three configuration of Fig. 7.9 and the OR configuration of Fig. 7.10.

Solution: The unavailability $Q_{s,1}(t)$ for Fig. 7.9 is given by (7.21):

$$Q_{s,1}(t) = \left\{ \begin{matrix} 3 \\ 2 \end{matrix} \right\} Q^2(1-Q) + \left\{ \begin{matrix} 3 \\ 3 \end{matrix} \right\} Q^3(1-Q)^0 = 3Q^2 - 2Q^3 \tag{7.22}$$

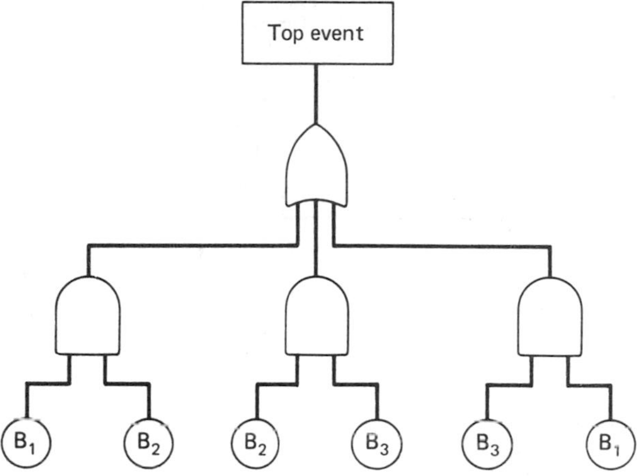

Figure 7.9. Fault tree for two-out-of-three system ($\Pr(B_i)=Q$).

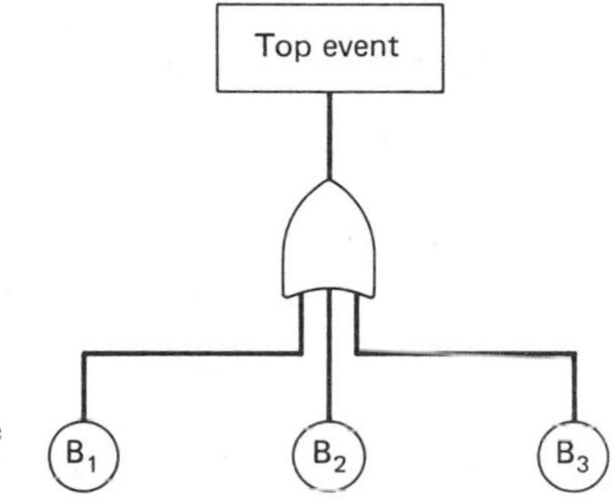

Figure 7.10. Gated OR fault tree ($\Pr(B_i)=Q$).

The unavailability $Q_{s,2}(t)$ for Fig. 7.10 is obtained from (7.16) or (7.19):

$$Q_{s,2}(t)=1-(1-Q)^3=3Q-3Q^2+Q^3 \tag{7.23}$$

Thus,

$$Q_{s,2}-Q_{s,1}=3Q(1-Q)^2>0, \quad \text{for } 0<Q<1$$

and we conclude that the safety system with a two-out-of-three configuration has a smaller probability of spurious shut-downs than the system with the simple OR configuration.

Example 2. Simple Combinations of Gates

Calculate the unavailability of the system described by the fault tree of Fig. 7.11, given the probabilities for the basic events shown in the tree.

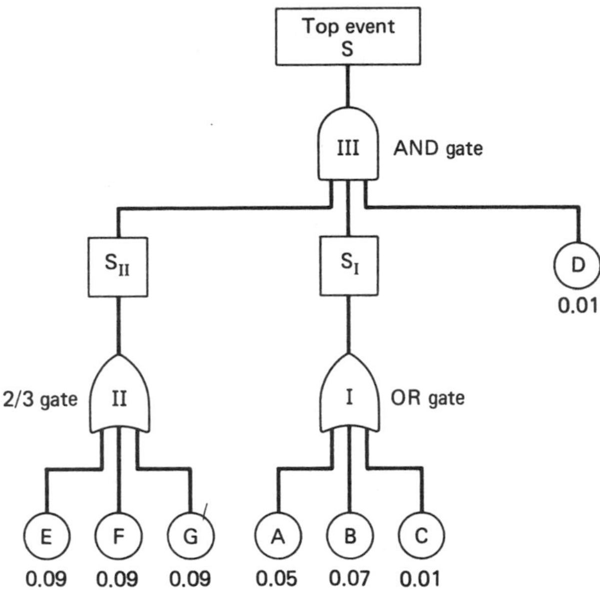

Figure 7.11. Simple combination of gates.

Solution: Using (7.16) for OR gate I:

$$\Pr(S_I) = 1 - (1 - 0.05)(1 - 0.07)(1 - 0.1) = 0.20485$$

For the two-out-of-three gate II, by (7.21):

$$\Pr(S_{II}) = \sum_{k=2}^{3} \left\{ \begin{matrix} 3 \\ k \end{matrix} \right\} 0.09^k (1 - 0.09)^{3-k}$$

$$= \left\{ \begin{matrix} 3 \\ 2 \end{matrix} \right\} (0.09)^2 (1 - 0.09) + \left\{ \begin{matrix} 3 \\ 3 \end{matrix} \right\} (0.09)^3 (1 - 0.09)^0$$

$$= 0.022842$$

Using (7.11) for AND gate III:

$$Q_s(t) = \Pr(S) = \Pr(S_I)\Pr(S_{II})\Pr(D)$$

$$= (0.20485)(0.022842)(0.01) = 4.68 \times 10^{-5}$$

Example 3. Tail Gas Quench and Clean Up System[2]

The system in Fig. 7.12 is designed to:

1) Decrease the temperature of a hot gas by a water quench.
2) Saturate the gas with water vapor.
3) Remove solid particles entrained in the gas.

[2]Caceres, S., and E. J. Henley, "Process Analysis by Block Diagrams and Fault Trees," *Ind. Engr. Chem.*, **15**, No. 2, 128, 1976.

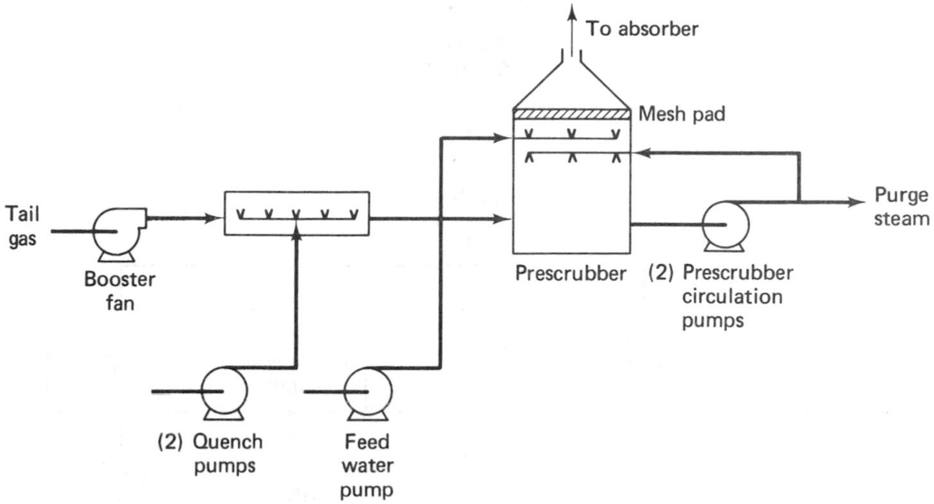

Figure 7.12. Schematic diagram of tail gas quench and clean up system.

A simplified fault tree is shown in Fig. 7.13. The booster fan (A), both of quench pumps (B and C), the feed water pump (D), both of circulation pumps (E and F), or the filter system (G) must fail for the top event S to occur. The top event expression for this fault tree is

$$S = A \cup (B \cap C) \cup D \cup (E \cap F) \cup G \tag{7.24}$$

Calculate the system unavailability $Q_s(t) = \Pr(S)$ using as data:

$$\begin{aligned} &\Pr(A)=0.9, \quad \Pr(B)=0.8, \quad \Pr(C)=0.7, \quad \Pr(D)=0.6 \\ &\Pr(E)=0.5, \quad \Pr(F)=0.4, \quad \Pr(G)=0.3 \end{aligned} \tag{7.25}$$

Solution: We proceed in a stepwise fashion:

$$\begin{aligned} &\Pr(B \cap C) = (0.8)(0.7) = 0.56 \\ &\Pr(A \cup (B \cap C)) = 0.9 + 0.56 - (0.9)(0.56) = 0.956 \\ &\Pr(A \cup (B \cap C) \cup D) = 0.956 + 0.6 - (0.956)(0.6) = 0.9824 \\ &\Pr(E \cap F) = (0.5)(0.4) = 0.2 \\ &\Pr(A \cup (B \cap C) \cup D \cup (E \cap F)) = 0.9824 + 0.2 - (0.9824)(0.2) = 0.98592 \\ &Q_s(t) = \Pr(S) = 0.98592 + 0.3 - (0.98592)(0.3) = 0.990144 \end{aligned} \tag{7.26}$$

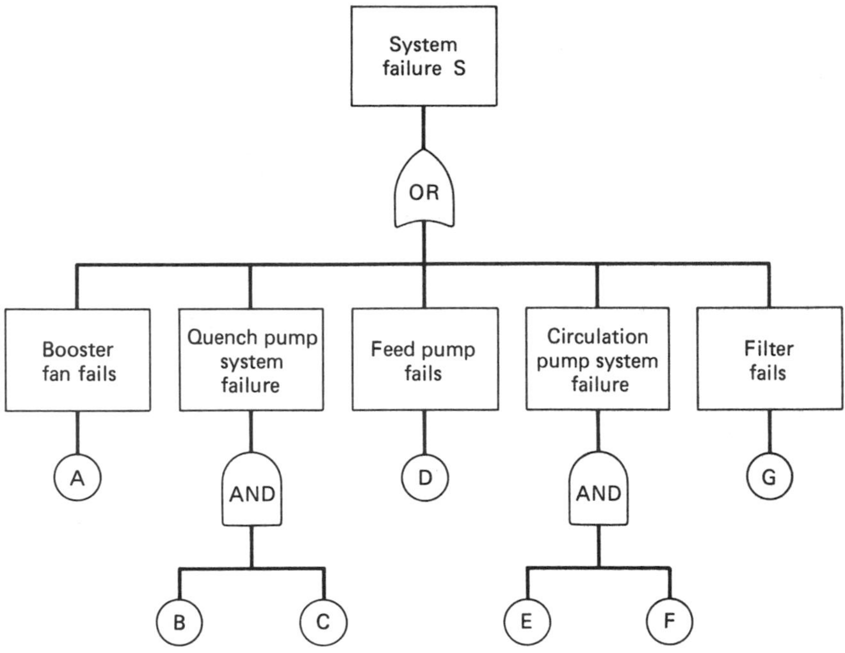

Figure 7.13. Fault tree for tail gas quench and clean up system.

7.2.5 Availability Calculation Based on Reliability Block Diagrams

Reliability *block diagrams* are an alternative way of representing events and gates, as are *success trees*, which are the mathematical duals of fault trees in which the top of the tree represents system success, and the events are success rather than failure status. The relationship between these three forms of system representation can best be shown by example.

Consider again the system of Fig. 7.12. The reliability block diagram is given as Fig. 7.14, where the booster fan ($\overline{A}$), either quench pump ($\overline{B}$ or $\overline{C}$), the feed water pump ($\overline{D}$), either circulation pump ($\overline{E}$ or $\overline{F}$) and the filter system ($\overline{G}$) must be operating successfully for the system to work.

Figure 7.15 is the success tree equivalent to the block diagram representation in Fig. 7.14. Boolean logic gates are used to indicate the parallel (OR) and the series (AND) connections in the block diagram. The expression for the success tree becomes

$$\overline{S}=\overline{A}\cap(\overline{C}\cup\overline{B})\cap\overline{D}\cap(\overline{F}\cup\overline{E})\cap\overline{G} \tag{7.27}$$

where $\overline{A},\ldots,\overline{G}$ are the events that components $\overline{A}-\overline{G}$ are functioning, and

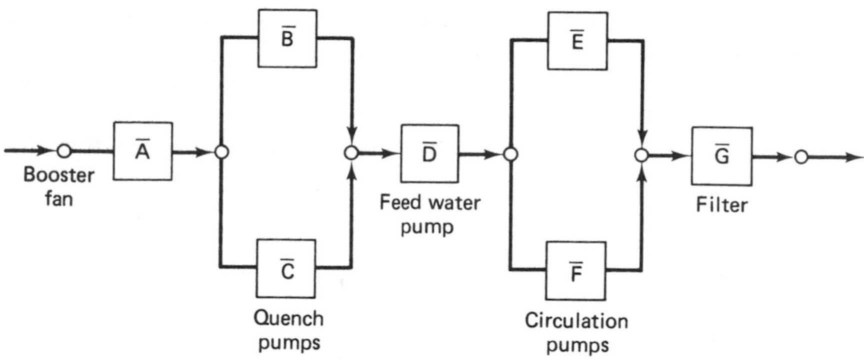

Figure 7.14. Reliability block diagram for tail gas quench and clean up system.

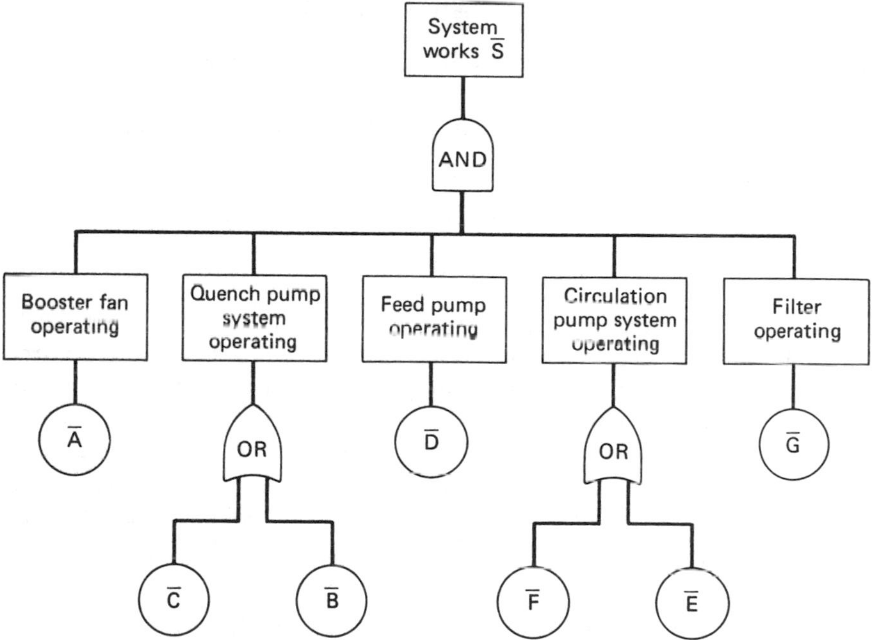

Figure 7.15. Success tree for tail gas quench and clean up system.

$\bar{S}$ is the event of the system functioning. The events $\bar{A},\ldots,\bar{G}$ are complements of the basic events $A,\ldots,G$ in the fault tree of Fig. 7.13.

Since the system is either functioning or failed at time t, the $\bar{S}$ of (7.27) should be the complement of the event S of (7.24). This complementary relation between (7.24) and (7.27) can also be stated in terms of *de Morgan's rule* which, for systems such as Figs. 7.13 and 7.15, states that if

$\bar{S}$ is complementary to S, we can obtain $\bar{S}$ from the negation of the Boolean expression for S, i.e., by interchanging AND's and OR's and replacing A by $\bar{A}, B$ by $\bar{B}$, etc. This is proven by examining (7.24) and (7.27) or Figs. 7.13 and 7.15.

The system availability $A_s(t)$ can be calculated by the probability $\Pr(\bar{S})$ in the following way:

From (7.25), in Example 3:

$$\Pr(\bar{A})=0.1, \quad \Pr(\bar{B})=0.2, \quad \Pr(\bar{C})=0.3, \quad \Pr(\bar{D})=0.4$$
$$\Pr(\bar{E})=0.5, \quad \Pr(\bar{F})=0.6, \quad \Pr(\bar{G})=0.7$$

Hence,

$$\begin{aligned} A_s(t) &= \Pr(\bar{S}) \\ &= (0.1)[0.3+0.2-(0.3)(0.2)](0.4) \\ &\quad \times[0.6+0.5-(0.6)(0.5)](0.7)=0.009856 \end{aligned}$$

This availability, and the unavailability in the preceding example, agree with identity (7.1):

$$A_s(t)+Q_s(t)=0.009856+0.990144=1$$

The foregoing examples show that:

1) A parallel reliability block diagram corresponds to a gated AND fault tree, and a series block diagram to the gated OR fault tree (Table 7.1). A parallel structure of components is called a *parallel system* and the series structure a *series system.*
2) The unavailability calculation methods for the fault tree can be extended directly to availability calculations for success trees when the basic events $B_1, \ldots, B_n$ are replaced by their complementary events $\bar{B}_1, \ldots, \bar{B}_n$ in (7.9) and (7.16):

$$\Pr(\bar{B}_1 \cap \bar{B}_2 \cap \ldots \cap \bar{B}_n)=\Pr(\bar{B}_1)\Pr(\bar{B}_2)\ldots\Pr(\bar{B}_n)$$
$$\Pr(\bar{B}_1 \cup \bar{B}_2 \cup \ldots \cup \bar{B}_n)=1-[1-\Pr(\bar{B}_1)][1-\Pr(\bar{B}_2)]\ldots[1-\Pr(\bar{B}_n)]$$

7.3 AVAILABILITY AND UNAVAILABILITY CALCULATIONS USING TRUTH TABLES

A *truth table* is a listing of all combinations of the states of basic events, the resulting occurrence or non-occurrence of a top event, and the corresponding probabilities for these combinations. A summation of a set of

TABLE 7.1 Reliability Block Diagram Versus Fault Tree

	Reliability Block Diagram	Fault Tree
Series System	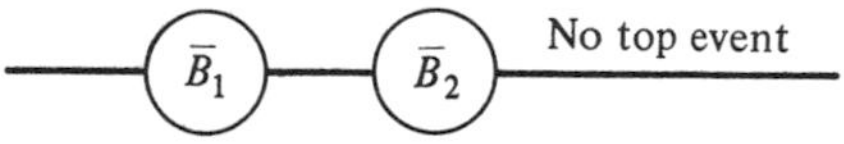 $\bar{B}_1$ $\bar{B}_2$ No top event $Pr(\bar{B}_1 \cap \bar{B}_2) = Pr(\bar{B}_1)Pr(\bar{B}_2)$	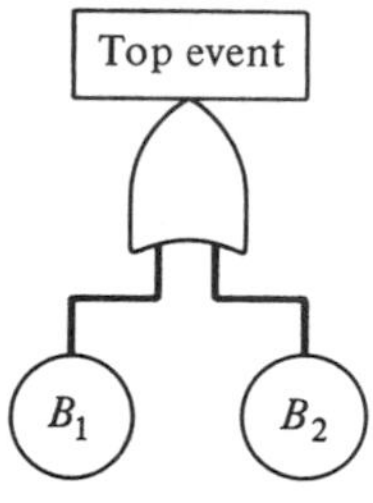 Top event B_1 B_2 $Pr(B_1 \cup B_2) = Pr(B_1) + Pr(B_2) - Pr(B_1)Pr(B_2)$
Parallel System	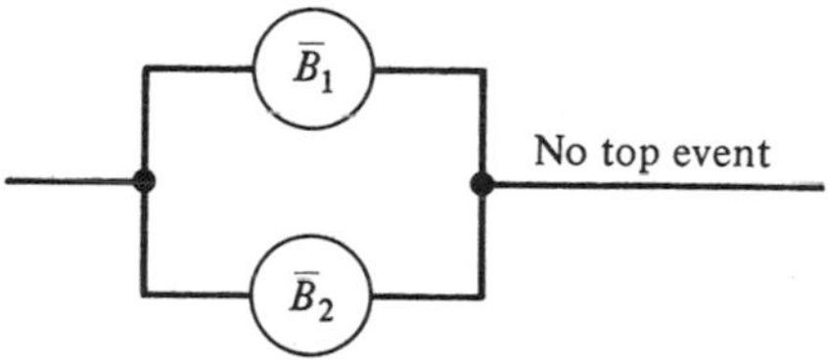 $\bar{B}_1$ $\bar{B}_2$ No top event $Pr(\bar{B}_1 \cup \bar{B}_2) = Pr(\bar{B}_1) + Pr(\bar{B}_2) - Pr(\bar{B}_1)Pr(\bar{B}_2)$	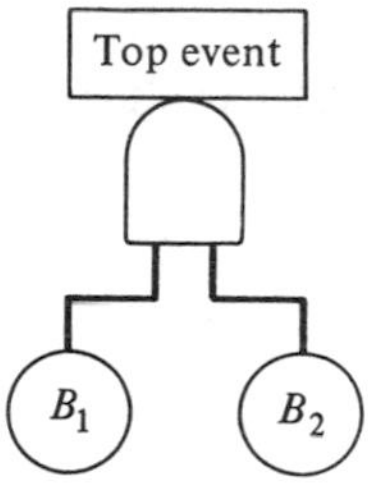 Top event B_1 B_2 $Pr(B_1 \cap B_2) = Pr(B_1)Pr(B_2)$

probabilities in the table yields the system unavailability $Q_s(t)$, and a complementary summation gives the system availability $A_s(t)$.

7.3.1 System with One AND Gate

Table 7.2 is a truth table for the system of Fig. 7.16. The system unavailability $Q_s(t)$ is given by row 1:

$$Q_s(t) = \Pr(B_1)\Pr(B_2)$$

TABLE 7.2 Truth Table for the Fault Tree in Fig. 7.16

	Basic Event B_1	Basic Event B_2	Top Event	Probability
1	Exists	Exists	Exists	$\Pr(B_1)\Pr(B_2)$
2	Exists	Not Exist	Not Exist	$\Pr(B_1)\Pr(\bar{B}_2)$
3	Not Exist	Exists	Not Exist	$\Pr(\bar{B}_1)\Pr(B_2)$
4	Not Exist	Not Exist	Not Exist	$\Pr(\bar{B}_1)\Pr(\bar{B}_2)$

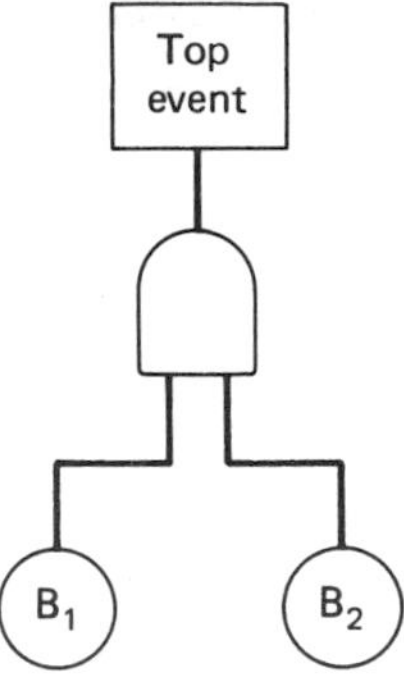

Figure 7.16. Gated AND fault tree.

7.3.2 System with One OR Gate

The system of Fig. 7.17 is represented by the truth table of Table 7.3. The unavailability $Q_s(t)$ is obtained by a summation of the probabilities of the mutually exclusive rows 1, 2, and 3.

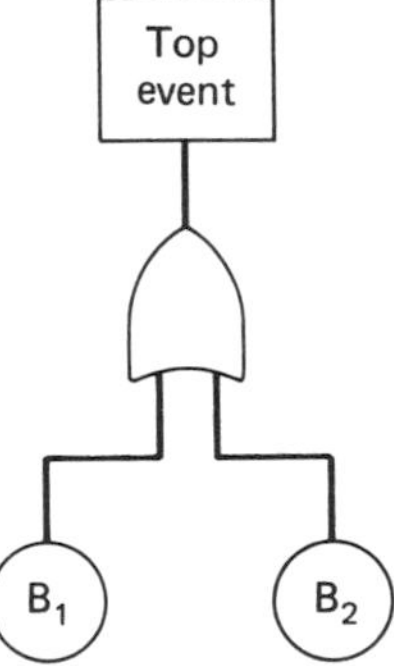

Figure 7.17. Gated OR fault tree.

$$\begin{aligned} Q_s(t) &= \Pr(B_1)\Pr(\bar{B}_2) + \Pr(\bar{B}_1)\Pr(B_2) + \Pr(B_1)\Pr(B_2) \\ &= \Pr(B_1)[1-\Pr(B_2)] + [1-\Pr(B_1)]\Pr(B_2) \\ &\quad + \Pr(B_1)\Pr(B_2) \\ &= \Pr(B_1) + \Pr(B_2) - \Pr(B_1)\Pr(B_2) \end{aligned} \tag{7.28}$$

This confirms equation (7.18).

Example 4.

A truth table provides a tedious but reliable technique for calculating the availability and unavailability for moderately complicated systems, as illustrated by the following example.†

†Courtesy of B. Bulloch, ICI Ltd., Runcorn, England.

TABLE 7.3 TRUTH TABLE FOR THE FAULT TREE IN FIG. 7.17

	Basic Event B_1	Basic Event B_2	Top Event	Probability
1	Exists	Exists	Exists	$\Pr(B_1)\Pr(B_2)$
2	Exists	Not Exist	Exists	$\Pr(B_1)\Pr(\bar{B}_2)$
3	Not Exist	Exists	Exists	$\Pr(\bar{B}_1)\Pr(B_2)$
4	Not Exist	Not Exist	Not Exist	$\Pr(\bar{B}_1)\Pr(\bar{B}_2)$

A plant has two, identical, parallel streams, A and B, consisting of one transfer pump and one rotary filter (Fig. 7.18). The failure rate of the pumps and filters are, respectively, 0.04 and 0.08 failures per day, whether equipment is in operation or standby. Assume MTTR's for the pumps and filters of 5 and 10 hr, respectively.

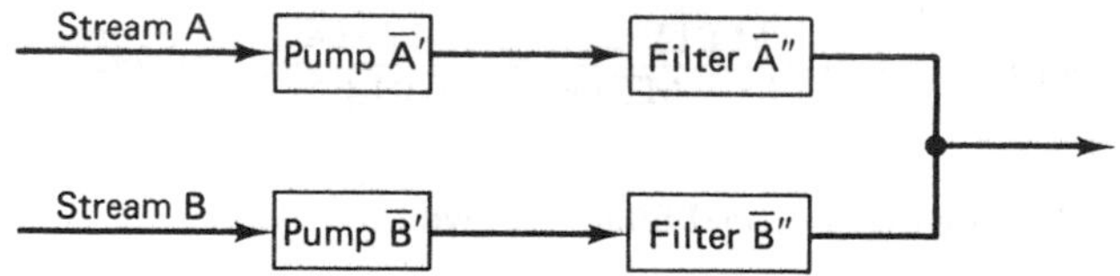

Figure 7.18. Two parallel process streams.

Two alternative schemes to increase plant availability are:

1) Add a third identical stream, C (Fig. 7.19).
2) Install a third transfer pump capable of pumping slurry to either filter (Fig. 7.20).

   Compare the effect of these two schemes on the ability of the plant to maintain: (a) full output; (b) *not less than* half output.

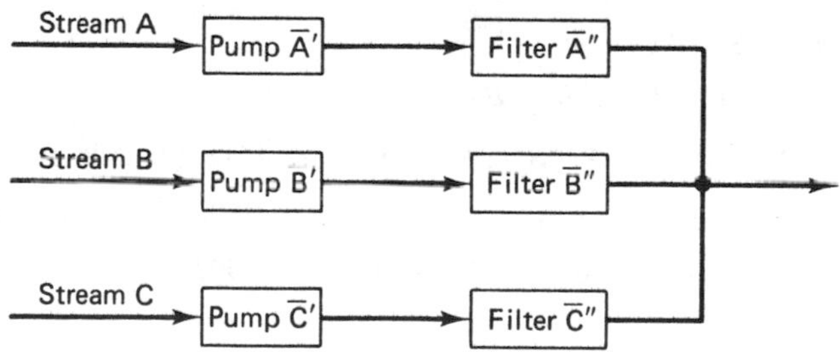

Figure 7.19. Three parallel process streams.

Solution:

1) Making the usual constant failure and repair rates assumption, the steady-state availabilities for the filter and the pump become (see Table 4.10, chapter 4).

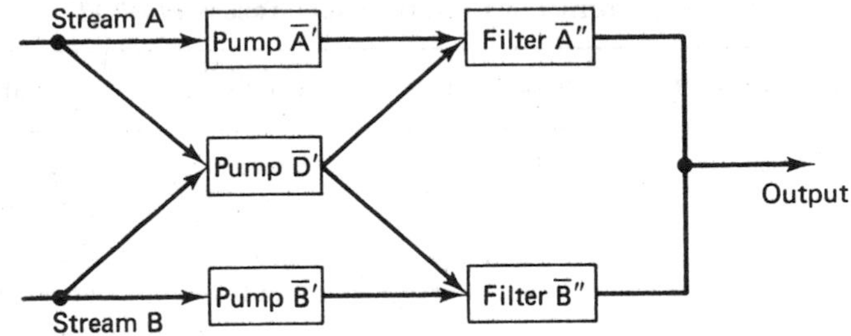

Figure 7.20. Additional spare pump $\bar{D}'$.

$$A\,(\text{filter}) = \frac{\text{MTTF}}{\text{MTTF}+\text{MTTR}} = \frac{1/0.08}{1/0.08+10/24} = 0.968$$

$$A\,(\text{pump}) = \frac{\text{MTTF}}{\text{MTTF}+\text{MTTR}} = \frac{1/0.04}{1/0.04+5/24} = 0.992$$

Thus, the steady-state event probabilities are given by

$$\Pr(\bar{A}'') = \Pr(\bar{B}'') = \Pr(\bar{C}'') = 0.968$$
$$\Pr(\bar{A}') = \Pr(\bar{B}') = \Pr(\bar{C}') = \Pr(\bar{D}') = 0.992$$

Considering the existing plant of Fig. 7.18, the availabilities for full output A_s(full) and for half output A_s(half) are

$$\begin{aligned}
A_s(\text{full}) &= \Pr(\bar{A}' \cap \bar{A}'' \cap \bar{B}' \cap \bar{B}'') \\
&= \Pr(\bar{A}')\Pr(\bar{A}'')\Pr(\bar{B}')\Pr(\bar{B}'') \\
&= 0.968^2 \times 0.992^2 = 0.922 \\
A_s(\text{half}) &= \Pr([\bar{A}' \cap \bar{A}''] \cup [\bar{B}' \cap \bar{B}'']) \\
&= \Pr(\bar{A}' \cap \bar{A}'') + \Pr(\bar{B}' \cap \bar{B}'') - \Pr(\bar{A}' \cap \bar{A}'')\Pr(\bar{B}' \cap \bar{B}'') \\
&= \Pr(\bar{A}')\Pr(\bar{A}'') + \Pr(\bar{B}')\Pr(\bar{B}'') - \Pr(\bar{A}')\Pr(\bar{A}'')\Pr(\bar{B}')\Pr(\bar{B}'') \\
&= 0.968 \times 0.992 + 0.968 \times 0.992 - 0.968^2 \times 0.992^2 \\
&= 0.9984
\end{aligned}$$

If a third stream is added, we have a two-out-of-three system for full production. Thus, using (7.21)

$$\begin{aligned}
A_s(\text{full}) &= 3 \cdot [\Pr(\bar{A}')\Pr(\bar{A}'')]^2 [1 - \Pr(\bar{A}')\Pr(\bar{A}'')] \\
&\quad + [\Pr(\bar{A}')\Pr(\bar{A}'')]^3 [1 - \Pr(\bar{A}')\Pr(\bar{A}'')]^0 \\
&= 0.9954
\end{aligned}$$

For half production we have three parallel units, thus:

$$A_s(\text{half}) = 1 - \left[1 - \Pr(\bar{A}')\Pr(\bar{A}'')\right]\left[1 - \Pr(\bar{B}')\Pr(\bar{B}'')\right]$$
$$\times\left[1 - \Pr(\bar{C}')\Pr(\bar{C}'')\right]$$
$$= 0.99994$$

2) Calculation of the reliability of the configuration shown as Fig. 7.20 represents a problem because this is a *bridged network* and cannot be reduced to a simple parallel system. A truth table is used to enumerate all possible component states and select the combinations which give full and half output (Table 7.4). The availability for full production is given by

$$A_s(\text{full}) = \Sigma\Pr(\text{rows } 1, 2, 5, 17)$$
$$= \Pr(\bar{A}')\Pr(\bar{A}'')\Pr(\bar{B}')\Pr(\bar{B}'')\Pr(\bar{D}')$$
$$+ \Pr(\bar{A}')\Pr(\bar{A}'')\Pr(\bar{B}')\Pr(\bar{B}'')\left[1 - \Pr(\bar{D}')\right]$$
$$+ \Pr(\bar{A}')\Pr(\bar{A}'')\left[1 - \Pr(\bar{B}')\right]\Pr(\bar{B}'')\Pr(\bar{D}')$$
$$+ \left[1 - \Pr(\bar{A}')\right]\Pr(\bar{A}'')\Pr(\bar{B}')\Pr(\bar{B}'')\Pr(\bar{D}')$$
$$= 0.93685$$

There are so many states leading to half production that it is easier to work with the unavailability Q_s(half).

$$Q_s(\text{half}) = \Sigma\Pr(\text{rows } 11, 12, 14, 15, 16, 20, 22, 24, 27, 28, 30, 31, 34)$$
$$= 0.001028$$

yielding

$$A_s(\text{half}) = 1 - Q_s(\text{half}) = 0.998972$$

The results are summarized in columns 2 and 3 of Table 7.5. The availabilities, when coupled with economic data on the equipment capital costs and the cost of lost production, permit economic assessments to be made. If, for example, the full cost of a pump (including maintenance, installation, etc.) is $15 per day and a filter costs $60 per day and the costs of full- and half-production lost are $10,000 per day and $2000 per day, respectively, then the expected loss can be calculated by the following formula.

$$\text{expected loss/day (dollars)} = n' \times 15 + n'' \times 60 + [1 - A_s(\text{half})] \times 10{,}000$$
$$+ [A_s(\text{half}) - A_s(\text{full})] \times 2000. \tag{7.29}$$

where n' = the number of pumps,

n'' = the number of filters.

The formula is illustrated by Fig. 7.21. Note that $1 - A_s$(half) is the proportion of the plant operation time expected to result in full-production lost, and A_s(half) $-$ A_s(full) is the proportion resulting in half-production lost. The expected costs are summarized in Table 7.5. We observe that the plant with the spare stream is the best choice.

TABLE 7.4 State Enumeration for System with Additional Spare Pump D' (W, working; F, failed)

State	$\bar{A}'$	$\bar{A}''$	$\bar{B}'$	$\bar{B}''$	$\bar{D}'$	Full Output	Half Output
1	W	W	W	W	W	Yes	Yes
2	W	W	W	W	F	Yes	Yes
3	W	W	W	F	W	No	Yes
4	W	W	W	F	F	No	Yes
5	W	W	F	W	W	Yes	Yes
6	W	W	F	W	F	No	Yes
7	W	W	F	F	W	No	Yes
8	W	W	F	F	F	No	Yes
9	W	F	W	W	W	No	Yes
10	W	F	W	W	F	No	Yes
11	W	F	W	F	W	No	No
12	W	F	W	F	F	No	No
13	W	F	F	W	W	No	Yes
14	W	F	F	W	F	No	No
15	W	F	F	F	W	No	No
16	W	F	F	F	F	No	No
17	F	W	W	W	W	Yes	Yes
18	F	W	W	W	F	No	Yes
19	F	W	W	F	W	No	Yes
20	F	W	W	F	F	No	No
21	F	W	F	W	W	No	Yes
22	F	W	F	W	F	No	No
23	F	W	F	F	W	No	Yes
24	F	W	F	F	F	No	No
25	F	F	W	W	W	No	Yes
26	F	F	W	W	F	No	Yes
27	F	F	W	F	W	No	No
28	F	F	W	F	F	No	No
29	F	F	F	W	W	No	Yes
30	F	F	F	W	F	No	No
31	F	F	F	F	W	No	No
32	F	F	F	F	F	No	No

TABLE 7.5 Comparison of Costs for Three Plants

Plant	A_s (full)	A_s (half)	Cost
Existing plant	0.922	0.9984	$318.8/day
Plant with spare stream	0.9954	0.99994	$234.7/day
Plant with spare pump	0.93685	0.99897	$299.5/day

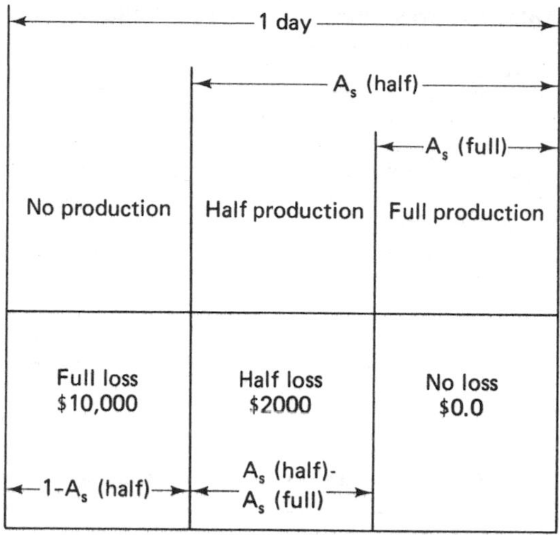

Figure 7.21. Illustration of expected loss per day, (7.29).

7.4 AVAILABILITY AND UNAVAILABILITY CALCULATIONS USING STRUCTURE FUNCTIONS

7.4.1 Structure Functions

It is possible to describe the state of the basic event or the system by a *binary indicator variable*. If we assign a binary indicator variable Y_i to the basic event i, then

$$Y_i = \begin{cases} 1, & \text{when the basic event is occurring} \\ 0, & \text{when the event is not occurring.} \end{cases}$$

Similarly, the top event is associated with a binary indicator variable $\psi(Y)$ related to the state of the system by

$$\psi(\mathbf{Y}) = \begin{cases} 1, & \text{when the top event is occurring} \\ 0, & \text{when the top event is not occurring} \end{cases}$$

Here $\mathbf{Y} = (Y_1, Y_2, \ldots, Y_n)$ is the vector of basic event states. The function $\psi(\mathbf{Y})$ is known as the *structure function* for the top event.

The event unions and intersections, $\cup$ and $\cap$, used in the set expressions to express relationships between events, correspond to the Boolean operators $\vee$ (OR) and $\wedge$ (AND), and to the usual algebraic operations $+$ and $\times$ as shown in Table 7.6. Note that $\Pr(B_i) = E(Y_i)$; thus, $E(\ \ )$ is an *expected number*, or probability.

TABLE 7.6 Event, Boolean, and Algebraic Operations

Event	Boolean	Algebraic	Note
B_i	$Y_i=1$	$Y_i=1$	Event i exists
$\bar{B}_i$	$Y_i=0$	$Y_i=0$	Event i does not exist
$B_i \cap B_j$	$Y_i \wedge Y_j=1$	$Y_i Y_j=1$	$\Pr(B_i \cap B_j)=E(Y_i \wedge Y_j)$
$B_i \cup B_j$	$Y_i \vee Y_j=1$	$1-[1-Y_i][1-Y_j]=1$	$\Pr(B_i \cup B_j)=E(Y_i \vee Y_j)$
$B_1 \cap \ldots \cap B_n$	$Y_1 \wedge \cdots \wedge Y_n=1$	$Y_1 \times \cdots \times Y_n=1$	$\Pr(B_1 \cap \cdots \cap B_n)=E(Y_1 \wedge \cdots \wedge Y_n)$
$B_1 \cup \cdots \cup B_n$	$Y_1 \vee \cdots \vee Y_n=1$	$1-[1-Y_1]\times \cdots \times[1-Y_n]=1$	$\Pr(B_1 \cup \cdots \cup B_n)=E(Y_1 \vee \cdots \vee Y_n)$

The operators $\vee$ and $\wedge$ can be manipulated in accordance with the rules of Boolean algebra. These rules and the corresponding algebraic representations are listed in Table 7.7.

TABLE 7.7 Rules for Boolean Manipulations

Laws	Algebraic Interpretation and Remark
Identities:	
$B \vee B=B$	$1-[1-Y][1-Y]=Y$
$B \wedge B=B$	$YY=Y$
Commutative laws:	
$B_1 \vee B_2=B_2 \vee B_1$	$1-[1-Y_1][1-Y_2]=1-[1-Y_2][1-Y_1]$
$B_1 \wedge B_2=B_2 \wedge B_1$	$Y_1 Y_2=Y_2 Y_1$
Associative laws:	
$B_1 \vee(B_2 \vee B_3)=(B_1 \vee B_2) \vee B_3$	Can be written as $B_1 \vee B_2 \vee B_3$
$B_1 \wedge(B_2 \wedge B_3)=(B_1 \wedge B_2) \wedge B_3$	Can be written as $B_1 \wedge B_2 \wedge B_3$
Distributive laws:	
$B_1 \wedge(B_2 \vee B_3)=(B_1 \wedge B_2) \vee(B_1 \wedge B_3)$	
$B_1 \vee(B_2 \wedge B_3)=(B_1 \vee B_2) \wedge(B_1 \vee B_3)$	
Absorption laws:	
$B_1 \wedge(B_1 \wedge B_2)=B_1 \wedge B_2$	$Y_1 Y_1 Y_2=Y_1 Y_2$
$B_1 \vee(B_1 \wedge B_2)=B_1$	$1-[1-Y_1][1-Y_1 Y_2]=Y_1$
Set definitions:	
$B \wedge 1=B$	$Y \cdot 1=Y$
$B \wedge 0=0$	$Y \cdot 0=0$
$B \vee 0=B$	$1-[1-Y][1-0]=Y$
$B \vee 1=1$	$1-[1-Y][1-1]=1$

7.4.2 System Representation in Terms of Structure Functions

The top event of the gated AND tree in Fig. 7.2 exists if and only if all basic events $B_1, \ldots, B_n$ exist. In terms of the system structure function,

$$\psi(\mathbf{Y})=\psi(Y_1, Y_2, \ldots, Y_n)=\bigwedge_{i=1}^{n} Y_i=Y_1 \wedge Y_2 \wedge \ldots \wedge Y_n. \tag{7.30}$$

where Y_i is an *indicator variable* for the basic event B_i.

The structure function can be expressed in terms of algebraic operators (see Table 7.6):

$$\psi(\mathbf{Y}) = \prod_{i=1}^{n} Y_i = Y_1 Y_2, \ldots, Y_n \tag{7.31}$$

The gated OR tree of Fig. 7.3 fails (the top event exists) if any of the basic events $B_1, B_2, \ldots, B_n$ are occurring. The structure function is

$$\psi(\mathbf{Y}) = \bigvee_{i=1}^{n} Y_i = Y_1 \vee Y_2 \vee \ldots \vee Y_n \tag{7.32}$$

and its algebraic form is

$$\psi(\mathbf{Y}) = 1 - \prod_{i=1}^{n} [1 - Y_i] \tag{7.33}$$

$$= 1 - [1 - Y_1][1 - Y_2] \cdots [1 - Y_n] \tag{7.34}$$

If the n in Fig. 7.3 is two, i.e., for a two event series structure

$$\psi(\mathbf{Y}) = Y_1 \vee Y_2 = 1 - [1 - Y_1][1 - Y_2] \tag{7.35}$$

$$= Y_1 + Y_2 - Y_1 Y_2 \tag{7.36}$$

This result is analogous to (7.18), where $Y_1 Y_2$ represent the probability of the intersected portion of the two events B_1 and B_2.

A slightly more sophisticated example is the two-out-of-three voting system of Fig. 7.8. The structure function is

$$\psi(\mathbf{Y}) = (Y_1 \wedge Y_2) \vee (Y_2 \wedge Y_3) \vee (Y_3 \wedge Y_1) \tag{7.37}$$

and its algebraic expression is obtained in the following way:

$$\begin{aligned}\psi(\mathbf{Y}) &= 1 - [1 - (Y_1 \wedge Y_2)][1 - (Y_2 \wedge Y_3)][1 - (Y_3 \wedge Y_1)] \\ &= 1 - [1 - Y_1 Y_2][1 - Y_2 Y_3][1 - Y_3 Y_1]\end{aligned} \tag{7.38}$$

This equation can be expanded and simplified by the *absorption law* of Table 7.7:

$$\begin{aligned}\psi(\mathbf{Y}) = 1 - [&1 - Y_1 Y_2 - Y_2 Y_3 - Y_3 Y_1 + Y_1 Y_2 Y_2 Y_3 + Y_2 Y_3 Y_3 Y_1 + Y_3 Y_1 Y_1 Y_2 \\ &- Y_1 Y_2 Y_2 Y_3 Y_3 Y_1]\end{aligned} \tag{7.39}$$

$$= Y_1 Y_2 + Y_2 Y_3 + Y_3 Y_1 - 2 Y_1 Y_2 Y_3 \tag{7.40}$$

where the absorption law

$$Y_1Y_2Y_2Y_3 = Y_2Y_3Y_3Y_1 = Y_3Y_1Y_1Y_2 = Y_1Y_2Y_2Y_3Y_3Y_1 = Y_1Y_2Y_3 \tag{7.41}$$

was used in going from (7.39) to (7.40).

Structure functions can be obtained in a stepwise way. For example, the structure function for Fig. 7.13 is given as follows:

$$\psi_1(Y) = Y_B \wedge Y_C = Y_B Y_C, \qquad \psi_2(Y) = Y_E \wedge Y_F = Y_E Y_F \tag{7.42}$$

where $\psi_1(Y)$ is a structure function for the first AND gate,
$\psi_2(Y)$ is a structure function for the second AND gate.

Here, Y_B is an indicator variable for basic event B, etc. The structure function for the fault tree is

$$\begin{aligned}\psi(\mathbf{Y}) &= Y_A \vee \psi_1(Y) \vee Y_D \vee \psi_2(Y) \vee Y_G \\ &= 1-[1-Y_A][1-\psi_1(Y)][1-Y_D][1-\psi_2(Y)][1-Y_G] \\ &= 1-[1-Y_A][1-Y_BY_C][1-Y_D][1-Y_EY_F][1-Y_G]\end{aligned} \tag{7.43}$$

7.4.3 Unavailability Calculations Using Structure Functions

It is of significance to recognize the probabilistic nature of expressions such as (7.36), (7.40), and (7.43). If we examine the system at some point in time, and the state of the basic event Y_i is assumed to be a Bernoulli random variable, then $\psi(\mathbf{Y})$ is also a Bernoulli random variable. The probability of occurrence of state $Y_i = 1$ is equal to the expected value of Y_i and to the probability of event B_i.

$$\Pr(Y_i = 1) = \Pr(B_i) = E(Y_i) \tag{7.44}$$

Note that this probability is the unavailability $Q_i(t)$, or existence probability depending on whether basic event B_i is a component failure or human error or environmental impact. The probability of the top event, i.e., the unavailability $Q_s(t)$, is the probability $\Pr(\psi(\mathbf{Y}) = 1)$, or expectation $E(\psi(\mathbf{Y}))$. An alternative way of stating this is as follows:

$$Q_s(t) = \Pr(\text{top event}) = \Pr(\psi(\mathbf{Y}) = 1) = E(\psi(\mathbf{Y})) \tag{7.45}$$

The next three examples demonstrate the use of structure functions in system analysis.

Example 5. Two-Out-of-Three System

Compare the unavailability for a two-out-of-three voting system with that of a two-component parallel system for

$$E(Y_1)=E(Y_2)=E(Y_3)=Q=0.6 \tag{7.46}$$

Solution: For the two-out-of-three system, according to (7.40)

$$Q_s(t)=E(\psi(\mathbf{Y}))$$
$$=E(Y_1Y_2)+E(Y_2Y_3)+E(Y_3Y_1)-2E(Y_1Y_2Y_3) \tag{7.47}$$
$$=E(Y_1)E(Y_2)+E(Y_2)E(Y_3)+E(Y_3)E(Y_1)-2E(Y_1)E(Y_2)E(Y_3) \tag{7.48}$$
$$=3\times 0.6^2-2\times 0.6^3=0.648 \tag{7.49}$$

Note that the expectation of the product of independent variables is equal to the product of expectations of these variables. This property is used in going from (7.47) to (7.48).

For the parallel system, from equation (7.36),

$$Q_s(t)=E(\psi(\mathbf{Y}))=E(Y_1)+E(Y_2)-E(Y_1)E(Y_2)$$
$$=2\times 0.6-0.6^2=0.84$$

Hence, a one-out-of-two system has an 84% chance of being in the top event state, and the two-out-of-three system has only a 64.8% probability.

Example 6. Tail Gas Quench and Clean Up System

Calculate the system unavailability $Q_s(t)$ for the fault tree of Fig. 7.13, assuming component unavailabilities of (7.25):

Solution: According to (7.43) we have

$$\begin{aligned}Q_s(t)&=E(\psi(\mathbf{Y}))\\&=1-E([1-Y_A][1-Y_BY_C][1-Y_D][1-Y_EY_F][1-Y_G])\end{aligned} \tag{7.50}$$

Each factor in the expected value operator E of this equation has different indicator variables, and these factors are independent, since the indicator variables are assumed to be independent. Thus, (7.50) can be written as

$$Q_s(t)=1-E(1-Y_A)E(1-Y_BY_C)E(1-Y_D)E(1-Y_EY_F)E(1-Y_G) \tag{7.51}$$
$$=1-[1-E(Y_A)][1-E(Y_B)E(Y_C)][1-E(Y_D)][1-E(Y_E)E(Y_F)]\times[1-E(Y_G)] \tag{7.52}$$

The component unavailabilities of (7.25) give

$$\begin{aligned}Q_s(t)&=1-[1-0.9][1-(0.8)(0.7)][1-0.6][1-(0.5)(0.4)]\times[1-0.3]\\&=0.990144\end{aligned} \tag{7.53}$$

This confirms the result of (7.26).

Contrary to (7.50), each indicator variable appears more than once in the products of (7.38), the structure function for a two-out-of-three system. For example, the variable Y_2 appears in $[1-Y_1Y_2]$ and $[1-Y_2Y_3]$. Here, we cannot proceed as we did in going from (7.50) to (7.51), since these factors are no longer independent. This is confirmed by the following derivation, which results in an incorrect result:

$$Q_s(t)=1-E(1-Y_1Y_2)E(1-Y_2Y_3)E(1-Y_3Y_1)$$

$$=1-[1-E(Y_1)E(Y_2)][1-E(Y_2)E(Y_3)][1-E(Y_3)E(Y_1)]$$

Substituting (7.46) into the above.

$$Q_s(t)=1-[1-0.6^2]^3=0.737856$$

This contradicts (7.49).

Example 7.

Draw a fault tree for the case of full production for the bridged circuit of Fig. 7.20, and calculate the system unavailability.

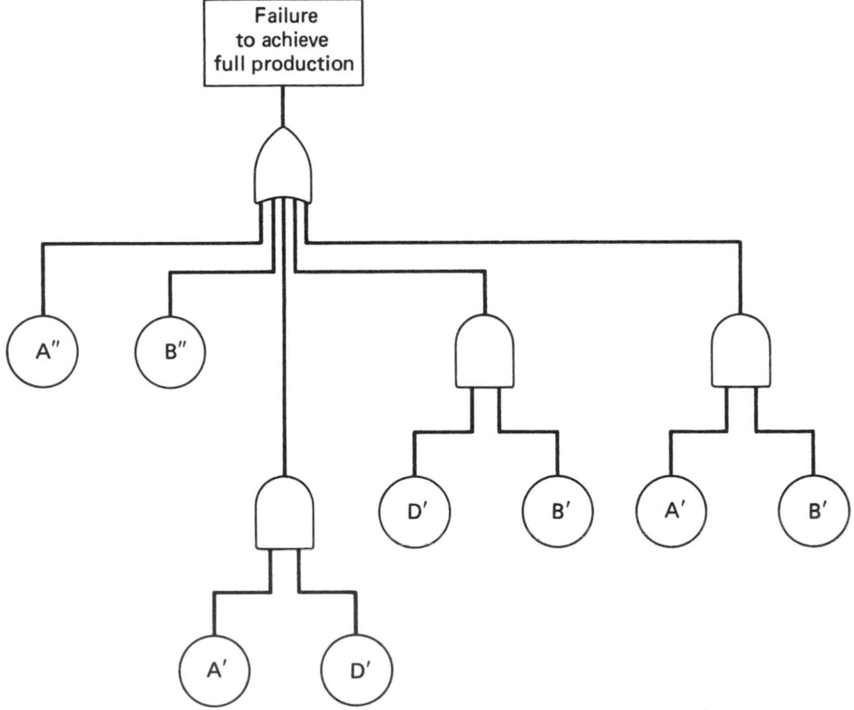

Figure 7.22. Condensed fault tree for full production.

Solution: The condensed fault tree is shown as Fig. 7.22.

$$\Pr(A'')=\Pr(B'')=0.032, \quad \Pr(A')=\Pr(B')=\Pr(D')=0.008$$

$$Q_s(t)=A''\cup B''\cup(A'\cap D')\cup(D'\cap B')\cup(A'\cap B') \tag{7.54}$$

$$A''\cup B''=0.032+0.032-(0.032)(0.032)=0.062976$$

The three pumps constitute a two-out-of-three voting system, so we can use the results of equation (7.40) and Example 1.

$$(A'\cap D')\cup(D'\cap B')\cup(A'\cap B')=3Q_i^2-2(Q_i)^3=3(0.008)^2-2(0.008)^3=0.000191$$

Thus,

$$Q_s(t)=(0.062976)+(0.00191)-(0.062976)(0.00191)=0.06315$$

This confirms the result of Table 7.5. Note that had we simply done the algebraic multiplications and additions in equation (7.54) without using the Boolean absorption laws, an incorrect answer would have resulted.

7.5 UNAVAILABILITY CALCULATIONS USING MINIMAL CUTS OR MINIMAL PATHS

7.5.1 Unavailability Calculations Using Minimal Cut Representations

The preceding section gave a method for constructing structure functions for calculating system unavailability. In this section another approach, based on minimal cut sets or path sets, is developed.

Consider a fault tree having the following m minimal cut sets.

$$\{B_{1,1}, B_{2,1}, \ldots, B_{n_1,1}\}: \quad \text{cut set } 1$$

$$\cdots\cdots\cdots\cdots\cdots\cdots\cdots\cdots$$

$$\{B_{1,j}, B_{2,j}, \ldots, B_{n_j,j}\}: \quad \text{cut set } j$$

$$\cdots\cdots\cdots\cdots\cdots\cdots\cdots\cdots$$

$$\{B_{1,m}, B_{2,m}, \ldots, B_{n_m,m}\}: \quad \text{cut set } m$$

Denote by $Y_{i,j}$ the indicator variable for the event $B_{i,j}$. The top event occurs if and only if all basic events in a minimal cut set occur simultaneously. Thus, the minimal cut set fault tree of Fig. 7.23 is equivalent to the fault tree. The structure function of this fault tree is

$$\psi(\mathbf{Y})=\bigvee_{j=1}^{m}\left[\bigwedge_{i=1}^{n_j} Y_{i,j}\right] \tag{7.55}$$

and its algebraic form is given by

$$\psi(\mathbf{Y}) = \bigvee_{j=1}^{n} \left[\prod_{i=1}^{n_j} Y_{i,j} \right] = 1 - \prod_{j=1}^{m} \left[1 - \prod_{i=1}^{n_j} Y_{i,j} \right] \tag{7.56}$$

Let $\kappa_j(Y)$ be a structure function for the AND gate G_j of Fig. 7.23:

$$\kappa_j(\mathbf{Y}) = \prod_{i=1}^{n_j} Y_{i,j} \tag{7.57}$$

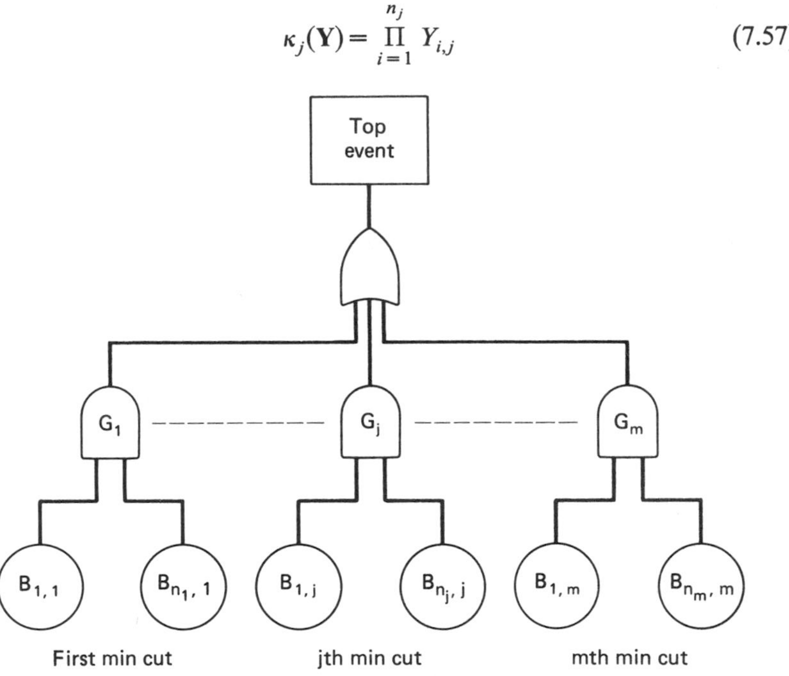

Figure 7.23. Minimal cut representation of fault tree.

The function $\kappa_j(\mathbf{Y})$ is the *jth minimal cut structure*. Equation (7.56) can be written as

$$\psi(\mathbf{Y}) = 1 - \prod_{j=1}^{m} \left[1 - \kappa_j(\mathbf{Y}) \right] \tag{7.58}$$

This equation is important because it gives a structure function of the fault tree in terms of minimal cut structures $\kappa_j(\mathbf{Y})$'s. The structure function $\psi(\mathbf{Y})$ can be expanded and simplified by the absorption law, resulting in a similar polynomial to (7.40). The system unavailability $Q_s(t)$ can be calculated by using (7.45), as is shown in the following example.

Example 8. Two-Out-of-Three System

Calculate the system unavailability for the two-out-of-three voting system in Fig. 7.8. Unavailabilities for the three components are as given by (7.46).

Solution: The voting system has three minimal cut sets:

$$\{B_1, B_2\}, \quad \{B_2, B_3\}, \quad \{B_3, B_1\} \tag{7.59}$$

The minimal cut structures $\kappa_1(Y)$, $\kappa_2(Y)$, $\kappa_3(Y)$, are

$$\kappa_1(Y) = Y_1 Y_2, \quad \kappa_2(Y) = Y_2 Y_3, \quad \kappa_3(Y) = Y_3 Y_1 \qquad (7.60)_1$$

Thus, the minimal cut representation of the structure function $\psi(\mathbf{Y})$ is

$$\psi(\mathbf{Y}) = 1 - [1 - Y_1 Y_2][1 - Y_2 Y_3][1 - Y_3 Y_1] \qquad (7.60)_2$$

which is identical to (7.38). This structure function is expanded, and (7.40) is obtained. The system unavailability is given by (7.49).

7.5.2 Unavailability Calculations Using Minimal Path Representations

Consider a fault tree with m minimal path sets:

$$\{B_{1,1}, B_{2,1}, \ldots, B_{n_1,1}\}: \quad \text{path set 1}$$
$$\{B_{1,j}, B_{2,j}, \ldots, B_{n_j,j}\}: \quad \text{path set } j$$
$$\{B_{1,m}, B_{2,m}, \ldots, B_{n_m,m}\}: \quad \text{path set } m$$

Denote by $Y_{i,j}$ the indicator variable for the event $B_{i,j}$. The top event occurs if and only if at least one basic event occurs in all minimal path sets. Thus, the original fault tree is equivalent to the fault tree of Fig. 7.24.

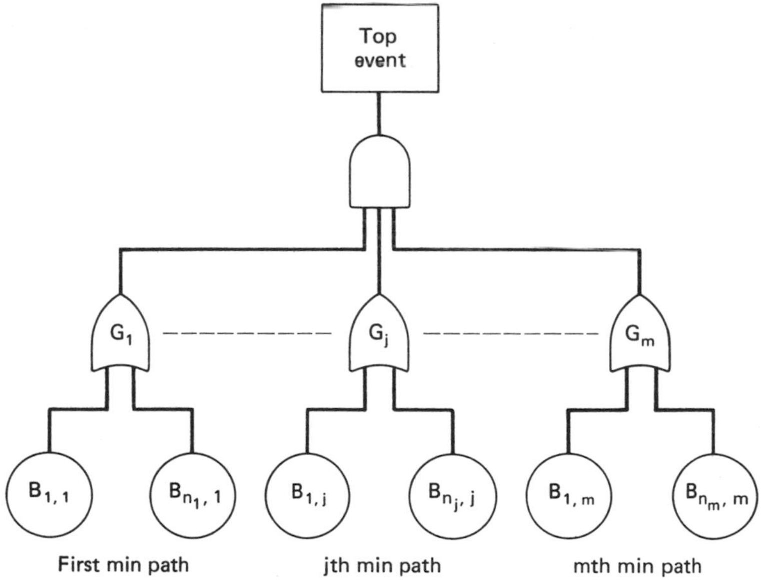

Figure 7.24. Minimal path representation of fault trees.

The structure function for this tree is

$$\psi(\mathbf{Y}) = \bigwedge_{j=1}^{m} \left[\bigvee_{i=1}^{n_j} Y_{i,j} \right] \tag{7.61}$$

An algebraic form for this function is

$$\psi(\mathbf{Y}) = \bigwedge_{j=1}^{m} \left[1 - \prod_{i=1}^{n_j} \left[1 - Y_{i,j}\right] \right] \tag{7.62}$$

$$= \prod_{j=1}^{m} \left[1 - \prod_{i=1}^{n_j} \left[1 - Y_{i,j}\right] \right] \tag{7.63}$$

Let $\rho_j(\mathbf{Y})$ be a structure function for the OR gate G_j of Fig. 7.24:

$$\rho_j(\mathbf{Y}) = 1 - \prod_{i=1}^{n_j} \left[1 - Y_{i,j}\right] \tag{7.64}$$

The structure function of (7.63) can be written as

$$\psi(\mathbf{Y}) = \prod_{j=1}^{m} \rho_j(\mathbf{Y}) \tag{7.65}$$

This $\psi(\mathbf{Y})$ is a minimal path representation, and the $\rho_j(\mathbf{Y})$ is the jth minimal path structure. The minimal path representation $\psi(\mathbf{Y})$ can be expanded and simplified via the absorption law. The system unavailability $Q_s(t)$ can be calculated by using (7.45), as shown in the next example.

Example 9. Two-Out-of-Three System

Calculate the system unavailability for the two-out-of-three voting system of Fig. 7.8. Unavailabilities for the three components are given by (7.46).

Solution: The voting system has three minimal path sets:

$$\{B_1, B_2\}, \quad \{B_2, B_3\}, \quad \{B_3, B_1\}$$

The minimal path structures are

$$\rho_1(\mathbf{Y}) = 1 - [1 - Y_1][1 - Y_2] = Y_1 + Y_2 - Y_1 Y_2$$

$$\rho_2(\mathbf{Y}) = 1 - [1 - Y_2](1 - Y_3] = Y_2 + Y_3 - Y_2 Y_3$$

$$\rho_3(\mathbf{Y}) = 1 - [1 - Y_3][1 - Y_1] = Y_3 + Y_1 - Y_3 Y_1$$

The minimal path representation of $\psi(\mathbf{Y})$ is

$$\psi(\mathbf{Y}) = [Y_1 + Y_2 - Y_1 Y_2][Y_2 + Y_3 - Y_2 Y_3][Y_3 + Y_1 - Y_3 Y_1] \tag{7.66}$$

The expansion of $\psi(\mathbf{Y})$ is,

$$\psi(\mathbf{Y}) = Y_1 Y_2 + Y_2 Y_3 + Y_3 Y_1 - 2 Y_1 Y_2 Y_3$$

which is identical to (7.40). The system unavailability is again given by (7.49).

7.5.3 Unavailability Calculations Using Partial Pivotal Decomposition

Some basic events appear in more than one minimal cut set. In this case, factors $[1-\kappa_j(\mathbf{Y})]$'s in (7.58) are no longer independent, and the equality

$$E(\psi(\mathbf{Y})) = 1 - \prod_{j=1}^{m} \left[1 - E(\kappa_j(\mathbf{Y})\right] \tag{7.67}$$

does not hold. For the same reason, (7.65) does not imply that

$$E(\psi(\mathbf{Y})) = \prod_{j=1}^{m} E(\rho_j(\mathbf{Y})) \tag{7.68}$$

One way of calculating $E(\psi(\mathbf{Y}))$ is to expand $\psi(\mathbf{Y})$ and simplify the results by the absorption law. However, this is a tedious process when the expansion contains a large number of terms. The process can be simplified by *partial pivotal decomposition.*

The structure function $\psi(\mathbf{Y})$ can be rewritten as

$$\psi(\mathbf{Y}) = Y_i \psi(1_i, \mathbf{Y}) + (1 - Y_i)\psi(0_i, \mathbf{Y}) \tag{7.69}$$

where $\psi(1_i, \mathbf{Y})$ and $\psi(0_i, \mathbf{Y})$ are binary functions obtained by setting the ith indicator variable Y_i to unity and zero, respectively. These binary functions can be pivoted around other indicator variables until the resulting binary functions consist only of independent factors; then $E(\psi(\mathbf{Y}))$ can be easily calculated. This technique is demonstrated by the following example.

Example 10. Two-Out-of-Three System

Consider the minimal path representation of (7.66). Then,

$$\begin{aligned} \psi(\mathbf{Y}) &= Y_1[Y_2 + Y_3 - Y_2 Y_3] \\ &\quad + [1 - Y_1]Y_2[Y_2 + Y_3 - Y_2 Y_3]Y_3 \\ &= Y_1[Y_2 + Y_3 - Y_2 Y_3] \\ &\quad + [1 - Y_1]Y_2 Y_3 \\ &\quad + [1 - Y_1][1 - Y_2]\cdot 0 \end{aligned} \qquad \begin{aligned} \psi(\mathbf{Y}) &= Y_1 \psi(1_1, \mathbf{Y}) \\ &\quad + [1 - Y_1]\psi(0_1, \mathbf{Y}) \\ &= Y_1 \psi(1_1, Y) \\ &\quad + [1 - Y_1]Y_2 \psi(0_1, 1_2, \mathbf{Y}) \\ &\quad + [1 - Y_1][1 - Y_2]\psi(0_1, 0_2, \mathbf{Y}) \end{aligned}$$

Thus,

$$\psi(\mathbf{Y}) = Y_1[Y_2 + Y_3 - Y_2Y_3] + [1 - Y_1]Y_2Y_3 \tag{7.70}$$

Note that Y_1 and $[Y_2 + Y_3 - Y_2Y_3]$ have different indicator variables. Similarly, $[1 - Y_1]$, Y_2, and Y_3 have no common variables. Thus, the factors of each of the two products in (7.70) are independent and the expected value $E(\psi(\mathbf{Y}))$ is given by

$$E(\psi(\mathbf{Y})) = E(Y_1)[E(Y_2) + E(Y_3) - E(Y_2)E(Y_3)] + [1 - E(Y_1)]E(Y_2)E(Y_3) \tag{7.71}$$

To confirm (7.71), we substitute (7.46).

$$\begin{aligned} Q(t) &= E(\psi(\mathbf{Y})) = (0.6)[0.6 + 0.6 - 0.6^2] + [1 - 0.6](0.6)^2 \\ &= 0.648 \end{aligned} \tag{7.72}$$

The results of (7.49) are obtained; thus, the methodology is confirmed.

7.5.4 Unavailability Calculations Using the Inclusion-Exclusion Principle

Define *event* d_i by

d_i = all basic events occurring in the ith minimal cut set at time t

The top event S can be expressed in terms of d_i as

$$S = \bigcup_{i=1}^{N_c} d_i \quad (N_c = \text{total number of minimal cuts}) \tag{7.73}$$

Thus,

$$Q_s(t) = \Pr\left(\bigcup_{i=1}^{N_c} d_i\right) \tag{7.74}$$

$$\begin{aligned} &= \sum_{i=1}^{N_c} \Pr(d_i) - \sum_{i=2}^{N_c} \sum_{j=1}^{i-1} \Pr(d_i \cap d_j) + \\ &\quad \cdots + (-1)^{N_c - 1} \Pr(d_1 \cap d_2 \cap \cdots \cap d_{N_c}) \end{aligned} \tag{7.75}$$

Equation (7.75) is an expansion of (7.74) obtained by the so-called *inclusion-exclusion principle*. The mth term on the right-hand side of (7.75) implies the contribution to $Q_s(t)$ from m out of N_c minimal cut sets being simultaneously failed at time t; i.e., all the basic events in these m minimal cut sets are occurring. A very useful property of (7.75) is that the top event probability is given in terms of intersections, which are easier to calculate than unions.

For small systems it is relatively easy to get exact values for $Q_s(t)$, and this is demonstrated by the following example.

Example 11. Two-Out-of-Three System

Calculate $Q_s(t)$ for the two-out-of-three voting system of Fig. 7.8 by assuming component unavailabilities of (7.46), using (7.75).

Solution: From the three minimal cut sets of the system, we have

$$d_1 = B_1 \cap B_2, \quad d_2 = B_2 \cap B_3, \quad d_3 = B_3 \cap B_1$$

The exact expression for $Q_s(t)$ from (7.75) is

$$\begin{aligned} Q_s(t) &= \sum_{i=1}^{3} \Pr(d_i) - \sum_{i=2}^{3} \sum_{j=1}^{i-1} \Pr(d_i \cap d_j) + \sum_{i=3}^{3} \sum_{j=2}^{i-1} \sum_{k=1}^{j-1} \Pr(d_i \cap d_j \cap d_k) \\ &= \Pr(d_1) + \Pr(d_2) + \Pr(d_3) - [\Pr(d_1 \cap d_2) + \Pr(d_2 \cap d_3) + \Pr(d_3 \cap d_1)] \\ &\quad - \Pr(d_1 \cap d_2 \cap d_3) \end{aligned} \tag{7.76}$$

$$\begin{aligned} &[A] \quad [\Pr(d_1) + \Pr(d_2) + \Pr(d_3)] - Q^2 + Q^2 + Q^2 = 1.08 \\ &[B] \quad [\Pr(d_1 \cap d_2) + \Pr(d_2 \cap d_3) + \Pr(d_3 \cap d_1)] - Q^3 + Q^3 + Q^3 = 0.648 \\ &[C] \quad \Pr(d_1 \cap d_2 \cap d_3) = Q^3 = 0.216 \end{aligned}$$

Thus,

$$Q_s(t) = [A] - [B] + [C] = 0.648 \tag{7.77}$$

This confirms (7.49).

7.6 LOWER AND UPPER BOUNDS FOR SYSTEM UNAVAILABILITY

For a large, complicated fault tree, calculations of the exact system unavailability by the methods of the preceding sections are time-consuming. When computing time becomes a factor, lower and upper bounds of the unavailability can be calculated by the *short-cut methods* explained in this section.

7.6.1 Lower and Upper Bounds Using the Inclusion-Exclusion Principle

Equation (7.75) can be bracketed by

$$\underbrace{\sum_{i=1}^{N_c} \Pr(d_i) - \sum_{i=2}^{N_c} \sum_{j=1}^{i-1} \Pr(d_i \cap d_j)}_{\text{lower bound}} \le Q_s(t) \le \underbrace{\sum_{i=1}^{N_c} \Pr(d_i)}_{\text{upper bound}} \tag{7.78}$$

Example 12. Two-Out-of-Three System

For Example 11 in the preceding section we have

$$Q_s(t)_{\min} = [A] - [B] = 0.432$$
$$Q_s(t)_{\max} = [A] = 1.08$$

The exact value of $Q_s(t)$ is 0.648, and these lower and upper bounds are not good. However, the brackets are usually within three significant figures of one another for practical problems, since component unavailabilities are usually much less than 1.

Example 13. Two-Out-of-Three System

Calculate $Q_s(t)$, $Q_s(t)_{\min}$, and $Q_s(t)_{\max}$ by assuming $Q = 0.001$ in Example 12.

Solution: From (7.76)

$$[A] = Q^2 + Q^2 + Q^2 = 3.0 \times 10^{-6}$$
$$[B] = Q^3 + Q^3 + Q^3 = 3.0 \times 10^{-9}$$
$$[C] = Q^3 = 1.0 \times 10^{-9}$$

Thus,

$$Q_s(t) = [A] - [B] + [C] = 2.998 \times 10^{-6}$$
$$Q_s(t)_{\min} = [A] - [B] = 2.997 \times 10^{-6}$$
$$Q_s(t)_{\max} = [A] = 3.0 \times 10^{-6}$$

We have tight lower and upper bounds.

7.6.2 Esary and Proschan Lower and Upper Bounds

We now restrict our attention to structure functions that are *coherent* (monotonic). The engineering interpretation of this is that, in a coherent system, the occurrence of a component failure always results in system degradation. Formally, $\psi(\mathbf{Y})$ is coherent if:[3]

1) $\psi(\mathbf{Y}) = 1$ if $Y = (1, 1, \ldots, 1)$,
2) $\psi(\mathbf{Y}) = 0$ if $Y = (0, 0, \ldots, 0)$,
3) $\psi(\mathbf{Y}) \geq \psi(\mathbf{X})$ if $Y_i \geq X_i$ for all i.
4) Each basic events appears in at least one minimal cut set. For a coherent structure function, the right-hand sides of (7.67) and (7.68)

[3]Esary, J. D., and F. Proschan, "Coherent Structures with Non-Identical Components," *Technometrics*, **5**, 191, 1963.

give upper and lower bounds for the system unavailability $Q_s(t)$:[4]

$$\underset{\text{lower bound}}{\prod_{i=1}^{N_p} E(\rho_i(\mathbf{Y}))} \leq Q_s(t) \leq \underset{\text{upper bound}}{1 - \prod_{i=1}^{N_c} \left[1 - E(\kappa_i(\mathbf{Y}))\right]} \tag{7.79}$$

where N_p = total number of minimal path sets.

Example 14. Two-Out-of-Three System

Calculate Esary and Proschan bounds for the problem in Example 13 of the preceding section.

Solution:

$$\kappa_1(\mathbf{Y}) = Y_1 Y_2, \quad \kappa_2(Y) = Y_2 Y_3, \quad \kappa_3(Y) = Y_3 Y_1$$
$$\rho_1(\mathbf{Y}) = Y_1 + Y_2 - Y_1 Y_2$$
$$\rho_2(\mathbf{Y}) = Y_2 + Y_3 - Y_2 Y_3$$
$$\rho_3(\mathbf{Y}) = Y_3 + Y_1 - Y_3 Y_1$$

Since $E(Y_1) = E(Y_2) = E(Y_3) = Q = 0.001$, we have

$$Q_s(t)_{\min} = [Q + Q - Q^2]^3 = 7.988 \times 10^{-9}$$
$$Q_s(t)_{\max} = 1 - [1 - Q^2]^3 = 3.0 \times 10^{-6}$$

This upper bound is as good as that obtained by the bracketing, while the lower bound is extremely conservative.

7.6.3 Lower and Upper Bounds Using Partial Minimal Cut Sets and Path Sets

Assume that there are N_c' minimal cut sets and N_p' minimal path sets. Here N_c' and N_p' are less than the actual number of minimal cut sets N_c and path sets N_p, respectively. The structure function $\psi_L(\mathbf{Y})$ with these N_c' cut sets is

$$\psi_L(\mathbf{Y}) = 1 - \prod_{i=1}^{N_c'} \left[1 - \kappa_i(\mathbf{Y})\right] \tag{7.80}$$

where $\kappa_i(\mathbf{Y})$ is the ith minimal cut structure removed. Similarly, the structure function $\psi_U(\mathbf{Y})$ with the N_p' path sets is

$$\psi_U(\mathbf{Y}) = \prod_{i=1}^{N_p'} \rho_i(\mathbf{Y}) \tag{7.81}$$

[4]Esary, J. D., and F. Proschan, "A Reliability Bound for Systems of Maintained and Independent Components," *Journal of the American Statistical Association*, **65**, 329–338, 1970.

Since the structure function $\psi_L(\mathbf{Y})$ has fewer minimal cut sets than $\psi(\mathbf{Y})$, we have

$$\psi_L(\mathbf{Y}) \le \psi(\mathbf{Y}) \tag{7.82}$$

Similarly,

$$\psi(\mathbf{Y}) \le \psi_U(\mathbf{Y}) \tag{7.83}$$

Thus,

$$E(\psi_L(\mathbf{Y})) \le E(\psi(\mathbf{Y})) \le E(\psi_U(\mathbf{Y}))$$

or

$$Q_s(t)_{\min} \equiv E(\psi_L(\mathbf{Y})) \le Q_s(t) \le E(\psi_U(\mathbf{Y})) \equiv Q_s(t)_{\max} \tag{7.84}$$

Example 15. Tail Gas Quench and Clean Up System

Calculate $Q_s(t)_{\min}$, $Q_s(t)$, and $Q_s(t)_{\max}$ for the fault tree of Fig. 7.13. Assume the component unavailabilities at time t to be 0.001.

Solution: The fault tree has five minimal cut sets:

$$\{A\},\quad \{D\},\quad \{G\},\quad \{B,C\},\quad \{E,F\} \tag{7.85}$$

and four minimal path sets:

$$\{A,D,G,C,F\},\quad \{A,D,G,C,E\},\quad \{A,D,G,B,F\},\quad \{A,D,G,B,E\} \tag{7.86}$$

Take only the cut sets $\{A\}$, $\{D\}$, and $\{G\}$ (ignore the higher-order ones) and only two path sets $\{A,D,G,C,F\}$ and $\{A,D,G,B,E\}$. Then,

$$\psi(\mathbf{Y}) = 1-[1-Y_A][1-Y_D][1-Y_G][1-Y_BY_C][1-Y_EY_F] \tag{7.87}$$

$$\psi_L(\mathbf{Y}) = 1-[1-Y_A][1-Y_D][1-Y_G] \tag{7.88}$$

$$\begin{aligned}\psi_U(\mathbf{Y}) &= [1-(1-Y_A)(1-Y_D)(1-Y_G)(1-Y_C)(1-Y_F)] \\ &\quad \times[1-(1-Y_A)(1-Y_D)(1-Y_G)(1-Y_B)(1-Y_E)] \\ &= 1-(1-Y_A)(1-Y_D)(1-Y_G)(1-Y_B)(1-Y_E) \\ &\quad -(1-Y_A)(1-Y_D)(1-Y_G)(1-Y_C)(1-Y_F) \\ &\quad +(1-Y_A)(1-Y_D)(1-Y_G)(1-Y_B)(1-Y_E)(1-Y_C)(1-Y_F)\end{aligned} \tag{7.89}$$

Thus,

$$\begin{aligned}Q_s(t) &= 1-(0.999)^3(0.999999)^2 = 2.998995\times10^{-3} \\ Q_s(t)_{\min} &= 1-(0.999)^3 = 2.997\times10^{-3} \\ Q_s(t)_{\max} &= 1-(0.999)^5-(0.999)^5+(0.999)^7 \\ &= 3.001\times10^{-3}\end{aligned} \tag{7.90}$$

Good upper and lower bounds are obtained.

As a first approximation, it is reasonable to include only the one or two event minimal cut sets in the N_c' cut sets. Similarly, we take as the N_p' path sets, minimal path sets containing the fewest number of basic events. Since fewer cut sets and path sets are involved, the calculation can be simplified. Further simplifications are possible if we pick out nearly disjoint cut sets or path sets, because the structure functions $\psi_U(\mathbf{Y})$ and $\psi_L(\mathbf{Y})$ can be expanded into polynomials with fewer terms, each of which consists of independent factors. The present method is used in "restricted-sampling Monte Carlo methods" described in Chapter 12.

7.7 SYSTEM QUANTIFICATION BY KITT

The previous sections covered availability and unavailability quantification methods for relatively simple systems. This section develops the theory and techniques germane to obtaining unavailabilities, availabilities, expected number of failures and repairs, and conditional failure and repair intensities, starting with minimal cut sets or path sets of large and complicated fault trees. We discuss, in some detail, the *KITT* (*kinetic tree theory*), and show how system parameters can be "guesstimated" by approximation techniques based on inclusion-exclusion principles. To be consistent with the previous chapters, we present here a revised version of Vesely's original derivation.

7.7.1 Overview of KITT[5]

The code is an application of kinetic tree theory and will handle independent basic events which are non-repairable or repairable, provided they have constant failure rates λ and constant repair rates μ. However, the exponential failure and/or repair distribution limitation can be circumvented by using the "phased mission" version of the program (KITT-2), which allows for tabular input of time-varying failure and repair rates. KITT requires also as input the minimal cut sets or the minimal path sets. Inhibit gates are also permitted.

Exact, time-dependent reliability parameters are determined for each basic event and cut set, but for the system as a whole the parameters are obtained by upper or lower bound approximations, or by bracketing. The upper and lower bounds are generally excellent approximations to the exact parameters. In the bracketing procedure the various upper and lower bounds can be obtained as close to each other as desired, and thus the exact value for system parameters are obtained if the user so chooses.

[5]Vesely, W. E., and R. E. Narum, *PREP and KITT: Computer Codes for Automatic Evaluation of Fault Trees*, Idaho Nuclear Corp., IN1349, 1970.

The probability characteristics, their definitions, the nomenclature, and the expected (mostly asymptotic) behavior of the variables are summarized in Tables 7.8 and 7.9. A flow sheet of the KITT computation is given as Fig. 7.25. The numbers on the flow sheet represent the equations used to obtain the parameters.

The program calculates $w(t)$ and $v(t)$ before $Q(t)$, using equation (4.64) of Chapter 4.

$$w(t)=f(t)+\int_0^t f(t-u)v(u)\,du \tag{7.91}$$

$$v(t)=\int_0^t g(t-u)w(u)\,du \tag{7.92}$$

In accordance with the definitions, the first term on the right-hand side of (7.91) is interpreted as the contribution to $w(t)$ from the first occurrence of the basic event, and the second term is the contribution to $w(t)$ from the failure repaired at time u, and then recurring at time t. A similar interpretation can be made for $v(t)$ in (7.92). If a rigorous solution of (7.91) is required, Laplace transform techniques can be used (Chapter 4, Section 4.4.3). KITT uses a numerical integration.

Before moving on to cut sets and system calculations, we demonstrate the use of these equations by a simple, one-component example. Component reliability parameters are unique and, as a first approximation, independent of the complexity of the system in which they appear. The calculation proceeds according to the flow chart of Fig. 4.17 of Chapter 4.

Example 16.

Using the flow chart of Fig. 4.17, calculate reliability parameters for non-repairable component 1 with $\lambda_1=1.0\times10^{-3}$ failures per hour, at $t=20$, $t=500$, and $t=1100$ hr. Repeat the calculations, assuming that the component is repairable with $\text{MTTR}=1/\mu_1=10$ hr.

Solution: Inspection of equations (7.91) and (7.92) reveals that, if the component is non-repairable, $g(t-u)$ and hence $v(t)$ are zero. Thus, the second term on the right-hand side of (7.91) also becomes zero. The first term, the expected number of failures per unit time to time t, for a constant failure rate, is simply $\lambda e^{-\lambda t}$ [see Chapter 4, equation (4.85)]. Thus,

$$w(t)=f(t)=\lambda e^{-\lambda t}, \qquad v(t)=0$$

and, as before [see Chapter 4, equation (4.70)]:

$$Q(t)=\int_0^t[w(u)-v(u)]\,du=\int_0^t\lambda e^{-u}\,du=1-e^{-\lambda t} \tag{7.93}$$

For the repairable case, the integrals of (7.91) and (7.92) must be calculated. The unavailability $Q(t)$ may be computed by (4.70). The values of $Q(t)$ and $w(t)$ (obtained by computer) are listed in Table 7.10.

TABLE 7.8 System Parameters Calculated by KITT

Symbols				Definition	Name
Component	Cut set	System	KITT Symbol		
$Q(t)$	$Q^*(t)$	$Q_s(t)$	Q	The probability of a failed state at time t	Unavailability
$w(t)$ $(v(t)$	$w^*(t)$ $v^*(t)$	$w_s(t)$ $v_s(t))^a$	W	The expected number of failures per unit time at time t	Unconditional failure intensity
$\lambda(t)$ $(\mu(t)$	$\lambda^*(t)$ $\mu^*(t)$	$\lambda_s(t)$ $\mu_s(t))^a$	L	The probability of a failure per unit time at time t, given no failures at time t	Conditional failure intensity
$W(0,t)$ $(V(0,t)$	$W^*(0,t)$ $V^*(0,t)$	$W_s(0,t)$ $V_s(0,t))^a$	WSUM	The expected number of failures during the time interval from 0 to t	Expected number of failures
$F(t)$	$F^*(t)$	$F_s(t)$		The probability of one or more failures in the time interval $(0,t]$	Unreliability[b]

[a]Analogous parameters for repair are shown within the parentheses.
[b]This cannot be obtained by KITT.

TABLE 7.9 Typical Behavior of System Parameters with Constant Rates $\lambda(t)=\lambda$ and $\mu(t)=1/\text{MTTR}$

COMPONENT		CUT SET		SYSTEM	
Repairable	Non-Repairable	Repairable	Non-Repairable	Repairable	Non-Repairable
Constant $Q(t)$ after $t>3\text{MTTR}$, $Q(t)\ll 1$	$Q(t)\to 1$ as $t\to\infty$, $Q(t)=F(t)=W(0,t)$ $=1-e^{-\lambda t}$	Constant $Q^*(t)$ after $t>3\text{MTTR}$, $Q^*(t)\ll 1$	$Q^*(t)\to 1$ as $t\to\infty$, $Q^*(t)=F^*(t)=W^*(0,t)$	Constant $Q_s(t)$ after $t>3\text{MTTR}$, $Q_s(t)\ll 1$	$Q_s(t)\to 1$ as $t\to\infty$, $Q_s(t)=F_s(t)=W_s(0,t)$
Constant $w(t)$ after $t>3\text{MTTR}$, $w(t)\simeq\lambda(t)$	Decreases with time, $w(t)=f(t)=\lambda e^{-\lambda t}$	Constant $w^*(t)$ after $t>3\text{MTTR}$, $w^*(t)\simeq\lambda^*(t)$	$w^*(t)$ increases, then decreases with time	Constant $w_s(t)$ after $t>3\text{MTTR}$, $w_s(t)\simeq\lambda_s(t)$	$w_s(t)$ increases, then decreases with time
Constant $\lambda(t)$ after $t>3\text{MTTR}$, $\lambda(t)\simeq w(t)$	Constant $\lambda(t)$	Constant $\lambda^*(t)$, $\lambda^*(t)\simeq w^*(t)$ after $t>3\text{MTTR}$	$\lambda^*(t)$ increases	Constant $\lambda_s(t)$ after $t>3\text{MTTR}$, $\lambda_s(t)\simeq w_s(t)$	$\lambda_s(t)$ increases
$W(0,t)\to\infty$ as $t\to\infty$	$W(0,t)\to 1$ as $t\to\infty$, $W(0,t)=F(t)=Q(t)$	$W^*(0,t)\to\infty$ as $t\to\infty$	$W^*(0,t)\to 1$ as $t\to\infty$, $W^*(0,t)=F^*(t)=Q^*(t)$	$W_s(0,t)\to\infty$ as $t\to\infty$	$W_s(0,t)\to 1$ as $t\to\infty$, $W_s(0,t)=F_s(t)=Q_s(t)$
$F(t)\to 1$ as $t\to\infty$	$F(t)\to 1$ as $t\to\infty$, $F(t)=W(0,t)=Q(t)$	$F^*(t)\to 1$ as $t\to\infty$	$F^*(t)\to 1$ as $t\to\infty$, $F^*(t)=W^*(0,t)=Q^*(t)$	$F_s(t)\to 1$ as $t\to\infty$	$F_s(t)\to 1$ as $t\to\infty$, $F_s(t)=W_s(0,t)=Q_s(t)$

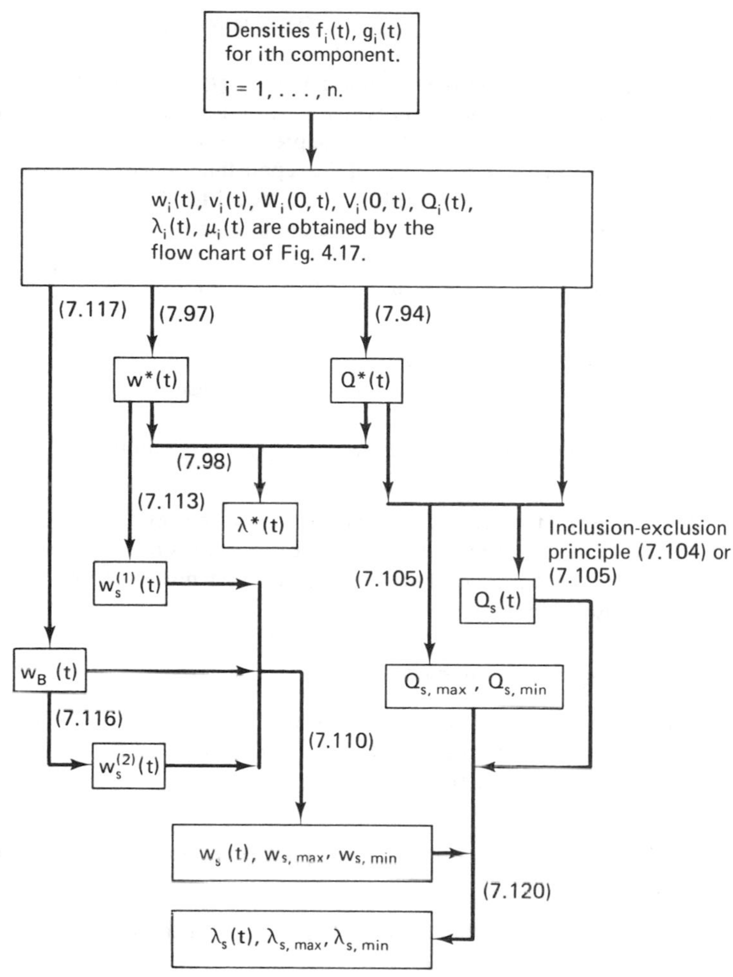

Figure 7.25. Flow sheet of KITT computations.

TABLE 7.10 Results of Example 16

	$\lambda_1 = 1\times10^{-3}$,	$\text{MTTR}_1 = \infty$	$\lambda_1 = 1\times10^{-3}$,	$\text{MTTR}_1 = 10$
t	$w(t)$	$Q(t)$	$w(t)$	$Q(t)$
20	9.80199×10^{-4}	1.98013×10^{-2}	9.91412×10^{-4}	8.58757×10^{-3}
500	6.06531×10^{-4}	3.93469×10^{-1}	9.90099×10^{-4}	9.90099×10^{-3}
1100	3.32871×10^{-4}	6.32121×10^{-1}	9.90099×10^{-4}	9.90099×10^{-3}
	$\lambda_1 = 1\times10^{-3}$,	$\text{MTTR}_1 = \infty$	$\lambda_1 = 1\times10^{-3}$,	$\text{MTTR}_1 = 10$
t	$W(0,t)$	$V(0,t)$	$W(0,t)$	$V(0,t)$
20	1.98013×10^{-2}	0.0	1.98870×10^{-2}	1.12994×10^{-2}
500	3.93469×10^{-1}	0.0	4.95148×10^{-1}	4.85247×10^{-1}
1100	6.67129×10^{-1}	0.0	1.08921×10^{0}	1.07931×10^{0}

At $t=1100$, 63.2% of the steady-state unavailability $Q(\infty)=1$ is attained for the non-repairable case, which is consistent with a mean time to failure of 1000 hr and infinite mean time to repair [T of (4.114) in Chapter 4 becomes MTTF]. In general, steady state is reached in a few multiples of the repair time, since MTTR is usually far smaller than MTTF and T of (4.114) is nearly equal to MTTR.

To calculate $W(0,t)$ and $V(0,t)$ we perform the integrations:

$$W(0,t)=\int_0^t w(u)\,du, \qquad V(0,t)=\int_0^t v(u)\,du$$

Numerical values for this example are shown in Table 7.10.

We observed previously that, for the non-repairable case,

$$W(0,t)=\int_0^t \lambda e^{-\lambda u}\,du=1-e^{-\lambda t}=F(t)=\text{unreliability}$$

This is not the case for the repairable system because the probability of the first failure is not the same as the expected number of failures, since the component is being repaired part of the time. The unreliability $F(t)$ must approach one, but $W(0,t)$ can be greater than one.

7.7.2 Minimal Cut Set Parameters

A cut set is *occurring* if all basic events in the cut set are occurring. The probability of cut set occurrence at time t, $Q^*(t)$, is obtained from the intersections of basic events [see equation (7.9)]:

$$Q^*(t)=\Pr(B_1\cap B_2\cap\ldots\cap B_n)=\prod_{j=1}^{n} Q_j(t) \tag{7.94}$$

where n is the number of cut set members, and $Q_j(t)$ the probability of the jth basic event existing at time t.

Examples 17, 18, and 19 demonstrate the procedure for calculating cut set parameters for a series, a parallel, and a two-out-of-three system.

Example 17.

Calculate the cut set reliability parameters for a three-component, repairable and non-repairable series system at $t=20$, $t=500$, and $t=1100$ hr, where the components have the following parameters

Component 1	Component 2	Component 3
$\lambda_1=10^{-3}, \mu_1=1/10,$	$\lambda_2=2\times10^{-3}, \mu_2=1/40,$	$\lambda_3=3\times10^{-3}, \mu_3=1/60$

Solution: For this configuration there are three cut sets, each component being a cut set. Thus, $Q_1^*(t)=Q_1(t)$, $Q_2^*(t)=Q_2(t)$, $Q_3^*(t)=Q_3(t)$, etc. The parameters for component one were calculated in Example 16. For the other two components (cut sets) we have Table 7.11.

TABLE 7.11 RESULTS OF EXAMPLE 17

t	$w^*(t)$	$Q^*(t)$	$W^*(0,t)=F(t)$
	$\lambda_2=2\times10^{-3}$, $\tau=0$ (non-repairable)		
20	1.92158×10^{-3}	3.92106×10^{-2}	3.92106×10^{-2}
500	7.35759×10^{-4}	6.32121×10^{-1}	6.32121×10^{-1}
1100	2.21606×10^{-4}	8.89197×10^{-1}	8.89197×10^{-1}
	$\lambda_2=2\times10^{-3}$, MTTR$=40$ (repairable)		
20	1.93818×10^{-3}	3.09075×10^{-2}	3.93265×10^{-2}
500	1.85185×10^{-3}	7.40741×10^{-2}	9.31413×10^{-1}
1100	1.85185×10^{-3}	7.40741×10^{-2}	2.04252×10^{0}
	$\lambda_3=3\times10^{-3}$, $\tau_3=0$ (non-repairable)		
20	2.82530×10^{-3}	5.82355×10^{-2}	5.82355×10^{-2}
500	6.69390×10^{-4}	7.76870×10^{-1}	7.76870×10^{-1}
1100	1.10650×10^{-4}	9.63117×10^{-1}	9.63117×10^{-1}
	$\lambda_3=3\times10^{-3}$, MTTR$=60$ (repairable)		
20	2.85118×10^{-3}	4.96062×10^{-2}	5.84145×10^{-2}
500	2.54240×10^{-3}	1.52534×10^{-1}	1.29445×10^{0}
1100	2.54237×10^{-3}	1.52542×10^{-1}	2.81988×10^{0}

An n-component parallel system has cut sets of the form $\{B_1, B_2, \ldots, B_n\}$. Thus, the calculation of $Q^*(t)$ represents no problem because

$$Q^*(t)=Q_1(t)Q_2(t)\ldots Q_n(t)$$

However, an extension of the theory must be made before $w^*(t)$ and $\lambda^*(t)$ can be obtained. Let us first examine $\lambda^*(t)$, which is defined by the probability of the occurrence of a cut set per unit time at time t, given no cut set failure at time t. Thus, $\lambda^*(t)\,dt$ is the probability that the cut set occurs during time interval $[t,t+dt)$, given that it is not active at time t:

$$\begin{aligned}\lambda^*(t)\,dt &= \Pr\left(C^*(t,t+dt)\,|\,\overline{C}^*(t)\right)\\ &= \frac{\Pr\left(C^*(t,t+dt)\cap\overline{C}^*(t)\right)}{\Pr\left(\overline{C}^*(t)\right)}=\frac{\Pr\left(C^*(t,t+dt)\right)}{\Pr\left(\overline{C}^*(t)\right)}\end{aligned}\tag{7.95}$$

where $C^*(t,t+dt)$ is the event *occurrence of the cut set during* $[t,t+dt)$, *and* $\bar{C}^*(t)$ is the event *non-existence of the cut set failure at time t.*

The last equation holds, since $C^*(t,t+dt)$ implies $\bar{C}^*(t)$; i.e., the cut set can occur during $[t,t+dt)$ only if it does not exist at time t. The numerator is equal to $w^*(t)dt$. As shown in Appendix 7.2 of this chapter, (7.95) can be written

$$\lambda^*(t)\,dt=\frac{\sum_{j=1}^{n} w_j(t)\,dt \prod_{\substack{l=1\\ l\neq j}}^{n} Q_l(t)}{1-Q^*(t)} \tag{7.96}$$

Each term of the summation in equation (7.96) is the probability of the jth basic event during $[t,t+dt)$ with the remaining basic events already existing at time t. At most, one basic event occurs during the small time interval $[t,t+dt)$, and the terms describing the possibility of two or more basic events has been neglected. The denominator on the right-hand side represents the probability of the non-existence of the cut set failure at time t.

The expected number of times the cut set occurs per unit time at time t, i.e., $w^*(t)$, is equal to the numerator of (7.95) divided by dt and, given by

$$w^*(t)=\sum_{j=1}^{n} w_j(t) \prod_{\substack{l=1\\ l\neq j}}^{n} Q_l(t) \tag{7.97}$$

Thus, $\lambda^*(t)$ in (7.96) is calculated from $w^*(t)$ and $Q^*(t)$:

$$\lambda^*(t)=\frac{w^*(t)}{[1-Q^*(t)]} \tag{7.98}$$

Similar equations hold for $\mu^*(t)$ and $v^*(t)$, i.e., $v^*(t)$ can be calculated by

$$v^*(t)=\sum_{j=1}^{n} v_j(t) \prod_{\substack{l=1\\ l\neq j}}^{n} [1-Q_l(t)] \tag{7.99}$$

and $\mu^*(t)$ is given by

$$\mu^*(t)=\frac{v^*(t)}{Q^*(t)} \tag{7.100}$$

The integral values $W^*(0,t)$ and $V^*(0,t)$ are, as before, obtained from the differentials $w^*(t)$ and $v^*(t)$:

$$W^*(0,t) = \int_0^t w^*(u)\,du \tag{7.101}$$

$$V^*(0,t) = \int_0^t v^*(u)\,du \tag{7.102}$$

These equations are applied in Example 18 to a simple parallel system, and in Example 19 to a two-out-of-three voting configuration.

Example 18.

Calculate the cut set parameters for a repairable and non-repairable parallel system consisting of components 1 and 2 of Example 17 at $t=20$, 500, and 1100.

Solution: Applying (7.94) to the two-component cut set $\{1,2\}$ we have

$$Q^*(t) = Q_1(t)Q_2(t)$$

For the repairable case, at 20 hr,

$$Q^*(t) = (8.58757\times 10^{-3})(3.09075\times 10^{-2}) = 2.65420\times 10^{-4}$$

From (7.97), $w^*(t) = w_1(t)Q_2(t) + w_2(t)Q_1(t)$; thus, for the repairable case, at 20 hr,

$$w^*(t) = (9.91412\times 10^{-4})(3.09075\times 10^{-2}) + (1.93818\times 10^{-3})(8.58757\times 10^{-3})$$
$$= 4.72864\times 10^{-5}$$

From (7.98), $\lambda^*(t) = w^*(t)/[1-Q^*(t)]$. At 20 hr, for the repairable case,

$$\lambda^*(t) = \frac{4.72864\times 10^{-5}}{1-2.65420\times 10^{-4}} = 4.72989\times 10^{-5}$$

Other differentials $v^*(t)$ and $\mu^*(t)$ are calculated by (7.99) and (7.100). The integral parameters $W^*(0,t)$ and $V^*(0,t)$ are readily obtained by equations (7.101) and (7.102). Part of the final results are listed in Table 7.12.

TABLE 7.12 RESULTS OF EXAMPLE 18

t	$Q^*(t)$	$w^*(t)$	$\lambda^*(t)$	$W^*(0,t)$
		Repairable System		
20	2.65420×10^{-4}	4.72864×10^{-5}	4.72989×10^{-5}	5.78476×10^{-4}
500	7.33405×10^{-4}	9.16758×10^{-5}	9.17431×10^{-5}	4.29513×10^{-2}
1100	7.33407×10^{-4}	9.16758×10^{-5}	9.17431×10^{-5}	9.79553×10^{-2}
		Non-Repairable System		
20	7.76421×10^{-4}	7.64839×10^{-5}	7.65434×10^{-5}	7.73524×10^{-4}
500	2.48720×10^{-1}	6.72899×10^{-4}	8.95670×10^{-4}	2.48685×10^{-1}
1100	5.93209×10^{-1}	4.43828×10^{-4}	1.09105×10^{-3}	5.93169×10^{-1}

As expected, the parallel (redundant) configuration is more reliable than the single-component system of Example 17. For the non-repairable case, $Q^*(t)$ equals $W^*(0,t)$ and, for the repairable case, $\lambda^*(t) \simeq w^*(t)$, since $Q^*(t) \ll 1$.

Example 19.

Calculate the cut set parameters for a repairable and non-repairable two-out-of-three voting system consisting of the three components of Example 17 at $t=20$, 500, and 1100 hr.

Solution: The fault tree for this system is given in Fig. 7.8, and the cut sets are easily identified as $K_1=\{1,2\}, K_2=\{2,3\}, K_3=\{3,1\}$. Reliability parameters for K_1 are obtained in Example 18. Parameters for the other two cut sets (from computer printouts) are listed in Table 7.13.

TABLE 7.13 RESULTS OF EXAMPLE 19

t	$Q^*(t)$	$w^*(t)$	$\lambda^*(t)$	$W^*(0,t)$
		Non-Repairable System $K_2(2,3)$		
20	2.28344×10^{-3}	2.22685×10^{-4}	2.23195×10^{-4}	2.26929×10^{-3}
500	4.91075×10^{-1}	9.94725×10^{-4}	1.95456×10^{-3}	4.90962×10^{-1}
1100	8.56400×10^{-1}	3.11822×10^{-4}	2.17147×10^{-3}	8.56291×10^{-1}
		Repairable System $K_2(2,3)$		
20	1.53321×10^{-3}	1.84269×10^{-4}	1.84552×10^{-4}	1.96798×10^{-3}
500	1.12988×10^{-2}	4.70796×10^{-4}	4.76176×10^{-4}	2.15316×10^{-1}
1100	1.12994×10^{-2}	4.70810×10^{-4}	4.76191×10^{-4}	4.97799×10^{-1}
		Non-Repairable System $K_3(3,1)$		
20	1.15314×10^{-3}	1.13027×10^{-4}	1.13157×10^{-4}	1.14742×10^{-3}
500	3.05674×10^{-1}	7.34580×10^{-4}	1.05798×10^{-3}	3.05619×10^{-1}
1100	6.42523×10^{-1}	3.94411×10^{-4}	1.10332×10^{-3}	6.42466×10^{-1}
		Repairable System $K_3(3,1)$		
20	4.25996×10^{-4}	7.36649×10^{-5}	7.36963×10^{-5}	8.22654×10^{-4}
500	1.51024×10^{-3}	1.76196×10^{-4}	1.76463×10^{-4}	8.03189×10^{-2}
1100	1.51032×10^{-3}	1.76204×10^{-4}	1.76471×10^{-4}	1.86040×10^{-1}

These results contain no surprises. The mean time to failure for the components is $\text{MTTF}_3 < \text{MTTF}_2 < \text{MTTF}_1$; thus, we would expect $w_2^*(t) < w_3^*(t) < w_1^*(t)$ for the non-repairable case, and that result is confirmed:† $3.11822\times10^{-4} < 3.94411\times10^{-4} < 4.43828\times10^{-4}$ (at 1100 hr). For the repairable case, the unavailabilities are a more significant parameter, and we see that $Q_2^*(t) > Q_3^*(t) > Q_1^*(t)$; $1.12994\times10^{-2} > 1.51032\times10^{-3} > 7.33407\times10^{-4}$. The longer repair times for the more reliable

†Suffix j attached to the cut set parameter refers to the jth cut set. Cut set $1=\{1,2\}$, Cut set $2=\{2,3\}$, cut set $3=\{3,1\}$.

components actually result in an increase in system unavailability. Another point to note is that a system composed of components having constant failure rates or constant conditional failure intensities λ will not necessarily have a constant conditional failure intensity λ^*. We see also that, unlike for a component where $w(t)$ decreases with time, $w^*(t)$ increases briefly, and then decreases.

The KITT program accepts as input path sets as well as cut sets, the calculations being done in much the same way. We do not discuss this option.

7.7.3 System Unavailability $Q_s(t)$

As in Section 7.5.4, we define event d_i as

d_i = all basic events in the ith minimal cut set exist at time t.
The ith minimal cut set failure exists at time t.

The expansion (7.75) in Section 7.5.4 was obtained by the inclusion-exclusion principle. Equation (7.75) can also be written as

$$Q_s(t)=\sum_{i=1}^{N_c}\Pr(d_i)-\sum_{i=2}^{N_c}\sum_{j=1}^{i-1}\Pr(d_i\cap d_j)+\cdots+(-1)^{m-1}\sum_{1\le i_1<i_2<\cdots<i_m<N_c}\Pr(d_{i_1}\cap d_{i_2}\cap\cdots\cap d_{i_m})+\cdots+(-1)^{N_c-1}\Pr(d_1\cap d_2\cap\cdots\cap d_{N_c}) \tag{7.103}$$

The mth term on the right-hand side of (7.103) is the contribution to $Q_s(t)$ from m minimal cut set failures existing simultaneously at time t. Thus, (7.103) can be rewritten as

$$Q_s(t)=\sum_{i=1}^{N_c}Q_i^*(t)-\sum_{i=2}^{N_c}\sum_{j=1}^{i-1}\prod_{i,j}Q(t)+\cdots+(-1)^{m-1}\sum_{1\le i_1<i_2<\cdots<i_m\le N_c}\prod_{i_1\ldots i_m}Q(t)+\cdots+(-1)^{N_c-1}\prod_{i_1\ldots i_{N_c}}Q(t) \tag{7.104}$$

where $\prod_{i_1\ldots i_m}$ is the product of $Q(t)$'s for the basic events in cut set i_1, or $i_2,\ldots,$ or i_m.

The lower and upper bounds of (7.78) can be rewritten as

$$\sum_{i=1}^{N_c} Q_i^*(t) - \sum_{i=2}^{N_c} \sum_{j=1}^{i-1} \prod_{i,j} Q(t) \le Q_s(t) \le \sum_{i=1}^{N_c} Q_i^*(t) \tag{7.105}$$

where $\prod_{i,j}$ is the product of $Q(t)$'s for the basic event which is a member of either cut set i or j. Since $Q(t)$ is usually much less than one, the brackets are within three significant figures of one another.

An alternative derivation based on a relationship derived by Esary and Proschan leads to an upper bound of (7.79), which can be written as

$$Q_s(t) \le 1 - \prod_{i=1}^{N_c} \left[1 - Q_i^*(t)\right] \tag{7.106}$$

This sometimes gives a conservative estimate of $Q_s(t)$. It is exact when the cut sets are disjoint sets of basic events.

Example 20.

Find the upper and lower brackets for $Q_s(t)$ at $t=20$ hr for a two-component, series, repairable system. The components are 1 and 2 of Example 17.

Solution: From Examples 16 and 17, the cut set and component values at 20 hr are

$$Q_1^*(t) = Q_1(t) = 8.58757 \times 10^{-3}$$
$$Q_2^*(t) = Q_2(t) = 3.09075 \times 10^{-2}$$

From (7.105),

$$\begin{aligned} Q_s(t)_{\max} &= Q_1^*(t) + Q_2^*(t) \\ &= 8.58757 \times 10^{-3} + 3.09075 \times 10^{-2} \\ &= 4.91114 \times 10^{-2} \\ Q_s(t)_{\min} &= Q_1(t) + Q_2(t) - Q_1(t)Q_2(t) \\ &= 3.92296 \times 10^{-2} \end{aligned}$$

The lower bound is the best and is the *last bracket*. It coincides with the exact system unavailability $Q_s(t)$ because all terms in the expansion are included.

Example 21.

Obtain the upper and lower brackets of $Q_s(t)$ for the parallel, two-component system of Example 18.

Solution: Here we have only one cut set, and so $Q_s(t)$ is exactly equal to $Q^*(t)$ and the upper and lower bounds are identical.

Example 22.

Find the upper and lower brackets for $Q_s(t)$ at 500 hr for the two-out-of-three system of Example 19 (non-repairable case), and compare the values with $Q_s(t)$ upper bound obtained from equation (7.106).

Solution: From Examples 16 and 17 at $t=500$ hr,

$$Q_1(t)=3.93469\times 10^{-1},\quad Q_2(t)=6.32121\times 10^{-1}$$

$$Q_3(t)=7.76870\times 10^{-1}$$

From Example 18 and 19,

$$Q_1^*(t)=2.48720\times 10^{-1},\quad Q_2^*(t)=4.91075\times 10^{-1},\quad Q_3^*(t)=3.05674\times 10^{-1}$$

The exact expression for $Q_s(t)$ from equation (7.103) is

$$Q_s(t)=\sum_{i=1}^{3}\Pr(d_i)-\sum_{i=2}^{3}\sum_{j=1}^{i-1}\Pr(d_i\cap d_j)+\sum_{i=3}^{3}\sum_{j=2}^{i-1}\sum_{k=1}^{j-1}\Pr(d_i\cap d_j\cap d_k)$$

$$=\Pr(d_1)+\Pr(d_2)+\Pr(d_3)-[\Pr(d_1\cap d_2)+\Pr(d_2\cap d_3)+\Pr(d_3\cap d_1)]$$
$$+\Pr(d_1\cap d_2\cap d_3)$$

$[A]$ $[\Pr(d_1)+\Pr(d_2)+\Pr(d_3)]=Q_1^*(t)+Q_2^*(t)+Q_3^*(t)=1.04547\times 10^{-1}$

$[B]$ $[\Pr(d_1\cap d_2)+\Pr(d_2\cap d_3)+\Pr(d_3\cap d_1)]=\prod_{1,2}Q(t)+\prod_{2,3}Q(t)+\prod_{3,1}Q(t)$
$=[Q_1(t)Q_2(t)Q_3(t)+Q_1(t)Q_2(t)Q_3(t)+Q_1(t)Q_2(t)Q_3(t)]=5.79669\times 10^{-1}$

$[C]$ $\Pr(d_1\cap d_2\cap d_3)=\prod_{1,2,3}Q(t)=Q_1(t)Q_2(t)Q_3(t)=1.93223\times 10^{-1}$
$Q_s(t)_{\min}=[A]-[B]=4.65800\times 10^{-1}$
$Q_s(t)_{\max}=[A]-[B]+[C]=6.59024\times 10^{-1}$

In this case $Q_s(t)_{\max}$ is the exact value and is the last bracket. As in the last example, all terms are included.

The upper bound obtained by (7.106) is

$$Q_s(t)_{\text{upper bound}}$$
$$=1-[1-4.91705\times 10^{-1}][1-3.05674\times 10^{-1}][1-2.48720\times 10^{-1}]$$
$$=7.34856\times 10^{-1}$$

We see that this upper bound is a conservative estimate compared to $Q_s(t)_{\max}$.

7.7.4 System Parameter $w_s(t)$

The parameter $w_s(t)$ is the expected number of times the top event occurs at time t, per unit time; thus, $w_s(t)\,dt$ is the expected number of times the top event occurs in t to $t+dt$. We now let

$e_i=$ the event that the ith cut set failure occurs in time t to $t+dt$; i.e., $\Pr(e_i)=w_i^*(t)\,dt$

For the top event to occur in the interval $[t, t+dt)$ none of the cut set failures can exist at time t, and then one (or more) of them must fail in time t to $t+dt$. Hence,

$$w_s(t)\,dt = \Pr\left(A \bigcup_{i=1}^{N_c} e_i\right) \tag{7.107}$$

where $\left(A \bigcup_{i=1}^{N_c} e_i\right)$ is $A \cap \left(\bigcup_{i=1}^{N_c} e_i\right)$,

$\left(\bigcup_{i=1}^{N_c} e_i\right)$ = the event that one or more of the cut set failures occur at time t,

A = the event of none of the cut sets failures existing at time t.

We have

$$\Pr\left(A \bigcup_{i=1}^{N_c} e_i\right) = \Pr\left(\bigcup_{i=1}^{N_c} e_i\right) - \Pr\left(B \bigcup_{i=1}^{N_c} e_i\right) \tag{7.108}$$

where $B = \bigcup_{j=1}^{N_c} d_j$ = the union of the events of the jth cut set failure existing at time t; i.e., $B = \bar{A}$.

Substitute (7.108) into (7.107); then

$$w_s(t)\,dt = \Pr\left(\bigcup_{i=1}^{N_c} e_i\right) - \Pr\left(B \bigcup_{i=1}^{N_c} e_i\right) \tag{7.109}$$

$$w_s(t) = w_s^{(1)}(t) - w_s^{(2)}(t) \tag{7.110}$$

The first right-hand term is the contribution from the event that one or more cut sets fail during $[t, t+dt)$. The second accounts for those cases in which one or more cut sets fail during $[t, t+dt)$ while other cut sets, already failed to time t, have not been repaired. It is a *second-order correction* term; hence, we label it $w_s^{(2)}(t)$. Expanding $w_s^{(1)}(t)$ in the same manner as (7.103) yields

$$\begin{aligned} w_s^{(1)}(t)\,dt &= \Pr\left(\bigcup_{i=1}^{N_c} e_i\right) \\ &= \sum_{i=1}^{N_c} w_i^*(t)\,dt - \sum_{i=2}^{N_c}\sum_{j=1}^{i-1} \Pr(e_i \cap e_j) + \\ &\quad \dots + (-1)^{N_c-1}\Pr(e_1 \cap e_2 \dots \cap e_{N_c}) \end{aligned} \tag{7.111}$$

The first summation, as in equation (7.103), is simply the contribution from cut set failures, whereas the second and following terms involve the simultaneous occurrence of two or more failures. The cut set failures considered in the particular combinations must not exist at time t and then must all simultaneously occur in t to $t+dt$.

The foregoing equations are adequate to obtain upper estimates of $w_s(t)$ for the simple series and parallel systems of Examples 17 and 18, since the expansion terms $\Pr(e_i \cap e_j)$ are zero because the cut sets do not have any common component.

Example 23.

Calculate $w_s^{(1)}(t)$ at 20 hr for a two-component ($\lambda_1 = 10^{-3}, \lambda_2 = 10^{-3}$), repairable ($\mu_1 = 1/10, \mu_2 = 1/40$), series system at 20 hr.

Solution: From Example 17,

$$w_2^*(20) = 1.93818 \times 10^{-3}, \quad w_1^*(20) = 9.91412 \times 10^{-4}$$

Noting that the second and the following terms on the right-hand side of (7.111) are equal to zero,

$$w_s^{(1)}(20) = \sum_{i=1}^{N_c} \frac{w_i^*(20)\,dt}{dt} = w_1^*(20) + w_2^*(20)$$

$$= 9.91412 \times 10^{-4} + 1.93818 \times 10^{-3} = 2.92959 \times 10^{-3}$$

This is maximum w_s, with $w_s^{(2)} = 0$ in (7.110). A non-repairable system would be treated in exactly the same way. For the two-component, parallel system, $w_s = w^*$, there being only one cut set.

The probability of one or more basic failures occurring in t to $t+dt$ is equal to $w(t)dt$ and, hence, is proportional to dt.† The simultaneous occurrence of two or more mode failures can thus only be caused by one basic failure occurring and, moreover, this basic failure must be a common member of all those mode failures which must occur simultaneously. Consider the general event $e_1 \cap e_2 \cap \dots \cap e_m$, i.e., the simultaneous occurrence of the m mode failures. Let there be k unique basic failures which are common members to all of the m mode failures: Each of these basic failures must be a member of every one of the mode failures $1, \dots, m$. If k is zero, then the event $e_1 \cap e_2 \cap \dots \cap e_m$ cannot occur, and its associated probability is zero. Assume, therefore, that k is greater than zero.

If one of these k basic failures does not exist at t and then occurs in t to $t+dt$, and all the other basic failures of the m mode failure exist at t

†The next few pages of this discussion closely follow Vesely's original derivation.

(including the $k-1$ common basic failures), then the event $e_1 \cap e_2 \cap \ldots \cap e_m$ will occur. The probability of the event $e_1 \cap e_2 \cap \ldots \cap e_m$ is,

$$\Pr(e_1 \cap \ldots \cap e_m) = w(t; 1,\ldots,m)\,dt \prod_{1,\ldots,m} Q(t) \tag{7.112}$$

The product symbol in equation (7.112) is defined such that

$\prod_{1,\ldots,m}$ = the product of $Q(t)$ for the basic event which is a member of at least one of the cut sets $1,\ldots,m$ but is not a common member of all of them

The product in equation (7.112) is, therefore, the product of the existence probabilities of those basic failures other than the k common primary failures. Also, a basic failure existence probability $Q(t)$ appears only once in the product even though it is a member of two or more mode failures (it cannot be a member of all m mode failures, since these are the k common basic failures).

The quantity $w(t; 1,\ldots,m)\,dt$ accounts for the k common basic failures and is defined such that

$w(t; 1,\ldots,m)$ = the unconditional failure intensity for a mode failure which has as its basic failures the basic failures which are common members to all the mode failures $1,\ldots,m$

If the m mode failures have no basic failures common to all of them, then $w(t; 1,\ldots,m)$ is defined to be identically zero:

$w(t; 1,\ldots,m) = 0$, no basic failures (events) common to all m mode (cut set) failures

The expression for a cut set failure intensity $w^*(t)$, equation (7.97), shows that the intensity consists of one basic failure occurring and the other basic failures already existing. This is precisely what is needed for the k common basic failures. Computation of $w(t; 1,\ldots,m)$ therefore consists of considering the k common basic events as being members of a cut set, and using equation (7.97) to calculate $w(t; 1,\ldots,m)$, the unconditional failure intensity for a cut set.

Computation of the probability of m mode failures simultaneously occurring by equation (7.112) is therefore quite direct. The unique basic events which are members of any of the m cut sets are first separated into two groups: those which are common to all m cut sets and those which are not common to all the cut sets. The common group is considered as a cut set in itself, and $w(t; 1,\ldots,m)$ is computed for this group directly from equation (7.97). If there are no basic events in this common group, then $w(t; 1,\ldots,m)$ is identically zero and computation need proceed no further ($\Pr(e_1 \cap \ldots \cap e_m) = 0$). For the non-common group, the product of the existence probabilities $Q(t)$ for the member basic events is computed. This product and $w(t; 1,\ldots,m)$ are multiplied and $\Pr(e_1 \cap \ldots \cap e_m)$ is obtained. The factor dt will "cancel out" in the final expression and will not be needed.

With the general term $\Pr(e_1 \cap \ldots \cap e_m)$ being determined, equation (7.111), which gives the first term for $w_s(t)\,dt$, is subsequently evaluated.

$$w_s^{(1)}(t)\,dt = \Pr\left(\bigcup_{i=1}^{N_c} e_i\right) = \sum_{i=1}^{N_c} w_i^*(t)\,dt - \sum_{i=2}^{N_c}\sum_{j=1}^{i-1} w(t; i,j)\,dt \prod_{i,j} Q(t)$$
$$+ \sum_{i=3}^{N_c}\sum_{j=2}^{i-1}\sum_{k=1}^{j-1} w(t; i,j,k)\,dt \prod_{i,j,k} Q(t) + \tag{7.113}$$
$$\ldots + (-1)^{N_c-1} w(t; 1,\ldots,N_c)\,dt \prod_{1,\ldots,N_c} Q(t)$$

The first term on the right-hand side of this equation is simply the sum of the unconditional failure intensities of the individual cut sets. Each product in the remaining terms consists of separating the common and uncommon basic events for the particular combination of cut sets and then performing the operations as described in the preceding paragraph. Moreover, each succeeding term on the right-hand side of equation (7.113) consists of combinations of a larger number of products of $Q(t)$. Therefore, each succeeding term rapidly decreases in value and the bracketing procedure is extremely efficient when applied to equation (7.113).

Equation (7.113) consequently determines the first term for $w_s(t)$ of (7.110), and the second term $w_s^{(2)}$ must now be determined. Expanding this second term yields

$$w_s^{(2)}(t) = \Pr\left(B \bigcup_{i=1}^{N_c} e_i\right)$$
$$= \sum_{i=1}^{N_c} \Pr(e_i \cap B) - \sum_{i=2}^{N_c}\sum_{j=1}^{i-1} \Pr(e_i \cap e_j \cap B) + \tag{7.114}$$
$$\ldots + (-1)^{N_c-1} \Pr(e_1 \cap e_2 \cap \ldots \cap e_{N_c} \cap B)$$

Consider a general term in this expression, $\Pr(e_1 \cap \ldots \cap e_m \cap B)$. This term is the probability of the m cut set failures simultaneously occurring in t to $t+dt$ with one or more of the other mode failures already existing at time t (event B). Let

$$w_B(t;1,\ldots,m)\,dt = \Pr(e_1 \cap \ldots \cap e_m \cap B) \tag{7.115}$$

where $w_B(t;1,\ldots,m)$ = the rate of occurrence of the m mode failures $1,\ldots,m$ simultaneously occurring at t with one or more of the other mode failures already existing at time t.

The term *rate of occurrence* simply means "probability per unit time." The term $w_B(t;1,\ldots,m)$ should not be confused with the term $w(t;1,\ldots,m)$. $w_B(t;1,\ldots,m)$ is simply used for ease of notation and refers to the entire event $e_1 \cap \ldots \cap e_m \cap B$ occurring, whereas $w(t;1,\ldots,m)$ refers to the common basic failures of the modes $1,\ldots,m$ occurring. With this notation, equation (7.114) may be rewritten as

$$\begin{aligned} w_s^{(2)}(t)\,dt = \sum_{i=1}^{N_c} w_B(t;i)\,dt - \sum_{i=2}^{N_c}\sum_{j=1}^{i-1} w_B(t;i,j)\,dt \\ + \ldots + (-1)^{N_c-1} w_B(t;1,\ldots,N_c)\,dt \end{aligned} \tag{7.116}$$

Since the event B involves a union, the general term in (7.116) may be expanded:

$$\begin{aligned} w_B(t;1,\ldots,m)\,dt = \sum_{i=1}^{N_c} \Pr(e_1 \cap \ldots \cap e_m \cap d_i) \\ - \sum_{i=2}^{N_c}\sum_{j=1}^{i-1} \Pr(e_1 \cap \ldots \cap e_m \cap d_i \cap d_j) + \\ \ldots + (-1)^{N_c-1} \Pr(e_1 \cap \ldots \cap e_m \cap d_1 \cap \ldots \cap d_{N_c}) \end{aligned} \tag{7.117}$$

where d_i is the event of the ith cut set failure existing at time t. Consider now a general term in this expansion, $\Pr(e_1 \cap \ldots \cap e_m \cap d_1 \cap \ldots \cap d_n)$. If this term is evaluated then $w_B(t;1,\ldots,m)\,dt$ will be determined and, hence, $w_s^{(2)}(t)\,dt$.

The event $e_1 \cap \ldots \cap e_m \cap d_1 \cap \ldots \cap d_n$ is similar to the event $e_1 \cap \ldots \cap e_m$ with the exception that now the mode failures $1,\ldots,n$ must also exist at time t. If a mode failure exists at time t all its basic failures must exist at time t, and these basic failures cannot occur in t to $t+dt$ since an occurrence calls for a non-existence at t and then an existence at $t+dt$. The expression for $\Pr(e_1 \cap \ldots \cap e_m \cap d_1 \cap \ldots \cap d_n)$ is, therefore, analogous to the previous expression for $\Pr(e_1 \cap \dots \cap e_m)$ [equation (7.112)] with one alteration. Those basic events common to all the m cut sets $1,\dots,m$, which are also in any of the n cut sets $1,\dots,n$, cannot contribute to $w(t; 1,\dots,m)$, since they must already exist at time t (for the event $d_1 \cap \dots \cap d_n$). Hence, these basic events, common to all m cut sets and also in any of the n cut sets, must be deleted from $w(t; 1,\dots,m)$ and must be incorporated in the product of basic failure existence probabilities appearing in equation (7.112).

For fault trees with a large number of cut sets, the bracketing procedure is an extremely efficient method of obtaining as tight an enveloping as desired for $w_s(t)$. In equations (7.113), (7.116), and (7.117), an upper bound can be obtained for $w_s^{(1)}(t)$, $w_s^{(2)}(t)$, or $w_B(t)$ by considering just the first terms in the respective right-hand expressions. Lower bounds can be obtained by considering the first two terms, etc. Various combinations of these successive upper and lower bounds will give successive upper and lower bounds for $w_s(t)$.

As an example of the application of the bracketing procedure, a first (and simple) upper bound for $w_s(t), w_s(t)_{\max}$, is given by the relations

$$w_s(t)_{\max} = w_s^{(1)}(t)_{\max} \tag{7.118}$$

where

$$w_s^{(1)}(t)_{\max} = \sum_{i=1}^{N_c} w_i^*(t) \tag{7.119}$$

This was done in Example 23.

The computer code based on these equations allows the user the luxury of determining how many terms in equations (7.113), (7.116), and (7.117) he wishes to use. This introduces a number of complications, since the terms are alternatively plus and minus. If one chooses, for example, to use two terms in equation (7.113), then $w_s^{(1)}(t)$ is a lower bound with respect to the first term, and the best solution is the lower bound. If three terms are considered, then the best solution is the upper bound. The same consideration applies to $w_s^{(2)}(t)$, so the final $w_s(t)$ brackets must be interpreted cautiously. For the KITT program the program input options are:

1) ISTOP=1 or ISTOP=2. If ISTOP=2, the bracketing procedure is used. If ISTOP=1, bracketing is not used [only (7.119) is used].
2) NBMAX=number of outer brackets to be obtained (assuming ISTOP=2).
3) If IFAG=2, the system failure correction term, equation (7.116), is used. If IFAG=1, system correction terms are not used.

4) NB2(N) = the number of inner brackets to be obtained for each outer bracket. Here N are the number of cut sets failing simultaneously, so $N \leq N_c$.

The overall system bounds are

$$w_s(t)_{\min} \leq w_s(t) \leq w_s(t)_{\max}$$
$$w_s(t)_{\min} = w_s^{(1)}(t)_{\min} - w_s^{(2)}(t)_{\max}$$
$$w_s(t)_{\max} = w_s^{(1)}(t)_{\max} - w_s^{(2)}(t)_{\min}$$

If one sets NBMAX equal to N_c, the number of cut sets, then all terms are calculated, and the exact values for the system parameters are produced. If all terms in the expansion exist

$$\left.\begin{aligned} Q_s(t) &= Q_s(t)_{\min} \\ w_s(t) &= w_s(t)_{\min} \\ \lambda_s(t) &= \lambda_s(t)_{\min} \end{aligned}\right\} \quad N_c\text{: even}$$

or

$$\left.\begin{aligned} Q_s(t) &= Q_s(t)_{\max} \\ w_s(t) &= w_s(t)_{\max} \\ \lambda_s(t) &= \lambda_s(t)_{\max} \end{aligned}\right\} \quad N_c\text{: odd}$$

Example 24, hopefully, will clarify the theory and equations.

Example 24.

Calculate w_s and the associated brackets for the two-out-of-three non-repairable voting system of Example 22 at 500 hr. The computer solution produced by KITT with ISTOP = 2, NBMAX = N_c = 3, IFAG = 2, and NB2(N) = 3, 2, 1 was:

	Computer Name for Variable	
Differential characteristics—upper bound	W	2.40220×10^{-3}
Differential characteristics—best brackets	WMIN	1.71296×10^{-3}
Failure rate contributions	WMAX	1.71296×10^{-3}
	W1MIN	1.71296×10^{-3}
	W1MAX	1.71296×10^{-3}
	W2MIN	0.00000
	W2MAX	0.00000
	WMIN	1.71296×10^{-3}
	WMAX	1.71296×10^{-3}
Differential characteristics—last brackets	W1LAST	1.72196×10^{-3}
	W2MIN-LAST	-6.89245×10^{-4}
	W2MAX-LAST	0.00000

Solution: With MBMAX$=N_c=3$ exact values should be obtained, and with N_c odd, these shoula be $w_s(t)_{\max}$. The system parameters at 500 hr (see Example 19) were:

$$w_3^*(3,1)=7.34580\times10^{-4},\quad w_2^*(2,3)=9.94724\times10^{-4},\quad w_1^*(1,2)=6.72899\times10^{-4}$$

$$Q_1=3.93469\times10^{-1},\quad Q_2=6.32121\times10^{-1},\quad Q_3=7.76870\times10^{-1}$$

$$w_1=6.06531\times10^{-4},\quad w_2=7.35759\times10^{-4},\quad w_3=6.69390\times10^{-4}$$

We proceed with a term-by-term evaluation of $w_s^{(1)}$, using equation (7.113).

A, first term:

$$\sum_{i=1}^{N_c} w_i^*(t)=w_1^*(t)+w_2^*(t)+w_3^*(t)=2.40220\times10^{-3}$$

B, second term:

$$\sum_{i=2}^{N_c}\sum_{j=1}^{i-1} w(t;i,j)\prod_{i,j} Q(t)=w_2Q_1Q_3+w_3Q_2Q_1+w_1Q_2Q_3=6.89245\times10^{-4}$$

C, third term:

$$\sum_{i=3}^{N_c}\sum_{j=2}^{i-1}\sum_{k=1}^{j-1} w(t;i,j,k)\prod_{i,j,k} Q(t)=0$$

The calculation of $w_s^{(2)}(t)$ is done by using equations (7.116) and (7.117).

D, first term:

$$\sum_{i=1}^{N_c} w_B(t;i)=\sum_{i=1}^{3}\left[\sum_{k=1}^{3}\Pr(e_i\cap d_k)-\sum_{k=2}^{3}\sum_{l=1}^{k-1}\Pr(e_i\cap d_k\cap d_l)\right.$$
$$\left.+\sum_{k=3}^{3}\sum_{l=2}^{k-1}\sum_{m=1}^{l-1}\Pr(e_i\cap d_k\cap d_l\cap d_m)\right]$$

Recall now that e_i is the event of the ith cut set failure occurring in t to $t+dt$, d_i is the event of the ith mode failure existing at t. The first term in the inner bracket is

$$i=1;\Pr(e_1\cap d_1)+\Pr(e_1\cap d_2)+\Pr(e_1\cap d_3)=0+w_1Q_2Q_3+w_2Q_1Q_3$$

If, for example, d_1 exists at time t, components 1 and 2 have failed, and $\Pr(e_1)=0$ (term 1). If d_2. exists, components 2 and 3 have failed, and only three can fail, etc. The second term in the inner bracket is zero, since if two cut sets have failed, all components have failed. This is true of term three also.

$$i=2;\Pr(e_2\cap d_1)+\Pr(e_2\cap d_2)+\Pr(e_2\cap d_3)=w_3Q_1Q_2+0+w_2Q_1Q_3$$

$$i=3;\Pr(e_3\cap d_1)+\Pr(e_3\cap d_2)+\Pr(e_3\cap d_3)=w_3Q_1Q_2+w_1Q_2Q_3+0$$

$$\sum_{i=1}^{3}=2[w_1Q_2Q_3+w_2Q_1Q_3+w_3Q_1Q_2]=1.37849\times10^{-3}$$

E, second term:

$$\sum_{i=2}^{3}\sum_{j=1}^{i-1}w_B(t;i,j)=\sum_{i=2}^{3}\sum_{j=1}^{i-1}\left[\sum_{k=1}^{3}\Pr(e_i\cap e_j\cap d_k)-\sum_{k=2}^{3}\sum_{l=1}^{k-1}\Pr(e_i\cap e_j\cap d_k\cap d_l)\right.$$
$$\left.+\sum_{k=3}^{3}\sum_{l=2}^{k-1}\sum_{m=1}^{l-1}\Pr(e_i\cap e_j\cap d_k\cap d_l\cap d_m)\right]$$

$$\begin{aligned}i=2,j=1;\ &\Pr(e_2\cap e_1\cap d_1)+\Pr(e_2\cap e_1\cap d_2)+\Pr(e_2\cap e_1\cap d_3)\\&-\Pr(e_2\cap e_1\cap d_2\cap d_1)-\Pr(e_2\cap e_1\cap d_3\cap d_1)-\Pr(e_2\cap e_1\cap d_3\cap d_2)\\&+\text{higher terms (all zero)}\\&=0+0+w_2Q_1Q_3-0-0-0+0\end{aligned}$$

$$\begin{aligned}i=3,j=1;\ &\Pr(e_3\cap e_1\cap d_1)+\Pr(e_3\cap e_1\cap d_2)+\Pr(e_3\cap e_1\cap d_3)\\&-\Pr(e_3\cap e_1\cap d_2\cap d_1)-\Pr(e_3\cap e_1\cap d_3\cap d_1)-\Pr(e_3\cap e_1\cap d_3\cap d_2)\\&+\text{(all terms zero)}\\&=0+w_1Q_2Q_3+0-0-0+0\end{aligned}$$

$$\begin{aligned}i=3,j=2;\ &\Pr(e_3\cap e_2\cap d_1)+\Pr(e_3\cap e_2\cap d_2)+\Pr(e_3\cap e_2\cap d_3)-\text{(all term zero)}\\&=w_3Q_1Q_2+0+0+0\end{aligned}$$

$$\sum_{i=2}^{3}\sum_{j=1}^{i-1}=w_2Q_1Q_3+w_1Q_2Q_3+w_3Q_1Q_2=6.89246\times10^{-4}$$

$$\begin{aligned}w_s&=w_s^{(1)}-w_s^{(2)}=[A-B+C]-[D-E]\\&=1.712955\times10^{-3}-6.89245\times10^{-4}=1.02371\times10^{-3}\end{aligned}$$

If we compare these calculations with the computer output we note that:

W = upper bound: This is the upper bound term $A=2.40220\times10^{-3}$ and is thus verified.

WMIN (best bracket): This should equal WMAX, since the calculation is exact. The equality is there, but the computer result should be 1.02371×10^{-3} instead of the 1.71296×10^{-3} computed by KITT.

W1MIN (failure rate contribution): Here the value of 1.71296×10^{-3} is correct $[A-B+C]$. It also equals W1MAX.

W2MIN, W1MAX: These numbers are incorrectly calculated by KITT as zero, thus accounting for the discrepancy in WMIN.

WMIN, WMAX: This is simply a duplicate printout of WMIN, WMAX above.

W1LAST: This is the exact value for $w_s^{(1)}$LAST and is correct.
W2MIN: This should equal W2MAX LAST and be 6.89246×10^{-4}. The computer output is incorrect.

It appears, therefore, that there is an error in the subroutine which produces $w_s^{(2)}$. In general, for large systems with diverse components, this error is trivial.

7.7.5 Other System Parameters Calculated by KITT

Once Q_s and w_s have been computed, it is comparatively easy to obtain the other system parameters, λ_s and W_s. Like $\lambda^*(t)\,dt$, its cut set analog, the probability that the top event occurs in time t to $t+dt$, given there is no top event failure at t, is related to $w_s(t)\,dt$ and $Q_s(t)$ by

$$w_s(t)\,dt = [1 - Q_s(t)]\lambda_s(t)\,dt \tag{7.120}$$

This is identical to equation (7.98), the cut set analog. For the failure to occur in t to $t+dt, (w_s(t)\,dt)$, it must not exist at time $t, (1-Q_s(t))$, and must occur in time t to $t+dt, (\lambda_s(t)\,dt)$.

Example 25.

Calculate $\lambda_{s,\max}$ and $\lambda_{s,\min}$ at 500 hr for the two-out-of-three non-repairable voting system of Example 22, using Q_s and w_s values from Examples 22 and 24.

Solution: Using the KITT w_s values from Example 24 and the $Q_{s,\max}$ and $Q_{s,\min}$ from Example 22:

$$\lambda_s(500)_{\max} = \frac{w_{s,\max}}{1 - Q_{s,\max}} = \frac{1.71296 \times 10^{-3}}{1 - 6.59024 \times 10^{-1}} = 5.02369 \times 10^{-3}$$

$$\lambda_s(500)_{\min} = \frac{w_{s,\min}}{1 - Q_{s,\min}} = \frac{1.02371 \times 10^{-3}}{1 - 4.65800 \times 10^{-1}} = 1.91634 \times 10^{-3}$$

The integral value $W_s(0,t)$ is, as before, obtained by from the differential $w_s(t)$ by

$$W_s(0,t) = \int_0^t w_s(u)\,du \tag{7.121}$$

7.7.6 Short-cut Calculation Methods

Back-of-the-envelope guesstimates have a time-honored role in engineering and will always be with us, computers notwithstanding. In this section we develop a modified version of a calculation technique originated by J. Fussell.[6] It requires as input failure and repair rates for each component, and minimal cut sets. It assumes exponential distributions of λ and μ and independence of component failures. We begin the

[6]Fussell, J., "How to Hand-Calculate System Reliability and Safety Characteristics," *IEEE Trans. on Reliability*, R-24, No. 3, 1975.

derivation by restating a few equations presented earlier in this chapter and previous chapters. We use Q_i as the symbol for unavailability. For non-repairable component i,

$$Q_i = (1 - e^{-\lambda_i t}) \simeq \lambda_i t \tag{7.122}$$

[see equation (7.93)].

If the component is repairable, by Markov analysis, Chapter 4, equation (4.120),

$$Q_i = \frac{\lambda_i}{\lambda_i + \mu_i}\left[1 - e^{-(\lambda_i + \mu_i)t}\right] \tag{7.123}$$

As t becomes large and if $\lambda_i / \mu_i \ll 0.1$,

$$Q_i \simeq \frac{\lambda_i}{\lambda_i + \mu_i} \simeq \frac{\lambda_i}{\mu_i} \tag{7.124}$$

These approximations overpredict Q_i, in general. To obtain the cut set reliability parameters, we write the familiar

$$Q_i^* = \prod_{j=1}^{n} Q_i \tag{7.125}$$

Equation (7.125), coupled with equation (7.124), gives the steady-state value for Q_i^* directly. To calculate the other cut-set parameters, further approximations need be made.

We start by combining equation (7.97) with (4.77) to get

$$w_i^*(t) = \sum_{j=1}^{n} \left[1 - Q_j(t)\right]\lambda_j(t) \prod_{\substack{l=1 \\ l \neq j}}^{n} Q_j(t) \tag{7.126}$$

Substituting equation (7.125) and making the approximation that $[1 - Q_j(t)] = 1$ we obtain,

$$w_i^*(t) \simeq Q_i^*(t) \sum_{j=1}^{n} \frac{\lambda_j}{Q_j}$$

Furthermore, we have, for λ_i^*,

$$\lambda_i^* = \frac{w_i^*}{[1 - Q_i^*]} \tag{7.127}$$

System parameters are readily approximated from cut set parameters by bounding procedures previously developed.

$$Q_s \simeq \sum_{i=1}^{N_c} Q_i^* \tag{7.128}$$

$$\lambda_s \simeq \sum_{i=1}^{N_c} \lambda_i^* \tag{7.129}$$

$$w_s \simeq \sum_{i=1}^{N_c} w_i^* \tag{7.130}$$

Some caveats apply to the use of these equations, which are summarized in Table 7.14. In general, the overprediction can become significant in the following cases:

1) Unavailability of a repairable component, system, or cut set is evaluated at less than twice the mean repair time $1/\mu_i$.
2) Unavailability of a non-repairable component, system, or cut set is evaluated at more than one-tenth the $\text{MTTF} = 1/\lambda_i$.
3) When cut set or system unavailabilities are greater than 0.1.

TABLE 7.14 SUMMARY OF SHORT-CUT CALCULATION EQUATIONS

Component	Cut Set	System
$Q_i = \lambda_i t$	$Q_i^* = \prod_{i=1}^{n} Q_i$	$Q_s = \sum_{i=1}^{N_c} Q_i^*$
(non-repairable, $\lambda_i t < 0.1$)	$w_i^* = Q_i^* \sum_{i=1}^{n} \frac{\lambda_i}{Q_i}$	$w_s = \sum_{i=1}^{N_c} w_i^*$
$Q_i = \lambda_i/\mu_i$ (repairable, $t > 2/\mu_i$)	$\lambda_i^* = \frac{w_i^*}{[1 - Q_i^*]}$	$\lambda_s = \sum_{i=1}^{N_c} \lambda_i^*$

We now test these equations by using them to calculate the reliability parameters for the two-out-of-three system at 100 hr. The input information is, as before:

Component	λ (hour^{-1})	μ (hour^{-1})
1	10^{-3}	1/10
2	2×10^{-3}	1/40
3	3×10^{-3}	1/60

Cut sets are $\{1,2\}, \{2,3\}, \{1,3\}$.

The calculations are summarized in Table 7.15. The test example is a particularly severe one since, at 100 hr, we are below the minimum time required by component 3 to come to steady state ($t=100=1.67\times 60$). We see that Q_s has been calculated conservatively to an accuracy of 25% or better.

TABLE 7.15 Repairable and Non-Repairable Cases of Short-Cut Calculations

Characteristic	Approximation	Numerical Result (short-cut)	Exact Value (computer)	Time Bounds to Ensure Small Overprediction	
				Minimum	Maximum
Q_1 (repairable)	λ_1/μ_1	10×10^{-3}	9.9×10^{-3}	20	∞
Q_2	λ_2/μ_2	80×10^{-3}	74×10^{-3}	80	∞
Q_3	λ_3/μ_3	180×10^{-3}	152×10^{-3}	120	∞
Q_1 (non-repairable)	$\lambda_1 t$	10^{-1}	0.95×10^{-1}	0	100
Q_2	$\lambda_2 t$	2×10^{-1}	1.8×10^{-1}	0	50
Q_3	$\lambda_3 t$	3×10^{-1}	2.6×10^{-1}	0	33
Q_1^* (repairable)	Q_1Q_2	8×10^{-4}	7.3×10^{-4}		
Q_2^*	Q_2Q_3	14.4×10^{-3}	11×10^{-3}		
Q_3^*	Q_1Q_3	1.8×10^{-3}	1.5×10^{-3}		
Q_1^* (non-repairable)	Q_1Q_2	2×10^{-2}	1.8×10^{-2}		
Q_2^*	Q_2Q_3	6×10^{-2}	4.7×10^{-2}		
Q_3^*	Q_1Q_3	3×10^{-2}	2.5×10^{-2}		
w_1^* (repairable)	$Q_1^*\Sigma(\lambda_i/Q_i)$	10×10^{-5}	9.2×10^{-5}		
w_2^*	$Q_2^*\Sigma(\lambda_i/Q_i)$	6×10^{-4}	4.2×10^{-4}		
w_3^*	$Q_3^*\Sigma(\lambda_i/Q_i)$	2.1×10^{-4}	1.8×10^{-4}		
w_1^* (non-repairable)	$Q_1^*\Sigma(\lambda_i/Q_i)$	4×10^{-4}	3.2×10^{-4}		
w_2^*	$Q_2^*\Sigma(\lambda_i/Q_i)$	12×10^{-4}	8.3×10^{-4}		
w_3^*	$Q_3^*\Sigma(\lambda_i/Q_i)$	6×10^{-4}	4.5×10^{-4}		
Q_s (repairable)	ΣQ_i^*	17×10^{-3}	13×10^{-3}		
Q_s (non-repairable)	ΣQ_i^*	11×10^{-2}	9×10^{-2}		
w_s (repairable)	Σw_i^*	9.1×10^{-4}	7×10^{-4}		
w_s (non-repairable)	Σw_i^*	2.2×10^{-3}	1.5×10^{-3}		

7.7.7 The Inhibit Gate

An *inhibit gate*, Fig. 7.26, represents an event which occurs with some fixed probability of occurrence.† It produces the output event only if its input event exists and the inhibit condition has occurred.

An example of a section of a fault tree containing an inhibit gate is shown in Fig. 7.27. The event "fuse cut" occurs if a primary or secondary fuse failure occurs. Secondary fuse failure can occur if an excessive current in the circuit occurs because an excessive current can cause a fuse to open.

†See also row 3, Table 2.1 of Chapter 2.

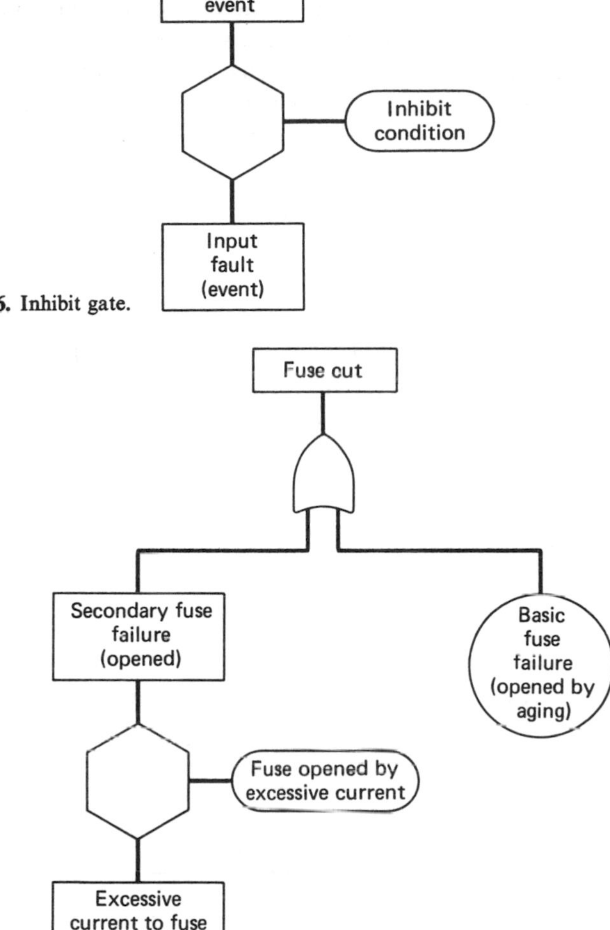

Figure 7.26. Inhibit gate.

Figure 7.27. Fault tree for fuse.

The fuse does not open, however, every time an excessive current is present in the circuit because all conditions of an excessive current do not result in sufficient overcurrent to open the fuse. The inhibit condition is then used as a weighting factor applied to all the fault events in the domain of the inhibit gate. Since the inhibit condition is treated as an AND logic gate in a probabilistic analysis, it is a probabilistic weighting factor. The inhibit condition has many variations of usage in fault tree analysis, but in all cases it represents a probabilistic weighting factor. A human operator, for example, is simulated by an inhibit gate, since his reliability or availability is a time-independent constant.

If, in the input to KITT, an event is identified as an inhibit condition, the cut set parameters Q^* and w^* are multiplied by the inhibit value. In the two-component parallel system of Fig. 7.28, a value of 0.1 is assigned to the inhibit condition, $Q_2 = 0.1$. The results of the computations are summarized in Table 7.16. We see that the effect of the inhibit gate in Fig. 7.28 is to yield a $Q^* = Q_1Q_2 = Q_1 \times 0.1$ and a $w^* = w_1 \times 0.1$, independent of time.

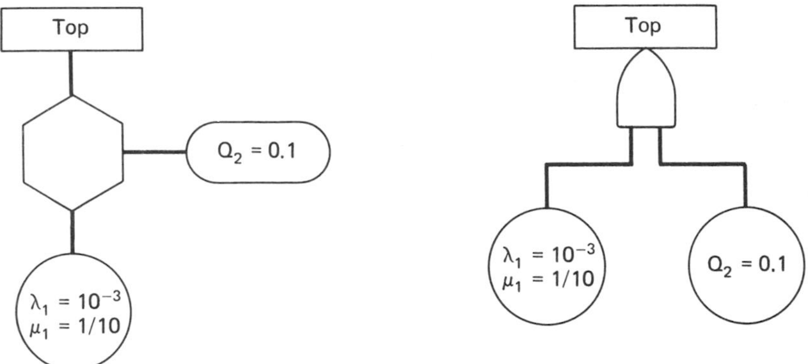

Figure 7.28. Example system with inhibit gate.

TABLE 7.16 COMPUTATION FOR INHIBIT GATE

Time	$Q_1 \times 10^3$	$w_1 \times 10^4$	$Q_s \times 10^4$	$w_s \times 10^5$
20	8.58757	9.91412	8.58757	9.91412
500	9.90099	9.90099	9.90099	9.90099

7.7.8 Remarks On Quantification Methods

1) *Component Unavailability for Age-Dependent Failure Rate*

Assume a component has an age-dependent failure rate $r(s)$, where s denotes age. The equation

$$r(t) = \frac{w(t)}{[1 - Q(t)]} \tag{7.131}$$

incorrectly quantifies unavailability $Q(t)$ at time t, for a given $r(t)$ and $w(t)$. This equation is correct if $r(t)$ is replaced by $\lambda(t)$, the conditional failure intensity of the component:

$$\lambda(t)=\frac{w(t)}{[1-Q(t)]} \tag{7.132}$$

However, it is difficult to use this equation for the quantification of $Q(t)$ because $\lambda(t)$ itself is an unknown parameter. One feasible approach is, as previously discussed in this chapter, to use (4.70) to quantify $Q(t)$.

2) *Cut Set or System Reliability*

$$R(t)=\text{Exp}\left[-\int_0^t \lambda(u)\,du\right] \tag{7.133}$$

This equation is not generally true. The correct equation is, as shown in Chapter 4, Section 4.3.1,

$$R(t)=\text{Exp}\left[-\int_0^t r(u)\,du\right] \tag{7.134}$$

Equation (7.133) is correct only in the case where the failure rate $r(t)$ is constant and, hence, coincides with the (constant) conditional failure intensity $\lambda(t)=\lambda$. For cut sets or systems, the conditional failure intensity is not constant, so we cannot use (7.133). In Chapter 9, we develop exact methods whereby the system reliability can be obtained.

APPENDIX 7.1 VENN DIAGRAMS

In Venn diagrams the set of all possible causes is denoted by rectangles, and a rectangle becomes a *universal set*. Some causes in the rectangle result in an event but others do not. Since the occurrence of the event is equivalent to the occurrence of its causes, the event is represented by a closed region, i.e., a *subset*, within the rectangle.

Example 1.

Assume an experiment where we throw a dice and observe its outcome. Consider the events A and B which are defined as

$$A=\{\text{outcome}\geq 3\}$$
$$B=\{2\leq\text{outcome}\leq 4\}$$

Represent these events by using a Venn diagram.

Solution: The rectangle (universal set) consists of six possible outcomes, 1, 2, 3, 4, 5, and 6. The event representation is shown in Fig. 7.29.

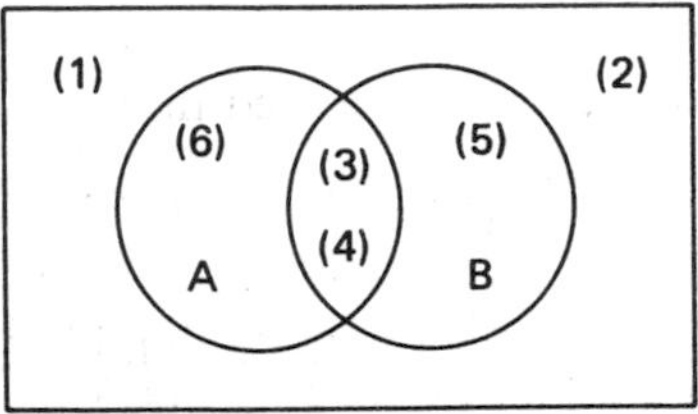

Figure 7.29. Venn diagram in Example 1.

TABLE 7.17 Relations between Venn Diagram, Event, Boolean Variable, and Probability

Venn Diagram	Event	Boolean Variable	Probability
A	A	$Y_A = \begin{cases} 1 \text{ in } A \\ 0, \text{ otherwise} \end{cases}$	$Pr(A) = S(S)$
A $A \cap B$ B	Intersection $A \cap B$	$Y_{A \cap B} \equiv Y_A \wedge Y_B$ $= \begin{cases} 1 \text{ in } A \cap B \\ 0, \text{ otherwise} \end{cases}$ $= Y_A Y_B$	$Pr(A \cap B)$ $= S(A \cap B)$
A B $A \cup B$	Union $A \cup B$	$Y_{A \cup B} \equiv Y_A \vee Y_B$ $= \begin{cases} 1 \text{ in } A \cup B \\ 0, \text{ otherwise} \end{cases}$ $= 1 - [1 - Y_A]\,[1 - Y_B]$	$Pr(A \cup B)$ $= S(A \cup B)$ $= S(A) + S(B) - S(A \cap B)$ $= P(A) + P(B) - P(A \cap B)$
$\bar{A}$ A	Complement $\bar{A}$	$Y_{\bar{A}} \equiv \bar{Y}_A$ $= \begin{cases} 1, \text{ in } \bar{A} \\ 0, \text{ otherwise} \end{cases}$ $= 1 - Y_A$	$Pr(\bar{A})$ $= S(\bar{A})$ $= 1 - S(A)$ $= 1 - Pr(A)$

Venn diagrams yield a visual tool for handling events, Boolean variables, and event probabilities; their use is summarized in Table 7.17.

Event Manipulations via Venn Diagrams. The intersection $A \cap B$ of events A and B is the set of points which belong to both A and B (column 1, row 2 in Table 7.17). The intersection is itself an event, and the common causes of events A and B become the causes of event $A \cap B$. The union $A \cup B$ is the set of points belonging to either A or B (column 1, row 3). Either causes of event A or B can create event $A \cup B$. The complement $\overline{A}$ consists of points outside event A.

Example 2.

Prove $A \cap (B \cup C) = (A \cap B) \cup (A \cap C)$ (1)

Solution: Both sides of the equation correspond to the shaded area of Fig. 7.30. This proves equation (1).

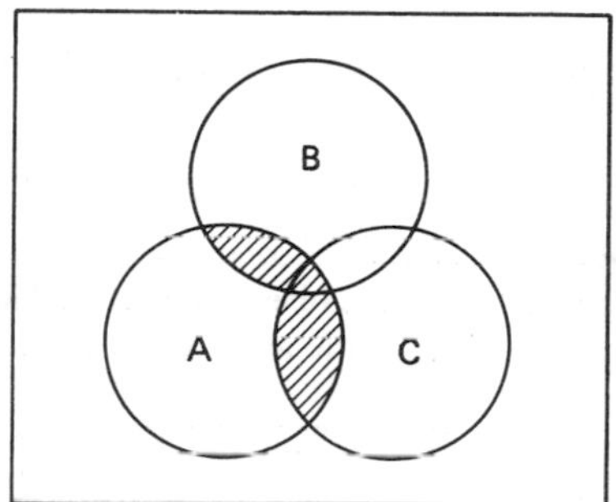

Figure 7.30. Venn diagram for $A \cap (B \cup C) = (A \cap B) \cup (A \cap C)$.

Boolean Variables and Venn Diagrams. The Boolean variable Y_A is an indicator variable for set A, as shown in column 3, row 1 in Table 7.17. Other variables such as $Y_{A \cap B}$, $Y_{A \cup B}$, $Y_{\overline{A}}$ are defined similarly. These three variables are often denoted by $Y_A \wedge Y_B$, $Y_A \vee Y_B$, and $\overline{Y}_A$, respectively. Algebraic equivalences shown in column 3 obviously hold.

Example 3.

Prove

$$\overline{Y_A \vee Y_B} = \overline{Y}_A \wedge \overline{Y}_B \quad \text{(de Morgan's law)}$$

Solution: By definition, $\overline{Y_A \vee Y_B} = \overline{Y_{A \cup B}}$ is the indicator for the set $\overline{A \cup B}$, whereas $\overline{Y}_A \wedge \overline{Y}_B = Y_{\overline{A}} \wedge Y_{\overline{B}} = Y_{\overline{A} \cap \overline{B}}$ is the indicator function for the set $\overline{A} \cap \overline{B}$. Both sets are the shaded region in Fig. 7.31 and de Morgan's law is proven.

Probability and Venn Diagrams. Let the rectangle have an area of unity. Denote by $S(A)$ the area of event A. Then, the probability of occurrence of event A is given by the area $S(A)$ (see column 4, row 1, Table 7.17):

$$\Pr(A) = S(A)$$

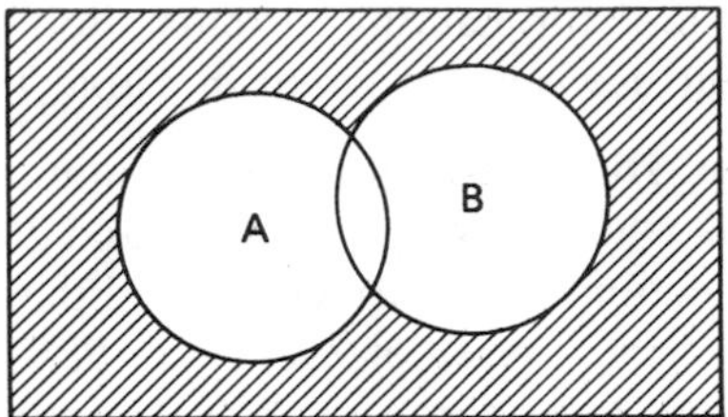

Figure 7.31. Venn diagram for de Morgan's law $\overline{A \cup B} = \bar{A} \cap \bar{B}$.

Other probabilities, $\Pr(A \cap B)$, $\Pr(A \cup B)$, and $\Pr(\bar{A})$ are defined by the areas $S(A \cap B)$, $S(A \cup B)$, and $S(\bar{A})$, respectively (column 4, Table 7.17). This definition of probabilities yields the relationship:

$$\Pr(A \cup B) = \Pr(A) + \Pr(B) - \Pr(A \cap B)$$

$$\Pr(\bar{A}) = 1 - \Pr(A)$$

Example 4.

Assume the occurrence of event A calls for the occurrence of event B. Then, prove that

$$\Pr(A \cap B) = \Pr(A) \tag{2}$$

Solution: Whenever event A occurs, event B must occur. This means that any cause of event A is also a cause of event B. Therefore, set A is included in set B as shown in Fig. 7.32. Thus, the area $S(A \cap B)$ is equal to $S(A)$, proving equation (2).

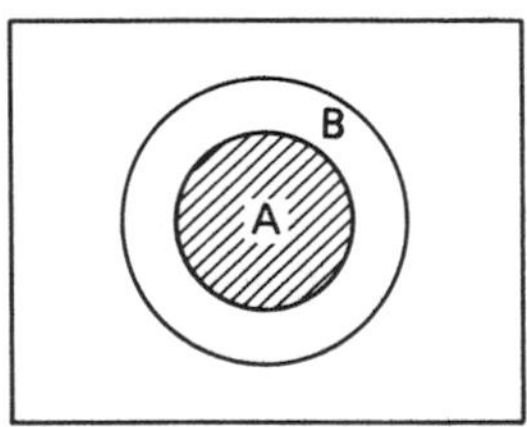

Figure 7.32. Venn diagram for $A \cap B$ when event A calls for event B.

Conditional probability, $\Pr(A|C)$, is defined by

$$\Pr(A|C) = \frac{S(A \cap C)}{S(C)}$$

In other words, the probability is the proportion of event A in the set C as is shown in Fig. 7.33.

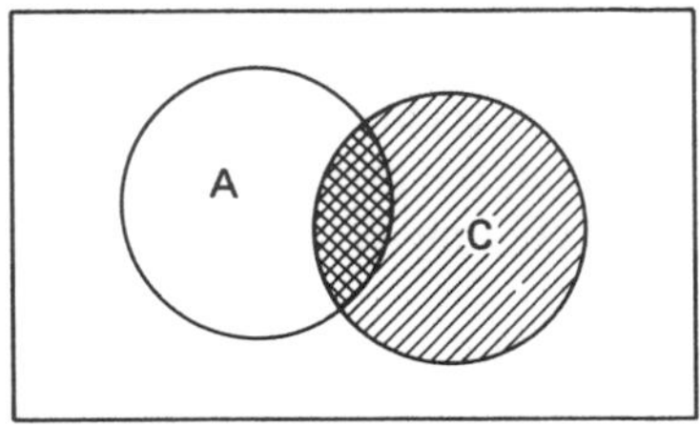

Figure 7.33. Venn diagram for conditional probability $\Pr(A|C)$.

Example 5.

Assume that event C calls for event B. Prove that

$$\Pr(A|B,C)=\Pr(A|C)$$

Solution:

$$\Pr(A|B,C)=\frac{S(A\cap B\cap C)}{S(B\cap C)}$$

Since set C is included in set B, as shown in Fig. 7.34, then

$$S(A\cap B\cap C)=S(A\cap C)$$

$$S(B\cap C)=S(C)$$

Thus,

$$\Pr(A|B,C)=\frac{S(A\cap C)}{S(C)}=\Pr(A|C)$$

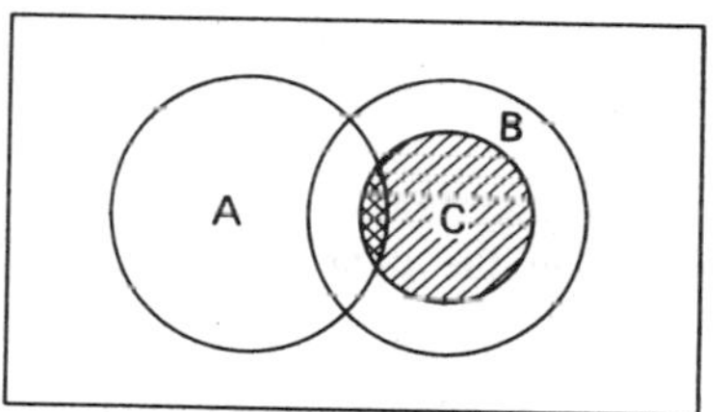

Figure 7.34. Venn diagram for $\Pr(A|B,C)=\Pr(A|C)$ when event C calls for event B.

This relation is intuitively obvious, since the additional observation of B brings no new information because it was already observed when event C happened.

APPENDIX 7.2 DERIVATION OF (7.96)

The cut set failure occurs if and only if one of the basic events in the cut set does not exist at t and then occurs in t to $t+dt$, and all other basic events exist at t. Thus,

$$\Pr(C^*(t,t+dt))=\sum_{j=1}^{n}\Pr(\text{event } j \text{ occurs in } t \text{ to } t+dt, \text{ and the other events exist at } t)$$

Since the basic events are mutually independent,

$$\Pr(C^*(t,t+dt)) = \sum_{j=1}^{n} \Pr(\text{event } j \text{ occurs in } t \text{ to } t+dt)\Pr(\text{the other events exist at } t)$$

$$= \sum_{j=1}^{n} W_j(t)\,dt \prod_{\substack{l=1 \\ l\neq j}} Q_l(t)$$

yielding (7.96).

PROBLEMS

7.1 Calculate the unavailability $Q_s(t)$ for the three-out-of-six voting system, assuming component unavailabilities of 0.1.

7.2. Calculate the system unavailability of the tail gas quench and clean up system of Fig. 7.12, using as data

$$\Pr(A)=\Pr(D)=\Pr(G)=0.01$$

$$\Pr(B)=\Pr(C)=\Pr(E)=\Pr(F)=0.1$$

7.3. Calculate the system availability $A_s(t)$ of the tail gas quench and clean up system, using the data in Problem 7.2 and the success tree of Fig. 7.15. Confirm that

$$A_s(t)+Q_s(t)=1$$

7.4. A safety system consists of three monitors. A plant requires shut-down with the probability of 0.2. If the safety system fails to shut-down, \$10,000 is lost. Each spurious shut-down costs \$4000. Determine the optimal m-out-of-three safety system, using the data

$$\Pr(\text{a monitor fails to shut down})=0.01$$

$$\Pr(\text{a monitor generates a spurious signal})=0.05$$

Assume statistically independent failures. Use the truth table method.

7.5. (a) Obtain the structure functions ψ_1, ψ_2, ψ_3, and ψ for the reliability block diagram of Fig. P7.5.

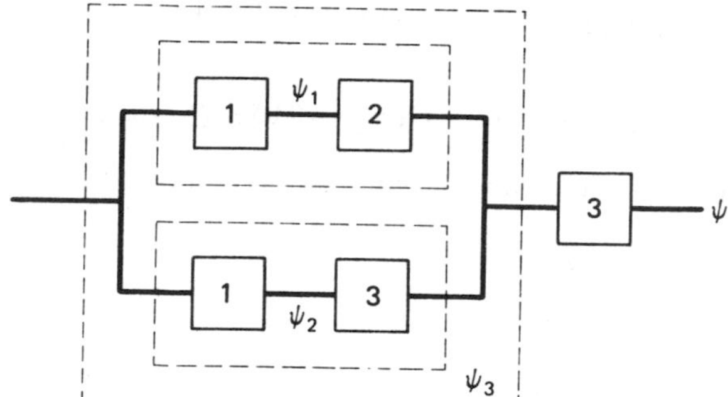

Figure P7.5.

(b) Calculate the system unavailability, using the component unavailabilities

$$Q_1 = 0.01, \quad Q_2 = 0.1, \quad Q_3 = 0.05$$

7.6. (a) Obtain by inspection, minimal cuts and minimal paths for the reliability block diagram of Fig. P7.6.

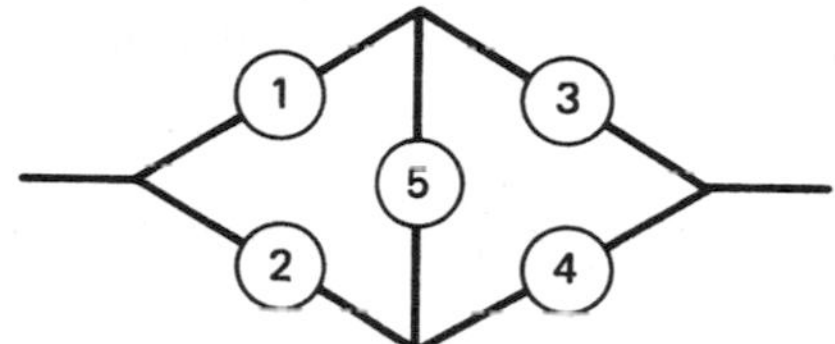

Figure P7.6.

(b) Determine the minimal cut and minimal path representation of the structure function.

(c) Calculate the system unavailability by expanding the two structure functions. Assume that

$$Q_1 = Q_2 = Q_3 = Q_4 = 0.05, \quad Q_5 = 0.01$$

7.7. Obtain the system unavailability by the partial pivotal decomposition of the minimal cut representation of Problem 7.6.

7.8. Calculate each bracket of the inclusion-exclusion principle for the reliability block diagram of Problem 7.6.

7.9 Obtain the successive lower and upper bounds of the bridge circuit of Problem 7.6, using the inclusion-exclusion principle. Obtain also the Esary and Proschan lower and upper bounds.

7.10. Obtain lower and upper bounds for the bridge circuit of Problem 7.6, using two event minimal cut sets and minimal path sets.

7.11. The components of the bridge circuit of Problem 7.6 have the stationary unavailabilities

$$Q_1 = Q_2 = Q_3 = Q_4 = 0.05, \quad Q_5 = 0.01$$

and the conditional failure intensities (failure rates)

$$\lambda_1 = \lambda_2 = \lambda_3 = \lambda_4 = \lambda_5 = 0.01$$

Obtain the conditional repair intensities (repair rates), $\mu_1, \ldots,$ and μ_5.

7.12. Obtain the component parameters, Q_i, W_i, and V_i at $t = 100$ for the bridge circuit with the rates

$$\lambda_1 = \lambda_2 = \lambda_3 = \lambda_4 = \lambda_5 = 0.001 = \lambda$$
$$\mu_1 = \mu_2 = \mu_3 = \mu_4 = \mu_5 = 0.01 = \mu$$

7.13. A bridge circuit has four minimal cut sets: $K_1 = \{1,2\}$, $K_2 = \{3,4\}$, $K_3 = \{1,4,5\}$, and $K_4 = \{2,3,5\}$. Calculate the cut set parameters Q_i^*, W_i^*, λ_i^*, V_i^*, and μ_i^* at $t = 100$, assuming the rates of Problem 7.12.

7.14. Obtain the upper bound of the system unavailability $Q_s(100)$ for the bridge circuit, using the result of Problem 7.13; $Q_1^*(100) = Q_2^*(100) = 3.6782 \times 10^{-3}$ $Q_3^*(100) = Q_4^*(100) = 2.2307 \times 10^{-4}$.

7.15. A bridge configuration has four minimal cut sets: $K_1 = \{1,2\}$, $K_2 = \{3,4\}$, $K_3\{1,4,5\}$, and $K_4 = \{2,3,5\}$. Define by:

e_1: cut set $\{1,2\}$ occurs at time t
e_2: cut set $\{3,4\}$ occurs at time t
e_3: cut set $\{1,4,5\}$ occurs at time t
e_4: cut set $\{2,3,5\}$ occurs at time t

Then $W_s^{(1)}(t)\,dt$ of equation (7.111) becomes

$$\begin{aligned} W_s^{(1)}(t)\,dt = {} & \Pr(e_1) + \Pr(e_2) + \Pr(e_3) + \Pr(e_4) \\ & - \Pr(e_2 \cap e_1) - \Pr(e_3 \cap e_1) - \Pr(e_3 \cap e_2) - \Pr(e_4 \cap e_1) \\ & - \Pr(e_4 \cap e_2) - \Pr(e_4 \cap e_3) \\ & + \Pr(e_1 \cap e_2 \cap e_3) + \Pr(e_1 \cap e_2 \cap e_4) + \Pr(e_1 \cap e_3 \cap e_4) \\ & + \Pr(e_2 \cap e_3 \cap e_4) - \Pr(e_1 \cap e_2 \cap e_3 \cap e_4) \end{aligned}$$

(a) Determine common members for each term on the right-hand side. Determine also $\prod_{1,\ldots,m} Q(t)$ and $W(t; 1, \ldots, m)$ of equation (7.112).

(b) Calculate lower and upper bounds for $w_s^{(1)}(100)$, using as data:

$$Q_i(100) = 6.0648 \times 10^{-2}, \quad w_i(100) = 9.3935 \times 10^{-4}, \quad i = 1, \ldots, 5$$
$$w_i^*(100) = w_2^*(100) = 1.1394 \times 10^{-4}, \quad w_3^*(100) = w_4^* = 1.0365 \times 10^{-5}$$

7.16. (a) If cut sets $K_1, K_2, \ldots,$ and K_m have no common members, then $\Pr(e_1 \cap \ldots \cap e_m \cap B)$ in equation (7.114) is zero. Noting this, prove for the bridge circuit, that

$$\begin{aligned} w_s^{(2)}(t) = {} & \Pr(e_1 \cap B) + \Pr(e_2 \cap B) + \Pr(e_3 \cap B) + \Pr(e_4 \cap B) \\ & - \Pr(e_3 \cap e_1 \cap B) - \Pr(e_3 \cap e_2 \cap B) - \Pr(e_4 \cap e_1 \cap B) \\ & - \Pr(e_4 \cap e_2 \cap B) - \Pr(e_4 \cap e_3 \cap B) \qquad (1) \end{aligned}$$

(b) Expand each term in the above equation, and simplify the results.
(c) Obtain lower and upper bounds for $w_s^{(2)}(100)$, using Q_i, w_i, and w_i^* in Problem 7.15.

7.17. (a) Obtain successive lower and upper bounds of $W_s(100)$, using the successive bounds:

$$w_s^{(1)}(100)_{\max,1} = 2.486 \times 10^{-4}$$

$$w_s^{(2)}(100)_{\min,1} = 2.4776 \times 10^{-4}$$

$$w_s^{(1)}(100)_{\max,2} = 2.4776 \times 10^{-4} = w_s^{(1)}(100)_{\min,2} = \text{last bracket}$$

$$w_s^{(2)}(100)_{\max,1} = 3.2765 \times 10^{-6}$$

$$w_s^{(2)}(100)_{\min,1} = 3.2130 \times 10^{-6}$$

$$w_s^{(2)}(100)_{\max,2} = 3.2130 \times 10^{-6} = w_s^{(2)}(100)_{\min,2} = \text{last bracket}$$

(b) Obtain an upper bound of $\lambda_s(100)$, using the results of Problem 7.14; $Q_s(100)_{\max} = 7.80254 \times 10^{-3}$.

7.18. Apply the short-cut calculation of the reliability parameters for the bridge circuit at $t = 500$ hr, assuming the rates

$$\lambda_1 = \lambda_2 = \lambda_3 = \lambda_4 = \lambda_5 = 0.001$$

$$\mu_1 = \mu_2 = \mu_3 = \mu_4 = \mu_5 = 0.01$$

8

SYSTEM QUANTIFICATION FOR DEPENDENT BASIC EVENTS

8.1 INTRODUCTION

So far, we have assumed statistical *independence* of basic events. *Dependent events* may appear in fault trees in the following cases.

1) *Standby Redundancies*. Equipment in *standby redundancy* is used to improve system availability and reliability. When an operating component fails, a standby component is put into operation, and the redundant configuration continues to function. Failure of an operating component thus causes a standby component to be more susceptible to failure because the latter component is now under load. This means that failure of one component affects the failure characteristics of the other components and component failures are not statistically independent.

2) *Common Causes*. As discussed in Chapter 3, a *common cause*, for example, fire, results in simultaneous failure of sets of components. Thus, under common causes, component failures are no longer independent.

3) *Components Supporting Loads*. Assume that a set of components supports loads such as stresses, currents, etc. A failure of one component increases the load supported by the other components. Consequently, the remaining components are more likely to fail, and we can not assume statistical independence of components.

4) *Mutually Exclusive Primary Events.* Consider basic events, "switch fails to close" and "switch fails to open." These two basic events are *mutually exclusive*, i.e., occurrence of one basic event precludes another. Thus, we encounter dependent basic events when a fault tree involves mutually exclusive basic events.

The *inclusion-exclusion principle*, when coupled with *Markov models*, enables us to quantify systems which include dependent basic events. A general procedure for system quantification is:

1) Represent system parameters by the inclusion-exclusion principle. For each term in the representation, examine whether it involves dependent basic events or not. If a term consists of independent events, quantify it by the methods in Chapter 7. Otherwise, proceed as follows:
2) Model dependent events by a *Markov transition diagram.* Obtain differential equations for state probabilities.
3) Quantify terms involving dependent basic events by the solutions of the differential equations.
4) Now each term in the inclusion-exclusion representation has been quantified. Determine the first and the second brackets, and obtain lower and upper bounds for the system parameter. If possible, calculate a complete expansion of the system parameter and obtain the exact value.

In this chapter, we discuss dependency created by standby redundancies or common causes. Other types of dependency can be treated in a similar manner.

8.2 QUANTIFICATION OF SYSTEMS, INCLUDING STANDBY REDUNDANCY

Consider the tail gas quench and clean up system of Fig. 8.1.† The corresponding fault tree is given by Fig. 8.2. There are two quench pumps, A and B; one is in *standby* and the other, which we call the *principal* pump, is in operation.

Assume that pump A is principal at a given time t, whereas pump B is in standby. If pump A fails, standby pump B takes the place of A, and the pumping continues. The failed pump A is repaired and is put into standby when the repair is completed. Standby pump A will replace principal pump B when it fails. The redundancy increases system or subsystem reliability.

†This system is also shown as Fig. 7.12.

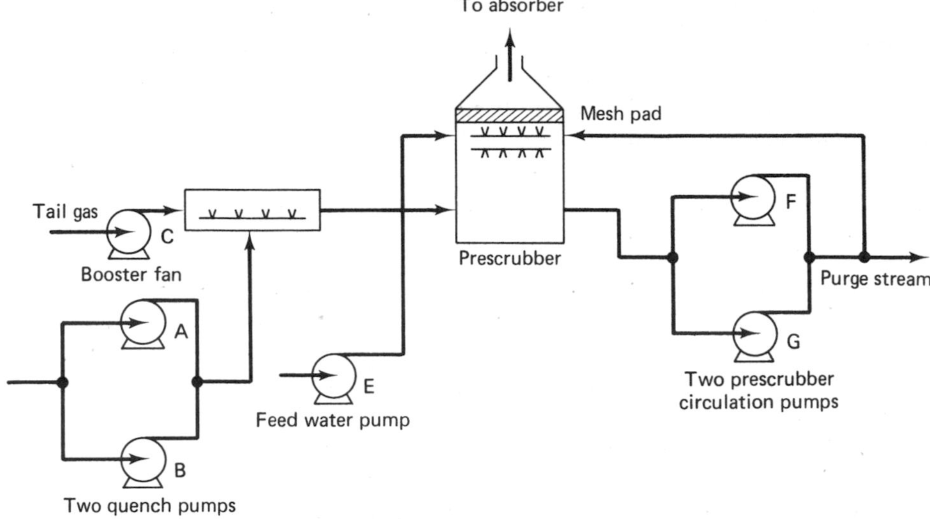

Figure 8.1. Schematic diagram of tail gas quench and clean up system.

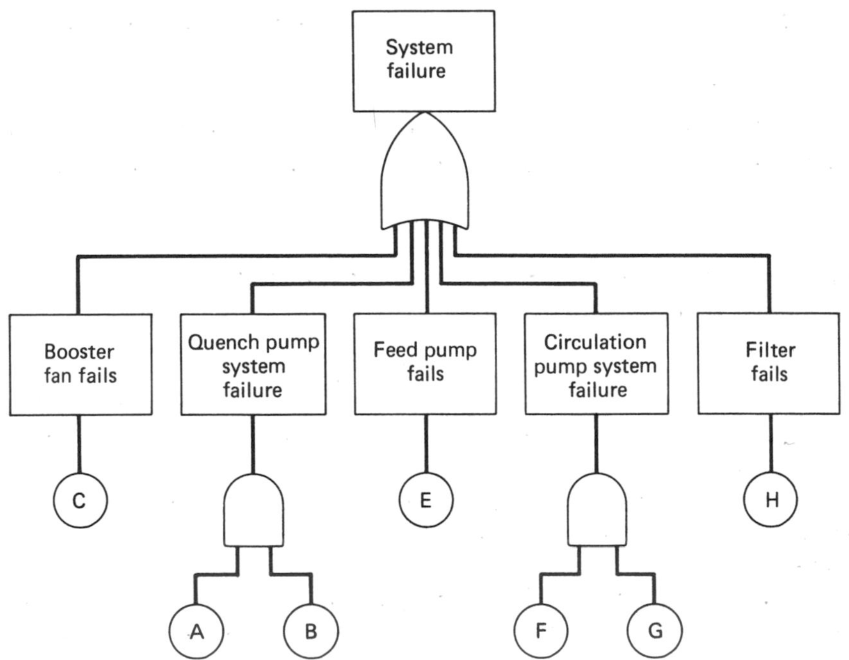

Figure 8.2. Fault tree for tail gas quench and clean up system.

The tail gas quench and clean up system has another standby redundancy consisting of the two circulation pumps F and G.

Each component in standby redundancy has three phases: *standby*, *operation*, and *repair*. Components can fail only when they are in operation or in standby. Depending on component failure characteristics during these phases, standby redundancy is classified into the following three types:

1) *Hot Standby*: Each component has the same failure rate regardless of whether it is in standby or in operation. Since the failure rate of one component is unique, and is not affected by the other components, the hot standby redundancy consists of statistically independent components.
2) *Cold Standby*: Components do not fail when they are in standby. The components have non-zero failure rates only when they are in operation. A failure of one principal component forces a standby component to start operating and to have a non-zero failure rate. Thus, failure characteristics of one component are affected by the other, and the cold standby redundancy results in mutually dependent basic events (component failures).
3) *Warm Standby*: A standby component can fail, but it has a smaller failure rate than the principal component. Failure characteristics of one component are affected by the other, and warm standby induces dependent basic events.

The fault tree in Fig. 8.2 has five minimal cut sets:

$$d_1=\{C\},\quad d_2=\{E\},\quad d_3=\{H\},\quad d_4=\{A,B\},\quad d_5=\{F,G\}$$

The inclusion exclusion principle of (7.103) gives the following lower and upper bounds for the system unavailability $Q_s(t)$.

$$\begin{aligned} Q_s(t)_{\max} &= \text{first bracket} \\ &= \Pr(C)+\Pr(E)+\Pr(H)+\Pr(A\cap B)+\Pr(F\cap G) \end{aligned} \tag{8.1}$$

$$\begin{aligned} Q_s(t)_{\min} &= Q_s(t)_{\max} - \text{second bracket} \\ &= Q_s(t)_{\max} - \Pr(C\cap E) - \Pr(C\cap H) - \Pr(C\cap A\cap B) \\ &\quad - \Pr(C\cap F\cap G) \\ &\quad - \Pr(E\cap H) - \Pr(E\cap A\cap B) - \Pr(E\cap F\cap G) \\ &\quad - \Pr(H\cap A\cap B) - \Pr(H\cap F\cap G) \\ &\quad - \Pr(A\cap B\cap F\cap G) \end{aligned} \tag{8.2}$$

The events C, E, H, $A\cap B$, and $F\cap G$ are mutually independent by assumption. Thus, (8.2) can be written as

$$
\begin{aligned}
Q_s(t)_{\min} = Q_s(t)_{\max} \\
& - \Pr(C)\Pr(E) - \Pr(C)\Pr(H) - \Pr(C)\Pr(A \cap B) - \Pr(C)\Pr(F \cap G) \\
& - \Pr(E)\Pr(H) - \Pr(E)\Pr(A \cap B) - \Pr(E)\Pr(F \cap G) \\
& - \Pr(H)\Pr(A \cap B) - \Pr(H)\Pr(F \cap G) \\
& - \Pr(A \cap B)\Pr(F \cap G)
\end{aligned} \tag{8.3}
$$

Note that the equalities

$$
\left.
\begin{aligned}
\Pr(A \cap B) = \Pr(A)\Pr(B) \\
\Pr(F \cap G) = \Pr(F)\Pr(G)
\end{aligned}
\right\} \tag{8.4}
$$

hold only for hot standby redundancies. Cold or warm standby do not satisfy these equalities.

Probabilities $\Pr(C)$, and $\Pr(E)$, and $\Pr(H)$ are component unavailabilities, and they can be computed by methods in Chapter 4 or 7. Probabilities $\Pr(A \cap B)$ and $\Pr(F \cap G)$ will be denoted by $Q_r(t)$, which is the unavailability in standby redundancy and can be calculated by the methods explained in the following paragraphs. In all cases we assume perfect switching and no failure to start.

8.2.1 Markov Model for Standby Redundancy

Figure 8.3 summarizes the behavior of a standby redundancy system consisting of components A and B. Each rectangle represents a state of the redundancy. The extreme left box in a rectangle is for a standby component, the middle box is for a principal component, and the extreme right box is for components under repair. Thus, rectangle 1 represents a state where component B is in standby and component A is in operation. Similarly, rectangle 4 expresses the event that component B is in operation and component A is under repair. Possible transitions of state are shown in the same figure. The warm or hot standby has transitions from state 1 to 3, or state 2 to 4, whereas the cold standby does not. For the warm or hot standby, the standby component is assumed to fail with a constant failure rate $\bar{\lambda}$. For hot standby, $\bar{\lambda}$ is equal to λ, the failure rate for principal components. For cold standby, $\bar{\lambda}$ is zero. The warm standby ($0 < \bar{\lambda} < \lambda$) has as its special cases the hot standby ($\bar{\lambda} = \lambda$) and the cold standby ($\bar{\lambda} = 0$). Each component has a constant repair rate μ. In all cases, the system fails when it enters state 5.

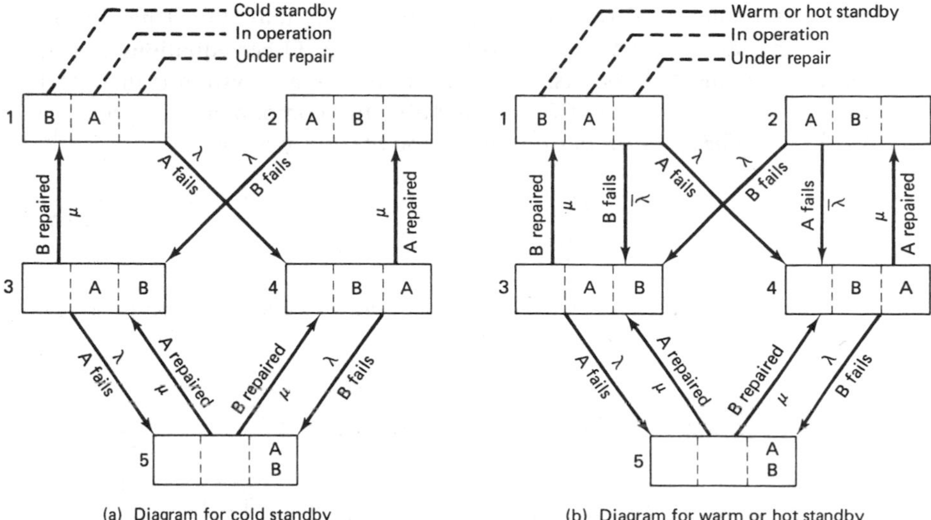

(a) Diagram for cold standby

(b) Diagram for warm or hot standby

Figure 8.3. Markov transition diagram for redundant configuration with cold, warm, or hot standby.

Denote by $P_i(t)$ the probability that the redundant system is in state i at time t. The derivative $\dot{P}_i(t)$ is given by

$$\begin{aligned}\dot{P}_i(t) &= (\text{inflow to state } i) - (\text{outflow from state } i)\\ &= \sum_j [\text{rate of transition to state } i \text{ from other states } j] \\ &\qquad\qquad \times [\text{probability of state } j] \\ &\quad - \sum_j [\text{rate of transition from state } i \text{ to other states } j] \\ &\qquad\qquad \times [\text{probability of state } i]\end{aligned} \tag{8.5}$$

This formula results in the differential equation set

$$\left.\begin{aligned}\dot{P}_1(t) &= -(\lambda+\bar{\lambda})P_1(t)+\mu P_3(t)\\ \dot{P}_2(t) &= -(\lambda+\bar{\lambda})P_2(t)+\mu P_4(t)\\ \dot{P}_3(t) &= \bar{\lambda}P_1(t)+\lambda P_2(t)-(\lambda+\mu)P_3(t)+\mu P_5(t)\\ \dot{P}_4(t) &= \bar{\lambda}P_2(t)+\lambda P_1(t)-(\lambda+\mu)P_4(t)+\mu P_5(t)\\ \dot{P}_5(t) &= \lambda P_3(t)+\lambda P_4(t)-2\mu P_5(t)\end{aligned}\right\} \tag{8.6}$$

The first equation of (8.6) is obtained by noting that state 1 has inflow rate μ from state 3 and two outflow rates λ and $\bar{\lambda}$. Other equations can be obtained in a similar way. Assume that the redundant system is in state 1 at time zero; i.e., component B is in standby and component A is principal at time zero. Then, the initial condition for (8.6) is given by

$$\begin{aligned} &P_1(0)=1 \\ &P_i(0)=0, \quad i=2,\ldots,5 \end{aligned} \tag{8.7}$$

Add the first equation of (8.6) to the second one and the third equation to the fourth one, respectively. Then,

$$\left.\begin{aligned} &\frac{d[P_1+P_2]}{dt}=-(\lambda+\bar{\lambda})[P_1+P_2]+\mu[P_3+P_4] \\ &\frac{d[P_3+P_4]}{dt}=(\lambda+\bar{\lambda})[P_1+P_2]-(\lambda+\mu)[P_3+P_4]+2\mu P_5 \\ &\frac{dP_5}{dt}=\lambda[P_3+P_4]-2\mu P_5 \end{aligned}\right\} \tag{8.8}$$

Define:

$$\left.\begin{aligned} &P_{(0)}=P_1(t)+P_2(t) \\ &P_{(1)}=P_3(t)+P_4(t) \\ &P_{(2)}=P_5(t) \end{aligned}\right\} \tag{8.9}$$

Then, (8.8) can be written as

$$\left.\begin{aligned} &\dot{P}_{(0)}=-(\lambda+\bar{\lambda})P_{(0)}+\mu P_{(1)} \\ &\dot{P}_{(1)}=(\lambda+\bar{\lambda})P_{(0)}-(\lambda+\mu)P_{(1)}+2\mu P_{(2)} \\ &\dot{P}_{(2)}=\lambda P_{(1)}-2\mu P_{(2)} \end{aligned}\right\} \tag{8.10}$$

with the initial condition

$$P_{(0)}(0)=1, \quad P_{(1)}(0)=0, \quad P_{(2)}(0)=0 \tag{8.11}$$

The differential equations of (8.10) are those for Fig. 8.4, the transition diagram consisting of states (0), (1), and (2). State (0) has outflow rate $\lambda+\bar{\lambda}$ and inflow rate μ. Meanings of states (0), (1), and (2) are shown in the same figure.

Equation (8.10) can be integrated numerically. If an analytical solution for $P_{(i)}$ is required, Laplace transformations may be used.

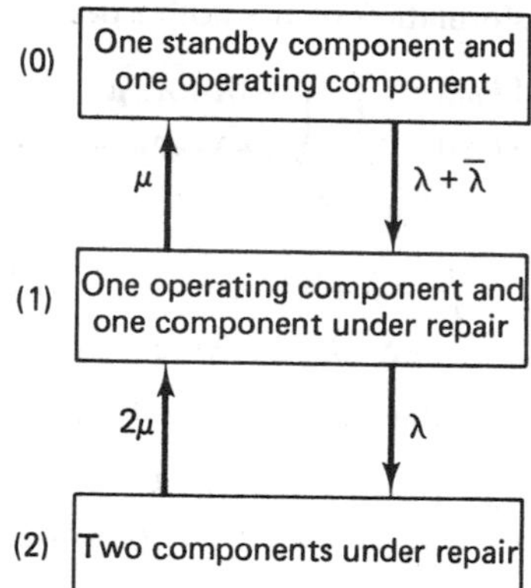

Figure 8.4. Simplified Markov transition diagram: redundant configuration.

8.2.2 Standby Redundancy Parameter $Q_r(t)$

The parameter $Q_r(t)=\Pr(A\cap B)$ is the unavailability of the standby redundancy $\{A,B\}$ and equals the probability that both components A and B are under repair at time t. Thus,

$$Q_r(t)=P_{(2)}(t) \tag{8.12}$$

Example 1. Warm Standby

Consider the redundant quench pumps A and B of Fig. 8.1. Assume the following failure and repair rates for each pump.

$$\lambda=10^{-3}(\text{hour}^{-1}),\quad \bar{\lambda}=0.5\times10^{-3}(\text{hour}^{-1}),\quad \mu=10^{-2}(\text{hour}^{-1})$$

Calculate parameter $Q_r(t)$ at $t=100$, 500, and 1000 hr.

Solution: Substitute $P_{(2)}=1-P_{(0)}-P_{(1)}$ into the second equation of (8.10). Then,

$$\begin{pmatrix}\dot{P}_{(0)}\\ \dot{P}_{(1)}\end{pmatrix}=\begin{pmatrix}-\lambda-\bar{\lambda}, \mu\\ \lambda+\bar{\lambda}-2\mu, -(\lambda+3\mu)\end{pmatrix}\begin{pmatrix}P_{(0)}\\ P_{(1)}\end{pmatrix}+\begin{pmatrix}0\\ 2\mu\end{pmatrix} \tag{8.13}$$

The Laplace transformation of $\dot{P}_{(i)}$ is related to $L[P_{(i)}]$ and $P_{(i)}(0)$ by

$$L[\dot{P}_{(i)}]=\int_0^\infty \dot{P}_{(i)}(t)e^{-st}\,dt=sL[P_{(i)}]-P_{(i)}(0) \tag{8.14}$$

Further, the Laplace transformation of the constant 2μ is given by

$$L[2\mu]=\int_0^\infty 2\mu e^{-st}\,dt=\frac{2\mu}{s} \tag{8.15}$$

Thus, the transformation of the both sides of (8.13) yields

$$\begin{pmatrix} sL[P_{(0)}]-1 \\ sL[P_{(1)}]-0 \end{pmatrix} = \begin{pmatrix} -(\lambda+\bar{\lambda}), \mu \\ \lambda+\bar{\lambda}-2\mu, -(\lambda+3\mu) \end{pmatrix} \begin{pmatrix} L[P_{(0)}] \\ L[P_{(1)}] \end{pmatrix} + \begin{pmatrix} 0 \\ 2\mu/s \end{pmatrix}$$

or

$$\begin{pmatrix} \lambda+\bar{\lambda}+s, -\mu \\ -\lambda-\bar{\lambda}+2\mu, \lambda+3\mu+s \end{pmatrix} \begin{pmatrix} L[P_{(0)}] \\ L[P_{(1)}] \end{pmatrix} = \begin{pmatrix} 1 \\ 2\mu/s \end{pmatrix} \tag{8.16}$$

Substitute the values of failure and repair rates into (8.16):

$$\begin{pmatrix} s+1.5\times 10^{-3}, -10^{-2} \\ 1.85\times 10^{-2}, s+3.1\times 10^{-2} \end{pmatrix} \begin{pmatrix} L[P_{(0)}] \\ L[P_{(1)}] \end{pmatrix} = \begin{pmatrix} 1 \\ 2\times 10^{-2}/s \end{pmatrix} \tag{8.17}$$

This is a linear simultaneous equation for $L[P_{(0)}]$ and $L[P_{(1)}]$ and can be solved:

$$L[P_{(0)}] = \frac{s+3.1\times 10^{-2}}{(s+a)(s+b)} + \frac{2\times 10^{-4}}{s(s+a)(s+b)}$$

$$L[P_{(1)}] = \frac{1.5\times 10^{-3}}{(s+a)(s+b)} + \frac{3\times 10^{-5}}{s(s+a)(s+b)}$$

where $a = 1.05436\times 10^{-2}$ and $b = 2.19564\times 10^{-2}$.

From standard tables of inverse Laplace transforms,

$$L^{-1}\left(\frac{K}{(s+a)(s+b)}\right) = \frac{K}{b-a}(e^{-at}-e^{-bt}) \tag{8.18}$$

$$L^{-1}\left(\frac{K(s+z)}{(s+a)(s+b)}\right) = \frac{K}{b-a}[(z-a)e^{-at}-(z-b)e^{-bt}] \tag{8.19}$$

$$L^{-1}\left(\frac{K}{s(s+a)(s+b)}\right) = \frac{K}{ab}\left(1-\frac{be^{-at}}{b-a}+\frac{ae^{-bt}}{b-a}\right) \tag{8.20}$$

These inverse transformations give

$$P_{(0)} = 0.863933 + 0.130340e^{-at} + 0.005727e^{-bt} \tag{8.21}$$

$$P_{(1)} = 0.129590 - 0.117879e^{-at} - 0.011711e^{-bt} \tag{8.22}$$

Thus, the cut set parameter $Q_r(t)$ is

$$\begin{aligned} Q_r(t) &= 1 - P_{(0)} - P_{(1)} \\ &= 0.006477 - 0.012461e^{-0.0105436t} + 0.005984e^{-0.0219564t} \end{aligned} \tag{8.23}$$

yielding

t	$Q_r(t)$
100	0.002801
500	0.006413
1000	0.006477

Example 2. Cold Standby

Assume the following failure and repair rates for quench pumps A and B.

$$\lambda = 10^{-3}(\text{hour}^{-1}),\quad \bar{\lambda} = 0.,\quad \mu = 10^{-2}(\text{hour}^{-1})$$

Calculate parameter $Q_r(t)$ at $T = 100$, 500, and 1000 hr.

Solution: Substitute the failure and the repair rates into (8.16).

$$\begin{pmatrix} s+0.001, -0.01 \\ 0.019, s+0.031 \end{pmatrix}\begin{pmatrix} L[P_{(0)}] \\ L[P_{(1)}] \end{pmatrix} = \begin{pmatrix} 1 \\ 0.02/s \end{pmatrix} \tag{8.24}$$

This has the solution

$$L[P_{(0)}] = \frac{s+0.031}{(s+a)(s+b)} + \frac{0.0002}{s(s+a)(s+b)} \tag{8.25}$$

$$L[P_{(1)}] = \frac{0.001}{(s+a)(s+b)} + \frac{0.00002}{s(s+a)(s+b)} \tag{8.26}$$

where $a = 0.0100839, b = 0.0219161$.

The inverse transformation of (8.18) to (8.20) gives

$$P_{(0)} = 0.904980 + 0.091488e^{-at} + 0.003532e^{-bt} \tag{8.27}$$

$$P_{(1)} = 0.090498 - 0.083109e^{-at} - 0.007389e^{-bt} \tag{8.28}$$

$$\begin{aligned} Q_r(t) &= 1 - P_{(0)} - P_{(1)} \\ &= 0.004522 - 0.008379e^{-0.0100839t} + 0.003857e^{-0.0219161t}, \end{aligned} \tag{8.29}$$

yielding

t	$Q_r(t)$
100	0.001896
500	0.004468
1000	0.004522

Example 3. Hot Standby

Let quench pump A and B have the failure and the repair rates

$$\lambda = \bar{\lambda} = 10^{-3}(\text{hour}^{-1}),\quad \mu = 10^{-2}(\text{hour}^{-1})$$

Calculate $Q_r(t)$ at $t = 100$, 500, and 1000 hr.

Solution: In this case, pump A and B are statistically independent and we can calculate $Q_r(t)$ without solving differential equations (8.13).

From (4.120) in Chapter 4,

$$\begin{aligned}\text{unavailability of pump } A &\\ &= \text{unavailability of pump } B \\ &= \frac{\lambda}{\lambda+\mu}(1-e^{-(\lambda+\mu)t}) \equiv Q(t)\end{aligned} \tag{8.30}$$

Thus, from (7.94)

$$Q_r(t) = Q(t)^2 = \left(\frac{\lambda}{\lambda+\mu}\right)^2(1-e^{-(\lambda+\mu)t})^2 \tag{8.31}$$

$$= \left(\frac{\lambda}{\lambda+\mu}\right)^2 - 2\left(\frac{\lambda}{\lambda+\mu}\right)^2 e^{-(\lambda+\mu)t} + \left(\frac{\lambda}{\lambda+\mu}\right)^2 e^{-2(\lambda+\mu)t} \tag{8.32}$$

$$= 0.008265 - 0.016529e^{-0.011t} + 0.008265e^{-0.022t} \tag{8.33}$$

Therefore,

t	$Q_r(t)$
100	0.003679
500	0.008198
1000	0.008265

From Examples 1 to 3 we see that:

$$\begin{aligned}\text{unavailability of hot standby} &\\ &> \text{unavailability of warm standby} \\ &> \text{unavailability of cold standby}\end{aligned}$$

Example 4. System Unavailability $Q_s(t)$

Consider the tail gas quench clean up system of Fig. 8.1. Assume the failure and repair rates in Example 1 for the quench pumps (warm standby), and the rates in Example 2 for the circulation pumps (cold standby). Assume further the following rates for booster fan C, feed pump E, and filter H:

$$\lambda^* = 10^{-4}, \quad \mu^* = 10^{-2}$$

Evaluate the system unavailability $Q_s(t)$ at $t = 100$, 500, and 1000 hr.

Solution: From Examples 1 and 2, we have

t	$Q_r'(t) = \Pr(A \cap B)$	$Q_r''(t) = \Pr(F \cap G)$
100	0.002801	0.001896
500	0.006413	0.004468
1000	0.006477	0.004522

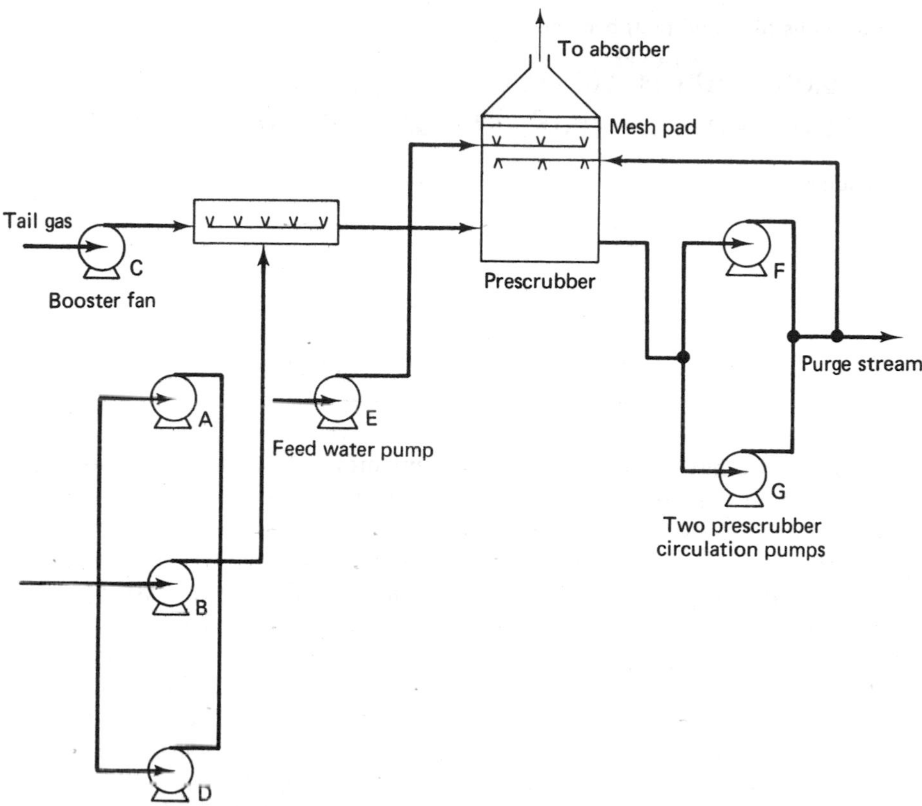

Figure 8.5. Schematic diagram of tail gas quench and clean up system with two-out-of-three quench pumps.

Further,

$$Q^*(t) = \Pr(C) = \Pr(E) = \Pr(H)$$
$$= \frac{\lambda^*}{\lambda^* + \mu^*}(1 - e^{-(\lambda^* + \mu^*)t})$$
$$= 0.009901[1 - e^{-0.0101t}]$$

Thus,

t	$Q^*(t)$
100	0.006295
500	0.009838
1000	0.009901

Equations (8.1) and (8.3) become

$$Q_s(t)_{\max} = 3Q^*(t) + Q_r'(t) + Q_r''(t)$$

$$Q_s(t)_{\min} = Q_s(t)_{\max} - 3Q^*(t)^2 - 3Q^*(t)Q_r'(t) - 3Q^*(t)Q_r''(t) - Q_r'(t)Q_r''(t)$$

yielding

t	$Q_s(t)_{\max}$	$Q_s(t)_{\min}$
100	0.025472	0.025259
500	0.040395	0.039755
1000	0.040702	0.040052

So far we have treated standby redundancy with two components. Let us now consider the tail gas quench and clean up system of Fig. 8.5, which has a two-out-of-three quench pumps system. Assume that each pump has failure rate λ when it is operating and failure rate $\bar{\lambda}$ when it is in standby. Assume further that only one pump at a time can be repaired. Possible transitions of states are shown in Fig. 8.6. State 1 means that pump A is in standby and pumps D and B are principal. State 13 shows that pumps A, D, and B are under repair, but only pump A is currently being repaired. Transition from state 7 to 13 occurs when pump B fails. The pump B is put into the last place of the repair queue in state 13. Transition from state 13 to 12 happens when the repair of pump A is complete. Pump D is being repaired in state 12. The other transitions can be explained in a similar way.

The states in Fig. 8.6 can be aggregated as shown in Fig. 8.7. The states in the first row of Fig. 8.6 can be regarded as substates of state (0) of Fig. 8.7. The state (0) implies one standby pump and two operating pumps. Similarly, the states in the second row of Fig. 8.6 are substates of state (1), which has two principal pumps and one pump under repair.

State (0) has substates 1, 2, and 3. Each substate goes into state (1) at the rate $2\lambda+\bar{\lambda}$. Thus, the inflow from state (0) to state (1) is given by

$$\begin{aligned}&(2\lambda+\bar{\lambda})P_1 + (2\lambda+\bar{\lambda})P_2 + (2\lambda+\bar{\lambda})P_3\\ &= (2\lambda+\bar{\lambda})(P_1+P_2+P_3)\\ &= (2\lambda+\bar{\lambda})P_{(0)}\end{aligned} \tag{8.34}$$

This means that the transition from state (0) to state (1) has the rate $(2\lambda+\bar{\lambda})$ as shown in Fig. 8.7. Rates of the other transitions can be obtained in a similar manner.

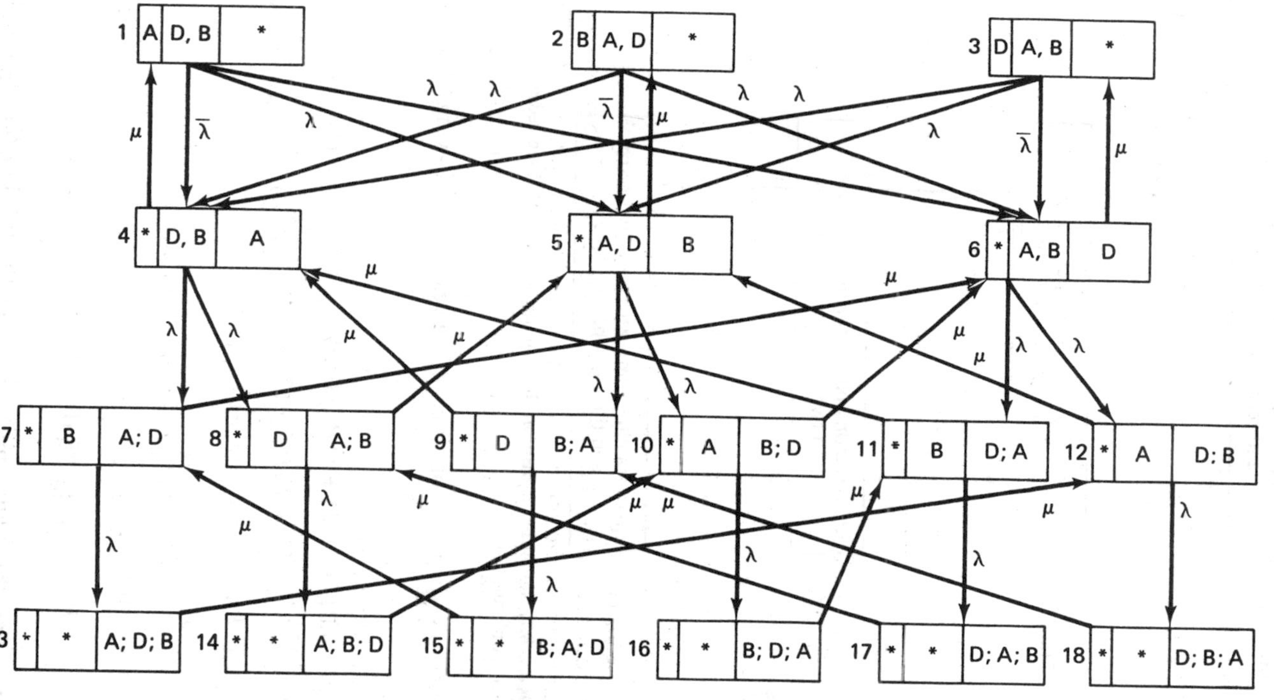

Figure 8.6. Markov transition diagram for two-out-of-three redundant configuration.

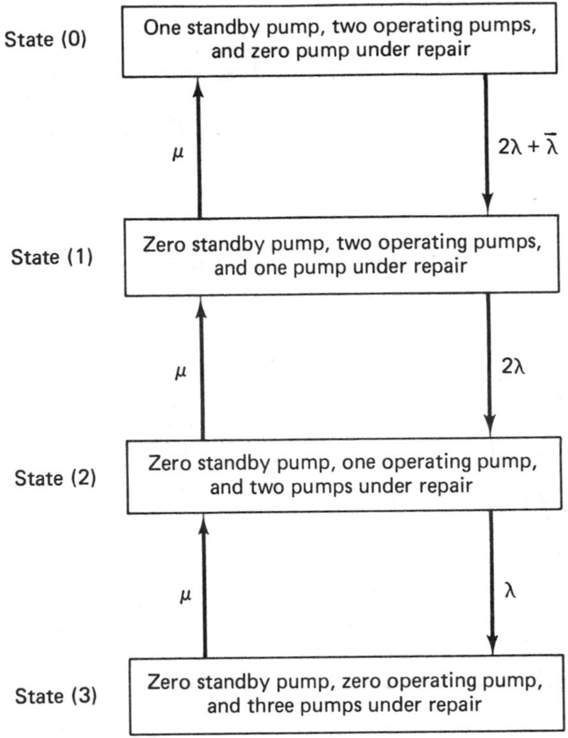

Figure 8.7. Simplified transition diagram for two-out-of-three redundant configuration.

The transition diagram of Fig. 8.7 gives the following differential equation.

$$\left.\begin{aligned} \dot{P}_{(0)} &= -(2\lambda+\bar{\lambda})P_{(0)} + \mu P_{(1)} \\ \dot{P}_{(1)} &= (2\lambda+\bar{\lambda})P_{(0)} - (2\lambda+\mu)P_{(1)} + \mu P_{(2)} \\ \dot{P}_{(2)} &= 2\lambda P_{(1)} - (\lambda+\mu)P_{(2)} + \mu P_{(3)} \\ \dot{P}_{(3)} &= \lambda P_{(2)} - \mu P_{(3)} \end{aligned}\right\} \tag{8.35}$$

with the initial condition

$$P_{(0)}(0)=1, \quad P_{(1)}(0)=P_{(2)}(0)=P_{(3)}(0)=0 \tag{8.36}$$

This can be integrated numerically and thus probabilities $P_{(i)}$ are obtained.

Two pumps must operate for the quench pump system of Fig. 8.5 to function. Thus, the redundancy parameter $Q_r(t)$ for the event "less than two pumps are operating" is given by

$$Q_r(t) = P_{(2)}(t) + P_{(3)}(t) \tag{8.37}$$

Example 5. Standby Redundancy with Three Pumps

Assume the failure and repair rates (hour^{-1}),

$$\lambda = 10^{-3}, \quad \bar{\lambda} = 0.5 \times 10^{-3}, \quad \mu = 10^{-2}$$

Calculate $Q_r(t)$ at time $t = 100$, 500, and 1000 hour.

Solution: Substitute the failure and repair rates into (8.35). The resulting differential equations can be integrated by the method in Appendix 8.1 of this chapter, yielding

t (hr)	$P_{(2)}(t)$	$P_{(3)}(t)$	$Q_r(t)$
100	0.011429	0.002716	0.014145
500	0.036325	0.003836	0.040161
1000	0.038238	0.003832	0.042070

Example 6. System Unavailability

A fault tree for the tail gas quench and clean up system of Fig. 8.5 is given by Fig. 8.8. Assume the failure and repair rates in Example 1 for quench pumps and the rates in Example 2 for circulation pumps. Assume further the rates in Example 4 for booster fan C, feed pump E, and filter H. Calculate lower and upper bounds for the system unavailability $Q_s(t)$ at $t = 1000$ hr.

Solution: The fault tree has minimal cut sets

$$\{C\}, \quad \{E\}, \quad \{H\}, \quad \{A,B\}, \quad \{B,D\}, \quad \{D,A\}, \quad \{F,G\}$$

The inclusion-exclusion principle of (7.103) gives the following upper and lower bounds for

$$Q_s(t) = \Pr(C \cup E \cup H \cup [(A \cap B) \cup (B \cap D) \cup (D \cap A)] \cup [F \cap G]) \tag{8.38}$$

$$\begin{aligned}
&\text{First bracket} = Q_s(t)_{\max}\\
&= \Pr(C) + \Pr(E) + \Pr(H) + \Pr((A \cap B) \cup (B \cap D) \cup (D \cap A)) + \Pr(F \cap G)\\
&\text{Second bracket} = Q_s(t)_{\min}
\end{aligned} \tag{8.39}$$

$$\begin{aligned}
&= Q_s(t)_{J_{\max}} - \Pr(C)\Pr(E) - \Pr(C)\Pr(H)\\
&- \Pr(C)\Pr((A \cap B) \cup (B \cap D) \cup (D \cap A))\\
&- \Pr(C)\Pr(F \cap G) - \Pr(E)\Pr(H) - \Pr(E)\Pr((A \cap B) \cup (B \cap D) \cup (D \cap A))\\
&- \Pr(E)\Pr(F \cap G) - \Pr(H)\Pr((A \cap B) \cup (B \cap D) \cup (D \cap A)) - \Pr(H)\Pr(F \cap G)\\
&- \Pr((A \cap B) \cap (B \cap D) \cup (D \cap A))\Pr(F \cap G)
\end{aligned} \tag{8.40}$$

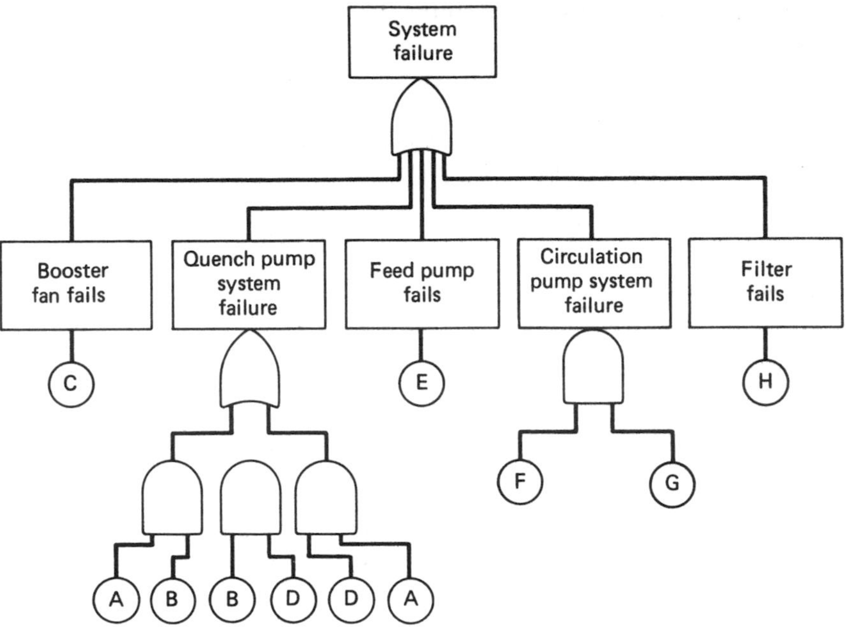

Figure 8.8. Fault tree for tail gas quench and clean up system.

The probability $\Pr((A \cap B) \cup (B \cap D) \cup (D \cap A))$ is given by $Q_r(t)$ in Example 5. Further, $\Pr(C) = \Pr(E) = \Pr(H)$ is calculated as $Q^*(t)$ in Example 4, and $\Pr(F \cap G)$ is equal to $Q_r''(t)$ in Example 4. Thus,

$$\begin{aligned} Q_s(t)_{\max} &= 3Q^*(t) + Q_r(t) + Q_r''(t) \\ &= 3 \times 0.009901 + 0.042070 + 0.004522 = 0.076295 \\ Q_s(t)_{\min} &= 0.076295 - 3Q^*(t)^2 - 3Q^*(t)Q_r(t) - 3Q^*(t)Q_r''(t) \\ &\quad - Q_r(t)Q_r''(t) \\ &= 0.076295 - 0.000294 - 0.001250 - 0.000134 - 0.000190 \\ &= 0.074427. \end{aligned}$$

Thus, the system unavailability is bracked by

$$0.074427 \le Q_s(t) \le 0.076295$$

As a general case, consider standby redundancies satisfying the following requirements.

1) The standby redundancy consists of n identical components.
2) The redundant configuration has $m(\le n)$ principal components.
3) At most, r components can be repaired at a time.

An aggregated transition diagram is shown in Fig. 8.9, and we have the differential equations

$$\left.\begin{aligned}
\dot{P}_{(0)} &= -\lambda_0 P_{(0)} + \mu_1 P_{(1)} \\
&\cdots\cdots\cdots\cdots\cdots\cdots\cdots\cdots \\
\dot{P}_{(k)} &= \lambda_{k-1} P_{(k-1)} - (\lambda_k + \mu_k) P_{(k)} + \mu_{k+1} P_{(k+1)} \\
&\cdots\cdots\cdots\cdots\cdots\cdots\cdots\cdots \\
\dot{P}_{(n)} &= \lambda_{n-1} P_{(n-1)} - \mu_n P_{(n)}
\end{aligned}\right\} \tag{8.41}$$

where

$$\left.\begin{aligned}
\lambda_k &= m\lambda + (n-m-k)\bar{\lambda} && \text{for } 0 \le k \le n-m \\
\lambda_k &= (n-k)\lambda && \text{for } n-m+1 \le k \le n-1 \\
\mu_k &= \min\{r,k\}\cdot\mu && \text{for } 1 \le k \le n
\end{aligned}\right\} \tag{8.42}$$

The parameter $Q_r(t)$ is given by

$$Q_r(t) = P_{(n-m+1)}(t) + \ldots + P_{(n)}(t) \tag{8.43}$$

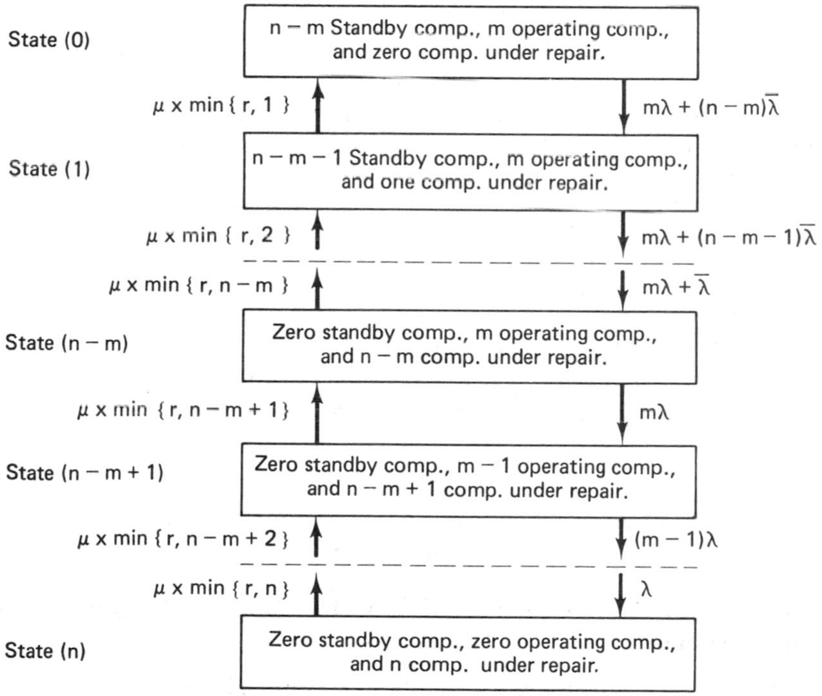

State (0)
n − m Standby comp., m operating comp., and zero comp. under repair.
μ x min { r, 1 }
mλ + (n − m)λ̄
State (1)
n − m − 1 Standby comp., m operating comp., and one comp. under repair.
μ x min { r, 2 }
mλ + (n − m − 1)λ̄
μ x min { r, n − m }
mλ + λ̄
State (n − m)
Zero standby comp., m operating comp., and n − m comp. under repair.
μ x min { r, n − m + 1 }
mλ
State (n − m + 1)
Zero standby comp., m − 1 operating comp., and n − m + 1 comp. under repair.
μ x min { r, n − m + 2 }
(m − 1)λ
μ x min { r, n }
λ
State (n)
Zero standby comp., zero operating comp., and n comp. under repair.

Figure 8.9. Transition diagram for *m*-out-of-*n* redundant configuration.

Equation (8.10) is a special case of (8.41), where $n=2$, $m=1$, and $r=2$. Similarly, equation (8.35) is obtained from (8.41) by setting $n=3$, $m=2$, and $r=1$.

8.2.3 Steady-State Value for Redundancy Parameter $Q_r(t)$

The steady-state solution of (8.41) satisfies

$$\left.\begin{aligned} 0&=-\lambda_0 P_{(0)}+\mu_1 P_{(1)} \\ &\cdots\cdots\cdots\cdots\cdots\cdots\cdots\cdots\cdots \\ 0&=\lambda_{k-1}P_{(k-1)}-(\lambda_k+\mu_k)P_{(k)}+\mu_{k+1}P_{(k+1)} \\ &\cdots\cdots\cdots\cdots\cdots\cdots\cdots\cdots\cdots \\ 0&=\lambda_{n-1}P_{(n-1)}-\mu_n P_{(n)} \end{aligned}\right\} \tag{8.44}$$

Define

$$\pi_k=-\lambda_k P_{(k)}+\mu_{k+1}P(_{k+1}),\quad k=0,\ldots,n-1 \tag{8.45}$$

Then, (8.44) can be written as

$$\left.\begin{aligned} &\pi_0=0 \\ &\pi_k-\pi_{k-1}=0,\quad k=1,\ldots,n-1 \\ &\pi_{n-1}=0 \end{aligned}\right\} \tag{8.46}$$

In other words,

$$\pi_0=\pi_1=\ldots=\pi_{n-1}=0$$

If $\mu_{k+1}\neq 0$ for $k=0,\ldots,n-1$, then

$$P_{(k+1)}=\frac{\lambda_0\lambda_1\cdots\lambda_k}{\mu_1\mu_2\cdots\mu_{k+1}}=\theta_{k+1}P_{(0)},\quad k=0,\ldots,n-1 \tag{8.47}$$

Since the sum of all probabilities is equal to unity, we have

$$P_{(k)}=\frac{\theta_k}{\sum_{k=0}^{n}\theta_k},\quad \theta_0=1 \tag{8.48}$$

The steady state $Q_r(\infty)$ can readily be obtained from (8.48).

Example 7. Standby Redundancy of Two Pumps

Calculate the steady-state unavailability $Q_r(\infty)$ for the pumping system of Examples 1, 2, and 3.

Solution: Note that $n=2$, $m=1$, and $r=2$. Equation (8.42) or (8.10) gives values for λ_0, λ_1, μ_1, and μ_2.

	Warm Standby	Cold Standby	Hot Standby
$\lambda_0=\lambda+\bar{\lambda}$	0.0015	0.001	0.002
$\lambda_1=\lambda$	0.001	0.001	0.001
$\mu_1=\mu$	0.01	0.01	0.01
$\mu_2=2\mu$	0.02	0.02	0.02

The values for θ_0, θ_1, and θ_2 are

	Warm Standby	Cold Standby	Hot Standby
$\theta_0=1$	1	1	1
$\theta_1=\frac{\lambda_0}{\mu_1}$	0.15	0.1	0.2
$\theta_2=\frac{\lambda_0\lambda_1}{\mu_1\mu_2}$	0.0075	0.005	0.01

Therefore, the probabilities $P_{(k)}$ and $Q_r(\infty)$ are given by

	Warm Standby	Cold Standby	Hot Standby
$\Sigma\theta$	1.1575	1.105	1.21
$P_{(0)}=\frac{\theta_0}{\Sigma\theta}$	0.863930	0.904977	0.826446
$P_{(1)}=\frac{\theta_1}{\Sigma\theta}$	0.129590	0.090498	0.165289
$P_{(2)}=\frac{\theta_2}{\Sigma\theta}$	0.006479	0.004525	0.008264
$Q_r(\infty)=P_{(2)}$	0.006479	0.004525	0.008264

We observe from Examples 1, 2, and 3, that the steady-state values of $Q_r(t)$ are attained at $t=1000$ within the accuracy of round-off error.

Example 8. Standby Redundancy with Three Pumps

Consider the pumping system of Example 5. Calculate the steady-state $Q_r(\infty)$ for the event "less than two pumps are operating."

Solution: We note that $n=3$, $m=2$, and $r=1$. Equation (8.35) or (8.42) gives

$$\lambda_0=2\lambda+\bar{\lambda}=0.0025, \quad \mu_1=\mu=0.01$$
$$\lambda_1=2\lambda=0.002, \quad \mu_2=\mu=0.01$$
$$\lambda_2=\lambda=0.001, \quad \mu_3=\mu=0.01$$

Values of θ_k are

$$\theta_0=1$$
$$\theta_1=\frac{\lambda_0}{\mu_1}=0.25$$
$$\theta_2=\frac{\lambda_0\lambda_1}{\mu_1\mu_2}=0.05$$
$$\theta_3=\frac{\lambda_0\lambda_1\lambda_2}{\mu_1\mu_2\mu_3}=0.005$$

Thus, (8.48) gives

$\Sigma\theta$	$P_{(0)}=\frac{\theta_0}{\Sigma\theta}$	$P_{(1)}=\frac{\theta_1}{\Sigma\theta}$	$P_{(2)}=\frac{\theta_2}{\Sigma\theta}$	$P_{(3)}=\frac{\theta_3}{\Sigma\theta}$
1.305	0.766284	0.191571	0.038314	0.003831

Hence, from (8.37),

$$Q_r(\infty)=0.038314+0.003831=0.042145$$

This steady-state value confirms $Q_r(t)$ at $t=1000$ in Example 5.

8.2.4 System Parameter $w_s(t)$

The parameter $w_s(t)$ is important in that its integration over a time interval is the expected number of system failures during the interval. As was shown by (7.118) and (7.119), an upper bound for $w_s(t)$ is

$$w_s(t)_{\max}=\sum_{i=1}^{N_c} w_i^*(t) \tag{8.49}$$

Let us consider the fault tree of Fig. 8.2. This tree has five minimal cut sets:

$$\begin{aligned} &d_1=\{C\}, \quad d_2\{E\}, \quad d_3=\{H\} \\ &d_4=\{A,B\}, \quad d_5=\{F,G\} \end{aligned} \tag{8.50}$$

Equation (8.49) becomes

$$w_s(t)_{\max}=w_1^*(t)+w_2^*(t)+w_3^*(t)+w_4^*(t)+w_5^*(t) \tag{8.51}$$

For the cut set $\{A,B\}$ to fail, either one of A and B should fail in t to $t+dt$ with the other remaining basic event already existing at time t. Thus, $w_4^*(t)\,dt$ is

$$\begin{aligned} w_4^*(t)\,dt = {} & \Pr\big(A \text{ fails during } (t,t+dt] \,|\, \bar{A}\cap B \text{ at time } t\big) \\ & \times \Pr(\bar{A}\cap B \text{ at time } t) \\ & + \Pr\big(B \text{ fails during } (t,t+dt] \,|\, A\cap \bar{B} \text{ at time } t\big) \\ & \times \Pr(A\cap \bar{B} \text{ at time } t) \end{aligned} \tag{8.52}$$

Assume failure rate λ' for pumps A and B. Then,

$$\begin{aligned} w_4^*(t)\,dt &= \lambda'\,dt\cdot\big[\Pr(\bar{A}\cap B) \text{ at time } t) + \Pr(A\cap\bar{B}) \text{ at time } t)\big] \\ &= \lambda'\,dt\cdot P'_{(1)}(t) \end{aligned} \tag{8.53}$$

where $P'_{(1)}(t)$ is the probability of one failed pump existing at time t, and is given by the solution of (8.10). Similarly,

$$w_5^*(t) = \lambda''\cdot P''_{(1)}(t) \tag{8.54}$$

where

λ'' = failure rate for pumps F and G,

$P''_{(1)}(t)$ = probability of either pump F or G, but not both, being failed at time t.

Thus, the upper bound $w_s(t)$ can be calculated by

$$\begin{aligned} w_s(t)_{\max} = {} & \lambda_1[1-Q_1(t)] + \lambda_2[1-Q_2(t)] + \lambda_3[1-Q_3(t)] \\ & + \lambda' P'_{(1)}(t) + \lambda'' P''_{(1)}(t) \end{aligned} \tag{8.55}$$

Example 9. Tail Gas Quench Clean Up System of Fig. 8.1

Calculate $w_s(t)_{\max}(1000)$ for the failure and repair rates of Example 4.

Solution: We have, from Example 4,

$$\lambda_1 = \lambda_2 = \lambda_3 = \lambda^* = 10^{-4} \tag{8.56}$$

$$\begin{aligned} Q_1(t) = Q_2(t) = Q_3(t) &= \frac{\lambda^*}{\lambda^*+\mu^*}\left(1-e^{-(\lambda^*+\mu^*)t}\right) \\ &= 0.009901 \text{ at } t = 1000 \end{aligned} \tag{8.57}$$

$$\lambda' = \lambda'' = 10^{-3}$$

Further, from (8.22) and (8.28), we have

$$P'_{(1)}(t) = 0.129590 - 0.117879e^{-0.0105436t} - 0.011711e^{-0.0219564t}$$
$$= 0.129587 \quad \text{at} \quad t = 1000 \tag{8.58}$$

$$P''_{(1)}(t) = 0.090498 - 0.083109e^{-0.0100839t} - 0.007389e^{-0.0219161t}$$
$$= 0.090495 \quad \text{at time } t = 1000 \tag{8.59}$$

Thus, from (8.55),

$$w_s(1000)_{\max} = 3 \times 10^{-4} \times [1 - 0.009901] + 10^{-3} \times 0.129587 + 10^{-3} \times 0.090495$$
$$= 0.000517 \text{ (times/hour)} \tag{8.60}$$

Let us next consider the fault tree of Fig. 8.8. This has seven minimal cut sets:

$$d_1 = \{C\}, \quad d_2 = \{E\}, \quad d_3 = \{H\}, \quad d_4 = \{A, B\}$$
$$d_5 = \{B, D\}, \quad d_6 = \{D, A\}, \quad d_7 = \{F, G\} \tag{8.61}$$

Denote by $w_r(t)$ the expected number of times that the quench pump system fails per unit time at time t. Then, similarly to (8.49), we have an upper bound

$$w_s(t)_{\max} = w_1^*(t) + w_2^*(t) + w_3^*(t) + w_r(t) + w_7^*(t) \tag{8.62}$$

For the redundant system to fail in time t to $t + dt$, one pump should fail during $(t, t + dt]$ with the redundant system already being in state (1) of Fig. 8.7. The rate of transition from state (1) to (2) is 2λ. Thus,

$$w_r(t) = 2\lambda \cdot P_{(1)}(t) \tag{8.63}$$

where $P_{(1)}(t)$ is the probability that the redundant system has one failed pump at time t, and is given by the solution of (8.35).

Example 10. Tail Gas Quench Clean Up System of Fig. 8.5

Calculate $w_s(t)_{\max}$ at $t = 1000$ for the failure and repair rates of Example 6.

Solution: We can use the results of Example 9 for $w_1^*(t) = w_2^*(t) = w_3^*(t)$ and $w_7^*(t)$. The parameter $w_r(t)$ is given by

$$w_r(t) = 2 \times 10^{-3} \times P_{(1)}(1000) = 2 \times 10^{-3} \times 0.1915061 = 0.000383$$

since the numerical integration of (8.35) yields

$$P_{(1)}(1000) = 0.1915061$$

Thus,

$$w_s(t)_{\max} = 3 \times 10^{-4} \times [1 - 0.009901] + 0.000383 + 10^{-3} \times 0.090495$$
$$= 0.000771$$

As a general case, consider an m-out-of-n redundant configuration with r repair crews. Let a fault tree have cut sets including component failures in the redundancy. The calculation of $w_s(t)_{max}$ can be reduced to evaluating $w_r(t)$ defined by

$$w_r(t) = \text{the expected number of times that the redundant configuration fails per unit time at time } t = m\lambda P_{(n-m)}(t) \tag{8.64}$$

where

$P_{(n-m)}(t)$ = the probability of $(n-m)$ components already being failed at time t, = the probability of state $(n-m)$ in Fig. 8.9

$m\lambda$ = the rate of transition from state $(n-m)$ to $(n-m+1)$.

8.3 QUANTIFICATION OF SYSTEMS SUBJECT TO COMMON CAUSE FAILURE

The tail gas quench and clean up system of Fig. 8.1 has two redundant configurations. Such redundancy is often used to improve system reliability. However, the configurations do not necessarily lead to sufficient improvement when common causes are involved. This point is discussed in Chapter 3.

Let us consider a cut set $\{A, B\}$ in a fault tree. Assume that the cut set suffers from a common cause C which occurs with the rate c. Let basic events A and B have failure rate λ_1 and λ_2 and repair rate μ_1 and μ_2, respectively, when no common causes are involved. Then, the behavior of the cut set can be expressed by the Markov transition diagram of Fig. 8.10. Here, the indicator variable 1 denotes the occurrence of the basic events, whereas the variable 0 the non-occurrence of the events. The cut set fails when it falls into state $(1,1)$. Common cause C creates the multiple-component transition from state $(0,0)$ to $(1,1)$.

There are a few methods for quantifying systems subject to common cause failure. B. B. Chu and D. P. Gaver use ordinary Markov models for the quantification.[1] W. E. Vesely proposed the Marshall-Olkin approach to common cause failures.[2] In this book, we propose a new approach which can handle any number of repairable basic events and common causes.

[1] Chu, B. B., and D. P. Gaver, "Stochastic Models for Repairable Redundant Systems Susceptible to Common Mode Failures," in *Nuclear Systems Reliability Engineering and Risk Assessment*, Fussell, J. B., and G. R. Burdick (eds.), *SIAM*, Philadelphia, p. 342, 1977.

[2] Vesely, W. E., "Estimating Common Cause Failure Probabilities in Reliability and Risk Analysis: Marshall-Olkin Specializations," ibid., p. 314.

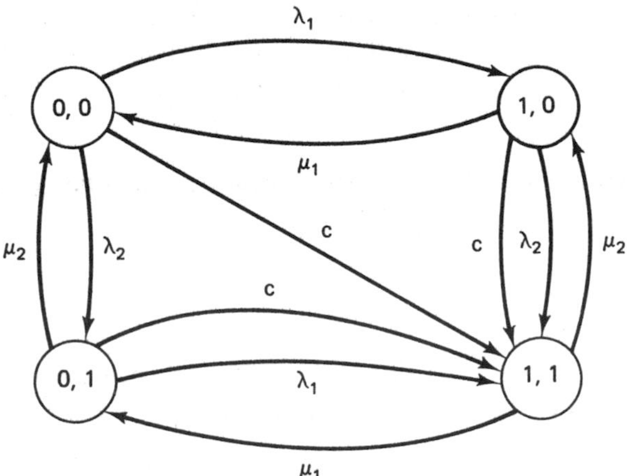

Figure 8.10. Markov transition diagram for components (A,B) subject to common cause C.

8.3.1 Quantification of Common Causes

Assume that a common cause has failure rate c, and repair rate b. The failure rate r denotes the rate of occurrence of the common cause, whereas the rate b refers to the disappearance of the cause. This is clarified in the following example.

Example 11. Rates for Common Causes

Assume that fire occurs once a year and that, when it happens, it continues for 4 hr on the average. Estimate rates c and b.

Solution: The reciprocal $1/c$ is equal to mean time to fire. Thus,

$$c = \frac{1}{(365 \times 24)} = 0.000114 \text{ (hour}^{-1}) \tag{8.65}$$

Similarly, $1/b$ is equal to mean time of the duration of the fire.

$$b = \frac{1}{4} = 0.25 \text{ (hour}^{-1}) \tag{8.66}$$

When a common cause occurs, its *common-mode events* occur simultaneously. Thus, the common-mode events are caused by the common cause with the frequency of $w_c(t)$, the expected number of common causes at time t per unit time. The parameter $w_c(t)$ is given by (4.103) in Chapter 4:

$$w_c(t) = \frac{cb}{c+b} + \frac{c^2}{c+b} e^{-(c+b)t} \tag{8.67}$$

Thus, its steady-state value $w_c(\infty)$ is

$$w_c(\infty)=\frac{cb}{c+b}=\frac{1}{(1/c)+(1/b)} \tag{8.68}$$

In most cases, mean time to common cause $1/c$ is much greater than the mean period of duration of the causes

$$\frac{1}{c}\gg\frac{1}{b} \tag{8.69}$$

This inequality means that the common cause occurs as a sequence of pulses; thus equation (8.68) becomes, simply,

$$w_c(\infty)=c \tag{8.70}$$

This relation gives the rate c in the Markov transition diagram of Fig. 8.10. Note that the state (0,0) changes to (1,1) when the common cause occurs.

8.3.2 System Unavailability $Q_s(t)$

Any sequences of common causes can be classified into one of the following two classes shown in Fig. 8.11.

1) No common causes to time t.
2) The last common cause occurs in time $u-du$ to u, and no common causes from time u to t.

The probability P_1 of class 1 is

$$P_1=e^{-ct} \tag{8.71}$$

whereas the probability P_2 of class 2 is

$$\begin{aligned} P_2=&\Pr(\text{common cause during } (u-du,u]) \\ &\times\Pr(\text{no common causes during } (u,t], \\ &\quad \text{given a common cause during } (u-du,u]) \end{aligned} \tag{8.72}$$

$$\begin{aligned} &=w_c(\infty)\,du\cdot e^{-c(t-u)} \\ &=ce^{-c(t-u)}\,du \end{aligned} \tag{8.73}$$

For class 1, the common-mode basic events A,B evolve according to the ordinary Markov transition diagram of Fig. 8.12(a), with the initial condition (0,0) at time zero. For class 2, the common-mode events A,B fall into the state (1,1) during $(u-du,u)$, and resume their Markov process

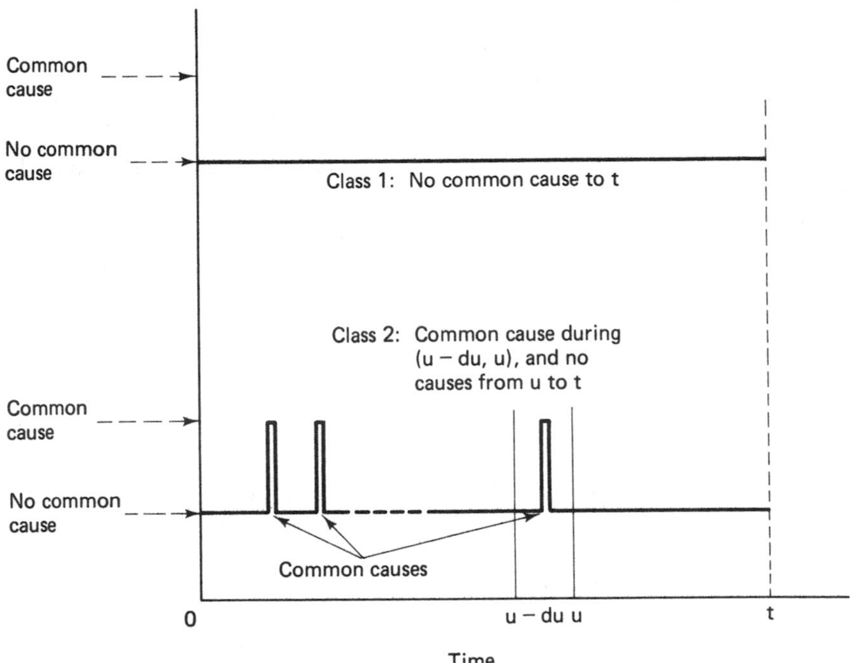

Figure 8.11. Classification of sequences of common causes.

according to the diagram of Fig. 8.12(b). This process has the initial condition (1, 1) at time u. Note that the state probabilities for these two diagrams can be calculated readily because basic events are independent, since no common causes are involved.

Consider the diagram of Fig. 8.12(a) for class 1. Both basic events A and B are failed state at time t with the probability

$$\prod_{i=1}^{2}\left(\frac{\lambda_i}{\lambda_i+\mu_i}\right)(1-e^{-(\lambda_i+\mu_i)t}) \tag{8.74}$$

On the other hand, consider the diagram of Fig. 8.12(b) for class 2. Note that the diagram has the initial condition (1, 1) at time u. Thus, the probability of both A and B being in failed states at time t is

$$\prod_{i=1}^{2}\left(\frac{\lambda_i}{\lambda_i+\mu_i}+\frac{\mu_i}{\lambda_i+\mu_i}e^{-(\lambda_i+\mu_i)(t-u)}\right) \tag{8.75}$$

Consequently, the probability of A and B being in the failed state at time t under the common cause can be calculated by a weighted sum of (8.74) and (8.75). The weighting factors are given by (8.71) and (8.73), respec-

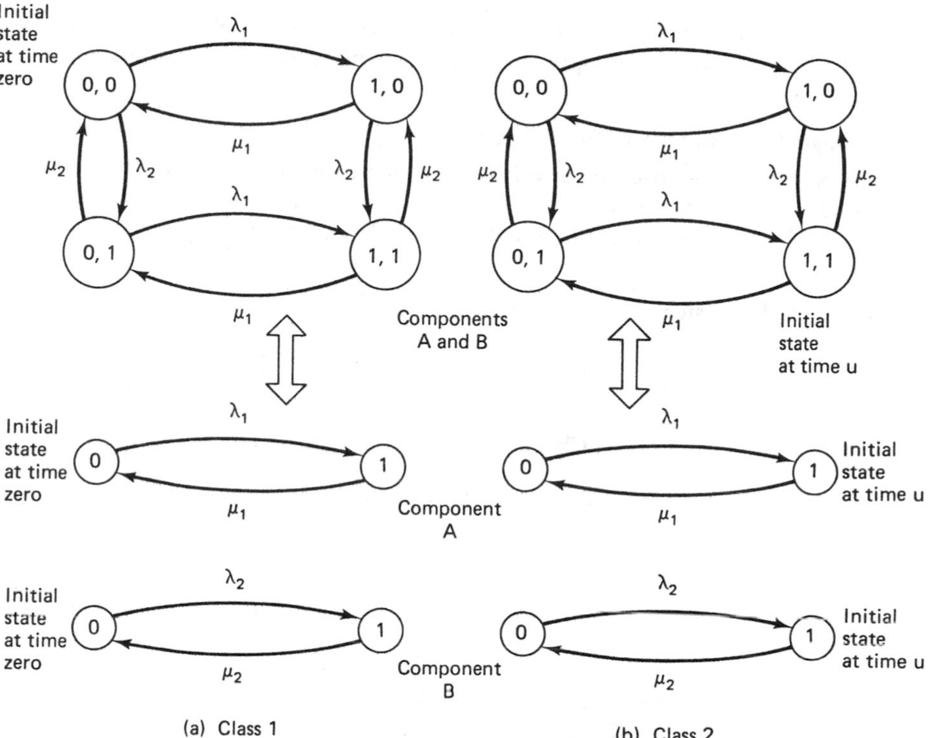

Figure 8.12. Markov transition diagrams for common causes in class 1 and class 2.

tively. Thus,

$$\begin{aligned}
&\Pr(A \cap B \text{ at time } t) \\
&= e^{-ct}\prod_{i=1}^{2}\left(\frac{\lambda_i}{\lambda_i+\mu_i}\right)(1-e^{-(\lambda_i+\mu_i)t}) \\
&\quad + \int_0^t ce^{-c(t-u)}\prod_{i=1}^{2}\left(\frac{\lambda_i}{\lambda_i+\mu_i}+\frac{\mu_i}{\lambda_i+\mu_i}e^{-(\lambda_i+\mu_i)(t-u)}\right)du
\end{aligned} \tag{8.76}$$

Equation (8.76) can be extended to the case of n common-mode basic events $A_1, \ldots, A_n$. Replacing 2 by n;

$$\begin{aligned}
&\Pr(A_1 \cap \cdots \cap A_n \text{ at time } t) \\
&= e^{-ct}\prod_{i=1}^{n}\left(\frac{\lambda_i}{\lambda_i+\mu_i}\right)(1-e^{-(\lambda_i+\mu_i)t}) \\
&\quad + \int_0^t ce^{-cs}\prod_{i=1}^{n}\left(\frac{\lambda_i}{\lambda_i+\mu_i}+\frac{\mu_i}{\lambda_i+\mu_i}e^{-(\lambda_i+\mu_i)s}\right)ds
\end{aligned} \tag{8.77}$$

where s is defined as $s=t-u$. This equation, when coupled with the inclusion-exclusion principle, gives lower and upper bounds for system unavailability $Q_s(t)$.

Example 12. Two-Out-of-Three System

Consider a two-out-of-three voting system consisting of components 1, 2, and 3. Assume components 1 and 2 suffer from common cause C. Evaluate system unavailability $Q_s(t)$ at $t=100$ by assuming $\lambda_1=\lambda_2=\lambda_3=0.001$, $\mu_1=\mu_2=\mu_3=0.004$, and $c=0.0005$.

Solution: The system has three cut sets:

$$d_1=\{1,2\},\quad d_2=\{2,3\},\quad d_3=\{3,1\} \tag{8.78}$$

Similarly to Example 11 in Chapter 7, we have the expansion

$$\begin{aligned}Q_s(t)=\Pr(d_1)+\Pr(d_2)+\Pr(d_3)-[\Pr(d_1\cap d_2)+\Pr(d_2\cap d_3)+\Pr(d_3\cap d_1)]\\ +\Pr(d_1\cap d_2\cap d_3)\end{aligned} \tag{8.79}$$

Note that components 1 and 2 are not independent, since they suffer from common cause C. Hence,

$$\begin{aligned}[A]\quad &[\Pr(d_1)+\Pr(d_2)+\Pr(d_3)]=Q_1^*(t)+Q_2^*(t)+Q_3^*(t)\\ &\quad=\Pr(1\cap 2)+\Pr(2)\Pr(3)+\Pr(3)\Pr(1)\end{aligned} \tag{8.80}$$

$$\begin{aligned}[B]\quad &[\Pr(d_1\cap d_2)+\Pr(d_2\cap d_3)+\Pr(d_3\cap d_1)]\\ &\quad=\Pr(1\cap 2)\Pr(3)+\Pr(1\cap 2)\Pr(3)+\Pr(1\cap 2)\Pr(3)=3\Pr(1\cap 2)\Pr(3)\end{aligned} \tag{8.81}$$

$$[C]\quad \Pr(d_1\cap d_2\cap d_3)=\Pr(1\cap 2)\Pr(3) \tag{8.82}$$

From the failure and repair rates,

$$\lambda_i+\mu_i=0.005,\qquad \frac{\lambda_i}{\lambda_i+\mu_i}=0.2,\qquad \frac{\mu_i}{\lambda_i+\mu_i}=0.8$$

Component 3 has no common causes. Thus,

$$\Pr(3)=0.2(1-e^{-0.005t})=0.078694$$

Components 1 and 2 are subjected to common cause C. Set in $n=1$ in (8.77). Then,

$$\begin{aligned}\Pr(1)&=\Pr(2)\\ &=e^{-0.0005t}\times 0.2\times(1-e^{-0.005t})\\ &\quad+\int_0^t 0.0005e^{-0.0005s}(0.2+0.8e^{-0.005s})\,ds\\ &=0.272727(1-e^{-0.0055t})=0.115377\end{aligned}$$

Set $n=2$ in (8.77). Then,

$$\Pr(1\cap 2)=e^{-0.0005t}\times[0.2\times(1-e^{-0.005t})]^2 + \int_0^t 0.0005e^{-0.0005s}(0.2+0.8e^{-0.005s})^2 ds$$
$$=0.099567-0.109091e^{-0.0055t}+0.009524e^{-0.0105t}$$
$$=0.039960$$

Consequently,

$$[A]=0.039960+2\times 0.115377\times 0.078694=0.058119$$
$$[B]=3\times 0.039960\times 0.078694=0.009434$$
$$[C]=0.039960\times 0.078694=0.003145$$
$$Q_s(t)_{\max}=[A]=0.058119$$
$$Q_s(t)_{\min}=[A]-[B]=0.048685$$
$$Q_s(t)=[A]-[B]+[C]=0.05183$$

Consider now cases where each one of m common causes $C_1,\dots,C_m$ affects some of the n basic events $A_1,\dots,A_n$. Denote by C_i occurrence of common cause C_i before time t, and designate by $\bar{C}_i$ non-occurrence of C_i to time t. Assume that the common cause is a low-probability event. Then we can restrict our attention to the following cases:

0) No common cause to time t; i.e., $\bar{C}_1\cap\dots\cap\bar{C}_m$.
1) Cause C_1 but no other causes to time t; i.e., $C_1\cap\bar{C}_2\cap\dots\cap\bar{C}_m$.
... ...
m) Cause C_m but no other causes to time t; i.e., $\bar{C}_1\cap\dots\cap\bar{C}_{m-1}\cap C_m$.

The other cases can be neglected because they imply occurrence of two or more common causes to time t. The probability $\Pr(A_1\cap\dots\cap A_n)$ at time t can be expressed by a weighted sum of the probabilities for cases $0,\dots,m$. This is illustrated by the following example.

Example 13. Two-Out-of-Three System

Assume λ_i and μ_i in Example 12. Assume further that components 1 and 2 suffer from common cause C_1, and components 2 and 3 are affected by common cause C_2. Rates for C_1 and C_2 are

$$c_1=c_2=0.0005$$

Evaluate system unavailability $Q_s(t)$ at $t=100$.

Solution:

$$[A]\quad [\Pr(d_1)+\Pr(d_2)+\Pr(d_3)]=\Pr(1\cap 2)+\Pr(2\cap 3)+\Pr(3\cap 1)$$

$$[B]\quad [\Pr(d_1\cap d_2)+\Pr(d_2\cap d_3)+\Pr(d_3\cap d_1)]=3\Pr(1\cap 2\cap 3)$$

$$[C]\quad \Pr(d_1\cap d_2\cap d_3)=\Pr(1\cap 2\cap 3)$$

In order to calculate $[A]$, $[B]$, and $[C]$, we must consider two common causes C_1 and C_2. We have the following three cases.

Case 0. $\bar{C}_1\cap\bar{C}_2$: In this case, no common causes occur to time t. Thus, basic events 1, 2, and 3 are independent and

$$\begin{aligned}\Pr(1\cap 2)&=\Pr(1)\Pr(2)=[0.2(1-e^{-0.005t})]^2=0.006193\\ \Pr(2\cap 3)&=0.006193\\ \Pr(3\cap 1)&=0.006193\\ \Pr(1\cap 2\cap 3)&=0.000487\end{aligned}$$

The weighting factor for case 0 is $\Pr(\bar{C}_1\cap\bar{C}_2)$ and is

$$\Pr(\bar{C}_1\cap\bar{C}_2)=e^{-c_1t}e^{-c_2t}=e^{-0.001t}=0.904837$$

Case 1. $C_1\cap\bar{C}_2$: In this case only common cause C_1 occurs to time t, and basic event 3 is independent of 1 or 2. Hence,

$$\begin{aligned}\Pr(2\cap 3)&=\Pr(2)\Pr(3)\\ \Pr(3\cap 1)&=\Pr(1)\Pr(3)\\ \Pr(1\cap 2\cap 3)&=\Pr(1\cap 2)\Pr(3)\end{aligned}$$

Since common cause C_2 does not occur for basic event 3,

$$\Pr(3)=0.2(1-e^{-0.005t})=0.078694$$

Basic events 1 and 2 suffer from common cause C_1. This common cause can be classified according to the time of its last occurrence, and $\Pr(1)$ or $\Pr(2)$ can be calculated by the second term ($n=1$) on the right-hand side of (8.77), the first term on the right-hand side being the contribution by "no common cause to time t."

$$\begin{aligned}\Pr(1)&=\Pr(2)\\ &=\int_0^t 0.0005e^{-0.0005s}(0.2+0.8e^{-0.005s})\,ds\\ &=0.272727-0.2e^{-0.0005t}-0.072727e^{-0.0055t}\\ &=0.040521\end{aligned}$$

Similarly,

$$\Pr(1\cap 2)=\int_0^t 0.0005e^{-0.0005s}(0.2+0.8e^{-0.005s})^2\,ds=0.034069$$

Thus,

$$\Pr(2\cap 3)=0.040521\times 0.078694=0.003189$$
$$\Pr(3\cap 1)=0.040521\times 0.078694=0.003189$$
$$\Pr(1\cap 2\cap 3)=0.034069\times 0.078694=0.002681$$

The weighting factor is $\Pr(C_1\cap \overline{C}_2)=\Pr(C_1)\Pr(\overline{C}_2)$. However, the factor $\Pr(C_1) = c_1e^{-c_1s}\,ds, s\in[0,t]$, has already been included as a weighting factor of the integral on the right-hand side of (8.77). Thus, only $\Pr(\overline{C}_2)=e^{-c_2t}=e^{-0.0005t}=0.951229$ is a multiplier in the final weighted sums.

Case 2. $\overline{C}_1\cap C_2$: Symmetrically to case 1,

$$\Pr(3\cap 1)=\Pr(1\cap 2)=0.003189$$
$$\Pr(2\cap 3)=0.034069$$
$$\Pr(1\cap 2\cap 3)=0.002681$$

The effective weighting factor is

$$\Pr(\overline{C}_1)=0.951229$$

Weighted sums:

$$\Pr(1\cap 2)=0.006193\times 0.904837+0.034069\times 0.951229+0.003189\times 0.951229$$
$$=0.041045$$
$$\Pr(2\cap 3)=\Pr(1\cap 2)=0.041045$$
$$\Pr(3\cap 1)=0.006193\times 0.904837+0.003189\times 0.951229+0.003189\times 0.951229$$
$$=0.011671$$
$$\Pr(1\cap 2\cap 3)=0.000487\times 0.904837+0.002681\times 0.951229+0.002681\times 0.951229$$
$$=0.005541$$

Unavailability:

$$[A]=\Pr(1\cap 2)+\Pr(2\cap 3)+\Pr(3\cap 1)$$
$$=2\times 0.041045+0.011671=0.093761$$
$$[B]=3\Pr(1\cap 2\cap 3)=3\times 0.005541=0.016623$$
$$[C]=\Pr(1\cap 2\cap 3)=0.005541$$
$$Q_s(t)_{\max}=[A]=0.093761$$
$$Q_s(t)_{\min}=[A]-[B]=0.077138$$
$$Q_s(t)=[A]-[B]+[C]=0.082679$$

Case 0 in this example gives as the availability for the no-common-cause situation

$$Q_s(t)_{\max}=0.018579$$
$$Q_s(t)_{\min}=0.017118$$
$$Q_s(t)=0.017605$$

We observe significant degradation of system availability by common causes C_1 and C_2.

8.3.3 System Parameter $w_s(t)$

Similar calculations are possible for system parameter $w_s(t)$ as shown by the following example.

Example 14. Two-Out-of-Three System

Calculate upper bound $w_s(t)_{\max}$ for the system of Example 12.

Solution:

$$w_s(t)_{\max} = w_1^*(t) + w_2^*(t) + w_3^*(t)$$

Let us first consider $w_1^*(t)$. Cut set $d_1 = \{1,2\}$ fails if:

1) both components 1 and 2 fail simultaneously by the common cause C when the cut set is in state $(0,0) = \bar{1} \cap \bar{2}$ or
2) component 1 fails by λ_1 or c when the cut set is in state $(0,1) = \bar{1} \cap 2$ or
3) component 2 fails by λ_2 or c when the cut set is in state $(1,0) = 1 \cap \bar{2}$.

Thus,

$$\begin{aligned} w_1^*(t) &= c \cdot \Pr(\bar{1} \cap \bar{2}) + (c+\lambda_1)\Pr(\bar{1} \cap 2) + (c+\lambda_2)\Pr(1 \cap \bar{2}) \\ &= c\left[\Pr(\bar{1} \cap \bar{2}) + \Pr(\bar{1} \cap 2) + \Pr(1 \cap \bar{2})\right] + \lambda_1 \Pr(\bar{1} \cap 2) + \lambda_2 \Pr(1 \cap \bar{2}) \\ &= 0.0005[1 - \Pr(1 \cap 2)] + 2 \times 0.001 \times \Pr(\bar{1} \cap 2) \end{aligned}$$

From Example 12,

$$\Pr(1 \cap 2) = 0.039960$$

Similarly to (8.77), we have the following equation for $\Pr(\bar{1} \cap 2)$:

$$\begin{aligned} \Pr(\bar{1} \cap 2) &= e^{-0.0005t} \times 0.2(1 - e^{-0.005t})(0.8 + 0.2e^{-0.005t}) \\ &\quad + \int_0^t 0.0005e^{-0.0005s} \times 0.2(1 - e^{-0.005s})(0.8 + 0.2e^{-0.005s})\,ds \\ &= 0.070915 \end{aligned}$$

where

1) the term $0.2(1 - e^{-0.005s})$ is the probability that component 1 is functioning at time t, given that it was failed at time $t-s$;
2) the term $(0.8 + 0.2e^{-0.005s})$ is the probability that component 2 is failing at time t, given that it was failed at time $t-s$;
3) $0.0005e^{-0.0005s}\,ds$ is the weighting factor, which shows that the common cause occurred in time $t-s-ds$ to $t-s$ and does not occur from time $t-s$ to t.

Hence,

$$w_1^*(t) = 0.0005[1 - 0.033960] + 2 \times 0.001 \times 0.070915 = 0.000622$$

Cut set parameter $w_2^*(t)$ becomes

$$\begin{aligned} w_2^*(t) &= (c + \lambda_2)\Pr(\bar{2} \cap 3) + \lambda_3 \Pr(2 \cap \bar{3}) \\ &= 0.0015 \Pr(\bar{2}) \Pr(3) + 0.001 \Pr(2) \Pr(\bar{3}) \\ &= 0.0015[1 - \Pr(2)] \Pr(3) + 0.001 \Pr(2)[1 - \Pr(3)] \end{aligned}$$

Use Pr(2) and Pr(3) in Example 12:

$$\begin{aligned} w_2^*(t) &= 0.0015[1 - 0.115377] \times 0.078694 + 0.001 \times 0.115377[1 - 0.078694] \\ &= 0.000211 \end{aligned}$$

From symmetry,

$$w_3^*(t) = w_2^*(t) = 0.000211$$

Consequently,

$$w_s(t)_{\max} = 0.000622 + 0.000211 + 0.000211 = 0.001044$$

APPENDIX 8.1 METHOD FOR NUMERICAL INTEGRATION OF LINEAR DIFFERENTIAL EQUATIONS

Consider a linear differential equation

$$\dot{\mathbf{P}}(t) = \mathbf{A}\mathbf{P}(t), \quad \mathbf{P}(0) = \mathbf{P}_0 \tag{1}$$

where $\mathbf{P} = n$-vector,
$\mathbf{A} = n \times n$ matrix.

Let Δ be a small time increment. Define a new $n \times n$ matrix, $\exp[\mathbf{A}\Delta]$ by the Taylor series,

$$\Phi(\Delta) = \exp[\mathbf{A}\Delta] = \mathbf{I} + \mathbf{A}\Delta + \frac{\mathbf{A}^2\Delta^2}{2} + \frac{\mathbf{A}^3\Delta^3}{3!} + \frac{\mathbf{A}^4\Delta^4}{4!} + \cdots \tag{2}$$

This matrix can be approximated by the finite Taylor series expansion including up to mth powers of Δ. The differential equation can be solved sequentially in time as

$$\left.\begin{aligned} \mathbf{P}(\Delta) &= \Phi(\Delta)\mathbf{P}(0) \\ \mathbf{P}(2\Delta) &= \Phi(\Delta)\mathbf{P}(\Delta) \\ \mathbf{P}(k\Delta) &= \Phi(\Delta)\mathbf{P}((k-1)\Delta) \end{aligned}\right\} \tag{3}$$

PROBLEMS

8.1. Let $\mathbf{P}(t)$ and $\mathbf{A}$ be an n-vector and an $n\times n$ matrix. It can be shown that the differential equation

$$\dot{\mathbf{P}}(t)=\mathbf{A}\mathbf{P}(t)$$

can be solved sequentially as

$$\mathbf{P}(\Delta)=\exp(\mathbf{A}\Delta)\mathbf{P}(\Delta)$$
$$\vdots$$
$$\mathbf{P}(k\Delta)=\exp(\mathbf{A}\Delta)\mathbf{P}([k-1]\Delta)$$

where

$$\Delta=\text{small length of time,}$$
$$\exp[\mathbf{A}\Delta]=\mathbf{I}+\mathbf{A}\Delta+\frac{\mathbf{A}^2\Delta^2}{2}+\frac{\mathbf{A}^3\Delta^3}{3!}+\frac{\mathbf{A}^4\Delta^4}{4!}+\cdots,$$
$$\mathbf{I}=\text{unit matrix.}$$

Calculate $\exp[\mathbf{A}\Delta]$ for the differential equation of the warm standby with $\lambda=0.001$, $\bar{\lambda}=0.005$, and $\mu=0.01$, considering up to the second-order terms of $\Delta=10$.

8.2. Obtain $Q_r(10)$ and $Q_r(20)$ for the warm standby, using $\exp[-\mathbf{A}\Delta]$ in Problem 8.1.

8.3. Obtain the exact $Q_r(t)$ for the warm standby in Problem 8.1, using Laplace transforms. Compare the results with $Q_r(10)$ and $Q_r(20)$ of Problem 8.2.

8.4. Consider the bridge configuration of Fig. P8.4. Unit 2 is a cold standby for unit 1, and unit 4 is a hot standby for unit 3. Assume that

$$Q_1=\Pr(1)=0.03,\quad Q_2=\Pr(2)=0.005$$
$$Q_r=\Pr(1\cap 2)=0.0003,\quad Q_3=\Pr(3)=Q_4=\Pr(4)=0.02,$$
$$Q_5=\Pr(5)=0.0002$$

Calculate the system unavailability Q_s.

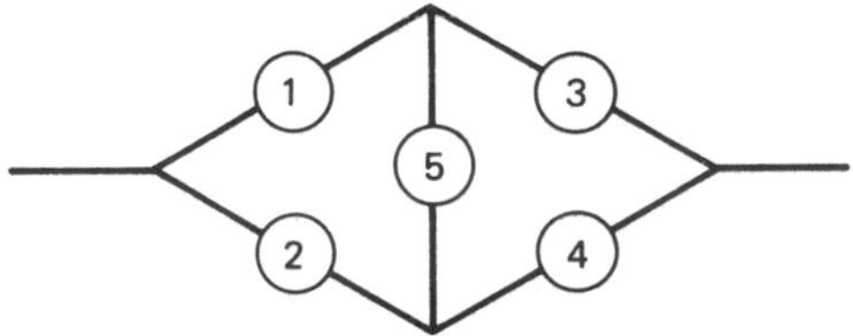

Figure P8.4.

8.5. Determine the differential equation for the two-out-of-three standby redundancy with three pumps, using as data

$$\lambda=0.001, \quad \bar{\lambda}=0.0001, \quad \mu=0.1$$

Obtain also the matrix $\exp[\mathbf{A}\Delta]=\mathbf{I}+\mathbf{A}\Delta$ for a small time length Δ. Calculate $P_{(i)}(3)$ by setting $\Delta=1$.

8.6. A standby redundancy consists of five identical components and has two principal components and repair crews. (a) Obtain the differential equation for the redundant configuration, using as data, $\lambda=0.001, \bar{\lambda}=0.0005, \mu=0.01$. (b) Calculate the steady-state unavailability $Q_r(\infty)$.

8.7. Prove (8.64) for the standby redundancy in Problem 8.6. Calculate the steady-state unconditional failure intensity $w_r(\infty)$.

8.8. Calculate the steady-state probabilities $\Pr(1)$, $\Pr(2)$, and $\Pr(1\cap 2)$ for the two-component redundant system suffering common cause C. [Use equation (8.77).]

8.9. (a) Give a Markov transition diagram for the components 1 and 2 in Example 12.

(b) Write a differential equation of the state probabilities for the components.

(c) Obtain the first-order approximation of $\exp[\mathbf{A}\Delta], (\Delta=1)$ for the differential equation.

(d) Calculate $\Pr(1\cap 2)$ at $t=1,2,3,4$, using the matrix $\exp[\mathbf{A}\Delta]$. Compare the results with the exact solution of Example 12:

$$\Pr(1\cap 2)=0.0995671-0.1090909e^{-0.0055t}+0.0095238e^{-0.0105}$$

8.10 Consider a two-out-of-three system consisting of identical components. Components 1 and 2 suffer common cause C_1, and components 2 and 3 suffer common cause C_2.

(a) Obtain the Markov differential equation.

(b) Using the above result, determine the differential equation for the system of Example 13.

9

SYSTEM QUANTIFICATION: RELIABILITY

W, the expected number of failures, and Q, the unavailability, refer to the *existence probability* of a system failure and are useful for predicting non-catastrophic faults such as production or service shutdowns. This kind of system failure does not result in complete destruction of the system and will occur many times during a system's lifetime. A fault which occurred and was repaired before time t can occur again and exist at time t.

The system reliability, on the other hand, is more useful for describing catastrophic *non-repairable system breakdowns* such as missile failures, explosions, etc. This kind of fault destroys the system; thus, the probability of its reoccurrence in the same system is meaningless. The system reliability $R_s(t)$ is the probability that the system suffers from no system hazard to time t, and is equivalent to the system survival probability. Its complement is system unreliability $F_s(t)$, which was defined as the probability of the system's suffering one hazard to time t. It is shown in Chapter 7 that equation (7.133) provides an incorrect calculation of the unreliability. Although the numerical values obtained by the equation are sufficiently accurate for practical purposes, this chapter presents methods by which exact values can be obtained.

9.1 ONE-COMPONENT SYSTEM

Consider a one-component system which has the failure rate λ and repair rate μ. A Markov transition diagram is shown as Fig. 9.1. The arrow from state 1 to state 0 denotes completion of a system repair. In reliability evaluation, however, we only consider the period starting at initial time zero and ending with the first system failure. Therefore, we remove this arrow from Fig. 9.1, obtaining Fig. 9.2. State 1 of Fig. 9.2 is an absorption state because it has only inflow from state 0 and has no outflow to state 0.

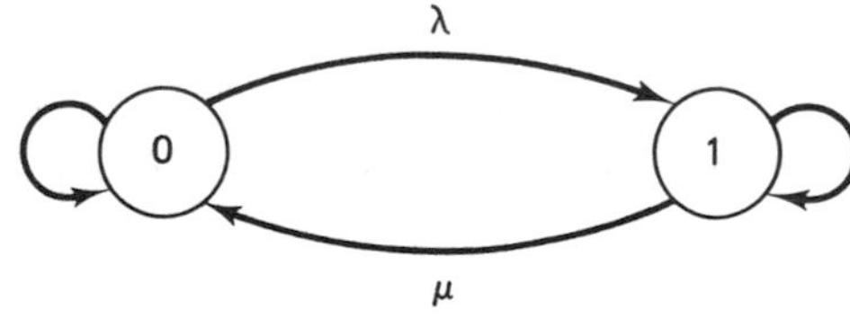

Figure 9.1. Transition diagram for one-component system.

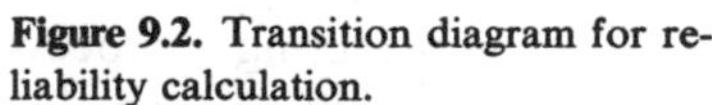

λ
0
1
Absorbtion state

Figure 9.2. Transition diagram for reliability calculation.

The differential equation which describes the state probability $P_0(t)$ of the component is

$$\dot{P}_0 = -\lambda P_0, P_0(0) = 1 \tag{9.1}$$

This has the solution

$$P_0(t) = e^{-\lambda t} \tag{9.2}$$

$P_0(t)$ is the probability of the system's continuing to function to time t, since there is no inflow from state 1 to state 0. Thus, the system reliability $R_s(t)$ is

$$R_s(t) = P_0(t) = e^{-\lambda t} \tag{9.3}$$

confirming (4.84) in Chapter 4. System unreliability $F_s(t)$ is equal to unity minus the reliability,

$$F_s(t) = 1 - e^{-\lambda t} \tag{9.4}$$

State probability $P_1(t)$, on the other hand, is described by

$$\dot{P}_1 = \lambda P_0 \tag{9.5}$$

or

$$P_1(t) = \int_0^t \lambda P_0(u)\, du$$

$P_1(t)$ is the probability of system failure before time t, since the diagram of Fig. 9.2 has no outflow from state 1. Hence, system unreliability $F_s(t)$ coincide with $P_1(t)$, and is given by

$$F_s(t) = P_1(t) = \int_0^t \lambda P_0(u)\,du = \int_0^t \lambda e^{-\lambda u}\,du = 1 - e^{-\lambda t} \tag{9.6}$$

which confirms (9.4). We see that the unreliability approaches unity as t gets larger. In other words, the system fails after a sufficiently large time interval.

9.2 TWO-COMPONENT SERIES SYSTEM

Consider a series system consisting of components 1 and 2. The corresponding Markov transition diagram is shown as Fig. 9.3. The system fails when it enters into state (1,0) or (0,1). Since we are considering a process ending in the first system failure, the diagram can be simplified, and Fig. 9.4 is obtained. In this simplified diagram, nodes (1,0) and (0,1) are absorption states.

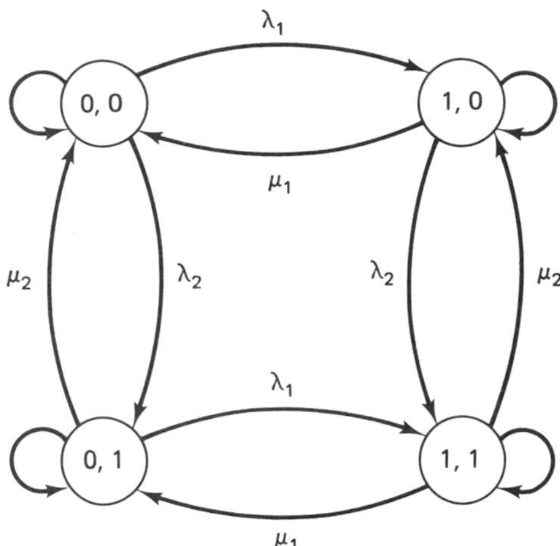

Figure 9.3. Transition diagram for a two-component system.

The differential equation for state probability $P_0(t)$ is

$$P_0 = -(\lambda_1 + \lambda_2)P_0,\ P_0(0) = 1 \tag{9.7}$$

yielding the solution

$$P_0(t) = e^{-(\lambda_1 + \lambda_2)t} \tag{9.8}$$

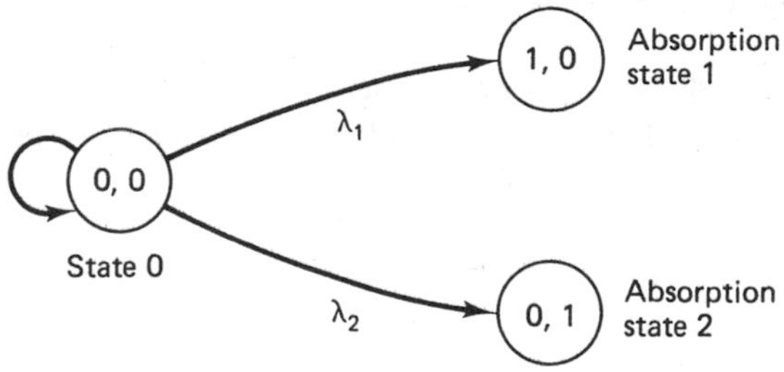

Figure 9.4. Transition diagram for reliability calculation of two-component series system.

The system reliability $R_s(t)$ is equal to $P_0(t)$:

$$R_s(t)=e^{-(\lambda_1+\lambda_2)t} \tag{9.9}$$

We see that the system reliability is a product of the component reliabilities $e^{-\lambda_1 t}$ and $e^{-\lambda_2 t}$. This property holds generally for n-component series systems. System unreliability $F_s(t)$ is

$$F_s(t)=1-R_s(t)=1-e^{-(\lambda_1+\lambda_2)t} \tag{9.10}$$

Another way of deriving (9.10) is to use differential equations for state probabilities $P_1(t)$ and $P_2(t)$:

$$\left.\begin{aligned} \dot{P}_1(t)&=\lambda_1 P_0(t)\\ \dot{P}_2(t)&=\lambda_2 P_0(t)\end{aligned}\right\}$$

This gives

$$\left.\begin{aligned} P_1(t)&=\int_0^t \lambda_1 P_0(u)\,du=\frac{\lambda_1}{\lambda_1+\lambda_2}\left(1-e^{-(\lambda_1+\lambda_2)t}\right)\\ P_2(t)&=\int_0^t \lambda_2 P_0(u)\,du=\frac{\lambda_2}{\lambda_1+\lambda_2}\left(1-e^{-(\lambda_1+\lambda_2)t}\right)\end{aligned}\right\}$$

Probability $P_1(t)$ is the probability that the system fails to time t by the first component failure. Similarly, $P_2(t)$ is the contribution to system unreliability due to the second component failure. System unreliability $F_s(t)$ is given as the sum of state probabilities $P_1(t)$ and $P_2(t)$:

$$F_s(t)=P_1(t)+P_2(t)=1-e^{-(\lambda_1+\lambda_2)t}$$

which confirms (9.10). Note that $F_s(t)$ is a sum of state probabilities over absorption states 1 and 2. In general, system unreliability is the total of all probabilities over all absorption states.

9.3 *n*-COMPONENT SERIES SYSTEM

Series systems consisting of n components are described by the transition diagram of Fig. 9.5, which is an extended version of Fig. 9.4. The resulting differential equation for state 0 is

$$\dot{P}_0 = -(\lambda_1 + \dots + \lambda_n) P_0, \; P_0(0) = 1 \tag{9.11}$$

System reliability $R_s(t)$ equals

$$R_s(t) = e^{-(\lambda_1 + \dots + \lambda_n)t} \tag{9.12}$$

We see that the reliability of the series system is the product of component reliabilities $e^{-\lambda_i t}$, $i = 1, \dots, n$. System unreliability $F_s(t)$ is

$$F_s(t) = 1 - e^{-(\lambda_1 + \dots + \lambda_n)t}. \tag{9.13}$$

Thus, the unreliability is equal to a sum of probability of absorption states $1, \dots, n$.

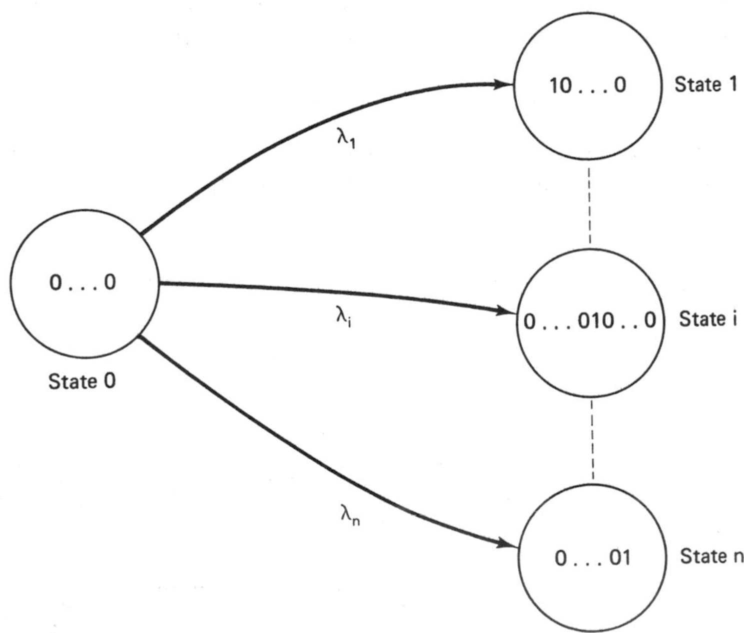

Figure 9.5. Transition diagram for reliability calculation of *n*-component series system.

9.4 PARALLEL SYSTEM

9.4.1 Two-Component Parallel System

Two-component parallel systems are described by the transition diagram shown in Fig. 9.3. Since we consider a period ending in the first system failure, the transitions from state (1, 1) can be removed, and Fig. 9.3 is reduced to Fig. 9.6. The resulting differential equations are

$$\left.\begin{aligned}\dot{P}_{0,0}&=-(\lambda_1+\lambda_2)P_{0,0}+\mu_1 P_{1,0}+\mu_2 P_{0,1}\\ \dot{P}_{1,0}&=\lambda_1 P_{0,0}-(\mu_1+\lambda_2)P_{1,0}\\ \dot{P}_{0,1}&=\lambda_2 P_{0,0}-(\mu_2+\lambda_1)P_{0,1}\\ \dot{P}_{1,1}&=\lambda_2 P_{1,0}+\lambda_1 P_{0,1}\end{aligned}\right\} \tag{9.14}$$

with the initial conditions

$$P_{0,0}(0)=1,\quad P_{1,0}(0)=P_{0,1}(0)=P_{1,1}(0)=0 \tag{9.15}$$

Note that the first three equations of (9.14) can be solved separately from the last one. The system functions as long as it is in state (0,0), (1,0), or (0,1). Thus, system reliability $R_s(t)$ is given by the sum

$$R_s(t)=P_{0,0}(t)+P_{1,0}(t)+P_{0,1}(t) \tag{9.16}$$

The state probabilities on the right-hand-side of (9.16) can be calculated by using the numerical integration method of Appendix 8.1.

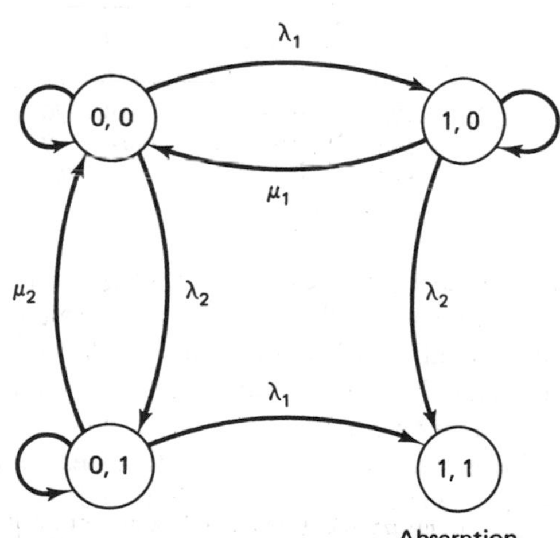

Figure 9.6. Transition diagram for reliability calculation of two-component parallel system.

System unreliability $F_s(t)$ is

$$F_s(t)=1-P_{0,0}(t)-P_{1,0}(t)-P_{0,1}(t) \tag{9.17}$$

This coincides with $P_{1,1}(t)$, the probability for absorption state (1, 1), since the total of all state probabilities is unity.

$$F_s(t)=P_{1,1}(t) \tag{9.18}$$

Example 1. Two-Component Parallel System

Assume the following failure and repair rates (hour^{-1}) for components 1 and 2.

Component 1	Component 2
$\lambda_1=1/1000$	$\lambda_2=2/1000$
$\mu_1=1/10$	$\mu_2=1/40$

Calculate $R_s(t)$ and $F_s(t)$ at $t=100$, 500, and 1000 hr, and compare the values with those of $Q_s(t)$, and $A_s(t)$.

Solution: The first three equations of (9.14) become

$$\dot{P}_{0,0}=-0.003P_{0,0}+0.1P_{1,0}+0.025P_{0,1}$$
$$\dot{P}_{1,0}=0.001P_{0,0}-0.102P_{1,0}$$
$$\dot{P}_{0,1}=0.002P_{0,0}-0.026P_{0,1}$$

By numerical integration, equations (9.16) and (9.17) yield

t	$R_s(t)$	$F_s(t)$	$A_s(t)$	$Q_s(t)$
100	0.993630	0.006370	0.999316	0.000684
500	0.959001	0.040999	0.999267	0.000733
1000	0.917232	0.082768	0.999267	0.000733

System availabilities $A_s(t)$ and unavailabilities $Q_s(t)$ listed above were calculated using the method of Example 17, Chapter 7. We observe that the reliability is less than the availability, and unreliability is larger than unavailability. Also, note that $F_s(t)$ continually increases, whereas $Q_s(t)$ approaches steady state.

When failure rates and repair rates are the same for all components, the system reliability can be calculated analytically by using the method of Appendix 9.1 or Laplace transformations. This is illustrated by the following example.

Example 2. Two-Component Parallel System with Common Rates

Assume a failure rate $\lambda=2$ (year^{-1}) and a repair rate $\mu=3$ (year^{-1}) for each component. Obtain analytical expressions for $R_s(t)$ and $F_s(t)$.

Solution: Since the two components have common failure and repair rates, the transition diagram of Fig. 9.6 simplifies to the one of Fig. 9.7. The corresponding differential equations are

$$\left.\begin{aligned}\dot{P}_0&=-2\lambda P_0+\mu P_1\\ \dot{P}_1&=2\lambda P_0-(\lambda+\mu)P_1\\ \dot{P}_2&=\lambda P_1\end{aligned}\right\}\tag{9.19}$$

Substituting $\lambda=2$ and $\mu=3$ into the first two equations of (9.19), we obtain:

$$\begin{pmatrix}\dot{P}_0\\ \dot{P}_1\end{pmatrix}=\begin{pmatrix}-4 & 3\\ 4 & -5\end{pmatrix}\begin{pmatrix}P_0\\ P_1\end{pmatrix}\tag{9.20}$$

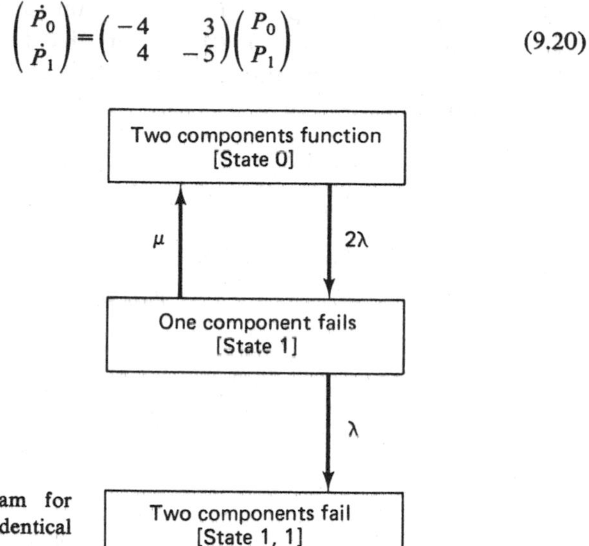

Figure 9.7. Transition diagram for parallel system with two identical components.

Equation (9.20) is identical to equation (10) in the example problem of Appendix 9.1. The solution is given as

$$P_0(t)=\tfrac{4}{7}e^{-t}+\tfrac{3}{7}e^{-8t}$$

$$P_1(t)=\tfrac{4}{7}e^{-t}-\tfrac{4}{7}e^{-8t}$$

Hence, system reliability $R_s(t)$ is

$$R_s(t)=P_0(t)+P_1(t)=\tfrac{8}{7}e^{-t}-\tfrac{1}{7}e^{-8t}$$

and system unreliability is

$$F_s(t) = 1 - \tfrac{8}{7}e^{-t} + \tfrac{1}{7}e^{-8t}$$

9.4.2 Parallel System with More Than Two Components

When a parallel system consists of n components, the corresponding transition diagram has 2^n states. The number of states increases exponentially with n, and it becomes difficult to calculate $R_s(t)$ or $F_s(t)$ for large values of n. Monte Carlo methods, which are discussed in Chapter 12, provide a feasible approach in this case.

9.5 TWO-OUT-OF-THREE SYSTEM

The transition diagram for a two-out-of-three system is shown in Fig. 9.8. State (1, 1, 1) does not appear in the diagram, since the first system failure occurs at either (1, 1, 0) or (0, 1, 1) or (1, 0, 1). These three states are absorption states for the transition diagram. Differential equations for the non-absorption states are

$$\left.\begin{aligned}
\dot{P}_{000} &= -(\lambda_1+\lambda_2+\lambda_3)P_{000} + \mu_1 P_{100} + \mu_2 P_{010} + \mu_3 P_{001} \\
\dot{P}_{100} &= \lambda_1 P_{000} - (\mu_1+\lambda_2+\lambda_3)P_{100} \\
\dot{P}_{010} &= \lambda_2 P_{000} - (\lambda_1+\mu_2+\lambda_3)P_{010} \\
\dot{P}_{001} &= \lambda_3 P_{000} - (\lambda_1+\lambda_2+\mu_3)P_{001}
\end{aligned}\right\} \quad (9.21)$$

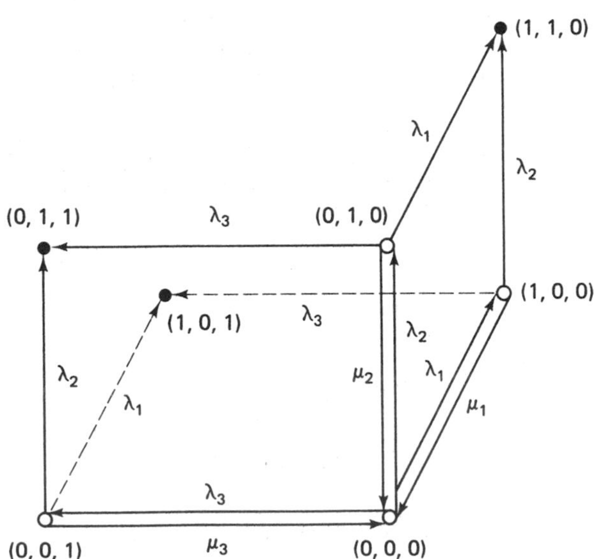

Figure 9.8. Transition diagram for reliability calculation of two-out-of-three system.

The initial conditions for (9.21) are,

$$P_{000}(0)=1, \quad P_{100}(0)=P_{010}(0)=P_{001}(0)=0 \tag{9.22}$$

The differential equations of set (9.21) can be solved numerically, by the method of Appendix 8.1. Reliability $R_s(t)$ is given by the sum of state probabilities for non-absorption states:

$$R_s(t)=P_{000}(t)+P_{100}(t)+P_{010}(t)+P_{001}(t) \tag{9.23}$$

Example 3. Two-Out-of-Three System

Assume the following failure and repair rates (hour^{-1}).

Component 1	Component 2	Component 3
$\lambda_1=1/1000$ $\mu_1=1/10$	$\lambda_2=2/1000$ $\mu_2=1/40$	$\lambda_3=3/1000$ $\mu_3=1/60$

Calculate the reliability $R_s(t)$ at $t=100$, 500, and 1000 hr.

Solution: The numerical integration method of Appendix 8.1 produced the results listed below:

t	P_{000}	P_{100}	P_{010}	P_{001}	$R_s(t)$
100	0.78713	0.00758	0.05354	0.10905	0.95730
500	0.59762	0.00573	0.04216	0.09427	0.73978
1000	0.43202	0.00414	0.03048	0.06815	0.53479

When the three components have common failure and repair rates, $R_s(t)$ can be obtained analytically in the same way as in Example 2. The results are[1]

$$R_s(t)=\frac{1}{k_2-k_1}[k_2\exp(-k_1\lambda t)-k_1\exp(-k_2\lambda t)] \tag{9.24}$$

where

$$b=\mu/\lambda,$$

$$2k_1=(b+5)-\sqrt{b^2+10b+1},$$

$$2k_2=(b+5)+\sqrt{b^2+10b+1}.$$

[1]Halperin, M. "Some Waiting Time Distributions for Redundant Systems with Repair," *Technometrics*, Vol. 6, pp. 27–40, 1964.

9.6 UPPER AND LOWER BOUNDS FOR SYSTEM UNRELIABILITY

For large systems it is difficult to obtain $F_s(t)$ [or $R_s(t)$] because the transition diagrams have large numbers of states. From an engineering point of view, therefore, it is desirable to obtain bounds for $F_s(t)$ to simplify the calculations. In this section we present an approach based on the inclusion-exclusion principle.

Assume that a fault tree has minimal cut sets $K_1,\dots,K_n$. Define

$$K_i = 1 \Leftrightarrow \text{cut set } K_i \text{ fails before time } t$$

$$K_i \equiv 0 \Leftrightarrow \text{cut set } K_i \text{ does not fail before time } t$$

The inclusion-exclusion principle gives,

$$\begin{aligned} F_s(t) &= \Pr(K_1=1 \cup K_2=1 \cup \dots \cup K_n=1) \\ &= \sum_{i=1}^{n} \Pr(K_i=1) - \sum_{i=2}^{n} \sum_{j=1}^{i-1} \Pr(K_i=1 \cap K_j=1) \\ &\quad + \sum_{i=3}^{n} \sum_{j=2}^{i-1} \sum_{k=1}^{j-1} \Pr(K_i=1 \cap K_j=1 \cap K_k=1) + \dots \\ &\quad + (-1)^{n-1} \Pr(K_1=1 \cap K_2=1 \cap \dots \cap K_n=1) \end{aligned} \tag{9.25}$$

Upper and lower bounds for $F_s(t)$ are

$$\begin{aligned} F_s(t)_{\max} &= \text{first bracket} \\ &= \sum_{i=1}^{n} \Pr(K_i=1) \end{aligned} \tag{9.26}$$

$$\begin{aligned} F_s(t)_{\min} &= F_s(t)_{\max} - \text{second bracket} \\ &= F_s(t)_{\max} - \sum_{i=2}^{n} \sum_{j=1}^{i-1} \Pr(K_i=1 \cap K_j=1) \end{aligned} \tag{9.27}$$

Three-or-more-event cut sets seldom fail simultaneously, and their contribution to the system unreliability is much smaller than that of the two- or one-event cut sets. Therefore, we remove these higher-event cut sets from (9.26) and (9.27), and assume that each cut set consists of, at most, two basic events.

A one-event cut set is regarded as a one-component system and two-event cut set as a two-component, parallel system. Thus, probability $\Pr(K_i=1)$ on the right-hand side of (9.26) can be calculated by the methods of Sections 9.1 or 9.4 of this chapter.

Consider now $\Pr(K_i=1\cap K_j=1)$ in (9.27). If either K_i or K_j is a one-event cut set, then the two cut sets have no common events. Hence, because of statistical independence,

$$\Pr(K_i=1\cap K_j=1)=\Pr(K_i=1)\Pr(K_j=1) \tag{9.28}$$

Equation (9.28) holds also for two-event cut sets K_i and K_j if they have no common events. Consider, therefore, the case where K_i and K_j have a common event b:

$$K_i=\{a,b\},\quad K_j=\{b,c\} \tag{9.29}$$

The following identity holds:

$$\begin{aligned}
&\Pr(K_i=1\cap K_j=1)\\
&=1-\Pr(K_i\equiv 0\cup K_j\equiv 0)\\
&=1-\left[\Pr(K_i\equiv 0)+\Pr(K_j\equiv 0)-\Pr(K_i\equiv 0\cap K_j\equiv 0)\right]
\end{aligned} \tag{9.30}$$

Probabilities $\Pr(K_i\equiv 0)$ and $\Pr(K_j\equiv 0)$ can be calculated by

$$\Pr(K_i\equiv 0)=1-\Pr(K_i=1)$$

$$\Pr(K_j\equiv 0)=1-\Pr(K_j=1)$$

Event $[K_i\equiv 0\cap K_j\equiv 0]$ occurs if and only if component state vector (a,b,c) remains in the set

$$\{(0,0,0),\quad (1,0,0),\quad (0,1,0),\quad (0,0,1),\quad (1,0,1)\}$$

Each state in the set is designated by a circle in the transition diagram of Fig. 9.9, and absorption states are denoted by black disks. The differential equations for the non-absorption states are

$$\left.\begin{aligned}
\dot{P}_{000}&=-(\lambda_1+\lambda_2+\lambda_3)P_{000}+\mu_1P_{100}+\mu_2P_{010}+\mu_3P_{001}\\
\dot{P}_{100}&=\lambda_1P_{000}-(\mu_1+\lambda_2+\lambda_3)P_{100}+\mu_3P_{101}\\
\dot{P}_{010}&=\lambda_2P_{000}-(\lambda_1+\mu_2+\lambda_3)P_{010}\\
\dot{P}_{001}&=\lambda_3P_{000}-(\lambda_1+\lambda_2+\mu_3)P_{001}+\mu_1P_{101}\\
\dot{P}_{101}&=\lambda_1P_{001}+\lambda_3P_{100}-(\mu_1+\lambda_2+\mu_3)P_{101}
\end{aligned}\right\} \tag{9.31}$$

with the initial condition

$$P_{000}(0)=1,\quad P_{100}(0)=P_{010}(0)=P_{001}(0)=P_{101}(0)=0 \tag{9.32}$$

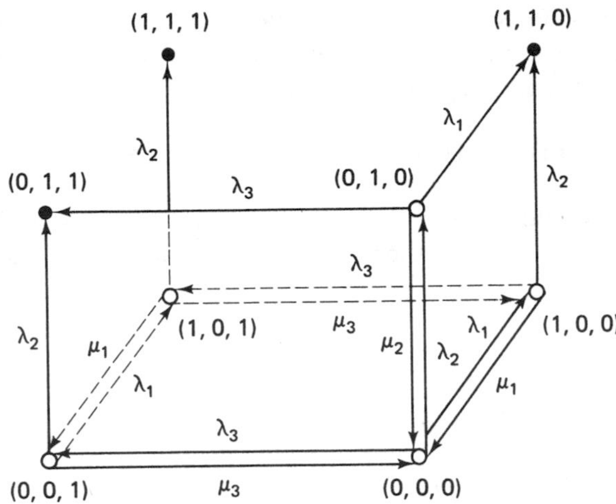

Figure 9.9. Transition diagram for event $K_i \equiv 0 \cap K_j \equiv 0$.

These differential equations can be solved by the method of Appendix 8.1, and $\Pr(K_i \equiv 0 \cap K_j \equiv 0)$ is the sum

$$\Pr\left(K_i \equiv 0 \cap K_j \equiv 0\right) = P_{000} + P_{100} + P_{010} + P_{001} + P_{101} \tag{9.33}$$

Equation (9.30) gives $\Pr(K_i = 1 \cap K_j = 1)$, and lower bound $F_s(t)_{\min}$ can be calculated by (9.27).

We note here that the upper bound $F_s(t)_{\max}$ of (9.26) can be improved by[2]

$$F_s(t)_{\text{upper}} = 1 - \prod_{i=1}^{n} \left[1 - \Pr(K_i = 1)\right] \tag{9.34}$$

Example 4. Two-Out-of-Three System

Consider the system of Example 3. Calculate $F_s(t)_{\max}$, $F_s(t)_{\text{upper}}$, and $F_s(t)_{\min}$ at $t = 100$, 500, and 1000 hr.

Solution: The minimal cut sets are

$$K_1 = \{1,2\}, \quad K_2 = \{2,3\}, \quad K_3 = \{1,3\}$$

For each cut set we obtain the differential equations shown in (9.14). Their solution

[2]Esary, J. D., and F. Proschan, "A Reliability Bound for Systems of Maintained, Interdependent Components," *Journal of the American Statistical Association*, Vol. 65, No. 329, pp. 329–338, 1970.

by (9.17) give:

t	$\Pr(K_1=1)$	$\Pr(K_2=1)$	$\Pr(K_3=1)$
100	0.006370	0.028260	0.010684
500	0.040999	0.184606	0.074842
1000	0.082768	0.346314	0.150170

Hence, equations (9.26), (9.27), and (9.34) give:

t	$F_s(t)_{max}$	$F_s(t)_{upper}$	$F_s(t)_{min}$	$F_s(t)$
100	0.045314	0.044766	0.041954	0.042710
500	0.300446	0.276560	0.252745	0.260222
1000	0.579253	0.490457	0.446733	0.465213

Exact unreliabilities $F_s(t)$ are listed for comparison.

9.7 HAND CALCULATION OF UPPER BOUNDS FOR SYSTEM UNRELIABILITY

Upper bound $F_s(t)_{max}$ in (9.26) requires the calculation of $\Pr(K_i=1)$. If K_i is a one-event cut set, we can readily calculate $\Pr(K_i=1)$.

$$\Pr(K_i=1)=1-e^{-\lambda t} \tag{9.35}$$

If, on the other hand, K_i is a two-event cut set, we must solve differential equation (9.14). This equation cannot be solved analytically; digital computers must be used. In this section, we present a hand-calculation method for (9.14) and provide a practical approach for obtaining $F_s(t)_{max}$.

Assume that cut set K_i consists of two components, 1 and 2. Differential equations (9.14) are based on the transition diagram of Fig. 9.6. Let us assume for each component that MTTR is much smaller than MTTF:

$$\frac{1}{\lambda_1} \gg \frac{1}{\mu_1}, \quad \frac{1}{\lambda_2} \gg \frac{1}{\mu_2} \tag{9.36}$$

Then, the transition from state (1,0) to state (0,0) is more likely to occur than the transition from (1,0) to (1,1), and we can neglect the latter transition. Similarly, the transition from state (0,1) to state (1,1) can be removed from the diagram of Fig. 9.6. The resulting diagram is shown as

Fig. 9.10, and that yields the following differential equations for the state probabilities.

$$\dot{P}_{00} = -(\lambda_1 + \lambda_2)P_{00} + \mu_1 P_{10} + \mu_2 P_{01} \tag{9.37}$$

$$\dot{P}_{10} = \lambda_1 P_{00} - \mu_1 P_{10} \tag{9.38}$$

$$\dot{P}_{01} = \lambda_2 P_{00} - \mu_2 P_{01} \tag{9.39}$$

These equations approximate the first three equations of (9.14).

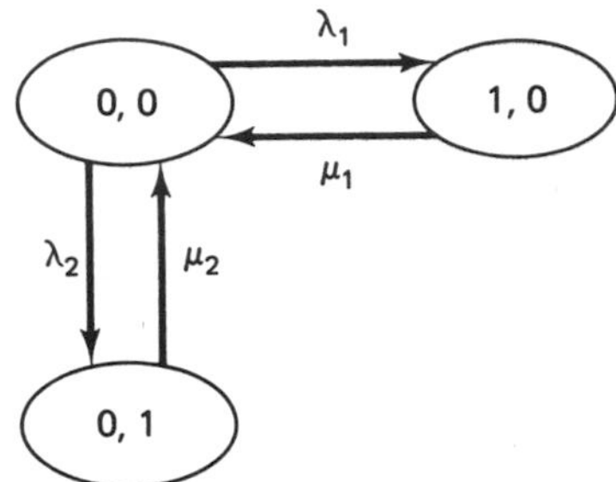

Figure 9.10. Transition diagram simplified by $1/\lambda_1 \gg 1/\mu_1$ and $1/\lambda_2 \gg 1/\mu_2$.

We note that the sum of all probabilities is unity:

$$P_{00} + P_{10} + P_{01} + P_{11} = 1. \tag{9.40}$$

Assume a time interval wherein each component remains sufficiently reliable. Then, probabilities P_{10}, P_{01}, and P_{11} are far smaller than P_{00}, and (9.40) becomes

$$P_{00} + P_{10} = 1 \tag{9.41}$$

$$P_{00} + P_{01} = 1 \tag{9.42}$$

Substitute P_{00} of (9.41) into (9.38). Then,

$$\dot{P}_{10} = \lambda_1 - (\lambda_1 + \mu_1)P_{10} \tag{9.43}$$

The initial condition of (9.15) gives

$$P_{10}(t) = \frac{\lambda_1}{\lambda_1 + \mu_1}\left(1 - e^{-(\lambda_1 + \mu_1)t}\right) \tag{9.44}$$

Similarly, equations (9.42) and (9.39) yield

$$P_{01}(t) = \frac{\mu_2}{\lambda_2 + \mu_2}\left(1 - e^{-(\lambda_2 + \mu_2)t}\right) \tag{9.45}$$

Equations (9.44) and (9.45) approximate $P_{10}(t)$ and $P_{01}(t)$ in (9.14).

Probability $\Pr(K_i=1)$ is equal to probability $P_{11}(t)$ in (9.14), since state (1, 1) is an absorption state for cut set K_1. Thus,

$$\Pr(K_i=1)=\int_0^t [\lambda_2 P_{10}(u)+\lambda_1 P_{01}(u)]\,du \tag{9.46}$$

Substitute (9.44) and (9.45) into (9.46). Then, $\Pr(K_i=1)$ can be calculated by

$$\begin{aligned}\Pr(K_i=1)=&\frac{\lambda_1\lambda_2}{\lambda_1+\mu_1}\left(t+\frac{1}{\lambda_1+\mu_1}e^{-(\lambda_1+\mu_1)t}-\frac{1}{\lambda_1+\mu_1}\right)\\&+\frac{\lambda_1\lambda_2}{\lambda_2+\mu_2}\left(t+\frac{1}{\lambda_2+\mu_2}e^{-(\lambda_2+\mu_2)t}-\frac{1}{\lambda_2+\mu_2}\right)\end{aligned} \tag{9.47}$$

Unreliability upper bound $F_s(t)_{\max}$ is obtained from (9.26) when $\Pr(K_i=1)$ is given. This upper bound is denoted by $F_s(t)_{\max,\text{ hand}}$.

Assume sufficiently small values of t. More precisely, let

$$(\lambda_i+\mu_i)t\le \tfrac{1}{10} \tag{9.48}$$

Equations (9.35) and (9.47) can be approximated by a Taylor series expansion of the exponential function.

$$\Pr(K_i=1)=\lambda t,\quad \text{for one-event cut set } K_i \tag{9.49}$$

$$\Pr(K_i=1)=\lambda_1\lambda_2 t^2,\quad \text{for two-event cut set } K_i \tag{9.50}$$

The resulting upper bound is designated as $F_s(t)_{\max,\text{ Taylor}}$.

Example 5. Two-Out-of-Three System

Consider the system of Example 3. Calculate $F_s(t)_{\max,\text{ hand}}$ and $F_s(t)_{\max,\text{ Taylor}}$ at $t=10$, 100, 200, 500, and 1000 hr.

Solution: The results are listed below. Exact upper bound $F_s(t)_{\max}$ and exact unreliabilities $F_s(t)$ are also given for comparison.

t	$F_s(t)_{\max,\text{ hand}}$	$F_s(t)_{\max,\text{ Taylor}}$	$F_s(t)_{\max}$	$F_s(t)$
10	0.000959	0.001100	0.000958	0.000949
100	0.046707	0.110000	0.045314	0.042710
200	0.117595	0.440000	0.110727	0.101381
500	0.338770	2.750000	0.300446	0.260222
1000	0.708114	11.00000	0.579253	0.465213

We see that $F_s(t)_{\max,\text{ hand}}$ gives good results up to 200 hr, whereas $F_s(t)_{\max,\text{ Taylor}}$ remains valid in a small interval around initial time zero.

APPENDIX 9.1 METHOD FOR SOLVING LINEAR DIFFERENTIAL EQUATIONS

Consider the differential equation

$$\dot{\mathbf{x}} = \mathbf{A}\mathbf{x}, \quad \mathbf{x}(0) = \mathbf{x}_0 \tag{1}$$

where $\mathbf{x}$ is an n-dimensional column vector and $\mathbf{A}$ is an $n \times n$ matrix. The differential equation can be solved by the steps stated below.

Step 1. Obtain eigenvalues $\gamma_1, \ldots, \gamma_n$ for matrix $\mathbf{A}$ by solving the characteristic equation

$$|\mathbf{A} - \gamma\mathbf{I}| = 0, \quad \mathbf{I}\text{: identity matrix} \tag{2}$$

We assume that the eigenvalues are distinct.

Step 2. Obtain eigenvector $\mathbf{v}_i$ for each eigenvalue γ_i by solving the linear equation

$$(\mathbf{A} - \gamma_i\mathbf{I})\mathbf{v}_i = 0 \tag{3}$$

Step 3. Construct the $n \times n$ matrix $\mathbf{T}$ defined by

$$\mathbf{T} = \{\mathbf{v}_1, \ldots, \mathbf{v}_n) \tag{4}$$

Obtain the inverse matrix $\mathbf{T}^{-1}$ of $\mathbf{T}$.

Step 4. Define the linear transformation from $\mathbf{x}$ to $\mathbf{x}^*$ by

$$\mathbf{x} = \mathbf{T}\mathbf{x}^* \tag{5}$$

Substitute (5) into (1). Then,

$$\dot{\mathbf{x}}^* = \mathbf{T}\mathbf{x}^*, \quad \mathbf{x}^*(0) = \mathbf{T}^{-1}\mathbf{x}_0 \tag{6}$$

where

$$\mathbf{T} \equiv \mathbf{T}^{-1}\mathbf{A}\mathbf{T} = \begin{bmatrix} \gamma_1 & & \mathbf{0} & \\ & \gamma_2 & & \\ \mathbf{0} & & \vdots & \\ & & & \gamma_n \end{bmatrix} : \text{diagonal matrix} \tag{7}$$

Step 5. Solve (6). Then,

$$\mathbf{x}_i^* = x_i^*(0) \exp[\gamma_i t] \tag{8}$$

where $x_i^*(0)$ is the ith component of $\mathbf{x}^*(0)$.

Step 6. Obtain $\mathbf{x}(t)$ from

$$\mathbf{x}(t)=\mathbf{T}\mathbf{x}^*(t) \tag{9}$$

Example. Solve the differential equation

$$\dot{\mathbf{x}}=\begin{pmatrix}-4 & 3\\ 4 & -5\end{pmatrix}\mathbf{x}, \quad \mathbf{x}(0)=\begin{pmatrix}1\\0\end{pmatrix} \tag{10}$$

Step 1. The characteristic equation is

$$\begin{vmatrix}-4-\gamma & 3\\ 4 & -5-\gamma\end{vmatrix}=(\gamma+4)(\gamma+5)-12=(\gamma+1)(\gamma+8)=0$$

Thus,

$$\gamma_1=-1, \quad \gamma_2=-8$$

Step 2. For $\gamma_1=-1$, equation (3) becomes

$$\begin{pmatrix}-3 & 3\\ 4 & -4\end{pmatrix}\begin{pmatrix}v_1\\v_2\end{pmatrix}=\begin{pmatrix}0\\0\end{pmatrix}$$

or

$$v_1=v_2$$

Hence, we have an eigenvector for γ_1:

$$\mathbf{v}_1=\begin{pmatrix}v_1\\v_2\end{pmatrix}=\begin{pmatrix}1\\1\end{pmatrix}$$

For $\gamma_2=-8$, equation (3) is

$$\begin{pmatrix}4 & 3\\ 4 & 3\end{pmatrix}\begin{pmatrix}v_1\\v_2\end{pmatrix}=\begin{pmatrix}0\\0\end{pmatrix}$$

or

$$4v_1+3v_2=0$$

Thus, we have an eigenvector for γ_2:

$$\mathbf{v}_2=\begin{pmatrix}3\\-4\end{pmatrix}$$

Step 3. Matrix **T** is given by

$$\mathbf{T}=\{\mathbf{v}_1,\mathbf{v}_2\}=\begin{pmatrix}1 & 3\\ 1 & -4\end{pmatrix}$$

The inverse matrix of **T** is

$$\mathbf{T}^{-1}=\begin{bmatrix}\frac{4}{7} & \frac{3}{7}\\ \frac{1}{7} & -\frac{1}{7}\end{bmatrix}$$

Step 4. Equation (6) is given by

$$\begin{pmatrix}\dot{x}_1^*\\ \dot{x}_2^*\end{pmatrix}=\begin{pmatrix}-1 & 0\\ 0 & -8\end{pmatrix}\begin{pmatrix}x_1^*\\ x_2^*\end{pmatrix}$$

and

$$\begin{pmatrix}x_1^*(0)\\ x_2^*(0)\end{pmatrix}=\begin{bmatrix}\frac{4}{7} & \frac{3}{7}\\ \frac{1}{7} & -\frac{1}{7}\end{bmatrix}\begin{pmatrix}1\\ 0\end{pmatrix}=\begin{bmatrix}\frac{4}{7}\\ \frac{1}{7}\end{bmatrix}$$

Step 5. The differential equation in step 4 has the solution

$$x_1^*(t)=\tfrac{4}{7}\exp[-t]$$
$$x_2^*(t)=\tfrac{1}{7}\exp[-8t]$$

Step 6. Equation (9) gives

$$\begin{pmatrix}x_1(t)\\ x_2(t)\end{pmatrix}=\begin{pmatrix}1 & 3\\ 1 & -4\end{pmatrix}\begin{pmatrix}x_1^*(t)\\ x_2^*(t)\end{pmatrix}=\begin{bmatrix}\frac{4}{7}e^{-t}+\frac{3}{7}e^{-8t}\\ \frac{4}{7}e^{-t}-\frac{4}{7}e^{-8t}\end{bmatrix} \tag{11}$$

When vector **x** has three or more columns, standard computer programs can be used for calculating eigenvalues, eigenvectors, and the inverse matrix $\mathbf{T}^{-1}$.

PROBLEMS

9.1. A system has three states:

State 1: Normal state
State 2: Partial degradation of the system
State 3: Complete degradation of the system.

A catastrophic hazard occurs if the system falls into state 3. The system is non-repairable and degrades consecutively. The transition rates from state 1 to 2 and from 2 to 3 are λ_1 and λ_2, respectively.
(a) Obtain a Markov transition diagram for the system.
(b) Calculate the system unreliability at $t=2$ years, using as data,

$$\lambda_1 = 10^{-1}/\text{year}, \quad \lambda_2 = 10^{-3}/\text{year}$$

9.2. Calculate the reliability of a two-component series system at $t=100$ days, using failure and repair rates, $\lambda_1=\lambda_2=0.001$ (hour^{-1}) and $\mu_1=\mu_2=0.1$ (hour^{-1}).

9.3. Calculate the reliability at $t=1$ year of a component with the rate of $\lambda=0.05$. Repeat the calculation for a 100-component series system consisting of the components.

9.4. Consider a series system consisting of n identical components. Prove that the system reliability at 1 year equals the component reliability at n years.

9.5. (a) Rewrite the differential equation of Example 1 of this chapter in the matrix form for equation (1) of Appendix 8.1.
(b) Obtain the fourth-order approximation of $\exp[/\mathbf{A}\Delta]$, where $\Delta=0.1$.
(c) Calculate $\mathbf{P}(0.1)$ and $\mathbf{P}(0.2)$, using the approximation of part (b).

9.6. (a) Calculate the conditional failure intensity $\lambda_s(t)$ for the parallel system of Example 2, using the KITT calculation in Chapter 8.
(b) Calculate $\overline{R}_s(t)=\exp[-\int_0^t\lambda_s(u)\,du]$. This corresponds to (7.133) in Chapter 7 and should differ from $R_s(t)$. Obtain graphs for $R_s(t)$ and $\overline{R}_s(t)$ for $0\le t\le 1.2$.

9.7. A two-out-of-three voting system consists of identical components with the failure and repair rates of

$$\lambda=0.001, \quad \mu=0.1$$

Obtain the system reliability at $t=100$.

9.8. Write a differential equation for calculating $\Pr(K_1\equiv 0\cap K_2\equiv 0)$, $\Pr(K_2\equiv 0\cap K_3\equiv 0)$, and $\Pr(K_1\equiv 0\cap K_3\equiv 0)$ of Example 4.

9.9. Consider a two-out-of-four voting system consisting of the failure and the repair rates of

$$\lambda=0.001, \quad \mu=0.1$$

(a) Hand-calculate the upper bound of the system unreliability at $t=100$.
(b) Determine the differential equation for calculating the exact system unreliability.
(c) Integrate the differential equation by a computer, using the algorithm in Appendix 8.1. Here, approximate $\exp[\mathbf{A}\Delta]$ by the fourth-order Taylor expansion with Δ being equal to 1.0. Compare the resulting $F_s(100)$ with the $F_s(100)_{\max}$.

10

IMPORTANCE

10.1 INTRODUCTION

A component or cut set's contribution to the top-event occurrence is termed its *importance*. It is a function of time, of failure and repair characteristics, and system structure. An importance analysis is akin to a *sensitivity analysis* and thus useful for system design, diagnosis, and optimization. For example, we can estimate possible variations of system availability caused by uncertainties in component reliability parameters. Inspection, maintainance, and failure detection can be carried out in their order of importance for components, and systems upgraded by improving components with relatively large importances.

A number of interesting and imaginative applications of importance concepts for improving system reliability, diagnosing failures, and generating repair checklists are developed by Dr. Howard Lambert in UCRL 51829, which also contains the IMPORTANCE code discussed in this chapter. The importance measures calculated by the code, their meaning, and the probabilistic expressions are shown in Table 10.1. Figure 10.1 is a flow sheet of the calculations performed.

We shall try to lead the reader through this maze by hand-calculating each of the importance measures for the same three systems used throughout this chapter: a two-component series, a two-component parallel, and a two-out-of-three voting system. These problems are also solved

TABLE 10.1 SUMMARY OF IMPORTANCE MEASURES

Symbol	Importance measure	Probabilistic Expression	Meaning
Δg_i	*Birnbaum* basic event importance	$g(1_i, \mathbf{Q}(t)) - g(0_i, \mathbf{Q}(t))$	Probability that the system is in a state in which the occurrence of event i is critical.
I_i^{CR}	*Criticality* basic event importance	$\dfrac{[g(1_i, \mathbf{Q}(t) - g(0_i, \mathbf{Q}(t))]Q_i(t)}{g(\mathbf{Q}(t))}$	The probability that event i has occurred and is critical to system failure.[a]
I_i^{FV}	*Fussell-Vesely* basic event importance	$\dfrac{g_i(\mathbf{Q}(t))}{g(\mathbf{Q}(t))}$	Probability that event i is contributing to system failure.[a]
I_i^{UF}	*Upgrading function* basic event importance	$\dfrac{\lambda_i(t)}{\partial(\lambda_i(t))} \times \dfrac{\partial g(\mathbf{Q}(t))}{g(\mathbf{Q}(t))}$	Fractional reduction in the probability of the top event when $\lambda_i(t)$ is reduced fractionally
I_i^{BP}	*Barlow-Proschan* basic event importance	$\int_0^t \{g(1_i, \mathbf{Q}(t)) - g(0_i, \mathbf{Q}(t))\} w_i(t)\,dt$	Expected number of failures caused by basic event i in $[0,t]$.
I_i^{SC}	*Contributory sequential* basic event importance	$\sum_j \int_0^t \{g(1_i, 1_j, \mathbf{Q}(t)) - g(1_i, 0_j, \mathbf{Q}(t))\} Q_i(t) w_j\,dt$ $j \neq i$ and j and $j \varepsilon^k$ for some l	The expected number of system failures in $[0,t]$ caused by min cut sets that contain basic event i with basic event i occurring prior to system failure.
$I^{*}{}_i^{\mathrm{VF}}$	*Fussell-Vesely* cut set importance	$\dfrac{Q_i^*(t)}{\mathbf{Q}(t)}$	Probability that min cut set K_i is contributing to system failure.[a]
$I_i^{*\mathrm{BP}}$	*Barlow-Proschan* cut set importance	$\sum_{i \in j} \int_0^t [1 - g(0_i, 1^{j-\{i\}}, \mathbf{Q}(t))] \prod_{\substack{l \neq i \\ l \in j}} Q_l(t)\,dw_i(t)\,dt\,\mathbf{Q}(t)$	Expected number of system failures caused by min cut set K_j.

[a]Given that system failure occurred.

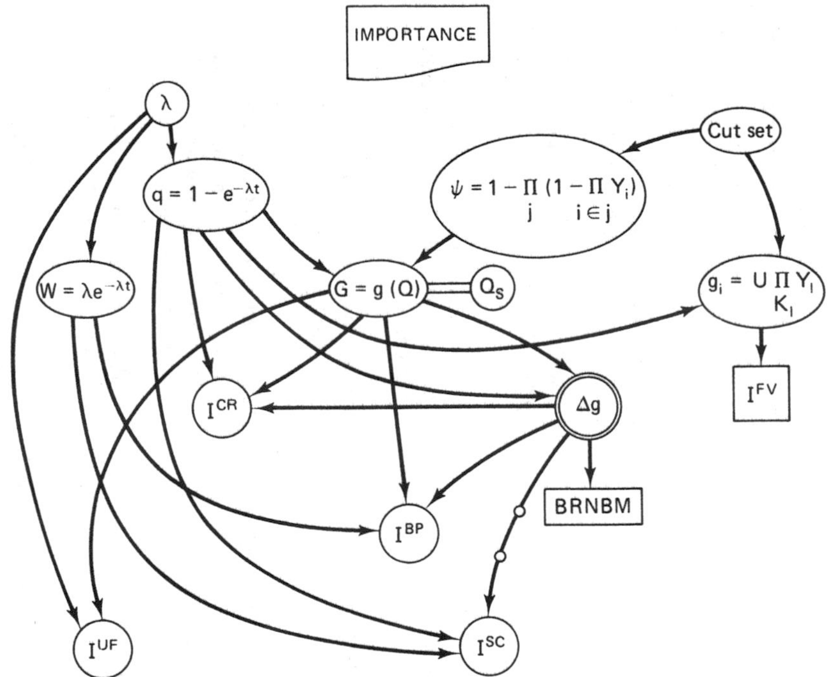

Figure 10.1. Flow sheet for IMPORTANCE calculation.

by using the IMPORTANCE code, with some interesting and, at times, surprising discrepancies. One general point which bears mentioning is an output called "probability of top event" which the code outputs. This is exactly equal to KITT's non-repairable Q_s if the components are not repairable, or the steady-state Q_s for repairable systems. The code assumes statistical independence of components, the unavailability being calculated from renewal theory which assumes that the time to failure and time to repair are exponential random variables.

Another general point which, as we shall see later, leads to difficulties is the structural representation of fault trees used in the code. In Section 7.4 of Chapter 7 we see that for a (series) two-event OR gate, this representation is

$$\psi(\mathbf{Y}) = Y_1 \vee Y_2 = 1 - (1 - Y_1)(1 - Y_2) \tag{10.1}$$

For a two-event AND gate (parallel structure) it is

$$\psi(\mathbf{Y}) = Y_1 \wedge Y_2 = Y_1 Y_2 \tag{10.2}$$

For a two-out-of-three voting system

$$\psi(\mathbf{Y}) = (Y_1 \wedge Y_2) \vee (Y_2 \wedge Y_3) \vee (Y_3 \wedge Y_1) \tag{10.3}$$

$$= 1 - [1 - Y_1 Y_2][1 - Y_2 Y_3][1 - Y_3 Y_1] \tag{10.4}$$

We shall return to this point shortly.

The nomenclature of Table 10.1 represents a slight elaboration of what we used previously:

$g[\mathbf{Q}]$ = a function that computes the probability of the top event in terms of the vector of basic event probabilities $\mathbf{Q}$

Thus,

$$E[\psi(\mathbf{Y})] = \Pr[\psi(\mathbf{Y}) = 1] = g[\mathbf{Q}]$$

are all ways of expressing the probability of top failure. We assume that the structure function ψ is coherent.

As before, we use Q_s for system unavailability and Q_i^* for cut set unavailability. For non-repairable systems, Q_s coincides with system unreliability. We assume all basic events (or component failures) to be independent.

10.2 BIRNBAUM'S STRUCTURAL IMPORTANCE, $\Delta g_i(t)$

This is the simplest of the importance criteria and is merely the partial derivative (the classical sensitivity) of the system Q_s with respect to a component's Q_i. As illustrated by (7.48) in Chapter 7, the system unavailability function $g[\mathbf{Q}]$ is a multiple linear function of $\mathbf{Q}$. Thus, the partial derivative is

$$\frac{\partial g(\mathbf{Q}(t))}{\partial \mathbf{Q}_i(t)} = g(1_i, \mathbf{Q}(t)) - g(0_i, \mathbf{Q}(t)) \tag{10.5}$$

$$\equiv \Delta g_i(t) \tag{10.6}$$

Mathematically, *structural importance* is the change in unavailability of the system with component i failed, minus the unavailability of the system with component i functioning. Another way of interpreting this importance is to rewrite (10.5) as

$$\Delta g_i(t) = E[\psi(1_i, \mathbf{y}(t)) - \psi(0_i, \mathbf{y}(t))]$$

$$= 1 \times \Pr[\psi(1_i, \mathbf{y}(t)) - \psi(0_i, \mathbf{y}(t)) = 1] \tag{10.7}$$

$$+ 0 \times \Pr[\psi(1_i, \mathbf{y}(t)) - \psi(0_i, \mathbf{y}(t)) = 0]$$

$$= \Pr[\psi(1_i, \mathbf{y}(t)) - \psi(0_i, \mathbf{y}(t)) = 1] \tag{10.8}$$

Thus, structural importance is the probability that the system is in a state in which the functioning of component i is critical.

Example 1.

Calculate Birnbaum's importance for the two-component non-repairable series, parallel, and two-out-of-three voting system at 20 hr by using the component failure data (hour^{-1}).

Component 1	Component 2	Component 3
$\lambda_1=0.001$	$\lambda_2=0.002$	$\lambda_3=0.003$

Solution: From (7.93) in Chapter 7, we have, at $t=20$,

$$Q_1=1.98013\times10^{-2}$$
$$Q_2=3.92106\times10^{-2}$$
$$Q_3=5.82355\times10^{-2}$$

For the series system at $t=20$, by equations (10.1) and (10.5),

$$g(\mathbf{Q})=1-(1-Q_1)(1-Q_2)$$
$$\Delta g_1(20)=[1-(1-1)(1-Q_2)]-[1-(1-0)(1-Q_2)]=1-Q_2$$
$$\Delta g_2(20)=[1-(1-Q_1)(1-1)]-[1-(1-Q_1)(1-0)]=1-Q_1$$
$$\Delta g_1(20)=1-3.92106\times10^{-2}=9.60789\times10^{-1}$$
$$\Delta g_2(20)=1-1.98013\times10^{-2}=9.80199\times10^{-1}$$

For the parallel system, $g(\mathbf{Q})=Q_1Q_2$; thus,

$$\Delta g_1(20)=Q_2-0=Q_2=3.92106\times10^{-2}$$
$$\Delta g_2(20)=Q_1-0=Q_1=1.98013\times10^{-2}$$

For the two-out-of-three system, we first expand the corresponding structure function.†

$$\psi(\mathbf{Y})=1-(1-Y_1Y_2)(1-Y_2Y_3)(1-Y_3Y_1)$$
$$=Y_1Y_2+Y_2Y_3+Y_3Y_1-2Y_1Y_2Y_3$$

Then, the unavailability function $g(\mathbf{Q})$ is obtained by

$$g(\mathbf{Q})=E[\psi(\mathbf{Y})]=Q_1Q_2+Q_2Q_3+Q_3Q_1-2Q_1Q_2Q_3$$

†There is an approximation in UCRL 51829 and the original IMPORTANCE code. It was assumed that

$$E[1-(1-Y_1Y_2)(1-Y_2Y_3)(1-Y_3Y_1)]=1-(1-Q_1Q_2)(1-Q_2Q_3)(1-Q_3Q_1)$$

In other words, factors $(1-Y_1Y_2)$, $(1-Y_2Y_3)$, and $(1-Y_3Y_1)$ were treated as statistically independent random variables. As a result, the calculations in UCRL 51829 are conservative in all cases when basic events are repeated, since, for example,

$$E[(1-Y_1Y_2)(1-Y_2Y_3)]=1-Q_1Q_2-Q_2Q_3+Q_1Q_2Q_3$$

whereas

$$(1-Q_1Q_2)(1-Q_2Q_3)=1-Q_1Q_2-Q_2Q_3+Q_1Q_2^2Q_3$$

Another way of obtaining $g(\mathbf{Q})$ is to rewrite the original $\psi(\mathbf{Y})$ by the partial pivotal decomposition described in Section 7.5.3 of Chapter 7 and then calculate the expected value of the resulting $\psi(\mathbf{Y})$.

From the expression of $g(\mathbf{Q})$, we have

$$\begin{aligned}
\Delta g_1 &= [Q_2+Q_2Q_3+Q_3-2Q_2Q_3]-[0+Q_2Q_3+0-0] \\
&= Q_2+Q_3-2Q_2Q_3 = 9.28792\times 10^{-2} \\
\Delta g_2 &= [Q_1+Q_3+Q_3Q_1-2Q_1Q_3]-[0+0+Q_3Q_1-0] \\
&= Q_1+Q_3-2Q_1Q_3 = 7.57305\times 10^{-2} \\
\Delta g_3 &= [Q_1Q_2+Q_2+Q_1-2Q_1Q_2]-[Q_1Q_2+0+0-0] \\
&= Q_2+Q_1-2Q_1Q_2 = 5.74591\times 10^{-2}
\end{aligned}$$

For the simple series system $\Delta g_2 > \Delta g_1$; i.e., component two is more likely to fail the system than component one which is to be expected since $\text{MTTF}_1 = 1000 > \text{MTTF}_2 = 500$ hr. The reverse result for the parallel system, $\Delta g_1 > \Delta g_2$, is because the Birnbaum importance is related to the probability that the system is in a state at time t in which the functioning of a component is critical. Since component two fails first, component one is more critical, according to this definition of importance. We will find the same "anomalous" result in parallel systems for the other importance measures used. Note also that the Birnbaum importance for one-event cut sets is always, and usually incorrectly, numerically equal to one. An example of this is shown in Chapter 13 (Example 4).

The result for the two-out-of-three system can be explained in the same way. We conclude from this that the Birnbaum measure is not a useful importance criteria except for simple series systems. Table 10.2 is a tabulation of the criticality rankings obtained by the various measures. The last column, "Expected Result," is the ordering of the expected component importances based on the simplistic (and partially incorrect) notion that importance is a direct measure of the weakest links in the system.

10.3 CRITICALITY IMPORTANCE $I_i^{\text{CR}}(t)$

The *criticality importance* considers the fact that it is more difficult to improve the more reliable components than to improve the less reliable components. The importance of component i is defined by

$$I_i^{\text{CR}}(t) = \lim_{\Delta Q_i \to 0} \frac{\left[\dfrac{\Delta g(\mathbf{Q}(t))}{g(\mathbf{Q}(t))}\right]}{\left[\dfrac{\Delta Q_i(t)}{Q_i(t)}\right]} \tag{10.9}$$

$$= \frac{\partial g(\mathbf{Q}(t))}{\partial Q_i(t)} \times \frac{Q_i(t)}{g(\mathbf{Q}(t))} \tag{10.10}$$

TABLE 10.2 IMPORTANCE RANKINGS (Non-repairable Case, 20 hr)

	Birnbaum	Criticality Importance	Fussell-Vesely	Upgrading Function	Barlow-Proschan	Sequential Contributory	Fussell-Vesely	Barlow-Proschan	Expected Result
Series system: Component 1	9.60790×10^{-1} 2	3.26736×10^{-2} 2	3.40182×10^{-1} 2	3.23630×10^{-1} 2	3.33400×10^{-1} 2	0 1			2
Component 2	9.80197×10^{-1} 1	6.59978×10^{-2} 1	6.73567×10^{-1} 1	6.47260×10^{-1} 1	6.66800×10^{-1} 1	0 1			1
Parallel system: Component 1	3.92105×10^{-2} 1	1.00000×10^{0} 1	1.00000×10^{0} 1	9.90571×10^{-1} 1	5.019390×10^{-1} 1	5.019390×10^{-1} 1			2
Component 2	1.98013×10^{-2} 2	1.00000×10^{0} 1	1.00000×10^{0} 1	9.80665×10^{-1} 2	4.98604×10^{-1} 2	4.98604×10^{-1} 2			1
Voting system: Component 1	9.28791×10^{-2} 1	4.37086×10^{-1} 3	4.58364×10^{-1} 3	4.52303×10^{-1} 3	2.26518×10^{-1} 3	2.21000×10^{-1} 3			3
Component 2	7.57305×10^{-2} 2	7.05713×10^{-1} 2	7.27632×10^{-1} 2	7.24381×10^{-1} 2	3.58721×10^{-1} 2	3.56000×10^{-1} 2			2
Component 3	6.05647×10^{-2} 3	9.08561×10^{-1} 1	8.16421×10^{-1} 1	7.93214×10^{-1} 1	4.03241×10^{-1} 1	4.03000×10^{-1} 1			1
Cut sets, voting system: cut set 1	—	—	—	—	—	—	1.84523×10^{-1} 3	1.77236×10^{-1} 3	3
cut set 2	—	—	—	—	—	—	5.42678×10^{-1} 1	5.36000×10^{-1} 1	1
cut set 3	—	—	—	—	—	—	2.74051×10^{-1} 2	2.67000×10^{-1} 2	2

Thus, from (10.6),

$$I_i^{CR}(t) = \frac{\Delta g_i \times Q_i(t)}{g(\mathbf{Q}(t))} \tag{10.11}$$

$$= \frac{[g(1_i, \mathbf{Q}(t)) - g(0_i, \mathbf{Q}(t))] Q_i(t)}{g(\mathbf{Q}(t))} \tag{10.12}$$

Equation (10.9) shows that the criticality importance of (10.12) is a fractional sensitivity. We now apply equation (10.12) to the three structures previously defined.

Example 2.

Calculate the criticality importance of the components to the three systems of Example 1 at $t=20$.

Solution: For the series system, by equations (10.1) and (10.12)

$$I_1^{CR}(20) = \frac{(1-Q_2)Q_1}{1-(1-Q_1)(1-Q_2)} = \frac{(1-3.92106\times10^{-2})(1.98013\times10^{-2})}{1-(1-1.98013\times10^{-2})(1-3.92106\times10^{-2})}$$

$$= 3.26689\times10^{-1}$$

Similarly,

$$I_2^{CR}(20) = \frac{(1-Q_1)Q_2}{1-(1-Q_1)(1-Q_2)} = 6.59979\times10^{-1}$$

For the parallel system, equations (10.2) and (10.12) produce

$$I_1^{CR}(20) = \frac{Q_2Q_1}{Q_2Q_1} = 1$$

$$I_2^{CR}(20) = \frac{Q_1Q_2}{Q_2Q_1} = 1$$

For the three-component, two-out-of-three system, by using equation (10.12) and the result of Example 1,

$$I_1^{CR}(20) = \frac{\Delta g_1 Q_1}{g(\mathbf{Q})} = \frac{(9.28792\times10^{-2})(1.98013\times10^{-2})}{4.12258\times10^{-3}} = 4.46111\times10^{-1}$$

$$I_2^{CR}(20) = \frac{\Delta g_2 Q_2}{g(\mathbf{Q})} = \frac{(7.57305\times10^{-2})(3.92106\times10^{-2})}{4.12258\times10^{-3}} = 7.20286\times10^{-1}$$

$$I_3^{CR}(20) = \frac{\Delta g_3 Q_3}{g(\mathbf{Q})} = \frac{(5.74591\times10^{-2})(5.82355\times10^{-2})}{4.12258\times10^{-3}} = 8.11666\times10^{-1}$$

Comparing these results with the expected result of Table 10.2 we see that the criticality importance rankings are reasonable with the usual caveat applying to the parallel component case.

10.4 FUSSELL-VESELY COMPONENT IMPORTANCE $I_i^{FV}(t)$

This measure was introduced by Vesely, and is used by Fussell in his hand-calculation procedure (Section 7.7.6, Chapter 7).[1] The basic idea is that a component can contribute to system failure without being critical, by its presence in one or more cut sets. By "not being critical," we mean that restoring the component will not change the system's overall state.

The probability of component i contributing to cut set failure $g_i(\mathbf{Q}(t))$, given that the system is failing at time t, $g(\mathbf{Q}(t))$, is the basis of this definition:

$$I_i^{FV}(t)=\frac{g_i(\mathbf{Q}(t))}{g(\mathbf{Q}(t))} \tag{10.13}$$

where $g_i(\mathbf{Q}(t)) = \Pr\left[\bigvee_j \bigwedge_{k\in K_j} Y_k=1\right]$,

$\bigvee_j \bigwedge_{k\in K_j} Y_k$ = Boolean indicator for the union of all cut sets, K_j's, that contain basic event i.

Example 3.

Compute $I_i^{FV}(20)$ for the three systems of the previous example by calculating exact values for $g(\mathbf{Q})$ and $g_i(\mathbf{Q})$.

Solution: For the series system,

$$I_1^{FV}(20)=\frac{Q_1}{Q_s}=\frac{1.98013\times 10^{-2}}{5.82354\times 10^{-2}}=3.40022\times 10^{-1}$$

$$I_2^{FV}(20)=\frac{Q_2}{Q_s}=\frac{3.92105\times 10^{-2}}{5.82354\times 10^{-2}}=6.73310\times 10^{-1}$$

For the parallel system,

$$I_1^{FV}(20)=\frac{Q_1Q_2}{Q_1Q_2}=1$$

$$I_2^{FV}(20)=\frac{Q_1Q_2}{Q_1Q_2}=1$$

[1]Fussell, B. J., "How to Hand-Calculate System Reliability Characteristics," *IEEE Trans. on Reliability*, R-24, No. 3, 1973.

For the two-out-of-three system,

$$I_1^{FV}(20) = \frac{Q_1Q_2 + Q_3Q_1 - Q_1Q_2Q_3}{Q_s} = 4.57079 \times 10^{-1}$$

$$I_2^{FV}(20) = \frac{Q_1Q_2 + Q_2Q_3 - Q_1Q_2Q_3}{Q_s} = 7.31254 \times 10^{-1}$$

$$I_3^{FV}(20) = \frac{Q_1Q_3 + Q_2Q_3 - Q_1Q_2Q_3}{Q_s} = 8.22634 \times 10^{-1}$$

The importance rankings produced by the Fussell-Vesely method relate closely to the criticality importance and produce the same rankings and almost the same numbers.

A point of digression here is that the Fussell-Vesely importance criteria is readily incorporated into the hand-calculation schemata developed in Section 7.7.6, Chapter 7. In terms of the short-cut calculation nomenclature, equation (10.13) is

$$I_i^{FV} \cong \frac{\sum_{k=1}^{m} Q_k^*}{Q_s} \tag{10.14}$$

where m is the number of minimal cut sets containing i. The cut set importance is defined simply by

$$I_k^{*FV} = \frac{Q_k^*}{Q_s} \tag{10.15}$$

Example 4.

Estimate the component and cut set importance for the system of Table 7.15 (which is the two-out-of-three system of the previous example, at 100 rather than 20 hr), using the Q_i, Q_s and Q_i^* values generated by short-cut calculations.

Solution:

	Cut sets (repairable)			*Cut sets (non-repairable)*		
	1	2	3	1	2	3
$I_i^{*FV}(100)$	4.7×10^{-2}	8.5×10^{-1}	1.1×10^{-1}	1.8×10^{-1}	5.5×10^{-1}	2.7×10^{-1}
	Components (repairable)			*Components (non-repairable)*		
	1	2	3	1	2	3
$I_i^{FV}(100)$:	1.6×10^{-1}	9.0×10^{-1}	9.6×10^{-1}	4.5×10^{-1}	7.3×10^{-1}	8.2×10^{-1}

We see, as expected, that $I_1^{FV}(100)$ for the non-repairable components is in excellent agreement with $I_i^{FV}(20)$ of Example 3. We show shortly that the cut set $I_i^{*FV}(100)$ is also in good agreement. The repairable component importances are in "logical" sequence.

10.5 UPGRADING FUNCTION I_i^{UF}

The *upgrading function* is of limited utility, since it applies to the non-repairable system only. It is defined as the fractional reduction in the probability of the top event when component failure rate λ_i is reduced fractionally.

$$I_i^{UF}(t) = \frac{\lambda_i\, \partial g(\mathbf{Q}(t))}{g(\mathbf{Q}(t))\, \partial \lambda_i}$$

This is not a very convenient measure to calculate, since derivatives are required.

Example 5.

Calculate $I_i^{UF}(20)$ for the three sample systems.

Solution: For the series system,

$$\frac{\partial g(\mathbf{Q}(t))}{\partial \lambda_1} = \frac{1 - e^{-\lambda_1 t} e^{-\lambda_2 t}}{\partial \lambda_1} = t e^{-\lambda_1 t} e^{-\lambda_2 t}$$

Thus,

$$I_1^{UF}(20) = \frac{\lambda_1 t e^{-(\lambda_1 + \lambda_2)t}}{g(\mathbf{Q}(20))} = 3.23433 \times 10^{-1}$$

Similarly,

$$I_2^{UF}(20) = \frac{\lambda_2 t e^{-(\lambda_1 + \lambda_2)t}}{g(\mathbf{Q}(20))} = 6.46867 \times 10^{-1}$$

For the parallel system,

$$I_1^{UF}(20) = \frac{\lambda_1 t e^{-\lambda_1 t}(1 - e^{-\lambda_2 t})}{g(\mathbf{Q}(20))} = 9.90033 \times 10^{-1}$$

$$I_2^{UF}(20) = \frac{\lambda_2 t e^{-\lambda_2 t}(1 - e^{-\lambda_1 t})}{g(\mathbf{Q}(20))} = 9.80133 \times 10^{-1}$$

For the two-out-of-three system,

$$g(\mathbf{Q}(t)) = Q_1 Q_2 + Q_2 Q_3 + Q_3 Q_1 - 2 Q_1 Q_2 Q_3 \tag{10.16}$$

Note that

$$g(\mathbf{Q}(t)) \neq 1-[1-Q_1Q_2][1-Q_2Q_3][1-Q_3Q_1] \tag{10.17}$$

$$\frac{\partial g(\mathbf{Q}(t))}{\partial \lambda_1} = \frac{\partial g}{\partial Q_1}\frac{\partial Q_1}{\partial \lambda_1} + \frac{\partial g}{\partial Q_2}\frac{\partial Q_2}{\partial \lambda_1} + \frac{\partial g}{\partial Q_3}\frac{\partial Q_3}{\partial \lambda_1} = \frac{\partial g}{\partial Q_1}\frac{\partial Q_1}{\partial \lambda_1} + 0 + 0$$

Thus, we calculate $\partial g/\partial \lambda_1$ as $(\partial g/\partial Q_1)(\partial Q_1/\partial \lambda_1) = \Delta g_1(\partial Q_1/\partial \lambda_1)$, where Δg_1 is Birnbaum's importance.

$$\begin{aligned}\Delta g_1 &= [Q_2+Q_2Q_3+Q_3-2Q_2Q_3]-[0+Q_2Q_3+0-0]\\ &= Q_2+Q_3-2Q_2Q_3 = 9.28792\times 10^{-2}\end{aligned}$$

Since

$$\frac{\partial Q_1}{\partial \lambda_1} = \frac{\partial}{\partial \lambda_1}[1-e^{-\lambda_1 t}] = te^{-\lambda_1 t}$$

we obtain, finally,

$$I_1^{UF}(20) = \frac{\lambda_1}{g(\mathbf{Q}(20))}[te^{-\lambda_1 t}][9.28792\times 10^{-2}] = 4.41666\times 10^{-1}$$

Similarly,

$$I_2^{UF}(20) = 7.05976\times 10^{-1}, \quad I_3^{UF}(20) = 7.87559\times 10^{-1}$$

The results of this rather laborious calculation give credible values for the two-out-of-three voting case, and ambivalent ones for the parallel system.

10.6 BARLOW-PROSCHAN IMPORTANCE, I_i^{BP}

Barlow and Proschan's analysis is for a system whose components fail sequentially in time, one at a time.[2] The analysis produces an expression for the expected number of failures caused by basic event i in the time interval 0 to t. The Barlow-Proschan, and the *sequential contributory importance* which follows, differ from the previously derived entities in that they consider the sequence of event failures that cause the system to fail in time and are, therefore, functions of past behavior rather than a point in time. The Barlow-Proschan equation is

$$I_i^{BP}(t) = \int_0^t [\, g(1_i, \mathbf{Q}(t)) - g(0_i, \mathbf{Q}(t)) \,] w_i(t)\, dt \tag{10.18}$$

[2]Barlow, R. E., and Proschan, F., "Importance of System Components and Fault Tree Analysis," Operations Research Center, University of California, Report ORC 74-3, 1974.

where $w_i(t)$ is the unconditional failure intensity, the expected number of failures per unit time at time t.

The sum of all component importances, when divided by Q_s, is unity. In essence I_i^{BP} is the probability of the system's failing because a critical cut set containing i fails, with i failing last.

Example 6.

Determine $I_i^{BP}(20)$ for the three test systems. Normalize the results by Q_s.

Solution: For the two-component series system, noting that $w_i = \lambda_i e^{-\lambda_i t}$ (see Chapter 4, equation (4.85)),

$$I_1^{BP}(20) = \int_0^{20} [1-(1-e^{-\lambda_2 t})]\lambda_1 e^{-\lambda_1 t}\,dt = \frac{-\lambda_1}{\lambda_1+\lambda_2}[e^{-(\lambda_1+\lambda_2)t}]_0^{20}$$
$$= 3.33333 \times 10^{-1}$$

Similarly,

$$I_2^{BP}(20) = 6.66667 \times 10^{-1}$$

In the case of the parallel system,

$$I_1^{BP}(20) = \int_0^{20} Q_2(t) w_1(t)\,dt = \int_0^{20} [1-e^{-\lambda_2 t}]\lambda_1 e^{-\lambda_1 t}\,dt$$
$$= -[e^{-\lambda_1 t}]_0^{20} + \frac{\lambda_1}{\lambda_1+\lambda_2}[e^{-(\lambda_1+\lambda_2)t}]_0^{20}$$
$$= 5.01666 \times 10^{-1}$$

Similarly,

$$I_2^{BP}(20) = 4.98333 \times 10^{-1}$$

The two-out-of-three voting system $I_1^{BP}(20)$ yields

$$I_1^{BP}(20) = \int_0^{20} [(Q_2 + Q_3 - Q_2 Q_3) - (Q_2 Q_3)] w_1\,dt$$
$$= \int_0^{20} \lambda_1 [e^{-(\lambda_1+\lambda_2)t} + e^{-(\lambda_1+\lambda_3)t} - 2e^{-(\lambda_1+\lambda_2+\lambda_3)t}]\,dt$$
$$= 9.39547 \times 10^{-4}$$

To normalize this result we divide by $g(\mathbf{Q}(20)) = 4.12258 \times 10^{-3}$, to get

$$I_1^{BP}(20) = 2.27903 \times 10^{-1}$$

Similar manipulations yield

$$I_2^{BP}(20) = 3.64408 \times 10^{-1} \quad \text{and} \quad I_3^{BP}(20) = 4.07688 \times 10^{-1}$$

These results are slightly different from the values given by the code because the original IMPORTANCE code uses (10.17) as $g(\mathbf{Q}(t))$.

We see that this measure, like the others, gives meaningful answers for the series and two-out-of-three cases.

10.7 SEQUENTIAL CONTRIBUTORY IMPORTANCE, I_i^{SC}

It is interesting to consider the role of the failure of component i, when another component j actually causes the system to fail. Two components, i and j, must be in at least one minimal cut set. For this case we have

$$I_i^{SC}(t) = \sum_{\substack{j \\ j \neq i}} \int_0^t [\, g(1_i, 1_j, \mathbf{Q}(t)) - g(1_i, 0_j, \mathbf{Q}(t)) \,] Q_i(t) w_j(t)\, dt \quad (10.19)^\dagger$$

Example 7.

Obtain the sequential contributory importance for the components of the three systems we have been considering.

Solution: For the series system, we see that

$$I_1^{SC}(20) = \int_0^{20} [1-1] Q_1 w_j\, dt = 0, \quad I_2^{SC}(20) = 0$$

The expression for the parallel configuration results in values for I_1^{SC} and I_2^{SC} exactly equal to the Barlow-Proschan importances I_2^{BP} and I_1^{BP}, respectively.

For the two-out-of-three system,

$$I_1^{SC}(20) = \int_0^{20} [\, g(1,1,Q_3) - g(1,0,Q_3)] Q_1 w_2\, dt + \int_0^{20} [\, g(1,Q_2,1) - g(1,Q_2,0)] Q_1 w_3\, dt$$

This results in a rather long algebraic expression. Final normalized values are

$$I_1^{SC}(20) = 2.254932 \times 10^{-1}, \quad I_2^{SC}(20) = 3.631900 \times 10^{-1}, \quad I_3^{SC}(20) = 4.113169 \times 10^{-1}$$

These results are extremely difficult to interpret except as a measure of the enabling events in the system.

10.8 TIME BEHAVIOR OF IMPORTANCES

All of the importance parameters discussed are time dependent. Furthermore, since they relate to the probability of a component's being critical to a system's functioning at a given point in time and/or past behavior, we

† In equation (3.10) in UCRL 51829 the differential is $dw_j(t)$, not $w_j(t)\,dt$ as in equation (10.19). The importance code, however, calculates $w_j(t)\,dt$.

should be prepared to observe some very complex behavior. It is possible that, for highly redundant systems at short times, the components which are more reliable will have the higher importances and, after long times, the less reliable components will be more important. The reverse is true for series systems, but could occur in complex systems. *Caveat emptor*.

We now examine some parameters developed for cut set analysis, even though importance ranking of cut sets is usually of considerably less engineering significance than the ranking of components, since it is difficult to look at cut sets as discrete entities when replications of basic events between cut sets occur.

10.8.1 Fussell-Vesely Cut Set Importance, I_i^{*FV}

We have already encountered this engagingly simple expression [equation (10.15)]:

$$I_i^{*FV}(t) = \frac{Q_i^*(t)}{Q_s(t)} \tag{10.20}$$

In Example 4 we obtained the non-repairable cut set importances for the voting system by using approximate Q_i, Q_i^*, and Q_s values.

For cut set one,

$$I_1{}^{*FV}(20) = \frac{Q_1 Q_2}{Q_s} = \frac{(1.98013\times 10^{-2})(3.92106\times 10^{-2})}{4.12258\times 10^{-3}}$$
$$= 1.88334\times 10^{-1}$$

Similarly,

$$I_2^{*FV}(20) = \frac{Q_2 Q_3}{Q_s} = 5.53888\times 10^{-1}$$

$$I_3^{*FV}(20) = \frac{Q_3 Q_1}{Q_s} = 2.79713\times 10^{-1}$$

The rank order is also in accord with expectations, since cut set two's components have MTTF's of 500 and 333 hr, cut set three has components whose MTTF's are 1000 and 333, whereas for cut set one, the MTTF's are 1000 and 500.

10.8.2 Barlow-Proschan Cut Set Importance, I_i^{*BP}

The Barlow-Proschan definition of cut set i's importance is the probability that it causes the system to fail. This means that some basic event in the set must fail, with all others having failed previously. The formal expression is

$$I_i^{*\mathrm{BP}}(t)=\frac{\sum_{j\in i}\int_0^t\left[1-g\left(0_j,1^{i-\{j\}},\mathbf{Q}(t)\right)\right]\prod_{\substack{k\neq j\\ k\in i}}Q_k(t)w_j(t)\,dt}{Q_s(t)} \tag{10.21}$$ †

where $1^{i-\{j\}}$ means Q_l equals one for each basic event $l\neq j$ contained in cut set i.

A numerical example will help to illustrate the meaning of the terms in the expression.

Example 8.

Obtain $I_i^{*\mathrm{BP}}$ for the three-component voting system at 20 hr.

Solution: According to (10.21) and (10.16),

$$I_1^{*\mathrm{BP}}(20)=\int_0^{20}[1-\{0\times1+1\times Q_3+Q_3\times0-2\times0\times1\times Q_3\}]Q_2w_1\,dt/Q_s$$
$$+\int_0^{20}[1-\{1\times0+0\times Q_3+Q_3\times1-2\times1\times0\times Q_3\}]Q_1w_2\,dt/Q_s$$

Substituting values of Q and w from previous examples and integrating, we have

$$I_1^{*\mathrm{BP}}(20)=1.80995\times10^{-1}$$

For the other two cut sets (by computer),

$$I_2^{*\mathrm{BP}}(20)=5.46603\times10^{-1},\quad I_3^{*\mathrm{BP}}(20)=2.72401\times10^{-1}$$

The rankings for the two-out-of-three system cut sets are therefore as anticipated. For the series system, equation (10.21) reduces to component values. An interesting point about the Barlow-Proschan treatment is raised by Lambert. If all components have equal failure probabilities, then cut set importances should be inversely proportional to *cut set order* (the number of basic events comprising a set). This result, unfortunately, is not always obtained.

There is another version of $I_i^{*\mathrm{BP}}$ and I_i^{BP} called the *limiting steady-state Barlow-Proschan unavailability* for repairable systems only, which can be obtained from equation (10.21) or (10.18) under the assumption that

$$\lim_{t\to\infty}Q_i(t)=\frac{\lambda_i}{\lambda_i+\mu_i}$$

† $w_j(t)\,dt$ is written $dw_j(t)$ in the original paper but evaluated as $w_j(t)\,dt$.

This parameter is also given as a subroutine in the IMPORTANCE code but will not be described here since the numbers obtained using this particular option are identical to those at large t's from equations (10.18) and (10.21).

10.9 REPAIRABLE SYSTEMS

Each of the importance functions, except for the upgrading functions, apply to repairable as well as non-repairable components and cut sets. Table 10.3 was computed for configurations of Examples 1 to 8, using the same λ's as before ($\lambda_1=0.001$, $\lambda_2=0.002$, $\lambda_3=0.003$), with repair times of $1/\mu_1=10$, $1/\mu_2=40$, $1/\mu_3=60$. We observe much the same behavior pattern as before. Some points of note are as follows:

Birnbaum. The 20 hr importance results for the parallel or voting system are counter-intuitive. This measure of importance should be used with caution, if at all.

Criticality importance. As for the non-repairable case, the values do not differ appreciably from the Fussell-Vesely values. At large t's, this will not necessarily be true.

Fussell-Vesely. This is, perhaps, the soundest of the measures and has the added virtue of being easy to calculate. It also closely approximates Lambert's upgrading function, which has not been discussed here but is described in detail in UCRL 51829.

Barlow-Proschan. This yields identical rankings as the Fussell-Vesely measure except for parallel systems.

Sequential contributory. Unless the components appear in the same cut sets this measure has no meaning. It is not applicable to series systems.

PROBLEMS

10.1. Calculate Birnbaum's importance for the three-component series system at $t=20$, using the failure data of Example 1.

10.2. Calculate Birnbaum's importance for the three-component parallel system at $t=20$, using the failure data of Example 1.

10.3. Prove that

$$\Delta g_i=(1-Q_1)\ldots(1-Q_{i-1})(1-Q_{i+1})\ldots(1-Q_n)$$

for n-component series systems

$$\Delta g_i=Q_1\ldots Q_{i-1}Q_{i+1}\ldots Q_n$$

for n-component parallel systems

TABLE 10.3 IMPORTANCE RANKING (Repairable case, 20 hr, by IMPORTANCE code)

	Birnbaum	Criticality Importance	Fussell-Vesely	Barlow-Proschan	Sequential Contributory	Fussell-Vesely	Barlow-Proschan	Expected Results
Series system:								
Component 1	9.69×10^{-1} 2	2.12×10^{-1} 2	2.19×10^{-1} 2	3.33×10^{-1} 2	0 1			2
Component 2	9.91×10^{-1} 1	7.81×10^{-1} 1	7.88×10^{-1} 1	6.67×10^{-1} 1	0 1			1
Parallel system:								
Component 1	3.09×10^{-2} 1	1.00×10^{0} 1	1.00×10^{0} 1	6.02×10^{-1} 1	3.98×10^{-1} 2			2
Component 2	8.59×10^{-3} 2	1.00×10^{0} 1	1.00×10^{0} 1	3.98×10^{-1} 2	6.02×10^{-1} 1			1
Voting system:								
Component 1	7.89×10^{-2} 1	3.05×10^{-1} 3	3.11×10^{-1} 3	2.50×10^{-1} 3	1.58×10^{-1} 3			3
Component 2	5.77×10^{-2} 2	8.03×10^{-1} 2	8.09×10^{-1} 2	3.68×10^{-1} 2	3.81×10^{-1} 2			2
Component 3	3.92×10^{-2} 3	8.75×10^{-1} 1	8.81×10^{-1} 1	3.82×10^{-1} 1	4.54×10^{-1} 1			1
Cut sets, voting system: cut set 1						1.19×10^{-1} 3	1.58×10^{-1} 3	3
cut set 2						6.90×10^{-1} 1	5.88×10^{-1} 1	1
cut set 3						1.92×10^{-1} 2	2.46×10^{-1} 2	2

10.4. Calculate criticality importance for the three-component series and parallel system of Problems 10.1 and 10.2.

10.5. Consider the tail gas quench and clean up system of Fig. 7.12, Chapter 7. Assume a failure rate $\lambda=10^{-3}$ (failures per hour) for each component. Calculate the criticality importance at $t=20$ (hours).

10.6. Prove that, if all components have the same failure rate, then component ranking by Birnbaum's importance coincides with that by criticality importance.

10.7. Calculate the Fussell-Vesely component importance at $t=20$ for the two-out-of-four voting system, using failure rates (hour^{-1}) of:

$$\lambda_1=0.001,\quad \lambda_2=0.002,\quad \lambda_3=0.003,\quad \lambda_4=0.004$$

Use the Esary and Proschan upper bound equation (7.79), Chapter 7, for the calculation of $g(\mathbf{Q})$ and $g_i(\mathbf{Q})$.

10.8. Calculate the Fussell-Vesely cut set importance at $t=100$ (hours) for the system of Example 4, using the exact Q_i, Q_s and Q_i^*. Compare the results with those of Example 4.

10.9. Prove that the upgrading function and the criticality importance yield the same component rankings if each component has the same failure rate λ.

10.10. Determine the Barlow-Proschan importance for the three-component parallel system at $t=20$ in Problem 10.2.

10.11. For the three-component parallel system of Problem 10.10, prove

$$I_1^{SC}(t)=I_2^{BP}(t)+I_3^{BP}(t),\quad I_2^{SC}(t)=I_1^{BP}(t)+I_3^{BP}(t),\quad I_3^{SC}=I_1^{BP}(t)+I_2^{BP}(t)$$

10.12. Obtain $I_i^{*FV}(2000)$ for the two-out-of-three voting system of Example 3. Use exact computation formulas for Q_i, Q_i^* and Q_s.

10.13. Express I_1^{*BP} by Q's and W's for the three-out-of-four voting system, where cut set 1 is $\{1,2,3\}$.

10.14. For the two-component series systems, prove the steady-state importances:

$$\Delta g_1=\frac{\mu_2}{\lambda_2+\mu_2},\quad \Delta g_2=\frac{\mu_1}{\lambda_1+\lambda_1}$$

$$I_1^{CR}=\frac{\lambda_1\mu_2}{\lambda_1\lambda_2+\lambda_1\mu_2+\lambda_2\mu_1},\quad I_2^{CR}=\frac{\lambda_2\mu_1}{\lambda_1\lambda_2+\lambda_1\mu_2+\lambda_2\mu_1}$$

$$I_1^{FV}=\frac{\lambda_1(\lambda_2+\mu_2)}{\lambda_1(\lambda_2+\mu_2)+\lambda_2\mu_1},\quad I_2^{FV}=\frac{\lambda_2(\lambda_1+\mu_1)}{\lambda_2(\lambda_1+\mu_1)+\lambda_1\mu_2}$$

11

STORAGE TANKS, PROTECTIVE SYSTEMS, COLD STANDBY, AND VOTING SYSTEMS

Fault tree methodology and associated analysis techniques were developed for aerospace and nuclear energy applications, where failures must be avoided at all costs. In the manufacturing industries in general, and the chemical industry, in particular, equipment failures are taken more or less for granted, and storage tanks, protective instrumentation and standby redundancy are used to smooth out production, reduce downtime to a minimum, and mitigate the effects of equipment failure. In this chapter we develop equations and special risk analysis techniques for process systems.

11.1 STORAGE TANKS

11.1.1 Introduction

Failures do not occur instantaneously: There is always a finite time lapse between cause and consequence, and failures occur if the cause is not repaired within the time lapse. In fault tree notation, this concept is shown as in Fig. 11.1.

The delay can be of mechanical or electrical origin, or it may represent storage, as in the case of a tank in a chemical plant. In general, delays uncouple units, thus blocking information transfer from upstream to downstream for a period of time. In Fig. 11.2 for example, if the upstream (SU) and downstream (SD) flow rates are at a steady rate of 100 liters per

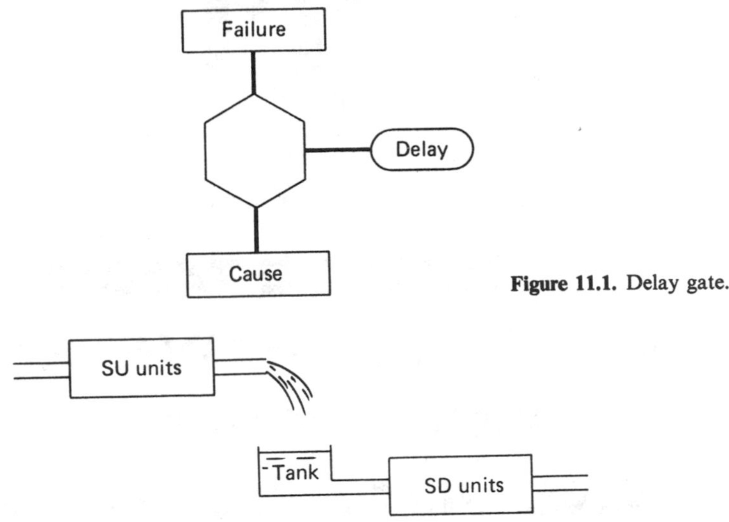

Figure 11.1. Delay gate.

Figure 11.2. System consisting of SU units, tank, and SD units.

hour and there is a storage tank which holds 100 liters between units, the downstream units will run for 1 hr after the upstream unit fails (provided the downstream units do not fail and the tank is full). Figure 11.3 shows the corresponding fault tree.

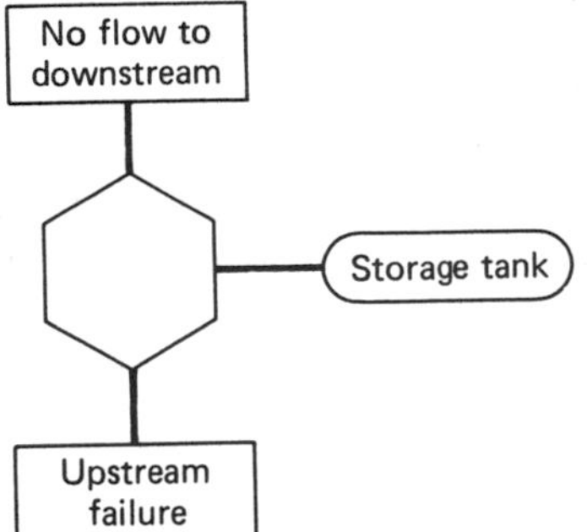

Figure 11.3. Storage tank.

Plant operating policies strongly depend on the reasons for having storage tanks. For example, imagine that the capacity of the SU and SD units are exactly the same, that we start with a full tank, and operating policy A in the event of a SU failure is as follows:

Policy A:

1) Run SD until the tank is empty or the SU failure is repaired.
2) Start up SU.

3) Fill the tank.
4) Start up SD.
5) Run until the next SD or SU failure.

This policy, for example, is used for plants where a sudden lack of feed to SD units poses a safety hazard: SU units may be providing a coolant, inert blanket, diluent, etc. We have time T of the tank capacity to cope with SU failures safely. Although the plant safety increases, the tank will have no effect on the availability of the plant, since the downtime interval of SU is simply shifted to that of SD (Fig. 11.4).

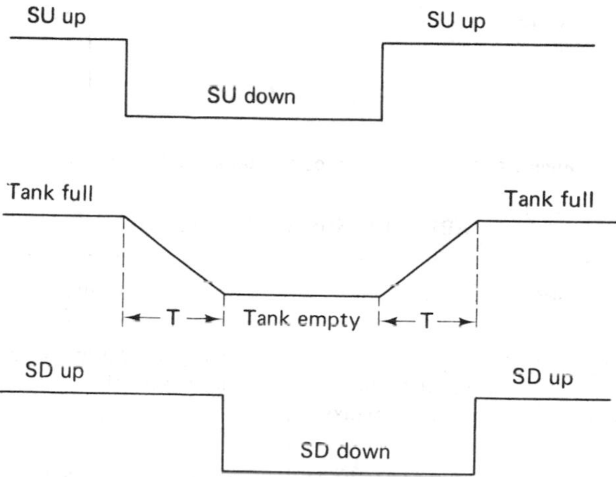

Figure 11.4. Time history of SU, tank, and SD.

As another example, starting with the empty tank and matched flows, imagine we adopt the following policy when SD fails.

Policy B:

1) Run SU until the tank is full or the SD failure is repaired.
2) Start up SD.
3) Empty the tank.
4) Start up SU.
5) Run until the next SD or SU failure.

This policy is useful for plants where SU units are difficult or costly to shut-down and/or overflow capability is required because of environmental pollution. The tank improves the plant safety but has no effect on the availability (Fig. 11.5).

Before attempting elaborate mathematical treatments of these and other possible operating policies, it is useful to consider the practical aspects of

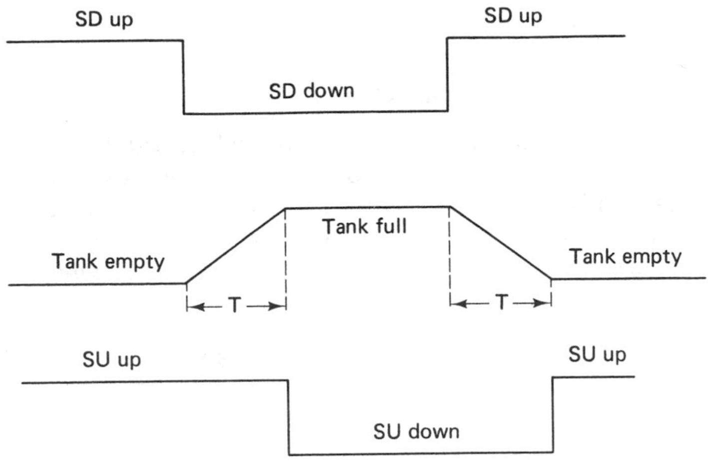

Figure 11.5. Time history of SD, tank, and SU for policy B.

TABLE 11.1 STORAGE TANK STRATEGY

Reason for Tank	Operating Policy
1. Lack of feed to SD units poses a safety hazard (i.e., SU units may be providing a coolant, inert blanket, dilutent, etc.).	1. Policy A. Always keep storage tank full if possible.
2. Quality of output from SU units is somewhat variable, and product must be blended to maintain specs.	2. Policy A. Always keep storage tank full if possible.
3. SD units are difficult or costly to shut down (cracking furnaces, catalytic reactions, etc.).	3. Policy A. Always keep storage tank full if possible.
4. SU units have excess capacity but are less reliable than SD units, and availability increase is desired.	4. Keep storage tank full. Fill *on-line* or when SD is down.
5. SD units have poor reliability, but excess capacity and availability increases are desired.	5. Keep storage tank empty. Empty *on-line* or when SU is down.
6. SD and SU units have balanced reliabilities, and capacity and availability increases are desired.	6. Keep tank half full, emptying and filling during shut-downs.
7. SU units are difficult or costly to shut down, and/or overflow capability is required because of environmental pollution.	7. Policy B. Keep storage tank empty if possible, filling when SD is down.

the situation. There are two major types of storage tanks used in chemical plants, and neither type is placed there strictly to obtain reliability or availability increases.

1) *Product and raw material tanks*. Raw material supply, shipping, delivery, schedule, and production rates are the primary variables dictating size and configuration.
2) *Intermediate storage tanks*. Here size, configuration, and operating policy depend on the purpose the tank is to serve. Table 11.1 details some typical storage tank functions. In policies 1, 2, 3, or 7 we fill or empty the tank immediately after plant startup, and the capacities SD and SU are matched. Policies 4, 5, and 6 result in an increase in availability, provided we fill or empty during breakdown periods and we are able to fill or empty or keep the tank half full during "normal" operation because of excess capacity SU or SD.

Table 11.1 does not, by any means, cover all of the reasons for having storage tanks. Tanks for the purpose of collecting overflows, effluents, off-spec materials, or streams from parallel or batch units also appear on plant sites, and each has a unique and optimal operating procedure.

11.1.2 Tanks Filled On-Line: Time Delays

We consider first Case 4, where the SU units have excess capacity, and we assume that the tank can be filled instantaneously, on-line. Thus, when SU fails the tank is always full and represents a pure time delay. Of course, when SD fails SU must shut down immediately. The fault tree for this situation is shown as Fig. 11.6. The top event is "no flow to SD." The SU failure and the SD failure in Fig. 11.6 can be traced backward to more basic events.

11.1.3 Unconditional Failure Intensity $w_{TE}(t)$ for "Empty Tank"

In the development of the theory, we neglect the event "tank fails" and adopt the following notation.

T: Time to empty the tank.

$w_{TE}(t)$: The expected number of events where the tank becomes empty at t per unit time due to SU failure at $(t-T)$. The subscript TE refers to "empty tank."

It is necessary to recognize that the tank becomes empty at t if and only if SU fails at $(t-T)$ and is not repaired by t, and that $w_{TE}(t)$ is not the same as the probability that the tank *is* empty. We now define:

$w_{SU}(t)$: The expected number of SU failures at t per unit time.

$\Pr(T)$: The probability of no upstream repair in time interval T.

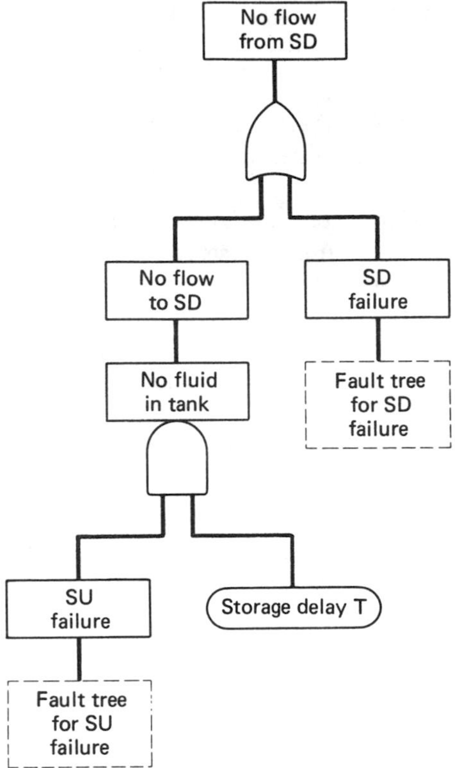

Figure 11.6. Fault tree for storage delay.

Thus, with reference to Fig. 11.7 and the above definitions,

$$w_{TE}(t) = w_{SU}(t - T)\,\mathrm{Pr}(T) \tag{11.1}$$

Note that w_{SU} can be obtained from computer code KITT described in Chapter 7.

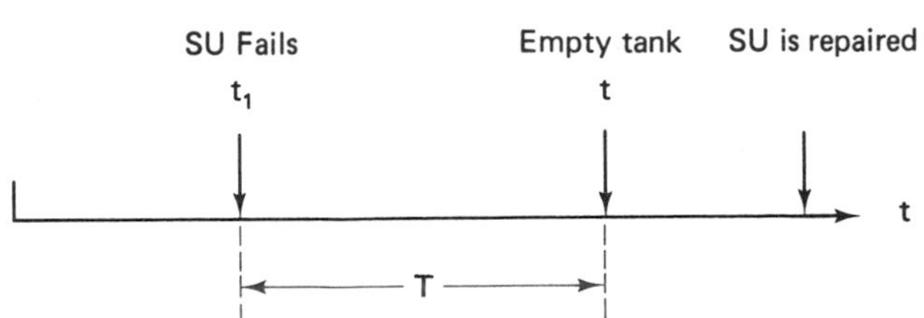

Figure 11.7. Tank capacity T.

We assume an exponential repair distribution of the upstream units; thus, the probability of no upstream repair in time T is

$$K \equiv \mathrm{Pr}(T) = \exp\left(\frac{-T}{\tau_{SU}}\right) \quad (\tau_{SU}\text{, upstream repair time}) \tag{11.2}$$

where τ_{SU} is obtained from the reciprocal of SU conditional repair intensity $\mu_{SU}(t-T)$. From (11.1) and (11.2), at steady state ($t \to \infty$),

$$w_{TE}(\infty) = e^{-T/\tau_{SU}} w_{SU}(\infty) = K w_{SU}(\infty) \tag{11.3}$$

11.1.4 Unavailability of Fluid from Tank

Recognizing now that the downstream units fail when the tank is empty, we define unavailability $Q_{TE}(t)$ of the tank as the probability of the tank being empty at t. We have, then,

$$Q_{TE}(t) = \int_0^{t-T} w_{SU}(t_1)\,\Pr(t-t_1)\,dt_1 \tag{11.4}$$

Figure 11.8 serves to clarify this equation. SU fails at t_1 and is not repaired by time t. An SU failure between $t-T$ and t does not influence the event of the tank being empty at t.

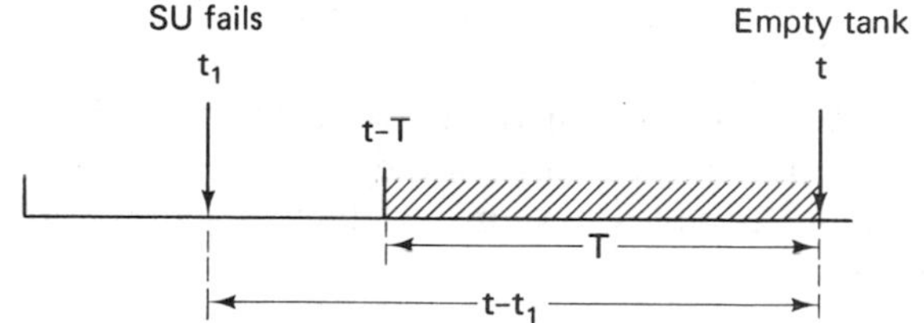

Figure 11.8. Notation for interval $t-t_1$.

With reference to Fig. 11.9, the unavailability of SU at t is described by a relationship similar to equation (11.4):

$$Q_{SU}(t) = \int_0^{t} w_{SU}(t_2)\,\Pr(t-t_2)\,dt_2 \tag{11.5}$$

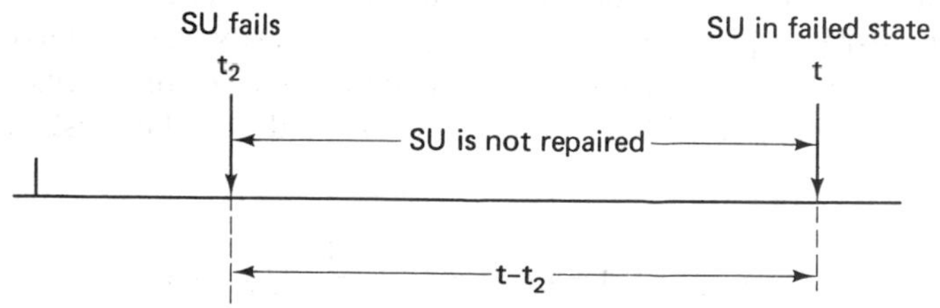

Figure 11.9. Notation for interval $t-t_2$.

Since, in general, for constant repair times τ_{SU},

$$\Pr(t) = \exp\left(\frac{-t}{\tau_{SU}}\right) \tag{11.6}$$

we substitute for $\Pr(t-t_1)$ in equation (11.4) and then, using equation (11.5) we obtain, for $Q_{TE}(t)$,

$$\begin{aligned} Q_{TE}(t) &= \int_0^{t-T} w_{SU}(t_1)\exp\left[\frac{-(t-t_1-T+T)}{\tau_{SU}}\right]dt_1 \\ &= \exp\left(\frac{-T}{\tau_{SU}}\right)\int_0^{t-T} w_{SU}(t_1)\exp\left[\frac{-(t-T-t_1)}{\tau_{SU}}\right]dt_1 \\ &= KQ_{SU}(t-T), \quad (t \geq T) \end{aligned} \tag{11.7}$$

This equation states that if SU is down at $t-T$ and not repaired in the interval T, the tank is empty at T. The failure of SU after $t-T$ has nothing to do with the empty tank at t because there is still some fluid left in the tank. At steady state, this becomes

$$Q_{TE}(\infty) = KQ_{SU}(\infty) \tag{11.8}$$

The above equations, combined with some relationships inherent in the definitions, are sufficient to solve steady- or unsteady-state availability problems involving complex systems containing storage which acts as a pure time delay.

11.1.5 Other Parameters

Mean Time to Repair of Upstream Units. Unavailability $Q_{SU}(t)$ and unconditional failure intensity $w_{SU}(t)$ for the above equations can be obtained from computer code KITT. This code also calculates conditional failure intensity Λ_{SU}, the probability of upstream failure occurring at time t per unit time, given that SU failure does not exist at time t. Under the assumptions of an exponential repair time distribution for SU, conditional repair intensity μ_{SU} is given by $1/\tau_{SU}$. At steady state, unconditional failure intensity $w_{SU}(t)$ is equal to unconditional repair intensity $v_{SU}(t)$. Equations (7.98) and (7.100) in Chapter 7 hold for the upstream units. Thus,

$$\Lambda_{SU}(\infty)[1-Q_{SU}(\infty)] = \frac{1}{\tau_{SU}}Q_{SU}(\infty) \tag{11.9}$$

or

$$\tau_{SU} = \text{MTTR}_{SU} = \frac{Q_{SU}(\infty)}{\Lambda_{SU}(\infty)[1-Q_{SU}(\infty)]} \tag{11.10}$$

This equation gives τ_{SU}, the mean time to repair of SU. For sufficiently large values of t, we may approximate τ_{SU} by

$$\tau_{SU} = \frac{Q_{SU}(t)}{\Lambda_{SU}(t)[1 - Q_{SU}(t)]} \tag{11.11}$$

Mean Time to Fill Tank. The tank continues to be empty if and only if the SU units fail for an interval larger than T. The SU units which are still in a failed condition at the end of interval T have an exponential repair time distribution with mean τ_{SU}. Thus, the mean time to fill the tank, τ_{TE}, is equal to τ_{SU}:

$$\tau_{TE} = \tau_{SU} \tag{11.12}$$

Mean Time to Empty Tank. At steady state, $[1 - Q_{TE}(\infty)]/Q_{TE}(\infty)$ is equal to the ratio of expected uptime to expected downtime. Thus,

$$\text{MTTF}_{TE}(\infty) = \frac{1 - Q_{TE}(\infty)}{Q_{TE}(\infty)} \text{MTTR}_{TE} \tag{11.13}$$

$$= \frac{1 - KQ_{SU}(\infty)}{KQ_{SU}(\infty)} \text{MTTR}_{TE} \tag{11.14}$$

Equation (11.12) yields

$$\text{MTTF}_{TE}(\infty) = \frac{1 - KQ_{SU}(\infty)}{KQ_{SU}(\infty)} \tau_{SU} \tag{11.15}$$

Similarly, for the upstream,

$$\text{MTTF}_{SU}(\infty) = \frac{1 - Q_{SU}(\infty)}{Q_{SU}(\infty)} \tau_{SU} \tag{11.16}$$

From (11.15) and (11.16),

$$\text{MTTF}_{TE}(\infty) = \frac{[1 - KQ_{SU}(\infty)]}{K[1 - Q_{SU}(\infty)]} \text{MTTF}_{SU}(\infty) \tag{11.17}$$

or, approximately,

$$\text{MTTF}_{TE}(\infty) \cong \frac{\text{MTTF}_{SU}(\infty)}{K} \tag{11.18}$$

We see that MTTF_{TE} is larger than MTTF_{SU} by a factor of $1/K$ at steady state.

Conditional Failure Intensity of Tank. Conditional failure intensities $\Lambda_{SU}(t)$ and $\Lambda_{TE}(t)$ satisfy

$$\Lambda_{SU}(t)=\frac{w_{SU}(t)}{1-Q_{SU}(t)} \tag{11.19}$$

$$\cong w_{SU}(t) \tag{11.20}$$

$$\Lambda_{TE}(t)=\frac{w_{TE}(t)}{1-Q_{TE}(t)} \tag{11.21}$$

$$=\frac{Kw_{SU}(t-T)}{1-KQ_{SU}(t-T)}\cong Kw_{SU}(t-T) \tag{11.22}$$

In (11.20) and (11.22), we obtain

$$\Lambda_{TE}(t)\cong K\Lambda_{SU}(t-T) \tag{11.23}$$

At steady state,

$$\Lambda_{TE}(\infty)\cong K\Lambda_{SU}(\infty) \tag{11.24}$$

Equations (11.23) and (11.24) calculate the conditional failure intensity of the tank.

So far we have not considered the SD failure of Fig. 11.6. Parameters for the top event, "no flow from SD," can be bounded in the following way (the subscript S refers to the system consisting of SU, tank, and SD).

$$\left.\begin{aligned} Q_S(t) &\le Q_{TE}(t)+Q_{SD}(t)\\ w_S(t) &\le w_{TE}(t)+w_{SD}(t)\\ \Lambda_S(t) &\le \Lambda_{TE}(t)+\Lambda_{SD}(t)\end{aligned}\right\} \tag{11.25}$$

We now apply these equations to test examples.

Example 1.

Obtain the MTTF_{TE} and Q_{TE} for a single SU unit, with $\lambda_{SU}=0.001$, $\tau_{SU}=10$, backed up by a storage tank, as a function of storage tank capacity in hours. Assume steady state.

Solution: At steady state, from (4.109) and (4.103) in Chapter 4,

$$Q_{SU}(\infty)=\frac{\Lambda_{SU}\tau_{SU}}{1+\Lambda_{SU}\tau_{SU}},\quad w_{SU}(\infty)=\frac{\Lambda_{SU}}{1+\Lambda_{SU}\tau_{SU}}$$

Substituting into (11.8)†

$$Q_{TE}(\infty)=KQ_{SU}(\infty)=\exp\left(\frac{-T}{\tau_{SU}}\right)Q_{SU}(\infty)$$

†This form of the steady-state equation was derived independently by G. Fussell, W. Johns, and ourselves.

Combining with (11.15)

$$\mathrm{MTTF}_{\mathrm{TE}}(\infty) = \frac{1 - KQ_{\mathrm{SU}}(\infty)}{KQ_{\mathrm{SU}}(\infty)}\tau_{\mathrm{SU}} = \frac{\exp\left(\dfrac{T}{\tau_{\mathrm{SU}}}\right)(1 + \Lambda_{\mathrm{SU}}\tau_{\mathrm{SU}}) - \Lambda_{\mathrm{SU}}\tau_{\mathrm{SU}}}{\Lambda_{\mathrm{SU}}}$$

Table 11.2 is obtained for $\Lambda_{\mathrm{SU}} = 0.001$ and $\tau_{\mathrm{SU}} = 10$. For a repairable system backed up by a tank which can be kept full, we note from Table 11.2 that the system availability and $\mathrm{MTTF}_{\mathrm{TE}}(\infty)$ increase dramatically with tank size.

TABLE 11.2 Reliability Parameters, Example 1

T	T/τ_{SU}	K	$\Lambda_{\mathrm{TE}} \times 10^3$	$\mathrm{MTTF} \times 10^{-3}$(hr)	$Q_{\mathrm{TE}} \times 10^3$
0	0	1	1	1	9.90099
10	1.0	0.36788	0.36557	2.73546	3.64236
20	2.0	0.13534	0.13417	7.45295	1.34121
40	4.0	0.01832	0.01814	55.13413	0.18151

Example 2.

The fault tree of Fig. 11.10 has the component characteristics:

λ (failures/hour)	τ (repair time, hours)
(1) 1.0×10^{-3}	10
(2) 2.0×10^{-3}	40
(3) 3.0×10^{-3}	60

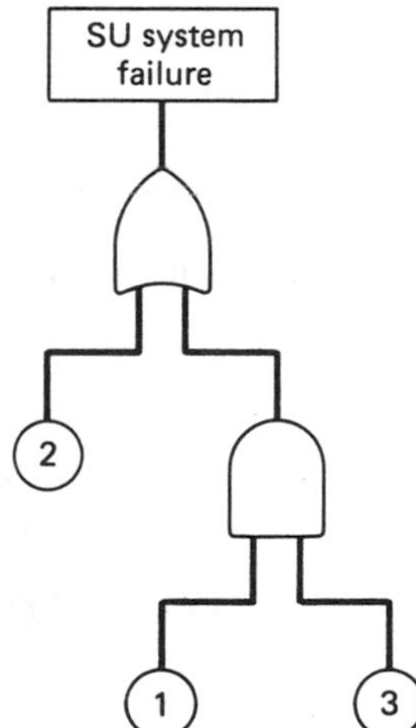

Figure 11.10. Fault tree of upstream in Example 2.

In view of the repair times, we would expect steady state at $t>200$ hr. The SU system parameters obtained by KITT are given in columns 2, 3, and 4 of Table 11.3, and τ_{SU} was calculated by equation (11.11).

TABLE 11.3 Reliability Parameters, Example 2

t	$Q_{SU}\times 10^2$	$w_{SU}\times 10^3$	$\Lambda_{SU}\times 10^3$	τ_{SU}
20	3.9756	2.00639	2.08946	19.8147
40	7.79010	1.98378	2.15139	39.2735
400	7.54726	2.02806	2.19362	37.21421

Let us now examine the effect of introducing tanks 1 and 2, representing time delays of 3.72 and 111.6 hr, respectively (Fig. 11.11). According to equation (11.2) this corresponds to $K_1=0.90485$ and $K_2=0.04979$. The computer output of Table 11.4 was obtained by using an inhibit gate as a delay gate, with inhibit conditions K_1 and K_2. The results are consistent with our expectations that the effect of the tanks is to increase the availability and reduce the number of system failures.

TABLE 11.4 Reliability Parameters, Example 2

t	K	$Q_{TE}\times 10^2$	$w_{TE}\times 10^3$	$\Lambda_{TE}\times 10^3$	τ_{TE}
20	0.90485	3.59755	1.81548	1.88232	19.39589
	0.04979	1.98041×10^{-1}	9.98922×10^{-2}	1.00090×10^{-1}	19.92547
40	0.90485	7.05043	1.79502	1.931184	39.27772
	0.04979	3.88295×10^{-1}	9.87665×10^{-2}	9.91515×10^{-2}	39.31444
400	0.90485	6.83011	1.83509	1.96962	37.2195
	0.04979	3.76285×10^{-1}	1.00971×10^{-1}	1.013523×10^{-2}	37.2665

Almost identical results are obtained if the SU system of Fig. 11.11 is replaced by one component with $\Lambda_{SU}(400)$ and $\tau_{SU}(400)$ from Table 11.3. One point of note is that the t's of Table 11.4 are really $(t-T)$.

11.1.6 Approximations Made in Example 2

When the upstream system consists of more than one unit, the upstream failure is a top event of a fault tree. The event does not follow the exponential repair distribution as was assumed in Example 2. Thus, the theory developed here holds only approximately. Equation (11.7), however, holds for the upstream units, since the tank is empty at time t if and only if SU is down at $t-T$ and not repaired in the time interval T. Thus,

$$Q_{TE}(t)=K(t-T)Q_{SU}(t-T) \tag{11.26}$$

where $K(t-T)$ is the probability of no upstream repair in the time interval

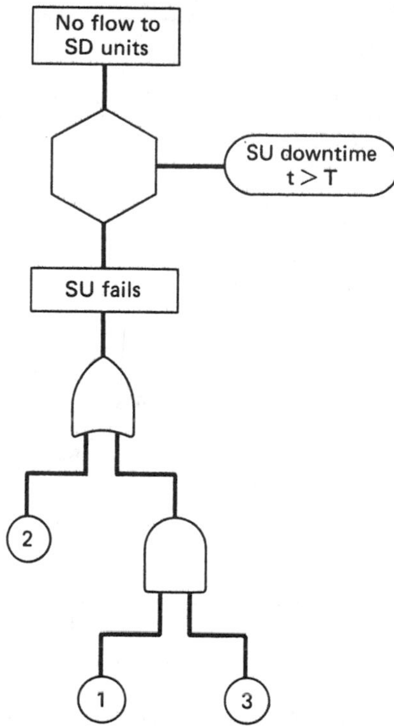

Figure 11.11. SU failure plus tank, Example 2.

T starting at $t-T$. We can evaluate $K(t-T)$ by solving the Markov differential equations as in Chapter 9, or by using the Monte Carlo methods of Chapter 12. Figure 11.12 shows the exact values of $K(\infty)$ obtained by Markov, and the exponential approximations. We see a good agreement.

Similarly, equation (11.1) can be extended to the case of non-exponential repair distributions:

$$w_{TE}(t) = K(t-T)w_{SU}(t-T) \tag{11.27}$$

Further, for reliable systems, unavailability Q is small, and conditional failure intensity Λ is approximately equal to unconditional failure intensity w. Thus, equation (11.23) also holds and, from (11.27),

$$\Lambda_{TE}(t) \cong K(t-T)\Lambda_{SU}(t-T) \tag{11.28}$$

11.1.7 Tanks Empty On-Line

Consider cases where the SD units have poor reliability but excess capacity, and an availability increase of the SU units is desired. Policy 5 of

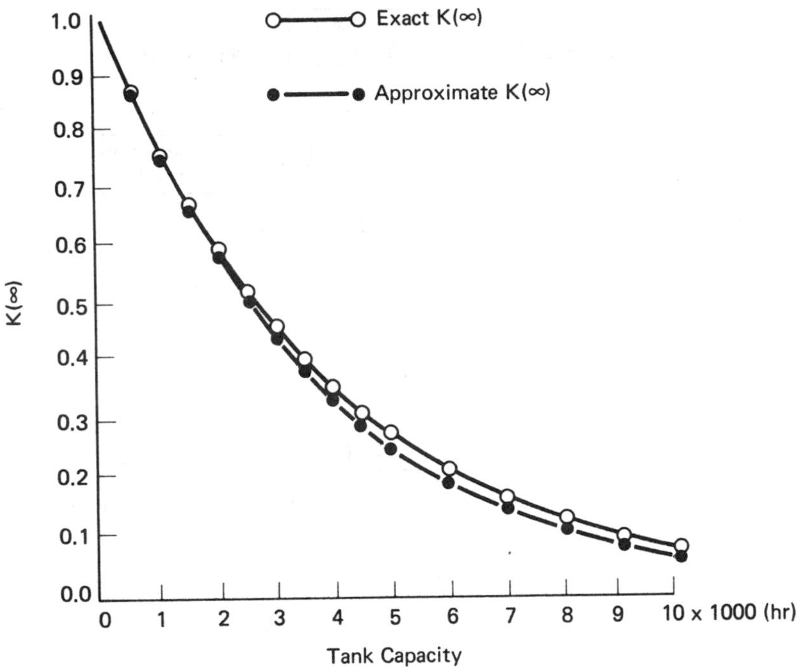

Figure 11.12. Comparison of exact and approximate $K(\infty)$ for upstream units in Example 2.

Table 11.1 is used for this purpose: The tank is empty on-line and filled during SD breakdown.

Let subscript TF refer to the full tank. Define probability K by

$$K \equiv \exp\left(\frac{-T}{\tau_{SD}}\right)$$

Similarly to the case of policy 4, we have

$$w_{TF}(t) = K w_{SD}(t-T), \quad Q_{TF}(t) = K Q_{SD}(t-T)$$

$$\tau_{SD} = \mathrm{MTTR}_{SD} = \frac{Q_{SD}(\infty)}{\Lambda_{SD}(\infty)[1-Q_{SD}(\infty)]}$$

$$\tau_{TF} = \tau_{SD}$$

$$\mathrm{MTTF}_{TF}(\infty) = \frac{1-KQ_{SD}(\infty)}{K[1-Q_{SD}(\infty)]}\mathrm{MTTF}_{SD}(\infty)$$

$$\cong K^{-1} \cdot \mathrm{MTTF}_{SD}(\infty)$$

$$\Lambda_{TF}(\infty) \cong K \Lambda_{SD}(\infty)$$

11.1.8 Other Tank Policies

We now examine the consequence of trying to keep the tank full, filling only while the downstreams unit is under repair, and assuming exactly matched upstream and downstream production capabilities. We make the assumption of steady state and constant failure rate and repair rate. The top event is "no flow to downstream units."

Case 1. Upstream Availability Greater Than Downstream Availability. The system shown in Fig. 11.13 has downstream availability A_{SD} of $20/23=0.87$ and the upstream availability A_{SU} of $10/11=0.91$. A good approximation to the system availability A_s without the storage tank is

$$A_s = A_{SD} A_{SU} = (0.87)(0.91) = 0.79$$

Since, however, we have excess upstream availability and the $MTTR_{SU}$ is small compared to the tank size, the system availability is nearly equal to $A_{SD}=0.87$, since the downstream unit seldom has to shut down because of lack of feed. Note that for a series system, to a first approximation, the system availability A_s is always bounded by

$$A_{SD} A_{SU} < A_s < A_{SD} \tag{11.29}$$

SU		SD
MTTF = 10	T = 2	MTTF = 20
MTTR = 1		MTTR = 3

Figure 11.13. Upstream availability greater than downstream availability.

Case 2. Downstream Availability Greater Than Upstream Availability. It is quite obvious that in this case the tank is relatively useless, since it will be empty most of the time. Thus, the system availability is simply the lower bound of A_s, $A_{SD} A_{SU}$.

Case 3. Equal Availability, Upstream and Downstream. The limits on the system availability are given by equation (11.29). One simple way of obtaining an exact answer for this unrealistic case is to make a graphical time history for the system, noting up and down times and obtaining the availability that way.

Case 4. Limiting Conditions in Terms of Capacity. Let C_{SU} and C_{SD} be the capacity of the upstream and downstream units in Fig. 11.2. Assume $C_{SU} > C_{SD}$ but $A_{SU} < A_{SD}$. This implies that the excess capacity of the upstream unit can be used to restore tank volume that is consumed due to the greater downtime of unit SU compared to unit SD.

Over a period of time Δt, there can exist four possible operating state combinations of units SU and SD.

Unit SU	Unit SD	Rate of Tank Fill	Time
Operating	Operating	$C_{SU}-C_{SD}$	$(1-Q_{SU})(1-Q_{SD})\Delta t$
Down	Operating	$-C_{SD}$	$Q_{SU}(1-Q_{SD})\Delta t$
Operating	Down	$+C_{SU}$	$(1-Q_{SU})Q_{SD}\Delta t$
Down	Down	0	$Q_{SU}Q_{SD}\Delta t$

Let $M(t)$ = expected value of tank mass at time t. At $t=0$,

$$M(0)=M_0 \tag{11.30}$$

A mass balance yields

$$M(t+\Delta t)=M(t)+(C_{SU}-C_{SD})(1-Q_{SU})(1-Q_{SD})\Delta t-C_{SD}Q_{SU}(1-Q_{SD})\Delta t + C_{SU}(1-Q_{SU})Q_{SD}\Delta t \tag{11.31}$$

To keep the tank volume M_0 at all t,

$$M(t+\Delta t)=M(t) \tag{11.32}$$

or

$$(C_{SU}-C_{SD})(1-Q_{SU})(1-Q_{SD})\Delta t-C_{SD}Q_{SU}(1-Q_{SD})\Delta t+ C_{SU}(1-Q_{SU})\,Q_{SD}\Delta t=0 \tag{11.33}$$

which reduces to

$$\frac{C_{SU}}{C_{SD}}=\frac{1-Q_{SD}}{1-Q_{SU}}=\frac{A_{SD}}{A_{SU}} \tag{11.34}$$

This means that for there to be a constant tank level M_0, the capacities C_{SU} and C_{SD} should satisfy

$$C_{SU}=\left(\frac{A_{SD}}{A_{SU}}\right)C_{SD} \tag{11.35}$$

11.2 PROTECTIVE SYSTEMS HAZARD ANALYSIS

11.2.1 Introduction

Every factory has instrumentation whose sole function is to protect the plant and personnel from destructive hazards due to equipment malfunctions, accidents, or other major risks. Fire alarms, sprinklers, reactor scram systems, quench systems, rupture discs, and pop-off valves are all examples of protective systems. It is assumed that as long as the protective system is working, the destructive hazard cannot occur.

Consider, for example, the hypothetical process of Fig. 11.14. The process consists of chlorinating a hydrocarbon gas in a glass-lined reactor. The possibility of an exothermic, runaway reaction occurs whenever the Cl_2/hydrocarbon gas ratio is too high, in which case a detonation occurs, since a source of ignition is always present.[1]

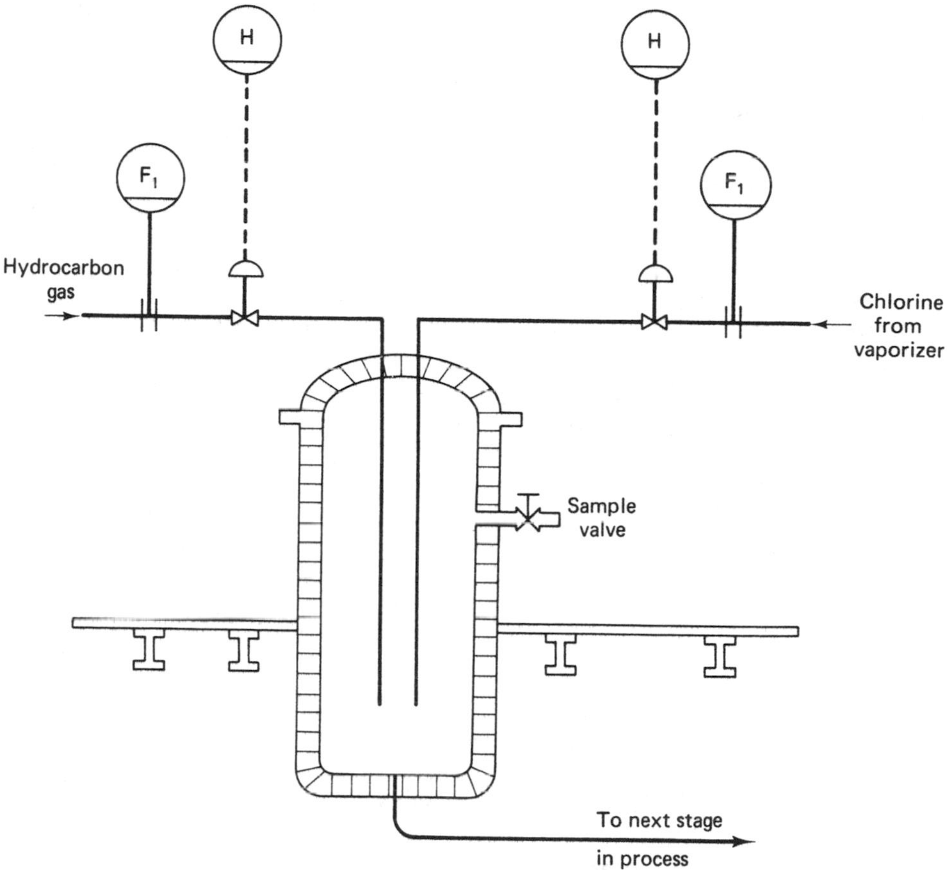

Figure 11.14. Line diagram of a chlorination process.

The fault tree for the process is shown as Fig. 11.15. The basic event probabilities such as "high pressure in vaporizer," which is postulated to occur 12 times per year, were obtained from plant operating data and are failure rates. Thus, the plant failure rate without protective instruments is

[1]Kumamoto, H., and E. J. Henley, "Protective Systems Hazards Analysis," *Ind. Engr. Chem.*, Vol. 13, No. 4, pp. 274–276, 1978.

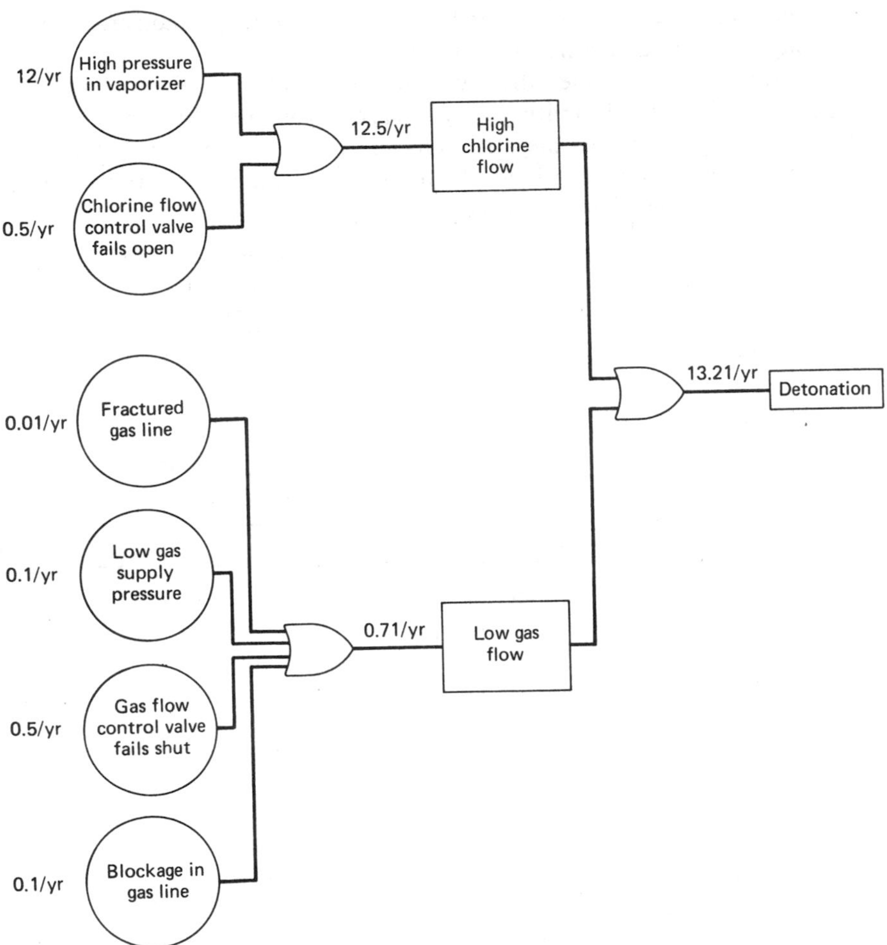

Figure 11.15. Fault tree for the plant of Fig. 11.14.

$D = 13.21/365 = 0.036\ \text{day}^{-1}$. The reliability, at 1 year, is

$$R(365) = e^{-(0.036)(365)} = 1.96 \times 10^{-6} \tag{11.36}$$

which is extremely low.

A logical way to reduce the probability of an explosion is to use a flow-ratio protective system which shuts off the chlorine whenever a high chlorine-to-gas ratio is detected. As long as the protective system functions, there is no hazard. A simplistic view of the process is that if the protective system is down for 27.6 days out of the year, the detonation will occur once a year, (13.21)(27.6/365)=1. In effect, the 13.21 is a "demand rate" on the protective system, and 27.6/365 is the fractional dead time of the protective system. Thus,

$$W = \text{ENF} = \text{demand rate} \times \text{fractional dead time}$$

A protective instrument system, like any other piece of equipment, has a certain failure rate λ. To treat safety systems as if they were simple AND gates, however, is incorrect, since the fractional dead time is a function not only of λ but also of how soon the fault is detected. This, in turn, is a function of T, the test (proof) interval, and D, the demand rate on it, since a demand is also a test. We assume the protective system is as good as new after testing.

11.2.2 Definition of Variables

(1) U: Life length of the plant (time). U is a random variable, and $U=t$ means that the destructive hazard occurs at time t.
(2) S: Life length of the protective system (time). S is a random variable.
(3) T: Inspection interval of the protection system (time).
(4) $W(0,T)$: Expected number of destructive hazards between 0 and T.
(5) D: Demand rate on the protective system (time^{-1}).
(6) $w(t)$: Expected number of destructive hazards at time t per unit time (time^{-1}).
(7) $A(t,t+dt]$: Event that the destructive hazard occurs in the interval $(t,t+dt]$; in other words, $t<U\le t+dt$.

11.2.3 Assumptions

1) *The protective system has n-redundantly connected devices.* The life length of the devices are mutually independent and they obey exponential distribution with failure rate λ. Therefore, for a given $t \ge 0$,

$$\Pr\{S \le t\} = (1-e^{-\lambda t})^n \tag{11.37}$$

2) *The probability distribution of U is dependent on the sample value of S:*

$$\Pr\{U \le t | S=u\} = \begin{cases} 0, & \text{if } t \le u & (11.38) \\ 1-e^{-D(t-u)}, & \text{if } t>u & (11.39) \end{cases}$$

Equation (11.38) means that no detonations occur as long as the protective system is functioning, whereas (11.39) represents that the occurrence of a destructive hazard after the protective system failure obeys an exponential distribution with failure rate D.

3) *Derivation of hazard rate w.* The following identities hold.

$$w(t)\,dt = \Pr\{A(t,t+dt]\} \tag{11.40}$$

$$W(0,T) = \int_0^T w(t)\,dt \tag{11.41}$$

The right-hand side of (11.40) can be written as

$$\Pr\{A(t,t+dt]\} = \Pr\{t<U\le t+dt\} \tag{11.42}$$

$$= \Pr\{t<U\le t+dt, S\le t\} + \Pr\{t<U\le t+dt, t<S\le t+dt\} + \Pr\{t<U\le t+dt, t+dt<S\} \tag{11.43}$$

The second term of (11.43) is of the order of $(dt)^2$, and the third term is zero because of (11.38).

$$\Pr\{A(t,dt]\} = \Pr\{t<U<t+dt, S\le t\} \tag{11.44}$$

$$= \int_0^t \Pr\{t<U\le t+dt, u<S\le u+du\} \tag{11.45}$$

Introducing the probability density function $p(U=t,S=u)$ defined by

$$p(U=t,S=u) = \lim_{\Delta\to 0} \frac{\Pr\{t<U\le t+\Delta, u<S\le u+\Delta\}}{\Delta^2} \tag{11.46}$$

Then, (11.45) can be rewritten as

$$\Pr\{A(t,t+dt]\} = dt\int_0^t p(U=t,S=u)\,du \tag{11.47}$$

The probability density of (11.46) satisfies the identity

$$p(U=t,S=u) = p(U=t|S=u)p(S=u) \tag{11.48}$$

where the densities $p(U=t|S=u)$ and $p(S=u)$ are defined by

$$p(U=t|S=u) = \lim_{\Delta\to 0} \frac{\Pr\{t<U\le t+\Delta|S=u\}}{\Delta} \tag{11.49}$$

$$p(S=u) = \lim_{\Delta\to 0} \frac{\Pr\{u<S\le u+\Delta\}}{\Delta} \tag{11.50}$$

From (11.38), (11.39), and (11.49), we have

$$p(U=t|S=u)=\frac{d}{dt}\Pr\{U\le t|S=u\} \tag{11.51}$$

$$=\begin{cases}0, & \text{if } t\le u\\ De^{-D(t-u)}, & \text{if } t>u\end{cases} \tag{11.52}$$

Further, from (11.37) and (11.50) we obtain

$$p(S=u)=\frac{d}{du}\Pr\{S\le u\} \tag{11.53}$$

$$=n\lambda e^{-\lambda u}(1-e^{-\lambda u})^{n-1} \tag{11.54}$$

Equations (11.48), (11.52), and (11.54) give

$$p(U=t,S=u)=\begin{cases}0, & \text{if } t\le u\\ De^{-D(t-u)}n\lambda e^{-\lambda u}(1-e^{-\lambda u})^{n-1}, & \text{if } t>u\end{cases} \tag{11.55}$$

Thus, from (11.47) we have

$$\Pr\{A(t,t+dt]\}=dt\,Dn\lambda\int_0^t e^{-D(t-u)}e^{-\lambda u}(1-e^{-\lambda u})^{n-1}du \tag{11.56}$$

Equations (11.40), (11.41) and (11.56) yield

$$w(t)=Dn\lambda e^{-Dt}\int_0^t e^{(D-\lambda)u}(1-e^{-\lambda u})^{n-1}du \tag{11.57}$$

$$W(0,T)=Dn\lambda\int_0^T e^{-Dt}\int_0^t e^{(D-\lambda)u}(1-e^{-\lambda u})^{n-1}du\,dt \tag{11.58}$$

The right-hand side of (11.58) is calculated in Appendix 11.1 of this chapter. The result is

$$W(0,T)=1+\sum_{i=0}^{n-1}\binom{n-1}{i}(-1)^i\frac{n}{D-(i+1)\lambda}\left(\lambda e^{-DT}-\frac{De^{-(i+1)\lambda T}}{i+1}\right) \tag{11.59}$$

The average hazard rate $\bar{w}$ is given by

$$\bar{w}=\frac{W(0,T)}{T} \tag{11.60}$$

11.2.4 Some Properties of $W(0, T)$

The following equations are derived in Appendix 11.2 of this chapter.

1) Let $\lambda T \ll 1$ and $DT \ll 1$. Then,

$$W(0,T) \cong \frac{D\lambda^n T^{n+1}}{n+1} \tag{11.61}$$

This result coincides with the formula derived by R. J. P. Briedly of the Petrochemical Division of ICI, Ltd.

2)
$$\lim_{D\to\infty} W(0,T) = (1-e^{-\lambda T})^n \tag{11.62}$$

$$\lim_{D\to 0} W(0,T) = 0 \tag{11.63}$$

(11.62) means that the plant has the fault characteristics of the protective system when DT is large.

3)
$$\lim_{\lambda\to\infty} W(0,T) = 1-e^{-DT} \tag{11.64}$$

$$\lim_{\lambda\to 0} W(0,T) = 0 \tag{11.65}$$

These equations are now applied to some practical examples.

Example 3.

The process shown in Fig. 11.14 is supervised by one process operator who spends 55 hr of his 2200 working hours per year near the reactor. At other times, the operator is out of the danger area. If the operator is near the unit when it detonates, there is a 0.1 chance of death.

Choose, from Table 11.5, the most suitable flow-trip device which will reduce the hazard to the operator to a FAFR (fatal accident frequency rate) of 1.25 (per 10^8 hours).

Solution: The fatal accident frequency rate is

$$\text{FAFR} = \bar{w}P(10^8)$$

where $\bar{w}$ = average hazard rate by equation (11.60),
P = probability of the hazard resulting in death.

To obtain P, we multiply the probability of death (0.1) by 55/2200:
$P = (0.1)(55/2200) = 0.0025.$

Substituting equation (11.60), assuming a monthly test interval, and $D = 13.21$ (see Fig. 11.15),

$$1.25 > \text{FAFR} = \frac{(0.0025)(10^8)}{(8760)}$$
$$\times \frac{1}{1/12}\left\{1+\sum_{i=0}^{n-1}\binom{n-1}{i}(-1)^i\frac{n}{D-(i+1)\lambda}\left[\lambda e^{-DT} - \frac{De^{-(i+1)\lambda T}}{i+1}\right]\right\}$$

TABLE 11.5

Function	Class	Dangerous Failure Rate/Yr[a]	Cost (dollars)
Flow trip: shuts off chlorine on detecting a high chlorine flow.	I	0.45, 0.45	10,000
	II	0.6, 0.6	1,250
	III	0.24	250
Flow trip: shuts off chlorine on detecting a low gas flow.	I	0.45, 0.45	10,000
	II	0.6, 0.6	1,250
	III	0.24	250
Flow-ratio trip: shuts off chlorine on detecting a high chlorine/gas ratio.	I	0.45, 0.45	24,000
	II	0.65, 0.65	2,350
	III	0.24	450

[a]Two failure rates are given when the shut-down system has two redundant channels.

We now substitute values of n and λ for the various instruments listed in Table 11.5 and choose the most appropriate. With a Class II flow-ratio trip, $n=2$, and $\lambda=0.6$, so the FAFR is

$$
\begin{aligned}
\text{FAFR} &= 0.2312 \\
&= \frac{(0.0025)(10^8)}{(8760)}\left\{12\left(1+\frac{2}{13.21-0.6}(0.6e^{-(13.21/12)}-13.21e^{-(0.6/12)})\right.\right. \\
&\quad \left.\left. -\frac{2}{13.21-(2)(0.6)}\left(0.6e^{-(13.21/12)}-\frac{13.21e^{-(2)(0.6/12)}}{2}\right)\right)\right\}
\end{aligned}
$$

This \$2,350 instrument is more than adequate.

Example 4.

The function of the non-return valve, NRV, in the tank-filling system in Fig. 11.16 is to protect the pump from being driven backwards by the liquid in Tank A in case of a pump trip-out or failure. Assume the demand rate on the NRV is 10 times per year, that it is inspected every year, and that it has a failure rate of 0.003 year^{-1}. Calculate the average hazard rate $\overline{w}$ for pump damage.

Solution: The hazard rate for $D=10$, $\lambda=0.003$ year^{-1}, $T=1$ year, $n=1$ is

$$
W(0,T)=1+\frac{1}{D-\lambda}(\lambda e^{-DT}-De^{-\lambda T})=0.0027
$$

Thus, the hazard occurs every 370 years. Note that this is a limiting case of $D\to\infty$, (11.62), where the plant has the characteristics of the protective system (whose mean time to failure is 333 years).

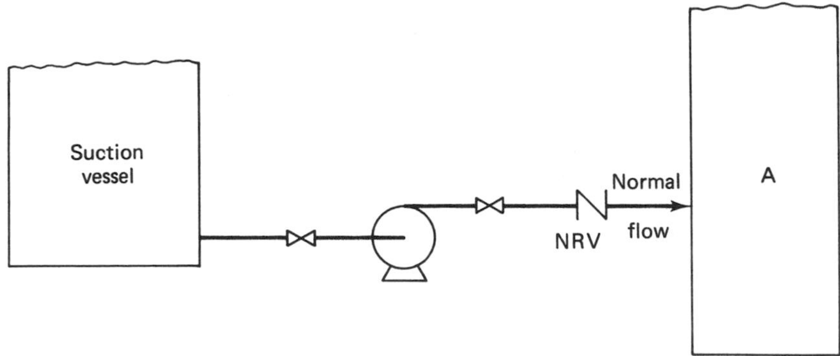

Figure 11.16. Pumping system.

11.3 HOT AND COLD REDUNDANCY, VOTING SYSTEMS

11.3.1 Hot Standby

If two or more units are in *hot redundancy*, and it is assumed that they all wear out at equal rates, independent of whether or not they are in service, the KITT computer code and the associated technology apply. Hot redundancy is commonly used in the nuclear, aerospace, and electric power industries, where failure to start poses a problem or hazard.

If the hot standby units are part of *voting system*, where r out of N units must function to have system functionality then, as shown in Chapter 7,

$$Q_N^* = 1 - \sum_{k=r}^{N} \frac{N!}{k!(N-k)!}(1-Q)^k Q^{N-k} \tag{11.66}$$

where N = total parallel units,

r = number of units required for system *functionality* $0 < r < N$,

Q_N^* = unavailability of parallel units (considered a cut set),

Q = unavailability of individual unit.

Note that this standby configuration corresponds to the $(N-r+1)$-out-of-N system of Chapter 7.

11.3.2 Cold Standby

In the chemical and other industries, standby (or spare) units are frequently in *cold standby*; i.e., they are not put into service unless primary units fail. Although either the Markov or Monte Carlo methods can be used to calculate exact reliability parameters for systems which include units in cold standby, the advantages of being able to calculate reliability parameters for systems which include cold standby units by the conventional kinetic tree theory methodology offers an attractive alternative.

For a voting system consisting of identical, parallel, and redundant units in cold standby,[2]

$$A_C = \sum_{j=0}^{N-r} \frac{a^r}{j!}(-r\ln a)^j \tag{11.67}$$

where N = total number of units available,
r = number of units required for operation $0<r<N$,
A_c = availability of the set of N redundant units,
a = availability of individual components.

Equation (11.67) is shown graphically in Fig. 11.17, where the ordinate, $z=-r\ln a$ is plotted versus $Q_c = 1 - A_c$. Equation (11.68) is an alternate form of (11.67), where the cold standby units are considered to be a cut set:

$$Q_C^* = 1 - \exp\left[\ln \sum_{j=0}^{N-r} \frac{z^j}{j!} - z\right] \tag{11.68}$$

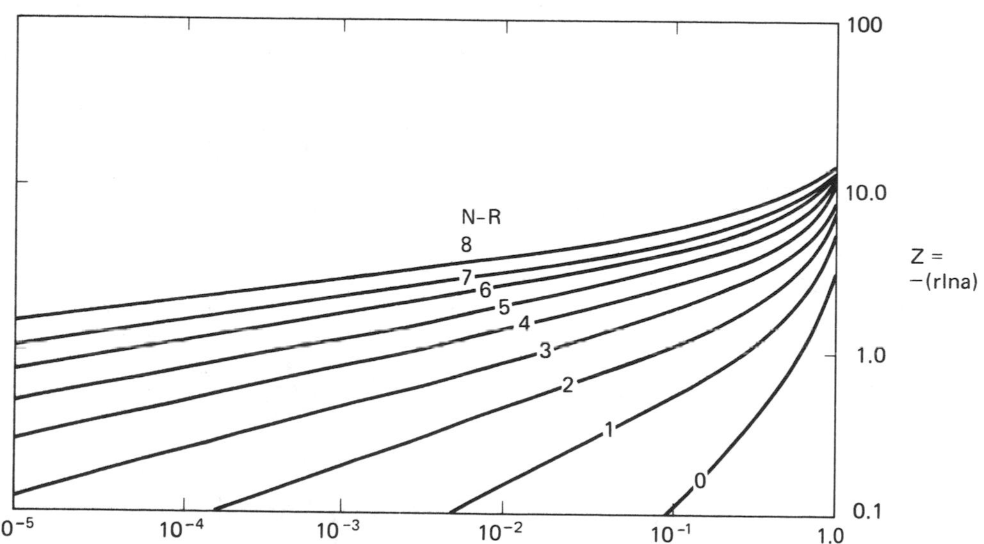

Figure 11.17. Effect of cold standby redundancy.

[2]Benning, C. J., "Reliability Prediction Formulas for Standby Redundant Structures," *IEEE Trans. on Reliability*, Vol. R-16, pp. 136–137, 1967.

11.3.3 Expected Number of Failures for Voting Systems

Given a system of N identical, redundant components, of which r must be operating for the system *not* to fail, we require a general formula for the failure rate. For $r=1$, by (7.97) of Chapter 7,

$$w_c^* = w^* = \sum_{j=1}^{N} w_j \prod_{\substack{l=1 \\ l \neq j}}^{N} Q_l \tag{11.69}$$

where w_c^* = unconditional failure intensity of the voting system,
w^* = unconditional failure intensity of cut set $\{1, 2, \ldots, N\}$.

Thus,

$$w_c^* = w_1 Q_2 + w_2 Q_1, \quad (N=2)$$
$$w_c^* = w_1 Q_2 Q_3 + w_2 Q_1 Q_3 + w_3 Q_1 Q_2, \quad (N=3)$$
$$w_c^* = w_1 Q_2 Q_3 Q_4 + w_2 Q_1 Q_3 Q_4 + w_4 Q_1 Q_2 Q_3, \quad (N=4)$$

Note that

$$w_c^* = \sum_{i=1}^{S} X_i Y_i$$

where X_i = product of unavailabilities of $(N-r)$ failed components, where a failure of one of r remaining components is critical to a cut set,
Y_i = unconditional failure intensity due to a failure of any
S = total number of critical states when cut set failures occur by a component failure. (11.70)

As an example, consider $r=2$, $N=4$. Then, we have cut sets

$$\{1,2,3\}, \{1,2,4\}, \{1,3,4\}, \{2,3,4\}$$

The critical states are

$$\{1,2\}, \{1,3\}, \{1,4\}, \{2,3\}, \{2,4\}, \{3,4\}$$

And the products X_i's are

$$\{X_i\} = \{Q_1 Q_2, Q_1 Q_3, Q_1 Q_4, Q_2 Q_3, Q_2 Q_4, Q_3 Q_4\} \tag{11.71}$$

The number of critical states S is

$$S = \binom{N}{N-r} = \binom{N}{r} = \frac{N!}{r!(N-r)!} = \frac{4!}{2!(4-2)!} = 6 \tag{11.72}$$

Thus, if

$$Q_1 = Q_2 = Q_3 = Q_4 = Q$$
$$Y_1 = Y_2 = Y_3 = Y_4 = Y$$

Then,

$$X_i = Q^{N-r} = Q^2$$

and

$$w_c^* = Y \sum_{i=1}^{S} Q^{N-r} = Y\binom{N}{r} Q^{N-r} \tag{11.73}$$

Consider now calculation of Y. For each critical state i, we see from (7.70) that

Y_i = unconditional failure intensity of a series system consisting of r identical components.

Thus, by Section 7.7.3 of Chapter 7, Y_i can be evaluated by $w_s^{(1)}$:

$$Y_i \cong w_s^{(1)} = rw \tag{11.74}$$

where w = unconditional failure intensity of each component.

Combining (11.73) and (11.74), we obtain

$$w_c^* = rw\binom{n}{r} Q^{n-r} \tag{11.75}$$

Equations (11.75), (11.66), and (11.68) are suitable for insertion into the KITT code in subroutine form. This has been accomplished by J. Ong[3] and the application of the modified KITT code, KITT-IT, is demonstrated by an example problem in Chapter 13.

APPENDIX 11.1 DERIVATION OF (11.59)

By expansion, we have

$$(1 - e^{-\lambda u})^{n-1} = \sum_{i=0}^{n-1} \binom{n-1}{i} (-1)^i e^{-\lambda i u} \tag{1}$$

[3]Ong, J., MS Thesis, Department of Chemical Engineering, University of Houston, June, 1978. Also Henley, E., and J. Ong, "Reliability Parameters For Chemical Process Systems," in *Synthesis and Analysis Methods for Risk, Safety, and Reliability Studies*, Apostolakis, G., S. Garriba, and G. Volta, (eds.), Plenum Press, N.Y., N.Y., 1980.

Substitution of (1) into (11.58) yields

$$W(0,T)=\sum_{i=0}^{n-1}\binom{n-1}{i}(-1)^i\frac{n}{D-(i+1)\lambda}\left(\lambda e^{-DT}-\frac{De^{-(i+1)\lambda T}}{i+1}\right) \quad (2)$$

$$+\sum_{i=0}^{n-1}\binom{n-1}{\lambda}(-1)^i\frac{n}{i+1} \quad (3)$$

It can be shown readily that the second term on the right-hand side of (3) is unity, and we have (11.59).

APPENDIX 11.2 DERIVATION OF ASYMPTOTIC PROPERTIES OF *W*(0, *T*)

1) *Derivation of (11.61)*: The approximation

$$e^{-Dt}e^{(D-\lambda)u}(1-e^{-\lambda u})^{n-1}\cong[1-Dt][1+(D-\lambda)u]\lambda^{n-1}u^{n-1} \quad (1)$$

$$\cong\lambda^{n-1}u^{n-1} \quad (2)$$

holds. The substitution of equation (2) into (11.58) yields (11.61).

2) *Derivation of (11.62) and (11.63):* From (11.59)

$$\lim_{D\to\infty}W(0,T)=1-\sum_{i=0}^{n-1}\binom{n-1}{i}(-1)^i\frac{n}{i+1}e^{-(i+1)\lambda T} \quad (3)$$

$$=1+\sum_{j=1}^{n}\binom{n}{j}(-1)^je^{-j\lambda T} \quad (4)$$

$$=1-1+\sum_{j=0}^{n}\binom{n}{j}(-1)^je^{-j\lambda T} \quad (5)$$

$$=(1-e^{-\lambda T})^n \quad (6)$$

On the other hand,

$$\lim_{D\to 0}W(0,T)=1-\sum_{i=0}^{n-1}\binom{n-1}{i}(-1)^i\frac{n}{i+1}=1-1=0 \quad (7)$$

3) *Derivation of (11.64) and (11.65):*

$$\lim_{\lambda\to\infty}W(0,T)=1-\sum_{i=0}^{n-1}\binom{n-1}{i}(-1)^i\frac{n}{i+1}e^{-DT}=1-e^{-DT} \quad (8)$$

$$\lim_{\lambda\to 0}W(0,T)=1-\sum_{i=0}^{n-1}\binom{n-1}{i}(-1)^i\frac{n}{i+1}=1-1=0 \quad (9)$$

PROBLEMS

11.1. The following reliability parameters apply to the plant of Problem 3.2 of Chapter 3.

		Mean Time to Failures (days)	Mean Repair Time (days)
1	Reactor	1800	180
3	Compressor I	450	75
4	Compressor II	210	75
5	Pump	915	65
2	Drier	810	72
6	Stripper	370	72

(a) Obtain, at steady state, Q_s, W_s, λ_s and MTTF's, using upper bound approximations.

(b) Investigate the effect of putting a storage tank between the stripper and the reactor. Assume a tank capacity of 50 days.

11.2. For the reactor system of the preceding problem, assume a tank capacity of 50 days, which stores the outflow from the drier and the stripper. Estimate A_s, assuming matched flows.

11.3. Consider the system of Fig. P11.3. Determine the capacity of the upstream unit to prevent the tank from being empty most of the time.

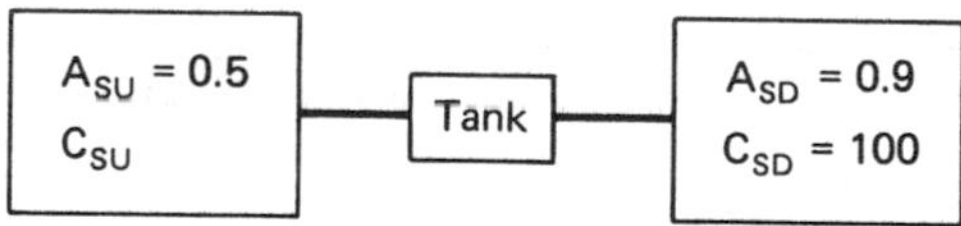

Figure P11.3

11.4. (a) Generalize (11.59) to the case where the plant is restored from the shut-down by an average reset time of $\tau = 1/\mu$

(b) Confirm (11.59), assuming $\mu^2 \gg D^2$.

(c) Obtain (11.59) when $\mu = 0$.

11.5. For the reactor system of Problem 11.1, assume the top-event failure is an explosion. If an operator is on duty at all times and there is a 10% chance that the operator will be killed in case of an accident, choose a protective instrument from Table 11.5 that will reduce the FAFR to 1.25. Assume that the instrumentation is pertinent and that there is a monthly proof test.

11.6. Consider the shut-down system shown in Fig. P11.6. Assume that the devices in the measurement and control sections have the same failure rate λ.

(a) Modify $\Pr\{S \leq t\}$ of (11.37).

(b) Prove that

$$W(0,T) = W_4(0,T) + W_2(0,T) - W_6(0,T) \tag{1}$$

where $W(0,T)$ = ENF of the plant of Fig. P11.6,
$W_n(0,T)$ = ENF of the plant containing n-redundant, connected, protective devices with failure rate λ.

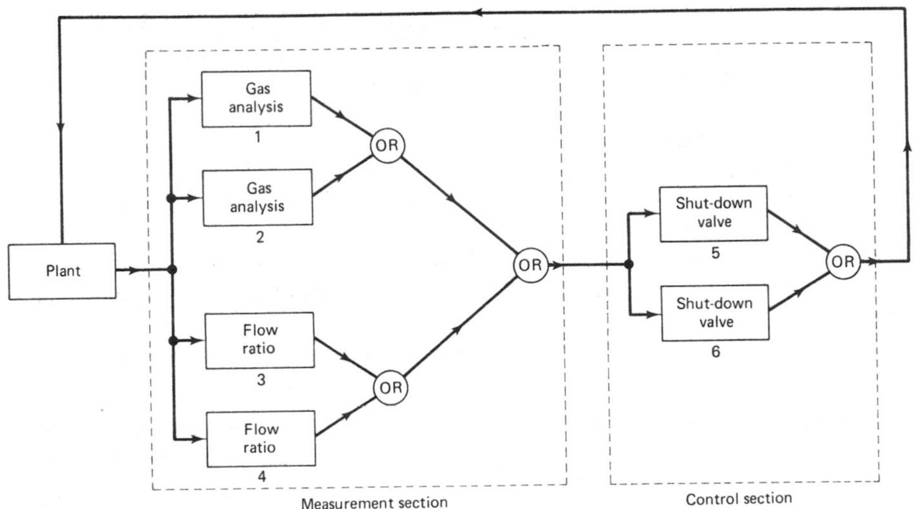

Figure P11.6. Automatic shutdown system.

11.7. Calculate the availability A_c for a shut-down system consisting of identical, parallel, and redundant units in cold standby, assuming that $N=3$, $r=2$, and $a=0.98$ in (11.67)

11.8. Obtain w_c^* of Section 11.33, assuming that $r=3$ and $N=5$.

12

MONTE CARLO METHODS FOR SYSTEM QUANTIFICATION

12.1 INTRODUCTION

Prior to the development of the kinetic tree theory, *Monte Carlo* methods were used extensively for fault tree and block diagram analysis. The technique consists of building, usually with a computer program, a probabilistic model of the system under investigation. A trial run of the model is repeated many times, and each time the performance of the synthesized system is recorded. For example, suppose one is interested in the reliability of a multicomponent system at 5000 hr. A simulation model of the system could be developed and run 100 times. Each simulation run is independent and, in effect, models the system anew. If 75 of the synthesized systems lasted longer than 5000 hr, but 25 failed prior to that time, one could conclude that the reliability of the system at 5000 hr is approximately 0.75. In general, Monte Carlo simulations are easy to carry out, and can be applied to systems that are too complex or too large to solve by deterministic methods. As is shown later in this chapter, the nature of reliability problems is such that *direct sampling* techniques are extremely wasteful of computer time, so new approaches such as *restricted sampling* and *dagger sampling* are developed.

12.2 UNIFORM RANDOM NUMBER GENERATOR

Monte Carlo simulation includes the chance variation inherent in most real-life problems, hence the descriptive name Monte Carlo. The device used to create this variation when the models are run on a digital computer is the *random number generator*. This generator usually is a subprogram that returns values from a uniform distribution of the intervals 0.0 to 1.0. Figure 12.1 is a listing of a simple, random number, generator function subprogram called RAN(NSEED) as written in FORTRAN. The single argument NSEED is the *seed* for the generator. This value is initially set by the user and determines the sequence of values that will be generated. The user, however, cannot select NSEED arbitrarily. For example, NSEED=0 results in a sequence of zeros. The subprogram has several recommended values of NSEED, such as 773311. Hence, if the user wants to carry out different sets of experiments, it would be safer to utilize disjoint subsequences generated from the subprogram with NSEED=773311, rather than to use different values of NSEED.

```
FUNCTION RAN(NSEED)
NSEED = IABS (NSEED*655393)
RAN = FLOAT (MOD(NSEED, 33554432))/FLOAT (33554432)
RETURN
END
```

Figure 12.1. Random number generator RAN(NSEED).

Figure 12.2 shows a FORTRAN program that calls on RAN(NSEED) 10 times and prints, each time, the value returned. The output from the program is also shown.

```
                                  0.4910020
                                  0.3085365
                                  0.3378267
      NSEED = 773311              0.2928143
      DO 100 I = 1, 10            0.5675969
      T = RAN(NSEED)              0.3361816
      WRITE (6,200) T             0.9255381
200   FORMAT (F12.5)              0.1872883
100   CONTINUE                    0.5499539
      STOP                        0.7460365
      END
                                  Output
```

Figure 12.2. Program and output of RAN(NSEED).

Figure 12.3 shows a histogram constructed from one thousand values generated from RAN(NSEED). As expected, each of the 10 intervals contains close to the same number of values. Function RAN(NSEED) is machine independent, and can run on any digital computer with a word size capable of storing a decimal integer equivalent to 2^{25} or larger.

The use of another uniform random number generator subprogram, RAND(L), is shown in Fig. 12.4. The figure shows a program that calls on

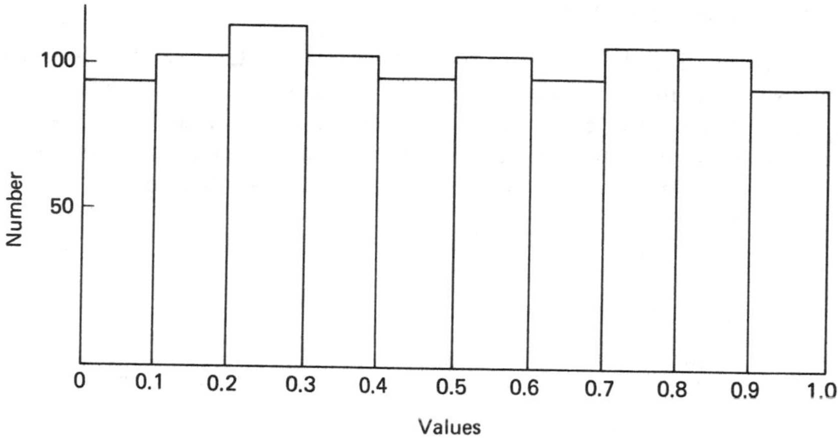

Figure 12.3. Histogram of values from RAN(NSEED).

Program

```
      L = 1
      DO 100 I = 1, 10
      T = RAND(L)
      WRITE (6,200)T
200   FORMAT (F12.5)
100   CONTINUE
      STOP
      END
```

Function RAND(L)

```
      FUNCTION RAND(L)
      DOUBLE PRECISION X1,X2,UU,VV
      IF (L.GT.1) GO TO 80
      X1 = 0.3785682200000000
      X2 = 0.0768029600000000
80    UU = X1*X2*1.0E+8
      X1 = X2
      VV = 1.0
      X2 = DMOD (UU,VV)
      L = L+1
      RAND = X2
      RETURN
      END
```

Output

```
0.41833
0.49172
0.45318
0.21663
0.42376
0.78475
0.32545
0.08142
0.21155
0.79538
```

Figure 12.4. Programs and output of RAND(L).

RAND(L) 10 times and prints, each time, the value returned. The output from the program is also shown. Function subprogram RAND(L) yields a good uniform distribution.[1]

12.3 GENERATORS FROM DISTRIBUTIONS

Algorithms are available to generate random variables from distributions by making transformations on one or several values generated from the uniform random number generator.

[1]Tsuda, T., "The Numerical Analysis of Multivariable Problems by the Use of Computers," Science Co., Ltd., Tokyo, 1973 (in Japanese).

12.3.1 Normal Distribution

Figure 12.5 is a listing of FUNCTION ANORM(EX, SD, NSEED), that returns a random value from a normal distribution with an expected value of EX and standard deviation SD. This technique is based on the central limit theorem,[2] and it calls on the uniform random number generator 12 times for a single value from the specified normal distribution.

```
      FUNCTION ANORM(EX, SD, NSEED)
      SUM = 0.0
      DO 100 I = 1, 12
      SUM = SUM + RAN(NSEED)
  100 CONTINUE
      ANORM = (SUM-6.0)*SD + EX
      RETURN
      END
```

Figure 12.5. FORTRAN function for normal distribution.

12.3.2 Exponential Distribution

Figure 12.6 is a listing of FUNCTION EXPO(AV, NSEED) that returns a random variable from an exponential distribution with expected value AV. This algorithm is based on the inverse of the cumulative distribution function.[2]

```
      FUNCTION EXPO(AV, NSEED)
      EXPO = -AV*ALOG(RAN(NSEED))
      RETURN
      END
```

Figure 12.6. FORTRAN function for exponential distribution.

12.3.3 Binary Random Variable

Function BINARY(Q, L) listed in Fig. 12.7 returns zero or one with probability Q or $1-Q$, respectively. Here, function RAND(L) is used to generate uniform random numbers. If a uniform random number falls in interval $[0, Q]$ then value zero is generated. Otherwise, value one is produced. Function BINARY(Q, L) realizes component failures occurring with probability Q.

```
      FUNCTION BINARY(Q, L)
      IF(Q.GT.RAND(L)) GO TO 100
      BINARY = 1.0
      RETURN
  100 BINARY = 0.0
      RETURN
      END
```

Figure 12.7. FORTRAN function for binary random variable.

[2]Naylor T. H., et al., *Computer Simulation Techniques*, John Wiley & Sons, Inc., New York, 1966.

12.4 DIRECT SIMULATION METHOD

A digital simulation model moves from one distinct state to another. The state of the model at any point in time is represented by a set of variables. An event in a simulation model causes the model to change its state, so whenever an event occurs one or more of the variables representing the model are changed. Most simulation models are programmed so that simulated time moves directly from one event to the next. For example, suppose that it is desired to find the percentage of downtime for a computer system. It has been determined that the time between failures for the computer follows an exponential distribution with an expected time of 320 hr. Repair time follows a normal distribution with an expected time of 25 hr and a standard deviation of 5 hr.[†] The simulation is to examine 30 independent periods of 5000 hr each and report, each time, the percentage of downtime. A flow chart of the model is shown as Fig. 12.8. The model and its output are shown as Fig. 12.9. The only two events in this model are the failure of the system and the completion of the repair. Simulation time, represented by the variable CLK, moves successively from one of these events to the other. Whenever CLK becomes equal to or greater than 5000, the percentage of downtime is printed, and the model is run again with CLK back to zero.

In the simple model just shown, only two types of events were involved, and the simulation moved alternately from one type to the other. In more complex models there are usually a greater number of event types, and the sequence of events is totally unregular. Most of the programming effort in developing one of these more difficult models is in keeping the events in proper chronological order and moving the model correctly from one event to the next. Figure 12.10 is a listing of three FORTRAN subprograms which are useful for event scheduling. The three routines are SCHED(INCOD,TMIN), RMV(NCODE,TMOUT), and CLEAR. When using SCHED the user furnishes a code for some event that is to occur (INCOD), and the time of occurrence for the event (TMIN). This information is stored by SCHED in a schedule. When subroutine RMV is called, it sets the first argument NCODE to the code of the event in the schedule that has the earliest occurrence time. The second argument TMOUT is set to the occurrence time of that event. The space that the earliest event occupied on the schedule will be made available. Subroutine CLEAR initializes the schedule, so it should be called before starting a new model.

[†]Note that, if TTF and TTR were exponentially distributed, the percentage of downtime would be (25)(100)/(320+25)=7.25. By specifying a normal distribution for TTR, we have complicated the problem considerably, yet the answer, 7.25% versus 7.39% (column 3, Fig. 12.9) is virtually the same.

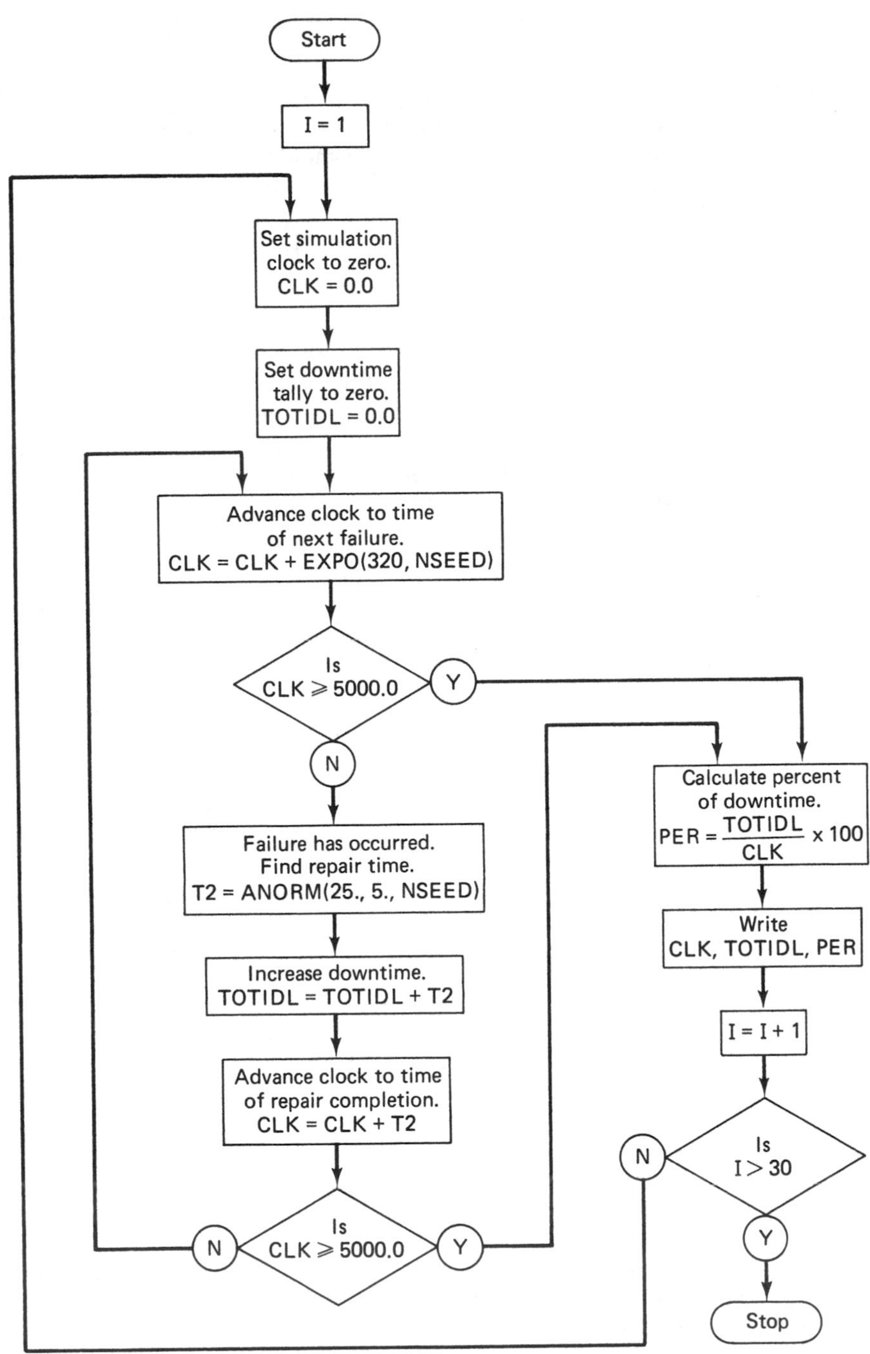

Figure 12.8. Flow chart for average down time of a computer.

```
      NSEED = 1234567
      DO 50 I = 1, 3
      CLK = 0.0
      TOTIDL = 0.0
C     ADVANCE CLOCK TO TIME OF NEXT FAILURE.
 5    CLK = CLK+EXPO (320 . , NSEED)
      IF (CLK .GE. 5000.0) GO TO 20
C     FIND REPAIR TIME.
      T2 = ANORM (25 . , 5 . , NSEED)
C     UPDATE DOWNTIME SUM.
      TOTIDL = TOTIDL+T2
C     ADVANCE CLOCK TO REPAIR COMPLETION.
      CLK = CLK+T2
      IF (CLK .GE. 5000.0) GO TO 20
      GO TO 5
C     CALCULATE PERCENT OF DOWNTIME.
 20   PER = TOTIDL/CLK*100.
      WRITE (6,21) CLK, TOTIDL, PER
 21   FORMAT (1H, 3F12.2)
 50   CONTINUE
      STOP
      END
```

	TIME	TOTAL DOWNTIME	% DOWNTIME
1	5175.90	367.49	7.10
2	5356.74	255.76	4.77
3	5103.91	437.96	8.58
4	5433.17	326.51	6.01
5	5263.39	297.01	5.64
6	5815.60	389.70	7.77
7	5363.89	336.61	6.28
8	5102.45	316.67	6.21
9	5475.78	564.51	11.12
10	5144.19	281.29	5.47
11	5759.50	431.19	7.49
12	5614.82	301.18	6.01
13	5127.52	421.92	8.13
14	5383.94	179.90	3.34
15	5600.68	367.49	7.35
16	5219.16	454.44	8.71
17	5759.46	365.07	7.22
18	5541.68	377.40	7.55
19	5914.13	403.49	8.05
20	5345.09	571.60	11.33
21	5330.34	359.92	6.75
22	5753.90	377.87	7.48
23	5877.18	375.23	6.98
24	5629.43	333.89	5.93
25	5449.64	476.79	7.46
26	5009.07	535.01	10.68
27	5104.13	339.36	6.65
28	5320.43	449.69	8.45
29	5138.72	412.44	8.73
30	5057.95	400.22	7.91

Figure 12.9. Example 1 model and output.

```
      SUBROUTINE SCHED(INCOD,TMIN)
      COMMON /SCBLK/ KODE(500),TM(500),NX(500),JP(500,JST,NEN,NCEL,LWC,
    1  NUMIN
      IF(NUMIN .EQ. NCEL) GO TO 5
      NUMIN = NUMIN+1
      J = LWC+1
    4 IF(J .GT. NCEL) J = 1
      IF(KODE(J) .EQ. 0) GO TO 10
      J = J+1
      GO TO 4
    5 WRITE(6, 7)
    7 FORMAT( ' SCHEDULE FILLED' )
      STOP
   10 TM(J) = TMIN
      KODE(J) = INCOD
      IF(JST .EQ. 0) GO TO 12
      IF(TMIN .GE. TM(NEN)) GO TO 15
      IF(JP(NEN) .EQ. 0) GO TO 18
      GO TO 25
   12 JST = J
   13 NEN = J
   14 LWC = J
      RETURN
   15 NX(NEN) = J
      JP(J) = NEN
      GO TO 13
   18 JP(NEN) = J
      NX(J) = NEN
   19 JST = J
      GO TO 14
   25 K = JP(NEN)
   26 IF(TMIN .GE. TM(K)) GO TO 30
      IF(K .EQ. JST) GO TO 27
      K = JP(K)
      GO TO 26
   27 NX(J) = K
      JP(K) = J
      GO TO 19
   30 JP(J) = K
      L = NX(K)
      JP(L) = J
      NX(J) = NX(K)
      NX(K) = J
      GO TO 14
      END
```

Figure 12.10. Subroutines SCHED, RMV and CLEAR.

```
      SUBROUTINE RMV(NCODE,TMOUT)
      COMMON /SCBLK/ KODE(500),TM(500),NX(500),JP(500),JST,NEN,NCEL,LWC,
     1  NUMIN
      IF(NUMIN .EQ. 0) GO TO 10
      NUMIN = NUMIN-1
      NCODE = KODE(JST)
      TMOUT = TM(JST)
      JX = JST
      JST = NX(JST)
      NX(JX) = 0
      JP(JX) = 0
      KODE(JX) = 0
      IF(JST .NE. 0) JP(JST) = 0
      RETURN
   10 WRITE(6, 11)
   11 FORMAT ( 'SCHEDULE EMPTY ' )
      RETURN
      END
```

```
      SUBROUTINE CLEAR
      COMMON /SCBLK/ KODE(500),TM(500),NX(500),JP(500),JST,NEN,NCEL,LWC,
     1  NUMIN
      DO 15 I = 1,500
      KODE(I) = 0
      NX(I) = 0
      JP(I) = 0
   15 CONTINUE
      JST = 0
      NEN = 0
      NCEL = 500
      LWC = 0
      NUMIN = 0
      RETURN
      END
```

Figure 12.10. (continued).

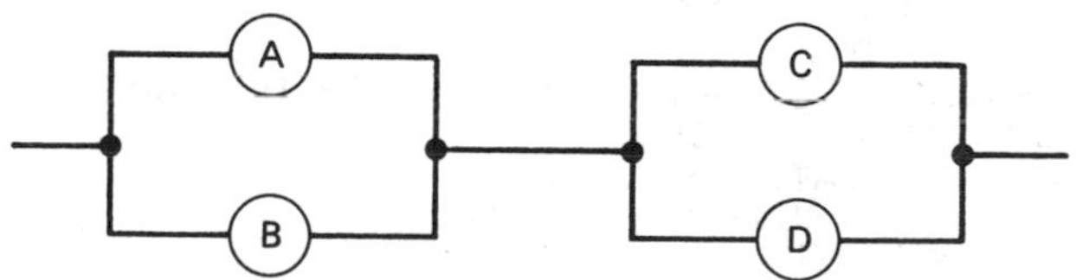

Figure 12.11. Reliability block diagram for four-component system.

To illustrate the use of these subprograms, consider the system shown as Fig. 12.11. Component *A* has a life distribution that is exponential with an expected time of 48 time units. Component *B* is a standby unit for *A* and has a life distribution that is normal with an expected time of 20 time units and a standard deviation of 4 units. Component *B* will only begin to function when *A* fails. Component *C* has an exponential life distribution with an expected value of 60 units. It has a standby component D, whose life distribution is normal with expected time of 15 units and a standard deviation of 3 units. Again, D will only begin to function when C fails. The program of Fig. 12.12 models this system, with a flow chart of the model shown as Fig. 12.13. There are four event types in this simulation. Code 1 indicates that component A has failed; 2, component B failure; 3, component C failure; and code 4 indicates the failure of component D. The output from the program is shown as Fig. 12.14.

```
      NSEED = 7654321
      SUM = 0.0
      SUMSQ = 0.0
      DO 100 I = 1,50
      CALL CLEAR
C   SCHEDULE COMPONENT A FAILURE.
      T = EXPO(48.0,NSEED)
      CALL SCHED(1, T)
C   SCHEDULE COMPONENT C FAILURE.
      T = EXPO(60.0,NSEED)
      CALL SCHED(3, T)
C   REMOVE EARLIEST EVENT IN SCHEDULE.
  2   CALL RMV(K,CLK)
C   BRANCH ACCORDING TO EVENT CODE.
      GO TO (10,20,30,40),K
C   EVENT IS COMPONENT A FAILURE, SCHEDULE B FAILURE.
 10   T = ANORM(20.0,4.0,NSEED) + CLK
      CALL SCHED(2, T)
      GO TO 2
C   EVENT IS COMPONENT B FIALURE, SYSTEM FAILS.
 20   WRITE(6,21) CLK
 21   FORMAT(1H ,F12.2,2H B)
      SUM = SUM+CLK
      SUMSQ = SUMSQ+CLK * CLK
      GO TO 100
C   EVENT IS COMPONENT C FAILURE, SCHEDULE D FAILURE.
 30   T = ANORM(15.0, 3.0, NSEED) + CLK
      CALL SCHED(4, T)
      GO TO 2
C   EVENT IS COMPONENT D FAILURE, SYSTEM FAILS.
 40   WRITE(6,41) CLK
 41   FORMAT(1H, F12.2, 2H D)
      SUM = SUM+CLK
      SUMSQ = SUMSQ+CLK * CLK
100   CONTINUE
C   CALCULATE MEAN AND VARIANCE FOR MODEL.
      AVG = SUM/50.0
      VAR = SUMSQ/50.0-AVG*AVG
      WRITE(6, 101) AVG,VAR
101   FORMAT (//, 2F16.2)
      STOP
      END
```

Figure 12.12. Program for four-component model.

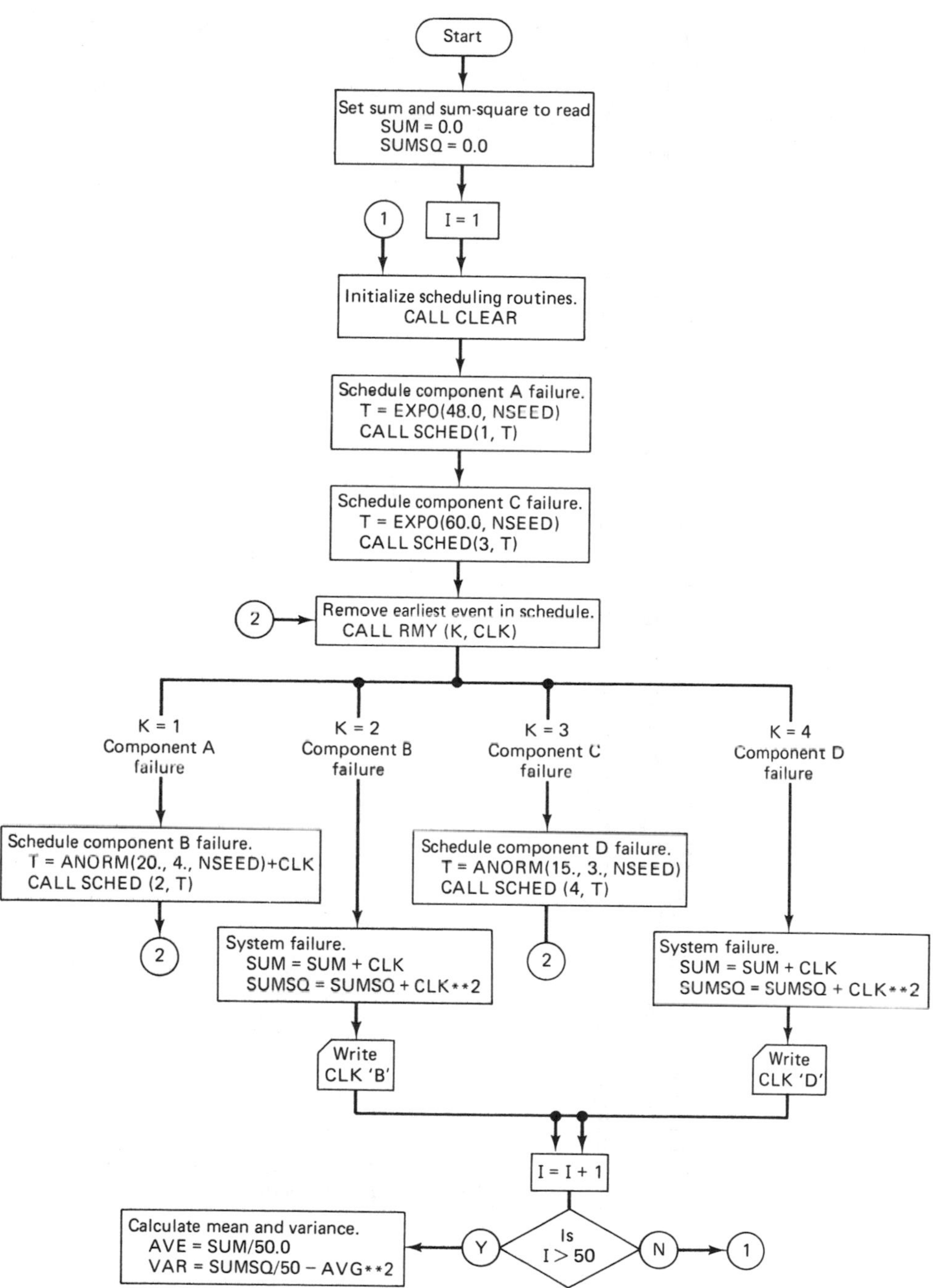

Figure 12.13. Flow chart for four-component model.

12.83	B
200.67	B
47.25	D
60.83	D
21.37	B
27.62	D
81.67	B
86.96	B
22.28	D
39.82	D
42.16	B
25.08	D
23.18	D
18.65	D
27.36	B
22.15	B
30.19	D
39.12	D
48.05	D
22.42	B
77.66	D
48.40	D
56.14	D
30.09	B
16.54	D
65.57	B
9.96	D
70.02	D
30.25	B
29.76	B
35.62	B
49.90	D
32.97	B
21.39	B
85.04	B
46.94	B
28.48	B
14.87	D
38.56	D
57.51	D
25.97	D
33.92	B
52.85	D
31.53	D
27.14	D
56.29	D
38.28	B
39.68	D
21.23	B
24.84	B

41.94 874.52

Figure 12.14. Output from four-component model.

These examples illustrate the Monte Carlo technique as applied to reliability problems that use programs written in FORTRAN. The functions and subroutines developed for these examples are general and are useful for any reliability modeling problem. There are other computer languages designed specifically for digital simulation. In these languages, the event scheduling routines are not as apparent to the user. One of the languages, GPSS (General Purpose Simulation System)[3,4] emphasizes the flow chart in its application. The model is flow-charted by using special GPSS symbols. Once the flow chart is completed the model is translated to program form, with each flow chart symbol equivalent to one program statement. A user can quickly learn the rudiments of GPSS and begin to flow chart and program simple simulation models after an hour or two of instruction. The emphasis on the flow charts makes for a clearer and quicker grasp of the models being simulated. There are over 50 flow chart symbols, but most models involve no more than a dozen different symbol types.

Another widely used simulation language is GASP (A General Activity Simulation Program).[5] GASP is a collection of FORTRAN subprograms which can be run on almost any computer system. In many ways it resembles the collection of FORTRAN routines described in this chapter. GASP, however, has a larger number of routines. The latest version, GASP IV, combines continuous and discrete modeling in a single package.

There are a variety of other digital simulation languages such as SIMSCRIPT,[6] SIMULA,[7] DYNAMO,[8] and CSMP.[9] Each has its adherents and detractors. Any of these languages can be used for reliability purposes. However, GPSS, GASP, or FORTRAN models are more common for reliability-type applications.

Prior to the development of the kinetic tree theory, Monte Carlo simulations were used routinely to obtain top-event probabilities for fault trees. Today, they are used primarily for reliability problems where the system under investigation is too complex or too large to solve realistically

[3]Gordon, G., *System Simulation*, 2nd ed., Prentice-Hall, Inc., Englewood Cliffs, N. J., 1978.

[4]Schriber, T. J., *Simulation Using GPSS*, John Wiley & Sons, Inc., New York, 1974.

[5]Pritsker, A. A., *The GASP IV Simulation Language*, John Wiley & Sons, Inc., New York, 1974.

[6]Kiviat, P. J., R. Villanueva, and H. M. Markowitz, *The SIMSCRIPT II Programming Language*, Prentice-Hall, Inc., Englewood Cliffs, N. J., 1969.

[7]Dahl, O. J., and K. Nygaard, *SIMULA: A Language for Programming and Description of Discrete Event Systems: Introduction and User's Manual*, Norwegian Computation Center, Oslo, Norway, 1967.

[8]Pugh, A. L., *DYNAMO User's Manual*, The M.I.T. Press, Cambridge, Mass., 1963.

[9]Gordon, G., *System Simulation*, 2nd ed., Prentice-Hall, Inc., Englewood Cliffs, N. J., 1978.

in any other manner. Electrical power generation systems represent examples of processes where Monte Carlo techniques provide the only practical approach to reliability analysis. Several recent papers have been published in this area.[10,11,12] These systems have large networks with on-line and standby generators at each node. Inspection of the equipment may be concurrent or on a staggered basis. The system may have covert or overt component failures. Shut-down may be scheduled or unscheduled, and repair times may vary due to component types, failure types, and availability of repair personnel. Clearly, any attempt to obtain reliability parameters for these kinds of problems by deterministic methods is virtually impossible.

Another advantage of the Monte Carlo technique is the ease with which the number and characteristics of components may be changed. Widawsky[13] shows how crew size and repair times in space missions may be varied to predict the probability of mission success. The critical components and factors in a system can quickly be determined by using the Monte Carlo technique. Most deterministic solutions only give expected values as results: for example, expected time between failures. Using Monte Carlo methods, a distribution of the results can be obtained. The distribution is usually described as a histogram, but in most cases that is sufficient; nothing more is gained by fitting a classic distribution to the results.

Another subtle advantage of the Monte Carlo technique is mentioned by Noferi et al.[11] The development and manipulation of the Monte Carlo model provides planning engineers with some "operating experience" with the system, as well as an insight into its structure and behavior. Often, that is of more value than the output from a deterministic model, since the simplifying assumptions required to solve a deterministic model are such that the true complexity of a large system is seldom understood.

The disadvantage of Monte Carlo technique is the time and expense involved in the development and execution of the models. To have reasonable confidence in a simulation result, a large number of iterations are required, thus necessitating a substantial amount of computer time.

[10]Cadwallader, G. J., et al. "Nuclear Reliability/Availability Monte Carlo Analysis," *Proceedings 1975 Annual Reliability and Maintainability Symposium.*

[11]Noferi, P. L., L. Paris, and L. Salvaderi, "Monte Carlo Methods for Power System Reliability Evaluations in Transmission and Generation Planning," *Proceedings 1975 Annual Reliability and Maintainability Symposium.*

[12]Henley, E., and R. Polk, "A Risk/Cost Assessment of Administrative Time Restrictions on Nuclear Power Plants," *Nuclear Systems Reliability Engineering and Risk Assessment*, Fussell, J., J. Burdick (eds.), *SIAM*, Philadelphia, pp. 495–518, 1977.

[13]Widawsky, W. H., *Reliability and Maintainability Parameters Evaluated with Simulation*, *IEEE Trans. on Reliability*, **R-20**, No. 3, p. 158, Aug. 1971.

12.5 MONTE CARLO SIMULATION FOR ERROR PROPAGATION THROUGH FAULT TREES

12.5.1 Introduction

To facilitate the quantitative analysis of fault trees, it is convenient to represent fault trees in mathematical form, and the structure functions described in Chapter 7 are an appropriate tool for this purpose. System unavailability $Q_s(t)$ may be obtained by methods such as complete expansion and partial pivotal decomposition of the structure function. For large and complex fault trees, the inclusion-exclusion principle can be used to approximate $Q_s(t)$. Figure 12.15 presents an AND and an OR gate (defining top event B), the corresponding structure function ψ, and system unavailability $Q_s(t)$.

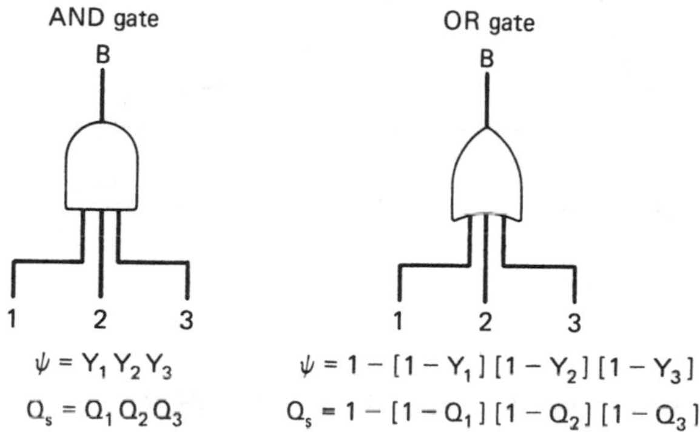

Figure 12.15. Schematic fault tree.

When fixed values are used for the failure rates and other data parameters, the system unavailability is referred to as a *point value*. However, because of the uncertainties and variations in the failure rates and data parameters, these quantities are treated as random variables and, since the system unavailability $Q_s(t)$ is a function of these random variables, it is itself a random variable. The term *error propagation* refers to the process of determining the system unavailability distribution in terms of the individual parameter distributions. Since a random variable can be viewed as having a range of values which arise due to uncertainties and population variations, a probability distribution, such as the log-normal, can be assigned to this range to obtain the likelihood of occurrence of any one particular value.

When the variation in data is characterized by factors or percentages, then the log-normal is the proper distribution to describe the data adequately. For example, if Q is a random variable which has a range between

$Q_L = Q_0/K$ and $Q_U = Q_0 K$, where Q_0 is some midpoint reference value, i.e., median, and K an error factor constant greater than one, then the log-normal is the natural distribution. When Q falls in the range $[Q_0/K, Q_0 K]$ with $100(1-2\alpha)$ percentile certainty, then the log-normal distribution has parameters $\mu = \log Q_0$ and $\sigma = (\log K)/L$, where L is the $100(1-\alpha)$ percentile of the normal distribution with a mean of zero and variance of unity. Figure 12.16 lists the characteristics of a log-normal distribution, where μ and σ are the location and scale parameters, respectively, and Q_L and Q_U are the $100\alpha\%$ and $100(1-\alpha)\%$ bounds, respectively.

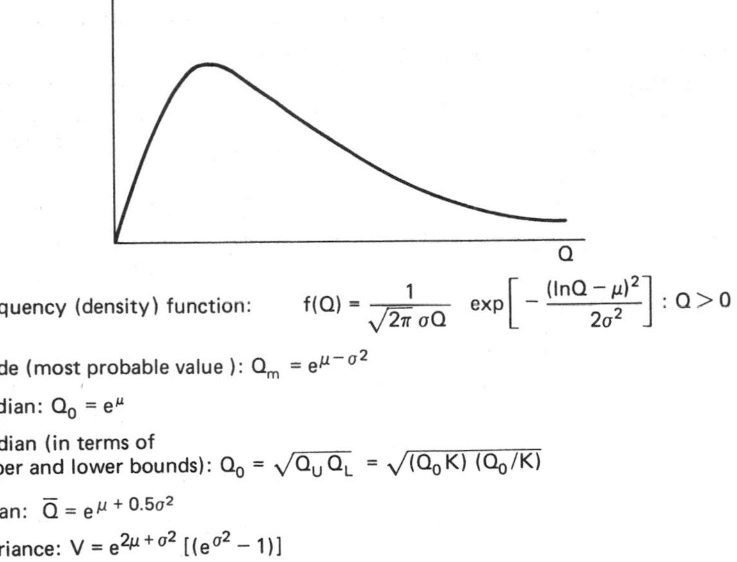

Figure 12.16. Characteristics of log-normal distribution.

With the log-normal distribution defined and assigned, the distribution may be propagated to obtain the system distribution and range. The error propagation is conveniently done by use of a computerized Monte Carlo technique. With this technique a probability is obtained (by a random sampling of the appropriate log-normal probability distribution) for the system unavailability, unreliability, etc. These probabilities are then used to compute a point value for top-event probabilities.

12.5.2 SAMPLE Program[14]

The SAMPLE program uses Monte Carlo sampling to obtain the mean, standard deviation, probability range, and distribution for a function $Y = f(X_1, X_2, \ldots, X_n)$. This function can be, for example, system unavailabil-

[14]WASH-1400, Appendix II, Vol. 1, p. 104.

ity Q_s in terms of component unavailabilities $Q_1, Q_2, \ldots, Q_n$. The function f could also be system reliability R_s, in terms of component failure and repair rates.

Given a function $Y=f(X_1,X_2,\ldots,X_n)$, values of the location and dispersion parameters of the independent variables, and a specific input distribution, SAMPLE obtains a Monte Carlo sampling $x_1,x_2,\ldots,x_n$ from the input variable distributions and evaluates the corresponding function $y=f(x_1,x_2,\ldots,x_n)$. The sampling is repeated N (input parameter) times, and the resultant estimates of Y are ordered in ascending values $y_1<y_2<\cdots<y_n$ to obtain the limits (percentiles) of the distribution of Y. The program currently has input options for three distributions: the normal, log-normal, and log-uniform.

The program was designed so that multiple functions using multiple data input descriptions can be processed during the same computer run. This is performed by the user-supplied function routine SAMPLE described in the next section.

The FUNCTION routine supplied by the user may contain one or more separate functions or algorithms for calculating Y. More than one data set may be used for the same function, or different functions. This is controlled by a computed "GO TO" subprogram.

Example 1. A SAMPLE Program

To illustrate the use of the SAMPLE program and propagation technique we consider the system for which a reliability block diagram and fault tree are shown in Fig. 12.17. The event data which contain the uncertainties for the above system is given below:

Component	Median Unavailability, Q_0	90% Error Factor K	Parameter $\mu=\log Q_0$	95% Percentile L	Parameter $\sigma=(\log K)/L$
1	9.90×10^{-3}	3.0	−4.615	1.605	0.6845
2	7.41×10^{-2}	3.0	−2.602	1.605	0.6845
3	1.53×10^{-1}	3.0	−1.877	1.605	0.6845

The system unavailability Q_s can be approximated as

$$Q_s = Q_2 + Q_1 \times Q_3$$

The input statements to SAMPLE for this system, in FORTRAN, are

```
FUNCTION SAMPLE (Q,IFLAG,NPROB)
DIMENSION Q(3)
SAMPLE=Q(2)+Q(1)*Q(3)
RETURN
END
```

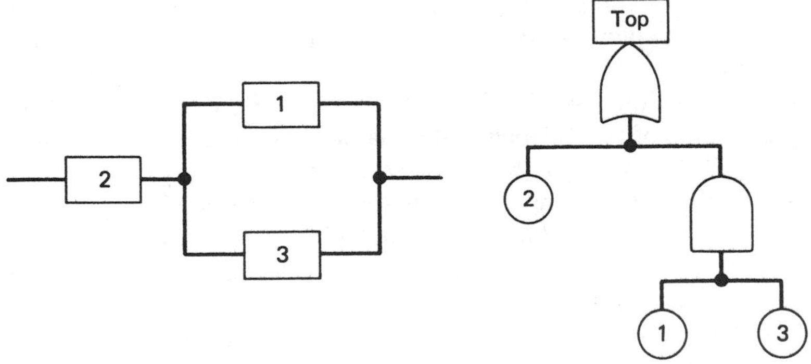

Figure 12.17. Reliability block diagram and fault tree for example problem.

The primary results of the computations are given in terms of probability confidence limits, the output being:

OUTPUT EVALUATIONS. SAMPLE SIZE = 1200
ACCURACY ON 95 PER CENT CONFIDENCE INTERVAL = 1.0 PERCENT
DISTRIBUTION PARAMETERS. MEAN = 9.4798 − 02 STANDARD DEVIATION = 6.7285 − 02

DISTRIBUTION CONFIDENCE LIMITS

CONFIDENCE (PER CENT)	FUNCTION VALUE
.5	1.7662 − 02
1.0	1.8778 − 02
2.0	2.2137 − 02
5.0	2.6577 − 02
10.0	3.3447b-02
20.0	4.4740 − 02
25.0	5.1134 − 02
30.0	5.6889 − 02
40.0	6.6473 − 02
50.0	7.6723 − 02
60.0	9.2312 − 02
70.0	1.0876 − 01
75.0	1.1929 − 01
80.0	1.3114 − 01
90.0	1.7735 − 01
95.0	2.2169 − 01
97.5	2.5573 − 01
99.0	3.1517 − 01
99.5	3.9093 − 01

Function values are the upper bounds of the indicated confidence limits. The 50% value is the median of the distribution, and the 95% and 5% values are the upper and lower bounds of the 90% probability intervals, respectively.

The SAMPLE program also generates frequency distribution functions in the form of the histogram of Fig. 12.18. The median point and the 90% confidence interval are indicated on the figure.

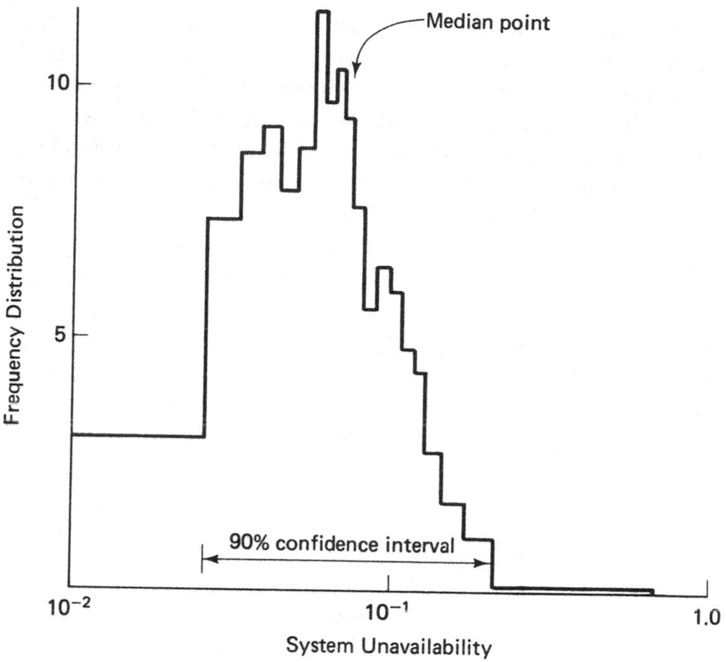

Figure 12.18. Confidence limits for top event.

12.6 SYSTEM UNAVAILABILITY EVALUATION BY MONTE CARLO METHOD WITH RESTRICTED SAMPLING

12.6.1 Introduction

References[15-21] offer further examples of the application of Monte Carlo techniques to reliability problems. Some of them use the direct sampling techniques described in the previous sections, and others employ more sophisticated ones. In the direct sampling cases a large number of trials are

[15]Kamat, S. J., and M. W. Riley, "Determination of Reliability Using Event-Based Monte Carlo Simulation," *IEEE Trans. on Reliability*, Vol. R-24, pp. 73–75, Apr. 1975.

[16]Kamat, S. J., and W. E. Franzmeier, "Determination of Reliability Using Event-Based Monte Carlo Simulation, Part II," *IEEE Trans. on Reliability*, Vol. R-25, pp. 254–255, Oct. 1976.

[17]Vesely, W. E., and R. E. Narum, "PREP and KITT: Computer Codes for the Automatic Evaluation of a Fault Tree," IN-1349, Aug., 1970. (Available from National Technical Information Service, Springfield, Va., 22161.)

[18]Becker, P. W., "Finding the Better of Two Similar Designs By Monte Carlo Techniques," *IEEE Trans. on Reliability*, Vol. R-23, pp. 242–246, Oct., 1974.

[19]Mazumdar, M., "Importance Sampling in Reliability Estimation, Reliability and Fault Tree Analysis," *SIAM*, Philadelphia, pp. 153–163, 1975.

[20]Van. Slyke, R., and H. Frank, "Network Reliability Analysis: Part I," *Networks*, Vol. 1, pp. 279–290, 1972.

[21]Van. Slyke, R., H. Frank, and A. Kershenbaum, "Network Reliability Analysis: Part II, Reliability and Fault Tree Analysis," Barlow, R. E., et al. (eds.), *SIAM*, Philadelphia, pp. 619–650, 1975.

required to obtain reasonably precise estimates of the unavailability. If, for example, the exact system unavailability is 1×10^{-5}, then 1,000,000 trials would yield about 10 system failures, whereas 10,000 trials could result in no system failure and might lead us to conclude that the system is completely reliable. On the average, we require at least 100,000 trials to have a failure, and about 10,000,000 trials are required to produce Monte Carlo estimates with one significant figure.

In this section, an improved Monte Carlo method is developed by applying variance-reducing techniques.[22] We begin by denoting by rectangle D of Fig. 12.19 as the area from which the direct Monte Carlo trials are sampled. Assume that the area of D is unity. Circle S designates the area in which trials result in system failures. The area of disk S corresponds to the system unavailability Q_s. Note that only a few trials result in the system failure; i.e., the diagram is not to scale.

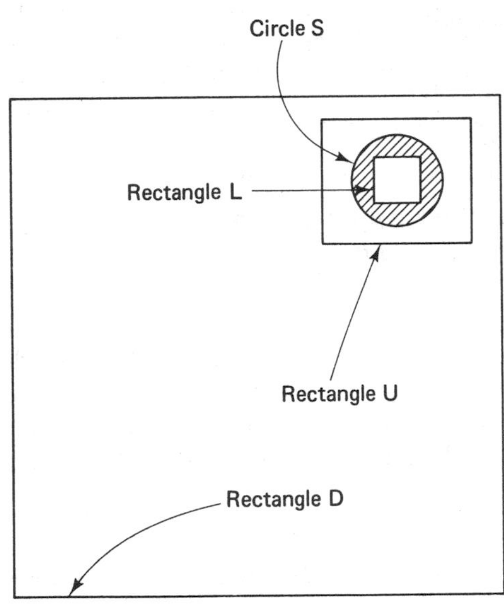

Figure 12.19. Venn diagram representation.

Disk S consists of various system failure modes. Assume now that we can analytically calculate contribution Q_L to system unavailability Q_s, Q_L being that part of the system failure modes denoted by rectangle L. Since the area of rectangle L is Q_L, the calculation of the system unavailability is

[22]Kumamoto, H., K. Tanaka, and K. Inoue, "Efficient Evaluation of System Reliability by Monte Carlo Method," *IEEE Trans. on Reliability*, Vol. R-26, No. 5, pp. 311–315, 1977.

reduced to carrying out the subtraction $Q_s - Q_L$, which is represented by the shaded area between circle S and rectangle L. In other words, if M-out-of-N trials fall in the shaded area, then the system unavailability Q_s is

$$Q_s = [Q_s - Q_L] + Q_L \tag{12.1}$$

$$\cong \frac{M}{N} + Q_L \tag{12.2}$$

It can proven that this Monte Carlo estimator has a smaller variance than the regular Monte Carlo estimator, since part of the system unavailability is calculated analytically. The technique of subtracting Q_L is regarded as an application of a variance-reducing technique called the *control variate method.*[23]

Consider now another rectangle, U, which encloses disk S. Assume that area U can be calculated analytically as Q_U. Let us restrict the Monte Carlo trials to the area between rectangle U and L. In other words, that area is a universal set. Assume that M-out-of-N Monte Carlo trials *from the area* fall on the shaded area between circle S and rectangle L. Then, ratio $[Q_s - Q_L]/[Q_U - Q_L]$ is estimated by

$$\frac{Q_s - Q_L}{Q_U - Q_L} \cong \frac{M}{N} \tag{12.3}$$

Thus, the difference $[Q_s - Q_L]$ is

$$Q_s - Q_L \cong [Q_U - Q_L] \times \frac{M}{N} \tag{12.4}$$

and the system unavailability Q_s of (12.1) is

$$Q_s \cong [Q_U - Q_s] \times \frac{M}{N} + Q_L \tag{12.5}$$

It can be proven that the Monte Carlo estimator on the right-hand side of (12.5) has smaller variance than that of (12.2), since the variance of M/N is compressed by factor $Q_U - Q_L$, which is smaller than unity. The following sections present a formal description of the Monte Carlo method based on these ideas.

[23]Hammersley, J. M., and D. C. Handscomb, *Monte Carlo Method,* Methuen and Co., Ltd., London, 1967.

12.6.2 Problem Statement

Assumptions:

1) The fault tree has k basic events, $1,\ldots,k$.
2) Basic events are statistically independent.
3) Several path and cut sets are known.

Notation:

$$x_i = \text{state of basic event } i, \quad \begin{cases} 1, & \text{if event } i \text{ is occurring} \\ 0, & \text{otherwise} \end{cases}$$

$\mathbf{x} = (x_1,\ldots,x_k)$ is a basic event state vector,

$\mathbf{b} = (b_1,\ldots,b_k)$ is a binary vector,

$$\psi(\mathbf{x}) = \text{binary function expressing the top event of the fault tree} = \begin{cases} 1, & \text{if the top event is occurring} \\ 0, & \text{otherwise} \end{cases}$$

N = sample size of Monte Carlo method.

We assume that every state vector is possible, i.e., for all $\mathbf{b}$, the inequality

$$0 < \Pr(\mathbf{x}=\mathbf{b}) = \prod_{i=1}^{k} \Pr(x_i = b_i) < 1 \tag{12.6, 12.7}$$

holds. We now need to calculate the system unavailability

$$Q_s \equiv \Pr(\psi(\mathbf{x}) = 1) \tag{12.8}$$

$$= \sum_{\mathbf{b}} \psi(\mathbf{b}) \Pr(\mathbf{x}=\mathbf{b}) \tag{12.9}$$

$$= E_{\mathbf{x}}(\psi(\mathbf{x})) \tag{12.10}$$

First, we consider how this is done by direct sampling, then by restricted sampling.

12.6.3 Direct Monte Carlo Method

We begin by generating N statistically independent samples $\mathbf{c}_1,\ldots,\mathbf{c}_N$ of $\mathbf{x}$. Figure 12.20 illustrates the generation of 100 samples for basic event 1. In general, $N \times k$ independent, uniform, random numbers are used for the generation of the N samples $\mathbf{c}_1,\ldots,\mathbf{c}_N$. Each element of sample vector $\mathbf{c}_\nu$ can be obtained by using function BINARY (Q, L) of section 12.3.3. Next, we evaluate Q_s by the unbiased binomial estimator $\hat{Q}_C$, where the subscript C stands for "direct sampling."

$$\hat{Q}_C \equiv N^{-1} \sum_{\nu=1}^{N} \psi(\mathbf{c}_\nu) \tag{12.11}$$

The variance of $\hat{Q}_C$ is equal to the sum of the variance of each $\psi(\mathbf{c}_\nu)$ divided by N^2, and is

$$\mathrm{Var}(\hat{Q}_C) = N^{-1} Q_s (1 - Q_s) \tag{12.12}$$

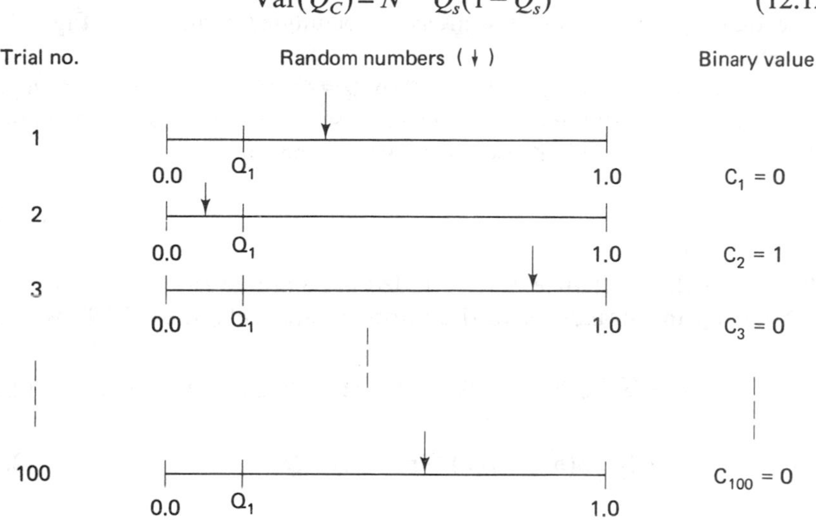

Figure 12.20. Generation of 100 samples for basic event 1.

12.6.4 Restricted-Sampling Monte Carlo Method

Let ψ_L and ψ_U be two binary functions satisfying (12.13) and (12.14).

$$\psi_L(\mathbf{b}) \le \psi(\mathbf{b}) \le \psi_U(\mathbf{b}), \quad \text{for all } \mathbf{b} \tag{12.13}$$

$$\psi_L(\mathbf{b}) \not\equiv 0, \quad \psi_U(\mathbf{b}) \not\equiv 1 \tag{12.14}$$

For any given $i, 0 \le i \le k$, define

$$Q_{L,i}(b_1, \ldots, b_i) = \sum_{b_{i+1}, \ldots, b_k} \psi_L(\mathbf{b}) \Pr(\mathbf{x} = \mathbf{b}) \tag{12.15}$$

$$Q_{U,i}(b_1, \ldots, b_i) = \sum_{b_{i+1}, \ldots, b_k} \psi_U(\mathbf{b}) \Pr(\mathbf{x} = \mathbf{b}) \tag{12.16}$$

Quantities $Q_{L,i}$ and $Q_{U,i}$ are used to generate random samples from the area between rectangles L and U of Fig. 12.19. Functions ψ_L and ψ_U and quantities $Q_{L,i}$ and $Q_{U,i}$ will be obtained by the method to be given in Section 12.6.5. Functions ψ_L and ψ_U are the indicator functions for rectangles L and U, respectively, and disk S has the indicator function ψ.

Quantities $Q_{L,0}$ and $Q_{U,0}$ are the unavailabilities of the fault trees with top events ψ_L and ψ_U, respectively, and are represented by Q_L and Q_U. The following inequality holds:

$$0<Q_L \leq Q_s \leq Q_U \leq 1 \tag{12.17}$$

Note that Q_L and Q_U are the areas of rectangles L and U in Fig. 12.19, respectively.

If the equality $Q_U=Q_L$ holds, then $Q=Q_L=Q_U$, and the problem is trivial: Q can be obtained without the use of the Monte Carlo method. Therefore, in the ensuing discussion we assume that

$$Q_U-Q_L>0 \tag{12.18}$$

This means that rectangle L is included in rectangle U.

Applying the straightforward control variate method to (12.9), we have

$$Q_s=\sum_{\mathbf{b}}[\psi(\mathbf{b})-\psi_L(\mathbf{b})]\Pr(\mathbf{x}=\mathbf{b})+\sum_{\mathbf{b}}\psi_L(\mathbf{b})\Pr(\mathbf{x}=\mathbf{b}) \tag{12.19}$$

$$=\sum_{\mathbf{b}}[\psi(\mathbf{b})-\psi_L(\mathbf{b})]\Pr(\mathbf{x}=\mathbf{b})+Q_L \tag{12.20}$$

This equation corresponds to (12.1).

We consider now the generation of random samples with probabilities different from $\Pr(\mathbf{x}=\mathbf{b})$, according to the importance sampling method. Define the sets

$$X\equiv\{\mathbf{b}|\psi(\mathbf{b})-\psi_L(\mathbf{b})=1\} \tag{12.21}$$

$$Y\equiv\{\mathbf{b}|\psi_U(\mathbf{b})-\psi_L(\mathbf{b})=1\} \tag{12.22}$$

Set X is the shaded area in Fig. 12.19, and set Y is the area between rectangle U and rectangle L. Since $X\subset Y$, we rewrite (12.20) as

$$Q_s\equiv\sum_{\mathbf{b}\in\mathbf{X}}[\psi(\mathbf{b})-\psi_L(\mathbf{b})]\Pr(\mathbf{x}=\mathbf{b})+Q_L \tag{12.23}$$

$$=\sum_{\mathbf{b}\in\mathbf{Y}}[\psi(\mathbf{b})-\psi_L(\mathbf{b})]\Pr(\mathbf{x}=\mathbf{b})+Q_L \tag{12.24}$$

Next, we introduce a new random vector $\mathbf{y}=(y_1,\ldots,y_k)$ in set Y, and define probability $\Pr(\mathbf{y}=\mathbf{b}), \mathbf{b}\in\mathrm{Y}$ by

$$\Pr(\mathbf{y}=\mathbf{b})\equiv\frac{\Pr(\mathbf{x}=\mathbf{b})}{[Q_U-Q_L]} \tag{12.25}$$

Note that this probability is defined on the area between rectangles U and L. Equation (12.24) can now be written as

$$Q_s = [Q_U - Q_L] \sum_{\mathbf{b} \in \mathbf{Y}} [\psi(\mathbf{b}) - \psi_L(\mathbf{b})] \Pr(\mathbf{y} = \mathbf{b}) + Q_L \tag{12.26}$$

Since $\psi_L(\mathbf{b}) \equiv 0$ for all $\mathbf{b} \in Y$,

$$Q_s = [Q_U - Q_L] \sum_{\mathbf{b} \in \mathbf{Y}} \psi(\mathbf{b}) \Pr(\mathbf{y} = \mathbf{b}) + Q_L \tag{12.27}$$

$$= [Q_U - Q_L] E_{\mathbf{y}}(\psi(\mathbf{y})) + Q_L \tag{12.28}$$

The expected value in (12.28) can be evaluated by random sampling; thus, (12.28) is a formal statement of the restricted-sampling Monte Carlo method. To show the relationship between the direct and the new method, consider N statistically independent samples $\mathbf{s}_1, \ldots, \mathbf{s}_N$ of $\mathbf{y}$ which belong to set Y. We evaluate Q_s by the unbiased binomial estimator $\hat{Q}_R$ (the subscript R standing for "restricted sampling"):

$$\hat{Q}_R \equiv [Q_U - Q_L] N^{-1} \sum_{\nu=1}^{N} \psi(\mathbf{s}_\nu) + Q_L \tag{12.29}$$

Random samples $\mathbf{s}_1, \ldots, \mathbf{s}_N$ can be generated readily by methods which are given in Section 12.6.6. Note that the estimator (12.29) corresponds to the right-hand side of (12.5).

The following theorem shows that the new restricted-sampling Monte Carlo method estimates the system unavailability with a smaller variance than the direct Monte Carlo.

Theorem: Let ψ_L and ψ_U satisfy (12.13) and (12.14). Assume $Q_U - Q_L > 0$ as in (12.18). Define $\hat{Q}_R$ as in (12.29). Then,

$$\text{Var}(\hat{Q}_R) = N^{-1}(Q_U - Q_s)(Q_s - Q_L) \tag{12.30}$$

$$< \text{Var}(\hat{Q}_C) = N^{-1} Q_s(1 - Q_s) \tag{12.31}$$

The theorem itself is proven in footnote reference 22.

12.6.5 Constructing $(\psi_L, Q_{L,i})$ and $(\psi_U, Q_{U,i})$

Take some, say m, cut sets $K_1, \ldots, K_m$ of a fault tree with top event ψ. Consider a new fault tree consisting of these m cut sets, and define ψ_L as the top event of this fault tree. Then, from (7.58), Chapter 7, we have

$$\psi_L(\mathbf{b}) = 1 - \prod_{j=1}^{m} \left[1 - \prod_{i \in K_j} b_i \right] \tag{12.32}$$

We now show that $\psi_L(\mathbf{b})$ satisfies (12.13) and (12.14). Suppose that $\psi_L(\mathbf{b})=1$. Then, from (12.32), at least one cut set among $K_1,\ldots,K_m$ is failed. Hence, $\psi(\mathbf{b})=1$, resulting in (12.13). Equation (12.14) is satisfied by (12.32).

$Q_{L,i}$ of (12.15) can be calculated by using the partial pivotal decomposition of Section 7.5.3 of Chapter 7, by rewriting (12.32) as

$$\psi_L(\mathbf{b})=\sum_{j=1}^{m^*} L_j(\mathbf{b})\equiv h_L(\mathbf{b}) \tag{12.33}$$

where each $L_j(\mathbf{b})$ consists of statistically independent factors. $Q_{L,i}$ can be obtained by

$$Q_{L,j}(b_1,\ldots,b_i)=h_L(b_1,\ldots,b_i,\Pr(x_{i+1}=1),\ldots,\Pr(x_k=1))\prod_{l=1}^{i}\Pr(x_l=b_l) \tag{12.34}$$

Take some, say n, path sets $P_1,\ldots,P_n$ of the original fault tree ψ. Consider another fault tree having these n path sets. Then, that fault tree has the top event $\psi_U(\mathbf{b})$ defined by

$$\psi_U(\mathbf{b})\equiv\prod_{j=1}^{n}\left[1-\prod_{i\in P_j}(1-b_i)\right] \tag{12.35}$$

In the same way, we see that ψ_U satisfies (12.13) and (12.14). We can obtain h_U like h_L, by partial pivotal expansions of ψ_U. The value $Q_{U,i}$ of (12.16) can thus be calculated by

$$Q_{U,i}(b_1,\ldots,b_i)=h_U(b_1,\ldots,b_i,\Pr(x_{i+1}=1),\ldots,\Pr(x_k=1))\prod_{l=1}^{i}\Pr(x_l=b_l) \tag{12.36}$$

We now show, by example, how this bounding process is applied.

Example 2. Two-Out-of-Three System

Consider a two-out-of-three system consisting of components 1, 2, and 3. The system has cut sets $\{1,2\}$, $\{2,3\}$, $\{1,3\}$ and path sets $\{1,2\}$, $\{2,3\}$, $\{1,3\}$. Represent cut sets $\{1,2\}$ and $\{2,3\}$ by K_1 and K_2 ($m=2$) and path sets $\{1,2\}$ and $\{1,3\}$ as P_1 and P_2 ($n=2$). Obtain ψ_L, ψ_U, h_L, h_U, $Q_{L,i}$, and $Q_{U,i}$.

Solution:

$$\psi_L(\mathbf{b})=1-[1-b_1b_2][1-b_2b_3],\quad h_L(\mathbf{b})=b_2[1-(1-b_1)(1-b_3)] \tag{12.37}$$

Denote by Q_i probability $\Pr(x_i=1)$. Then,

$$Q_L = Q_{L,0} = Q_2[1-(1-Q_1)(1-Q_3)]$$
$$Q_{L,1}(b_1) = Q_2[1-(1-b_1)(1-Q_3)]\Pr(x_1=b_1)$$
$$Q_{L,2}(b_1,b_2) = b_2[1-(1-b_1)(1-Q_3)]\Pr(x_1=b_1)\Pr(x_2=b_2)$$
$$Q_{L,3}(b_1,b_2,b_3) = b_2[1-(1-b_1)(1-b_3)]\Pr(x_1=b_1)\Pr(x_2=b_2)\Pr(x_3=b_3)$$

Similarly,

$$\psi_U(\mathbf{b}) = [1-(1-b_1)(1-b_2)][1-(1-b_1)(1-b_3)]$$
$$h_U(\mathbf{b}) = b_1+(1-b_1)b_2b_3$$
$$Q_U = Q_{U,0} = Q_1+(1-Q_1)Q_2Q_3$$
$$Q_{U,1}(b_1) = [b_1+(1-b_1)Q_2Q_3]\Pr(x_1=b_1)$$
$$Q_{u,2}(b_1,b_2) = [b_1+(1-b_1)b_2Q_3]\Pr(x_1=b_1)\Pr(x_2=b_2)$$
$$Q_{U,3}(b_1,b_2,b_3) = [b_1+(1-b_1)b_2b_3]\Pr(x_1=b_1)\Pr(x_2=b_2)\Pr(x_3=b_3)$$

Example 3.

Obtain the state vectors in set Y for ψ_L and ψ_U of Example 1.

Solution:

b_1	b_2	b_3	$\psi_U(\mathbf{b})$	$\psi_L(\mathbf{b})$	
0	0	0	0	0	No
0	0	1	0	0	No
0	1	0	0	0	No
0	1	1	1	1	No
1	0	0	1	0	Ok
1	0	1	1	0	Ok
1	1	0	1	1	No
1	1	1	1	1	No

Therefore,

$$Y = \{(1,0,0),(1,0,1)\}$$

12.6.6 Generation of Samples $\mathbf{s}_1,\ldots,\mathbf{s}_N$

Probability $\Pr(\mathbf{y}=\mathbf{b})$ can be represented as (see chain rule in Section 4.12 of Chapter 4)

$$\Pr(\mathbf{y}=\mathbf{b}) = \Pr(y_1=b_1)\Pr(y_2=b_2|y_1=b_1)\times\ldots \times\Pr(y_k=b_k|y_1=b_1,\ldots,y_{k-1}=b_{k-1}) \tag{12.38}$$

This identity shows that the generation of state vector $\mathbf{s}_\nu=(s_{1,\nu},\ldots,s_{k,\nu})$ reduces to the sequential generation of $s_{1,\nu},\ldots,s_{i,\nu},\ldots,s_{k,\nu}$ with probabilities $\Pr(y_i|y_1=s_{1,\nu},\ldots,y_{i-1}=s_{i-1,\nu})$, $i=1,\ldots,k$, respectively.

Assume now that the first $(i-1)$ elements of $\mathbf{s}_\nu$ have already been generated. Assume further that we can calculate $\Pr(y_i|y_1=s_{1,\nu},\ldots,y_{i-1}=s_{i-1,\nu})$ for $y_i=1$ and 0. Then, element $i,s_{i,\nu}$, can be generated by using BINARY (Q,L) in the following way.

$$s_{i,\nu}=\begin{cases}1, & \text{with probability } \Pr(y_i=1|y_1=s_{1,\nu},\ldots,y_{i-1}=s_{i-1,\nu})\\ 0, & \text{with probability } \Pr(y_i=0|y_1=s_{1,\nu},\ldots,y_{i-1}=s_{i-1,\nu})\end{cases}$$

For the general term, $\Pr(y_i=b_i|y_1=b_1,\ldots,y_{i-1}=b_{i-1})$, the following identity holds.

$$\begin{aligned}\Pr(y_i=b_i|y_1=b_1,\ldots,y_{i-1}=b_{i-1})&\\ =\frac{\Pr(y_1=b_1,\ldots,y_i=b_i)}{\Pr(y_1=b_1,\ldots,y_{i-1}=b_{i-1})}&\\ =\frac{\sum_{b_{i+1},\ldots,b_k}[\psi_U(\mathbf{b})-\psi_L(\mathbf{b})]\Pr(y_1=b_1,\ldots,y_k=b_k)}{\sum_{b_i,\ldots,b_k}[\psi_U(\mathbf{b})-\psi_L(\mathbf{b})]\Pr(y_1=b_1,\ldots,y_k=b_k)}&\end{aligned} \tag{12.39}$$
†

Substituting $\Pr(\mathbf{y}=\mathbf{b})$ of (12.25) into the denominator and numerator of (12.39), and using (12.15) and (12.16), we have

$$\begin{aligned}\Pr(y_i=b_i|y_1=b_1,\ldots,y_{i-1}=b_{i-1})&\\ =\frac{Q_{U,i}(b_1,\ldots,b_i)-Q_{L,i}(b_1,\ldots,b_i)}{Q_{U,i-1}(b_1,\ldots,b_{i-1})-Q_{L,i-1}(b_1,\ldots,b_{i-1})}&\end{aligned} \tag{12.40}$$

Example 4.

Evaluate (12.40) for ψ_L and ψ_U of Example 2.

Solution: From Example 2, we calculate $Q_{U,i}-Q_{L,i}$; thus,

$$\Pr(y_1=b_1)=\frac{b_1(1-Q_2)\Pr(x_1=b_1)}{Q_1(1-Q_2)}=\frac{b_1\Pr(x_1=b_1)}{Q_1}$$

$$\Pr(y_2=b_2|y_1=b_1)=\frac{(1-b_2)\Pr(x_2=b_2)}{(1-Q_2)}$$

$$\Pr(y_3=b_3|y_1=b_1,y_2=b_2)=\Pr(x_3=b_3)$$

†Set Y, the area between rectangles L and U, has indicator function $\psi_U(\mathbf{b})-\psi_L(\mathbf{b})$.

Hence, we have the probabilities

$\Pr(y_1=0)=0$

$\Pr(y_1=1)=1$ → $\Pr(y_2=0|y_1=1)=1$ → $\Pr(y_3=0|y_1=1,y_2=0)=1-Q_3$

$\Pr(y_3=1|y_1=1,y_2=0)=Q_3$

$\Pr(y_1=1)=1$ → $\Pr(y_2=1|y_1=1)=0$

We observe that any sample vector belongs to set $Y=\{(1,0,0), (1,0,1)\}$.

Example 5.

The system to be analyzed can be represented by the reliability block diagram of Fig. 12.21, since the new Monte Carlo method also applies to block diagrams. The unavailabilities of the components are $\Pr(x_i=1)=0.1$ for all i.

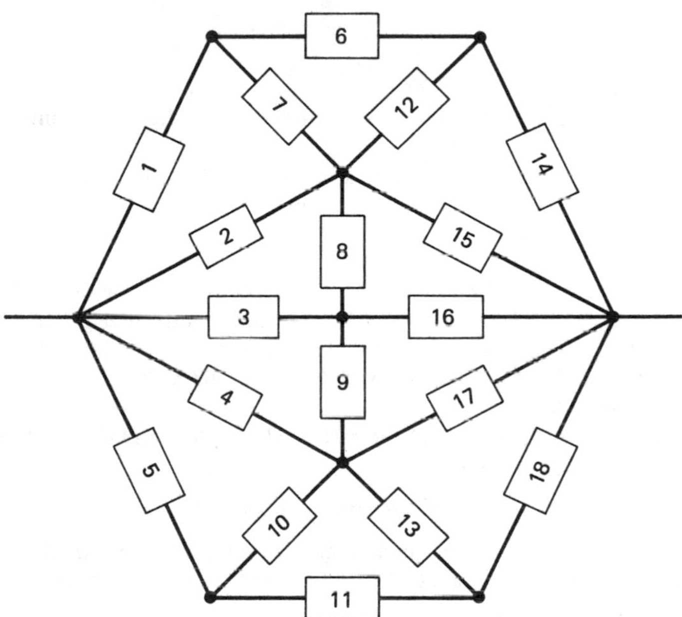

Figure 12.21. Reliability block diagram for example problem.

Solution: Consider the cut set $(m=1)$,

$$K_1=\{1,2,3,4,5\}$$

and the path sets $(n=6)$,

$$P_1=\{2,15\},\quad P_2=\{3,16\},\quad P_3=\{4,17\}$$
$$P_4=\{1,6,14\},\quad P_5=\{5,11,18\},\quad P_6=\{1,7,15\}$$

Then,

$$\psi_L(\mathbf{b}) = h_L(\mathbf{b}) = b_1 b_2 b_3 b_4 b_5 \tag{12.41}$$

$$\begin{aligned}\psi_U(\mathbf{b}) = &[1-(1-b_2)(1-b_{15})][1-(1-b_3)(1-b_{16})][1-(1-b_4)1-b_{17})] \\ &\times[1-(1-b_1)(1-b_6)(1-b_{14})][1-(1-b_5)(1-b_{11})(1-b_{18})] \\ &\times[1-(1-b_1)(1-b_7)(1-b_{15})]\end{aligned} \tag{12.42}$$

The partial pivotal decomposition of (12.42) gives

$$\begin{aligned}h_U(\mathbf{b}) = &[1-(1-b_3)(1-b_{16})][1-(1-b_4)(1-b_{17})][1-(1-b_5)(1-b_{11})(1-b_{18})] \\ &\times\{b_1[1-(1-b_2)(1-b_{15})] \\ &+(1-b_1)[1-(1-b_6)(1-b_{14})][b_{15}+(1-b_{15})b_2 b_7]\}\end{aligned} \tag{12.43}$$

From (12.41) and (12.43), we obtain

$$Q_L = 10\times10^{-6} \tag{12.44}$$

$$Q_U = 368\times10^{-6} \tag{12.45}$$

Inequality (12.17) ensures that the system unavailability lies in the interval $[10\times10^{-6}, 368\times10^{-6}]$.

Unavailability Q_s of (12.9) is the sum of $2^{18} = 262{,}144$ terms. A termwise calculation gives the exact system unavailability as

$$Q_s = 29.1\times10^{-6} \tag{12.46}$$

We see from (12.12) and (12.30) that estimators $\hat{Q}_C$ and $\hat{Q}_R$ with $N = 3000$ have standard deviations of 98.5×10^{-6} and 1.47×10^{-6}, respectively. The standard deviation of $\hat{Q}_R$ is much less than the standard deviation of $\hat{Q}_C$ and also much less than 358×10^{-6}, the length of the interval $[10\times10^{-6}, 368\times10^{-6}]$ of the unavailability obtained from (12.17).

Figure 12.22 shows the results of the restricted sampling and direct Monte Carlo methods. For $N = 3000$, we have

$$\hat{Q}_R = 27.9\times10^{-6}, \quad \hat{Q}_C = 0.0 \tag{12.47}$$

Estimator $\hat{Q}_C$ gave zero as the unavailability, since all the samples $\psi(\mathbf{c}_\nu), \nu = 1,\ldots,3000$ became zero. Figure 12.22 includes the 95% upper and lower statistical confidence limit of $\hat{Q}_N$ for different values of N.

12.7 SYSTEM UNAVAILABILITY EVALUATION BY A MONTE CARLO METHOD WITH DAGGER SAMPLING

12.7.1 Introduction

For the fault trees of Section 12.6, the direct Monte Carlo method requires $N\times k$ uniform random numbers to generate N sample states $\mathbf{c}_1,\ldots,\mathbf{c}_N$. As is shown in Section 12.7.4, most of the computation time for the direct method is spent in generating uniform random numbers. We

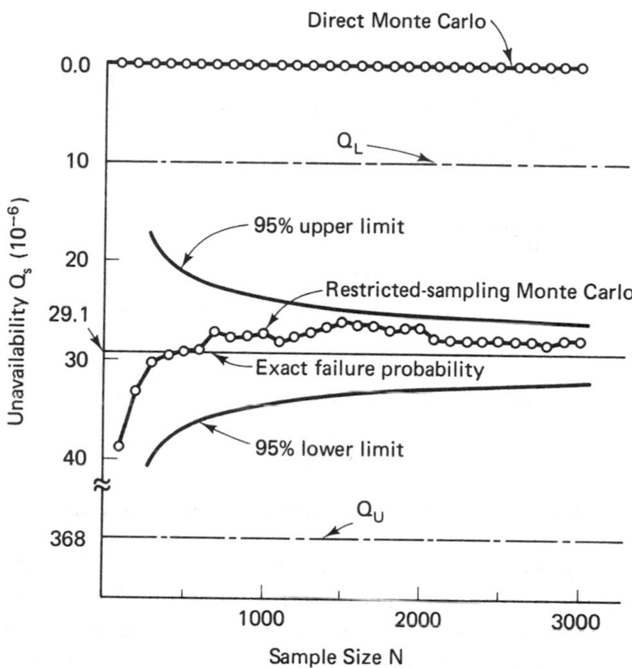

Figure 12.22. Result of restricted-sampling and direct Monte Carlo methods.

present here a Monte Carlo method which significantly decreases random number generations and, hence, saves computation time. The proposed method also decreases the variance of the Monte Carlo estimator for given sample size N.

12.7.2 Dagger Sampling

As an example, consider the case where each basic event occurs with probability 0.01. The proposed method generates 100 samples for event 1 in the manner shown in Fig. 12.23.

A group of 100 intervals between 0 and 1 is introduced for basic event 1. The ith interval is used for generating event 1 in the ith trial and has subinterval $[(i-1)\times 0.01, i\times 0.01)$, whose length is equal to probability 0.01 of the occurrence of the basic event. Thus, the first interval has subinterval $[0, 0.01)$ and the last interval $[0.99, 1)$.

Only *one uniform random number* is generated for the group. Assume that the random number falls in the subinterval of the ith interval. Then, basic event 1 is assumed to occur in the ith trial and not to occur in the other 99 trials. For example, random number 0.4256 determines that event

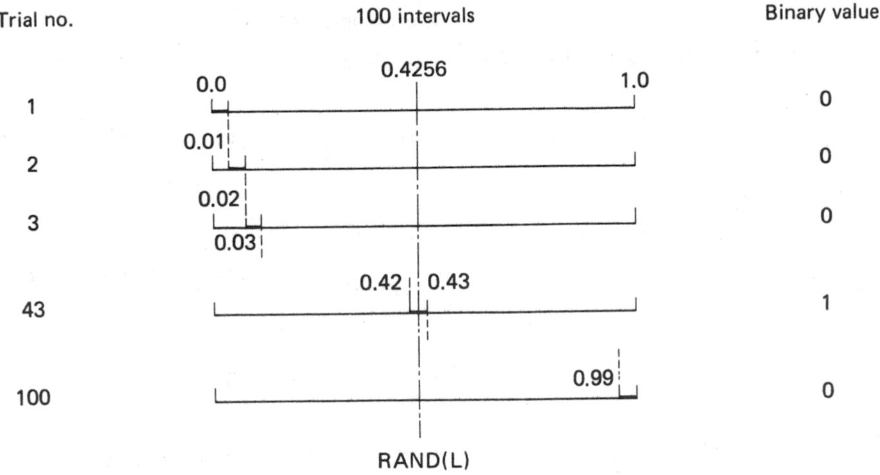

Figure 12.23. Generation of 100 binary values for event 1.

1 occurs in the 43rd trial and does not occur in the other 99 trials. It is evident that the basic event occurs with probability 0.01 in each trial.

Other basic events are generated similarly by using one uniform random number for each group of 100 intervals. Any number of event-state vectors can be sampled by independent repetitions of each 100 generations. One uniform random number pierces 100 intervals and determines 100 trials for a basic event: hence the term *dagger sampling*.

The proposed method generates $m \times 100$ trials by $k \times m$ uniform random numbers, whereas the direct method generates m trials by the same $k \times m$ numbers. Thus, we need only one-hundredth of the random numbers of the direct Monte Carlo method. This means that the proposed method can generate 500,000 trials within the same computational effort as 5000 direct Monte Carlo trials. Dagger sampling can be extended naturally to the cases where basic events have different probabilities of occurrence.

12.7.3 Estimator Based on Dagger Sampling

Let $\mathbf{z}_1, \ldots, \mathbf{z}_N$ be N sample states generated by dagger sampling. The system unavailability Q_s can be estimated by the unbiased binomial estimator $\hat{Q}_D$, the subscript D standing for "dagger".

$$\hat{Q}_D \equiv N^{-1} \sum_{\nu=1}^{N} \psi(\mathbf{z}_\nu) \tag{12.48}$$

The variance of $\hat{Q}_D$ is equal to the sum of variances of $\psi(\mathbf{z}_\nu)$ and covariances of different $\psi(\mathbf{z}_\nu)$'s divided by N^2.

$$\mathrm{Var}(\hat{Q}_D)=N^{-2}\left[\sum_{i=1}^{N}\mathrm{Var}(\psi(\mathbf{z}_i))+\sum_{i\neq j}\mathrm{Cov}(\psi(\mathbf{z}_i),\psi(\mathbf{z}_j))\right] \quad (12.49)$$

The first summation on the right-hand side of (12.49) corresponds to the variance of the direct estimator of (12.11), since $\mathrm{Var}(\psi(\mathbf{z}_i))=\mathrm{Var}(\psi(\mathbf{c}_j))$. Thus,

$$\mathrm{Var}(\hat{Q}_D)=\mathrm{Var}(\hat{Q}_C)+\sum_{i\neq j}\mathrm{Cov}(\psi(\mathbf{z}_i),\psi(\mathbf{z}_j)) \quad (12.50)$$

In dagger sampling, two sample vectors $\mathbf{z}_i$ and $\mathbf{z}_j$ are not independent, but correlated, since a small number of uniform random numbers generate a number of sample vectors. As seen from Fig. 12.23, if a basic event occurs during a trial, then the basic event does not occur in other trials for the same group. Thus, the correlation of two sample vectors $\mathbf{z}_i$ and $\mathbf{z}_j$ is negative as long as the vectors have some elements in the same group. Assume that top event $\psi(\mathbf{b})$ is a coherent function of $\mathbf{b}$. Then, since ψ is monotonically increasing, the negative correlation between $\mathbf{z}_i$ and $\mathbf{z}_j$ also applies to $\psi(\mathbf{z}_i)$ and $\psi(\mathbf{z}_j)$, and we have, from (12.50), that

$$\mathrm{Var}(\hat{Q}_D)\leq \mathrm{Var}(\hat{Q}_C) \quad (12.51)^{\dagger}$$

We already know that a Monte Carlo method based on dagger sampling required fewer random numbers than the direct Monte Carlo method. Inequality (12.51) shows another characteristic of this method: It yields a more accurate estimate for the system unavailability than the direct method, for a given sample size N.

12.7.4 Calculation of ψ(Sample Vector)

Most sample state vectors involve the simultaneous occurrence of, at most, two basic events. For such state vectors, we can calculate the value of ψ by table reference.

$$\psi(\text{Sample Vector})=\begin{cases}1, & \text{if the vector corresponds to a cut set}\\ 0, & \text{otherwise}\end{cases}$$

†A formal proof for (12.51) is possible, although it is not presented here.

For a sample vector involving the occurrence of three or more basic events, ψ is calculated by simulating the fault tree for the state vector. The probability of occurrence is relatively small, and we can significantly reduce the amount of computation required for the calculation of ψ. Since, in most cases, the major computational effort consists of generating uniform random numbers.

Example 6.

Consider the fault tree of Fig. 12.24, where each basic event occurs with probability 0.01. The termwise calculation of (12.9) gives the exact system unavailability as

$$Q_s = 3.72 \times 10^{-3} \tag{12.52}$$

Figure 12.25 shows the results of the dagger-sampling and direct Monte Carlo methods. We observe that good estimates of the system unavailability are obtained by the dagger-sampling Monte Carlo method.

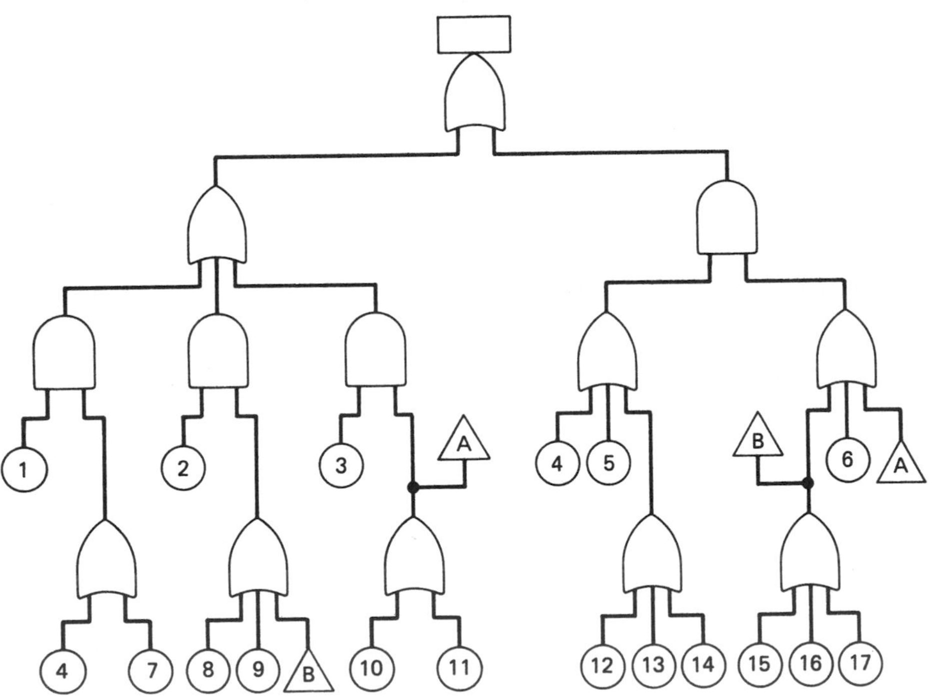

Figure 12.24. Fault tree for example problem.

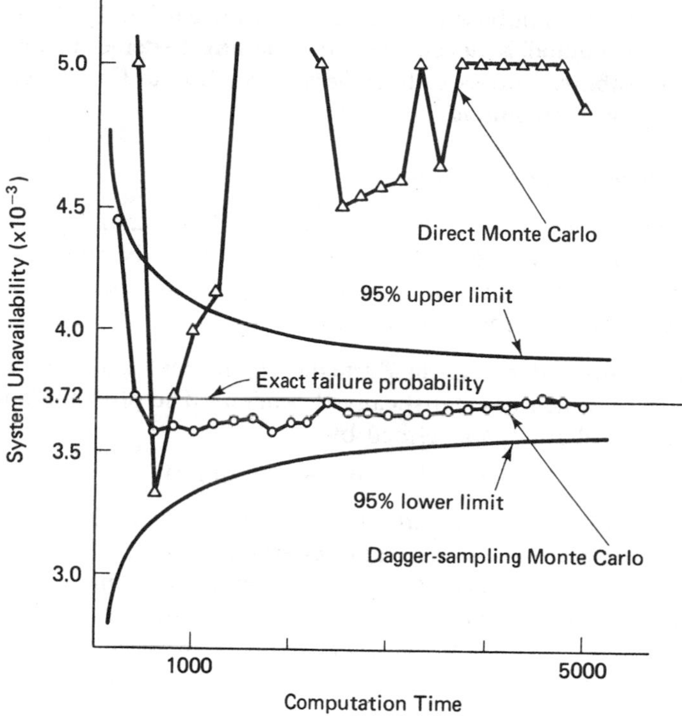

Figure 12.25. Result of direct Monte Carlo method and dagger sampling.

12.8 SYSTEM UNRELIABILITY EVALUATION BY STATE-TRANSITION MONTE CARLO METHOD

12.8.1 Introduction

In previous sections of this chapter we demonstrated applications of Monte Carlo methods to situations in which component unavailabilities were known, and top-event probabilities for large, complex systems were required at a specific instant of time. The previously developed restricted-sampling and dagger methods apply only to unavailabilities, and in this section a comparable method specific to unreliability is explained.

As discussed in Chapter 9, system unreliability (or reliability) can be calculated by differential equations associated with Markov transition diagrams. However, the size of the differential equations increases exponentially with the number of basic events. In Chapter 9 we developed a feasible approximation based on inclusion and exclusion. In this section, a Monte Carlo method is developed for obtaining better estimates for the system unreliability. First we describe a direct Monte Carlo method and then an improved technique.

12.8.2 Direct Monte Carlo Method

If a time profile for unreliabilities is required and λ and μ are constants, one can sample each time point by using a FORTRAN program, such as shown in Fig. 12.6, which samples uniformly from exponential distributions, and apply the direct Monte Carlo method.

12.8.2.1 Direct Monte Carlo Estimator. Consider a fault tree involving k basic events which are statistically dependent. The problem is to calculate system unreliability as defined by

$$F_s(t) = \Pr(\text{top event to time } t) \tag{12.53}$$

The fault tree has associated with it an appropriate Markov transition diagram. Each state in the diagram corresponds to a k-dimensional state vector for k basic events. The direct Monte Carlo method simulates sequences of state transitions to time t, by using random numbers.

Assume that M-out-of-N sequences result in the occurrence of the top event. Then, the system unreliability $F_s(t)$ is evaluated by the Monte Carlo estimator $\hat{F}_C(t)$ as

$$\hat{F}_C(t) \equiv \frac{M}{N} \tag{12.54}$$

12.8.2.2 Generation of State Transitions. Consider now the generation of a state transition from a given state A_0. Assume that the system jumped into state A_0 at time zero, and that the transition occurs to one of m states $A_1, \ldots, A_m$. Denote by γ_i the constant transition rate from A_0 to descendant A_i. This situation is summarized in Fig. 12.26.

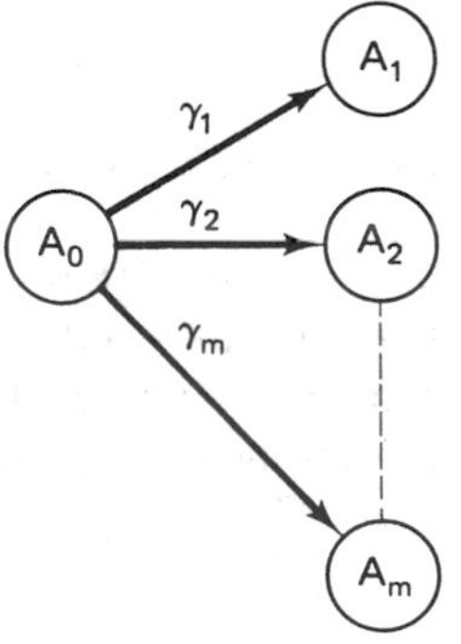

Figure 12.26. Transition from state A_0.

The differential equations for the state probabilities in the diagram are

$$\dot{P}_0 = -(\gamma_1 + \ldots + \gamma_m)P_0, \quad P_0(0) = 1 \tag{12.55}$$

$$\dot{P}_i = \gamma_i P_0, \quad P_i(0) = 0, \quad i = 1, \ldots, m \tag{12.56}$$

These equations have the solution,

$$P_0(T) = \exp[-\Gamma T] \tag{12.57}$$

$$P_i(T) = \frac{\gamma_i}{\Gamma}(1 - \exp[-\Gamma T]), \quad i = 1, \ldots, m \tag{12.58}$$

where parameter Γ is defined by

$$\Gamma \equiv \gamma_1 + \ldots + \gamma_m \tag{12.59}$$

Equation (12.57) shows that the time to the first transition is distributed exponentially with the expected time of $1/\Gamma$. Note that this distribution is determined uniquely by rates γ_i's. As is shown in Fig. 12.6, time T is generated by

$$T = -\frac{1}{\Gamma}\log\xi \tag{12.60}$$

where ξ = uniform random number RAN(NSEED) or RAND (L).

Equation (12.58) indicates that a transition occurs to state A_i with probability γ_i/Γ. Thus, if a uniform random number η falls in the interval between $[\gamma_1 + \ldots + \gamma_{i-1}]/\Gamma$ and $[\gamma_1 + \ldots + \gamma_i]/\Gamma$, then the transition to state A_i is sampled. (Fig. 12.27).

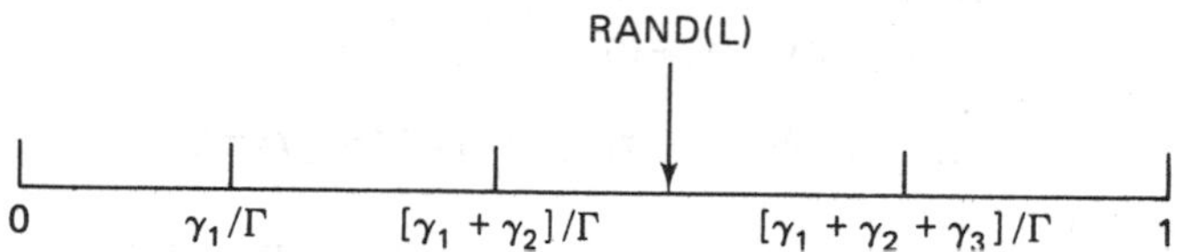

State A_3 is sampled as a descendant.

Figure 12.27. Selection of a descendant from A_1, A_2, A_3, *or* A_4.

Since probabilities $\gamma_i/\Gamma, i = 1, \ldots, m$ are functionally independent of the time to transition T, so is the selection of a descendant. Therefore, we can reverse the order for generating T and the selection of descendant A_i, and we have the following method for simulating a transition from state A_0:

Step 1. Select a descendant out of $A_1, \ldots, A_m$ with probabilities $\gamma_1/\Gamma, \ldots,$ and γ_m/Γ, respectively. Denote by B_0 the descendant selected.

Step 2. Generate time T to the transition to B_0 according to the exponential distribution, $1-\exp[-\Gamma T]$.

Assume that state B_0 is a non-absorption state and, hence, has several candidates as its descendant. The transition from state B_0 can be simulated by two independent random numbers as in the case of state A_0. A descendant from B_0 and the time to transition are determined. On the other hand, assume absorption state B_0. In this case no transition is made from state B_0, and the time to transition is infinite.

Example 7. Parallel System

Consider the transition diagram of Fig. 12.28. The absorption state is (1, 1). Parameters λ_1 and λ_2 are failure rates for components 1 and 2, and rate c indicates that the system suffers from common cause C. Repair rates are denoted by μ_1 and μ_2 for components 1 and 2, respectively. Develop sampling methods for the transitions from (0,0) or (1,1).

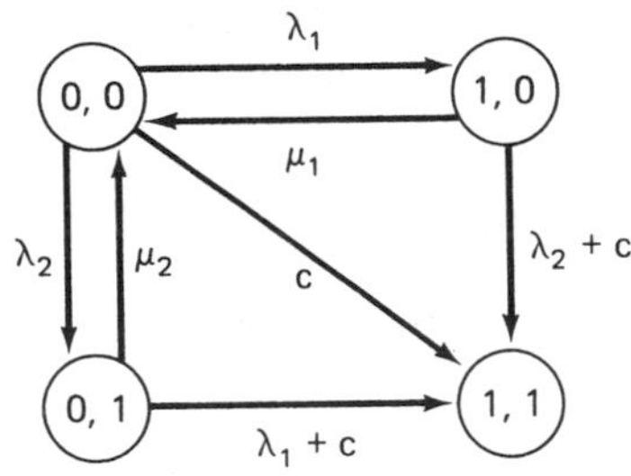

Figure 12.28. Transition diagram for two-component parallel system suffering from common cause C.

Solution:

1) Transition from state (0,0):

$$A_1=(1,0), \quad A_2=(0,1), \quad A_3=(1,1)$$

$$\Gamma=\lambda_1+\lambda_2+c, \quad \gamma_1=\lambda_1, \quad \gamma_2=\lambda_2, \quad \gamma_3=c$$

Select (1,0) or (0,1) or (1,1) with probabilities γ_1/Γ, γ_2/Γ, and γ_3/Γ, respectively. Generate the transition time according to the distribution $1-\exp[-\Gamma T]$.

2) Transition from state (1,1):

Since this is an absorption state, no transition is made, and the time to transition is infinite.

12.8.2.3 Reformulation of the Direct Monte Carlo Method. Step 1 of section 12.8.2.2 depends only on current transition rates, γ_i's and is functionally independent of previous times to transition. Thus, we can generate a sequence of states independent of the corresponding transition times. Since system unreliability $F_s(t)$ approaches unity as t gets larger, any sampled sequence of states reaches an absorption state in a finite number of transitions. Denote by $\mathbf{S}=(S_0,\ldots,S_n)$ any of the state sequences, where

$$S_0=(0,\ldots,0)=\text{system initial state}$$
$$S_1,\ldots,S_{n-1}=\text{non-absorption states}$$
$$S_n=\text{absorption state}$$

Note that n, the number of transitions to state S_n, is itself a random variable.

Let T_i be the time to transition from state S_{i-1} to state S_i. Denote by $\mathbf{T}=(T_1,\ldots,T_n)$ the transition time associated with $\mathbf{S}$. Both $\mathbf{S}$ and $\mathbf{T}$ are random variables.

If the sum of times comprising $\mathbf{T}$ is less than t, then pair $(\mathbf{T},\mathbf{S})$ contributes to the first occurrence of the top event before time t. Thus, the system unreliability $F_s(t)$ is

$$F_s(t)=\sum_{\mathbf{S}}\int D(\mathbf{T};t)\Pr(\mathbf{T},\mathbf{S})\,d\mathbf{T} \tag{12.61}$$

where

$$D(\mathbf{T};t)=\begin{cases}1, & \text{if } \sum_{i=1}^{n} T_i<t\\ 0, & \text{otherwise}\end{cases} \tag{12.62}$$

By using the above notation we can restate the Monte Carlo estimator of (12.54) as

$$\hat{F}_C(t)=N^{-1}\sum_{\nu=1}^{N} D(\mathbf{T}_\nu;t) \tag{12.63}$$

where $\mathbf{T}_1,\ldots,$ and $\mathbf{T}_N$ are N sample sequences of transition times associated with N sample sequences of states $\mathbf{S}_1,\ldots,$ and $\mathbf{S}_N$, respectively. Sequence $\mathbf{S}_\nu$ can be generated by step 1 of Section 12.8.2.2, and $\mathbf{T}_\nu$ for a given $\mathbf{S}_\nu$ by step 2.

12.8.3 State-Transition Monte Carlo Method

Suppose that $F_s(t)$ equals 10^{-5} for some t. On the average, the direct estimator requires at least 10^5 trials to obtain a sample sequence $\mathbf{T}_\nu$ such that $D(\mathbf{T}_\nu;t)=1$. This means that the direct Monte Carlo method requires a large number of trials for practical systems, which are highly reliable. Thus, we require a Monte Carlo method that yields a better estimate with fewer trials.

Equation (12.61) can be written as

$$F_s(t) = \sum_{\mathbf{S}} \left[\int D(\mathbf{T}; t) p(\mathbf{T}|\mathbf{S}) \, d\mathbf{T} \right] \Pr(\mathbf{S}) \tag{12.64}$$

The sum within the brackets is a function of **S**. Denote by $B(\mathbf{S}; t)$ the function:

$$B(\mathbf{S}; t) = \int D(\mathbf{T}; t) p(\mathbf{T}|\mathbf{S}) \, d\mathbf{T} \tag{12.65}$$

Equation (12.64) can be written as

$$F_s(t) = \sum_{\mathbf{S}} B(\mathbf{S}; t) \Pr(\mathbf{S}) \tag{12.66}$$

$$= E_{\mathbf{S}}(B(\mathbf{S}; t)) \tag{12.67}$$

From (12.67), we have the following Monte Carlo estimator with subscript N standing for "state transition":

$$\hat{F}_N(t) = N^{-1} \sum_{\nu=1}^{N} B(\mathbf{S}_\nu; t) \tag{12.68†}$$

Note that we generate only state sequences $\mathbf{S}_\nu$'s. The transition time sequences are processed analytically by $B(\mathbf{S}_\nu; t)$.

The following theorem proves that the state-transition Monte Carlo method estimates the system unreliability with a smaller variance than the direct Monte Carlo method.

Theorem: Consider a system which is in functioning state $(0, \ldots, 0)$ at time zero. Assume $t > 0$ and $F_s(t) \neq 0$. Then,

$$\mathrm{Var}(\hat{F}_N(t)) < \mathrm{Var}(\hat{F}_C(t)) \tag{12.69}$$

This theorem holds because the random variation due to the transition time sequences is zero in the state-transition Monte Carlo method. A formal proof is possible, but it is not presented here.

The state-transition Monte Carlo method requires calculation of $B(\mathbf{S}; t)$ for sample sequence **S**. Let T_i in **T** follow an exponential distribution $1 - \exp[-\Gamma_i T]$. Equation (12.62) shows that

$$B(\mathbf{S}; t) = \Pr\left(\sum_{i=1}^{n} T_i < t | \mathbf{S} \right) \tag{12.70}$$

†For actual calculations, a maximum length for the state-transition sequences $\mathbf{S}_\nu$ is given by the user. Although some sequences may not end the in absorption state and the corresponding $B(\mathbf{S}; t)$'s are set to be zero, the Monte Carlo output will change little, since $B(\mathbf{S}; t)$ is very small for the long sequences.

The probability on the right-hand side of (12.70) is the distribution of the sum of n independent random variables $T_1, \ldots, T_n$. Thus, $B(\mathbf{S}; t)$ is given by the integral of convolution of the probability density functions $f_i(u) = \Gamma_i \exp[-\Gamma_i u]$:

$$B(\mathbf{S}; t) = \int_0^t f_1 * f_2 * \ldots * f_n \, du \tag{12.71}$$

where symbol $*$ denotes a convolution integral.

The Laplace transformation of (12.71) is

$$L[B(\mathbf{S}; t)] = \frac{\Gamma_1 \ldots \Gamma_n}{(s+\Gamma_0)(s+\Gamma_1)\ldots(s+\Gamma_n)}, \quad \Gamma_0 \equiv 0 \tag{12.72}$$

Assume that parameters Γ_i's are distinct. Then, the inverse transformation of (12.72) is

$$B(\mathbf{S}; t) = \sum_{i=0}^{n} W_i \exp[-\Gamma_i t] \tag{12.73}$$

$$W_i \equiv \Gamma_1 \ldots \Gamma_n / \prod_{j \neq i} [\Gamma_j - \Gamma_i] \tag{12.74}$$

When some Γ_i's have the same value, we can calculate $B(\mathbf{S}; t)$ by substituting slightly different values into (12.73) or by applying an inverse transformation to (12.72). The transformation is not difficult, but the result is not presented here.

Example 8. Three Component Parallel System

Consider the system of Example 3 of Chapter 9. Assume that sequence 000→100→110→111 has been sampled. Calculate $B(\mathbf{S}, t)$ for this sequence.

Solution:

$$\Gamma_0 = 0$$

$$\Gamma_1 = \frac{1}{1000} + \frac{2}{1000} + \frac{3}{1000} = \frac{6}{1000}$$

$$\Gamma_2 = \frac{1}{10} + \frac{2}{1000} + \frac{3}{1000} = \frac{105}{1000}$$

$$\Gamma_3 = \frac{1}{10} + \frac{1}{40} + \frac{3}{1000} = \frac{128}{1000}$$

$$\Gamma_1\Gamma_2\Gamma_3 = 8.064 \times 10^{-5}$$

$$W_0 = \frac{8.065 \times 10^{-5}}{(0.006-0)(0.105-0)(0.128-0)} = 1$$

$$W_1 = \frac{8.065 \times 10^{-5}}{(0-0.006)(0.105-0.006)(0.128-0.006)} = -1.11295$$

$$W_2 = \frac{8.065 \times 10^{-5}}{(0-0.105)(0.006-0.105)(0.128-0.105)} = 0.3373277$$

$$W_3 = \frac{8.065 \times 10^{-5}}{(0-0.128)(0.006-0.128)(0.105-0.128)} = -0.2245467$$

$$B(\mathbf{S};t) = 1 - 1.11295e^{-0.006t} + 0.3373277e^{-0.105t} - 0.2245467e^{-0.128t}$$

Example 9. Two-Out-of-Three System

Consider the system of the preceeding example. Evaluate the system unreliability $F_s(t)$ by the state-transition and the direct Monte Carlo methods.

Solution: The results with the sample size of $N=1000$ are shown in Fig. 12.29. Better results are achieved by the new Monte Carlo method.

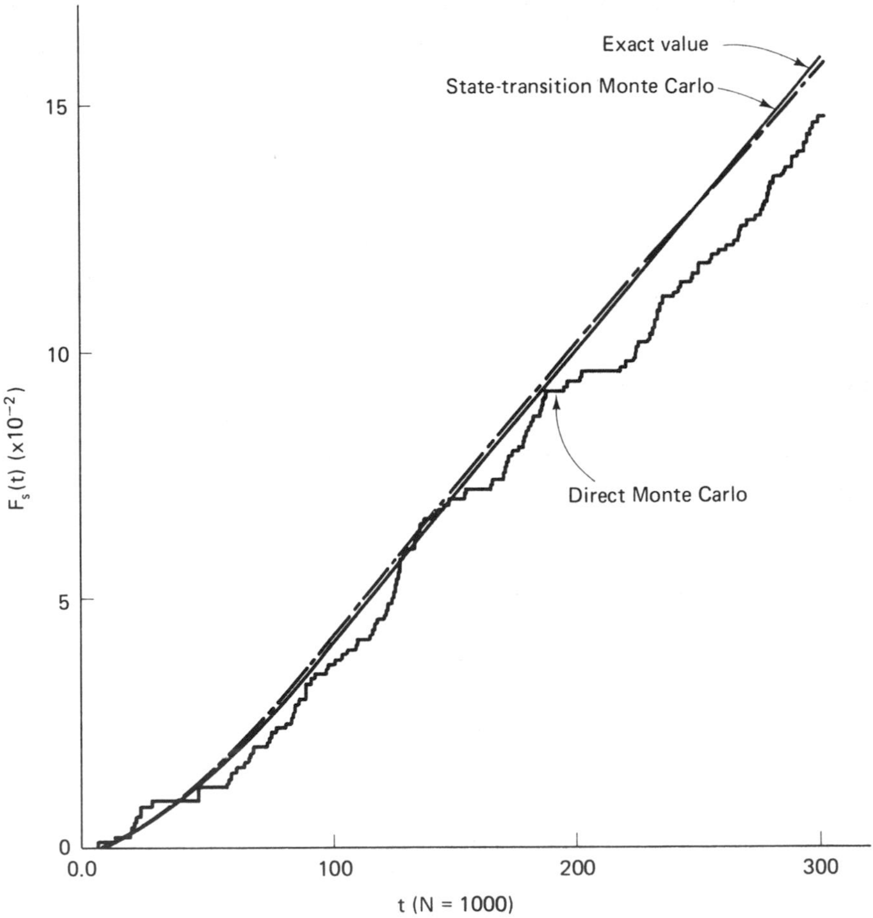

Figure 12.29. Results of state-transition and direct Monte Carlo methods.

12.8.4 Unavailability Calculation by the State-Transition Monte Carlo Method

The state-transition Monte Carlo method for the reliability calculation can be used to calculate system unavailability. Assume independent basic events. The steady-state unavailability of basic event i is

$$Q_i(\infty) = \frac{\lambda_i}{\lambda_i + \mu_i} \tag{12.75}$$

Let t be a time instant. Introduce new failure rate γ_i such that

$$1 - \exp[-\gamma_i t] = Q_i(\infty) \tag{12.76}$$

or

$$\gamma_i = \frac{-1}{t}\log[1 - Q_i(\infty)] \tag{12.77}$$

Consider a new *non-repairable* system having failure rates γ_i's. Then, the unreliability of this system coincides with the steady-state unavailability of the original (repairable) system, provided that it is coherent. The unavailability calculation reduces to the unreliability calculation, and the state-transition Monte Carlo method can apply.

Example 10. Two-Out-of-Three System

Consider the two-out-of-three voting system of Section 12.8.3. Construct the new non-repairable system by setting $t = 1$, and develop possible state transitions for the state-transition Monte Carlo method.

Solution:

$$\lambda_1 = \frac{1}{1000}, \qquad \lambda_2 = \frac{2}{1000}, \qquad \lambda_3 = \frac{3}{1000}$$

$$\mu_1 = \frac{1}{10}, \qquad \mu_2 = \frac{1}{40}, \qquad \mu_3 = \frac{1}{60}$$

$$Q_i(\infty) = 0.009901, \quad Q_2(\infty) = 0.074074, \quad Q_o(\infty) = 0.152542$$

$$\gamma_1 = 0.00995, \qquad \gamma_2 = 0.07696, \qquad \gamma_3 = 0.16551$$

Possible state transitions are:

(0,0,0) → (1,0,0), (0,1,0), (0,0,1)

(1,0,0) → (1,1,0), (1,0,1)

(0,1,0) → (1,1,0), (0,1,1)

(0,0,1) → (0,1,1), (1,0,1)

The transition rates are:

1) From (0,0,0) to (1,0,0): $\dfrac{\gamma_1}{\gamma_1+\gamma_2+\gamma_3}=0.039418$

to (0,1,0): $\dfrac{\gamma_2}{\gamma_1+\gamma_2+\gamma_3}=0.304889$

to (0,0,1): $\dfrac{\gamma_3}{\gamma_1+\gamma_2+\gamma_3}=0.655693$

2) From (1,0,0) to (1,1,0): $\dfrac{\gamma_2}{\gamma_2+\gamma_3}=0.317400$

to (0,1,1): $\dfrac{\gamma_3}{\gamma_2+\gamma_3}=0.682600$, etc.

Since the new system is non-repairable, Monte Carlo state transition $\mathbf{S}_\nu$ is obtained making new components fail one after another; thus, we can reach absorption states readily. For the two-out-of-three-voting system, for example, the absorption state can be reached in two state transitions. Further, $B(\mathbf{S};t)$ of (12.71) can be calculated as (12.73), since the parameter Γ_i's are distinct. It can be shown that this state-transition Monte Carlo method coincides with the sequential destruction method[24] when each basic event has the same steady-state unavailability.

Example 11.

Consider the fault tree of Fig. 12.24. The result of the state-transition Monte Carlo method is shown in Fig. 12.30. A result as good as the dagger sampling is obtained.

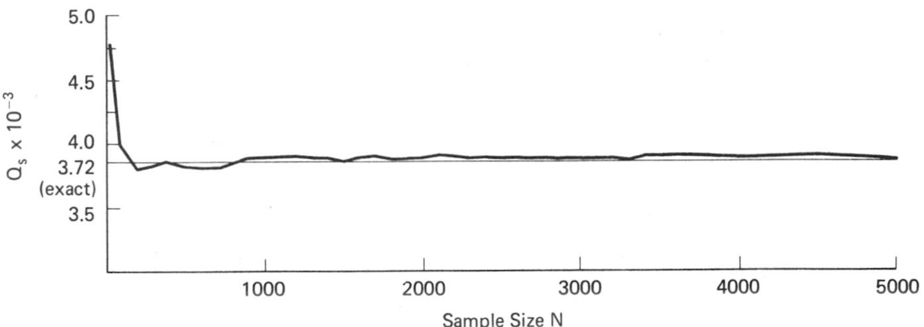

Figure 12.30. System unavailability by the state-transition Monte Carlo method.

[24]Easton, Malcolm C., and C. K. Wong, "The Sequential Destruction Method for Monte Carlo Evaluation of System Reliability," Research Report of IBM, Thomas J. Watson Research Center, Computer Science, RC7337 (#31441), pp. 1–16, Sept. 1978.

PROBLEMS

12.1. Generate five points from the domain $\{0 \le x \le 1, 0 \le y \le 1\}$, using the uniform random numbers of Fig. 12.4.

12.2. Generate five TTF's for a component with a MTTF of 100 hr, assuming exponential distributions. Use the uniform random numbers of Fig. 12.2.

12.3. Assume that a component has an unavailability of 0.3 at 1000 hr. Generate 10 binary values of the indicator variable y for the component at 1000 hr. Use the random numbers of Fig. 12.4.

12.4. Consider the system shown in Fig. 12.11. Components A and C have life distributions with MTTF's of 50 and 75 days, respectively. Component B is a cold standby unit for A and has a normal life distribution with an MTTF of 20 days and a standard deviation of 5 days. Component C has a hot standby unit D whose life distribution is normal with MTTF=65 days and a standard deviation of 4 days. Generate five TTF's for the system, and evaluate the system MTTF. Use the random numbers listed in the following table.

	TTF (Comp. A)	TTF (Comp. B)	TTF (Comp. C)	TTF (Comp. d)
1	35.57	24.82	81.76	63.06
2	58.80	17.67	5.80	66.21
3	54.26	18.03	125.63	67.54
4	61.41	11.17	44.84	71.37
5	28.32	21.45	21.97	62.74

12.5. (a) Generate five random numbers from the log-normal distribution with μ of -4.615 and σ of 0.6845. This distribution is the one for component 1 of Example 1 of this chapter. Repeat similar generations for components 2 and 3. Use the following normal random numbers with the mean zero and the standard deviation of unity.

1	2.026	6	−0.811	11	0.963
2	−0.350	7	0.763	12	−0.465
3	−0.122	8	−0.205	13	−0.393
4	−0.464	9	−0.422	14	−1.766
5	−0.830	10	−0.784	15	0.290

(b) Generate five system unavailabilities for the system in Fig. 12.17, using the component unavailabilities obtained in Problem 12.5(a).

12.6. Consider the two-out-of-three voting system in Example 2, Section 12.6.5. Assume that

$$Q_1 = Q_2 = Q_3 = \tfrac{1}{3}$$

Perform five direct Monte Carlo trials, using random number of Figs. 12.2 and 12.4. Compare the result with the exact system unavailability.

12.7. Carry out five restricted-sampling trials for the voting system of Problem 12.6, using ψ_L and ψ_U of Example 2, Section 12.6.5. Use the random numbers of Fig. 12.4.

12.8. Perform 15 dagger-sampling Monte Carlo trials for the voting system of Problem 12.6, using the first 15 random numbers of Figs. 12.2 and 12.4.

12.9. Consider a two-out-of-three voting system consisting of components with the following failure data:

$$\lambda_1 = \lambda_2 = \lambda_3 = 0.1 \equiv \lambda$$

$$\mu_1 = \mu_2 = \mu_3 = 0.2 \equiv \mu$$

Generate sequences of state transitions ending in an absorption state. Generate also the corresponding transition times. Use the random numbers of Figs. 12.4 and 12.2 to sample the state transitions and the transition times, respectively. Obtain TTF's of the system, and estimate the system reliability.

12.10. In Problem 12.9 sampled state transitions are

$$\begin{aligned}
&\text{A:} \quad (0,0,0) \overset{0.3}{\rightarrow} (0,0,1) \overset{0.4}{\rightarrow} (1,0,1)\\
&\text{B:} \quad (0,0,0) \overset{0.3}{\rightarrow} (1,0,0) \overset{0.4}{\rightarrow} (1,0,1)\\
&\text{C:} \quad (0,0,0) \overset{0.3}{\rightarrow} (0,0,1) \overset{0.4}{\rightarrow} (0,0,0) \overset{0.3}{\rightarrow} (1,0,0) \overset{0.4}{\rightarrow} (1,1,0)
\end{aligned}$$

Calculate $B(\mathbf{S};t)$ for the sequences A and B.

12.11. Calculate $B(\mathbf{S};t)$ for the sequence C of Problem 12.10. Use the slightly different values of Γ_i's.

$$\gamma_0 = 0, \quad \Gamma_1 = 0.3, \quad \Gamma_2 = 0.4, \quad \Gamma_3 = 0.300001, \quad \Gamma_4 = 0.400001$$

Obtain $B(\mathbf{S};t)$ at $t=5$.

12.12. Calculate exact $B(\mathbf{S};t)$ for the sequence C. Note that $B(\mathbf{S};t)$ is obtained by solving the differential equation:

$$\begin{aligned}
\dot{P}_1 &= -\Gamma_1 P_1, & P_1(0) &= 1\\
\dot{P}_2 &= \Gamma_1 P_1 - \Gamma_2 P_2, & P_2(0) &= 0\\
\dot{P}_3 &= \Gamma_2 P_2 - \Gamma_3 P_3, & P_3(0) &= 0\\
\dot{P}_4 &= \Gamma_3 P_3 - \Gamma_4 P_4, & P_4(0) &= 0\\
\dot{P}_5 &= \Gamma_4 P_4, & P_5(0) &= 0
\end{aligned}$$

This equation represents state probabilities of the following Markov transition diagram, and $B(\mathbf{S};t)$ of (12.70) is given by $P_5(t)$.

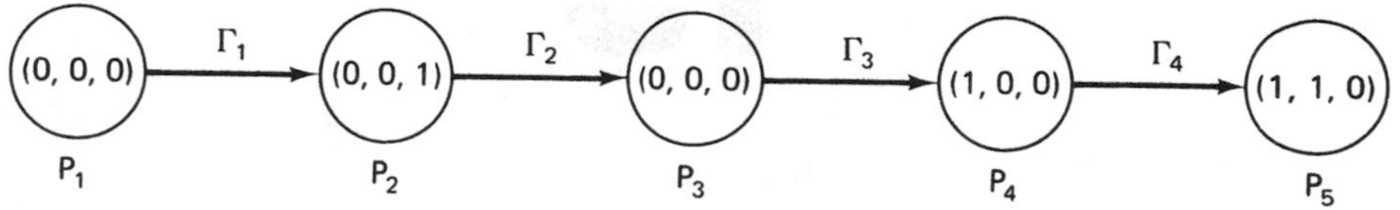

Figure P12.12.

12.13. Obtain the state-transition Monte Carlo estimate $\hat{F}_N(t)$ based on the sequences A, B, and C. Calculate $\hat{F}_N(t)$ at $t=0$, 5, and 10.

12.14. (a) Obtain γ_i's for the state-transition Monte Carlo method which evaluates the steady-state unavailability of the voting system in Problem 12.9. Use $t=1$ in (12.77).

(b) Generate three sequences of state transitions for the voting system, using the random numbers of Fig. 12.4.

(c) Evaluate the unavailability based on the three sequences. Compare the result with the exact unavailability.

13

CASE STUDIES AND APPLICATIONS: SOME WORKED EXAMPLES

This chapter consists of a (non-intersecting) set of worked examples which demonstrate risk and reliability applications and methodology. Many of these were developed by Prof. J. B. Fussell for use as example problems in industrial courses. They are not arranged in any "logical" order, or even in their order of complexity, so this chapter can be read in a "random sampled" fashion. The examples and what they are intended to illustrate are as follows:

Example 1. *A Risk Assessment Using Cause-Consequence Technology (after J. B. Fussell)*

Cause-consequence (CC) diagrams, which were introduced in Chapter 1, are combined event trees and fault trees. The event tree consequences are also included in a CC diagram, and so a risk assessment leading to a "Farmer curve" of accident probability versus consequence can be fashioned. In this example, the initiating event is a fire, the probability of which is obtained by fault tree analysis, the various consequences (financial losses) being contingent on the operability of various fire protection devices.

Cause-consequence diagrams were developed at the RISØ National Laboratory in Denmark, where a large group headed by Bob Taylor is currently studying their automated construction.

Example 2. *A Redundancy Study (after J. B. Fussell)*

Redundancy can be added on a component-by-component basis or by replicating entire systems or subsystems. This example explores both strategies and also teaches that every time one adds an additional protective shut-down device the probability of having spurious (unnecessary) shut-downs increases. Short-cut calculation methods are used.

Example 3. *Analysis of a Chemical Reactor Safety System Problem (after R. L. Browning, Hydrocarbon Processing, 253, Sept. 1975, and Chem. Engr. Prog., 72, June, 1976.)*

Construction of the fault tree for this system is discussed in Chapter 2, and the tree is shown as Fig. 2.26. We proceed now to demonstrate how the MOCUS, KITT, and IMPORTANCE codes are used to compute cut set and top-event reliability parameters. This problem also includes an operator failure as a basic event, and we show how an inhibit gate is used for this calculation. Failure data for the components were taken from WASH 1400 and are somewhat low for chemical processes, but the principles and trends are valid.

Example 4. *System Trade-Off Analysis Using Short-Cut Calculation Methods (after J. B. Fussell)*

The system in question, which is introduced in Chapter 2, Fig. 2.23, consists of a tank, a pump, a timer, and an operator, the problem being to prevent tank rupture due to overfilling. In addition to demonstrating fault tree construction and system definition principles, this example explains:

1) How to include an operator in a short-cut calculation.
2) That extreme care must be taken in choosing importance criteria.
3) That reliability techniques can be used to make economic analyses. In this case, the economics of construction of a repair facility are considered.

Example 5. *Pressure Tank Rupture*

This system is introduced in Chapter 2, and the fault tree is shown as Fig. 2.27. We now extend the analysis to demonstrate how top-event reliability parameters are calculated by using cut set information and the KITT program. The full spectrum of importance parameters are computed.

Example 6. *Validity of the Constant Failure Rate Model (after B. Bulloch and S. B. Gibson, Imperial Chemicals Industries Ltd.)*

When using plant failure data to obtain generic failure rates, engineering judgment and experience should always be used. Blind application of the statistical methods of Chapter 4 can lead to serious errors and inconsistencies.

This particular example illustrates that failure data are not necessarily random nor do they necessarily exhibit simple wearout characteristics.

Example 7. *Spares Inventory Calculations (after B. Bulloch and S. B. Gibson, Imperial Chemicals Industries, Ltd.)*

Since no applications of a discrete Poisson distribution have previously been discussed in this text, we include this example for the sake of completeness.

Example 8. *Availability Study of White River Shale Oil Plant (after J. Ong, Texaco Corporation)*

This analysis of the raw materials preparation portion of the White River Shale Oil Plant demonstrates how reliability techniques can be used to size units consisting of parallel, cold-standby equipment and storage facilities, with the objective of meeting a specified production goal. The calculations were done by using a modified KITT code which includes storage tank and cold-standby analysis capabilities.

Example 1. A Risk Assessment Example Using Cause-Consequence Technology

We show here how a cause-consequence diagram can be used to construct a Farmer curve of the probability of an event versus its consequence.

Consider the cause-consequence diagram in Fig. 13.1. The fault tree corresponding to the top event, "motor overheats," has an expected number of failures of 0.088 per 6 months, the time between motor overhauls. There is a probability of 0.02 that the overheating results in a fire. The consequences of a fire are C_0 to C_4, ranging from a loss of \$1000 if there is equipment damage [with probability $P_0(1-P_1)$] to $\$5\times10^7$ if the plant burns down (probability $P_0P_1P_2P_3P_4$). The downtime loss is estimated at \$1000 per hour; thus, the total loss for

$$C_0=\$1000+(2)(\$1000)=\$3000,\ C_1=\$15{,}000+\$24{,}000=\$39{,}000,\ \text{etc.}$$

Given the following parameters, we now calculate the possible consequence of each event, plot it versus the probability of its occurrence, and show the constant \$300 risk line on the plot.

Probalistic Event	Parameter
P_0	Probability is 0.088/6 months
P_1	Probability is 0.02
P_2	Operature failure$=0.1$
	Fire extinguisher, $\lambda=10^{-4}$, $T=$test period$=730$
P_3	Fire extinguisher control, $\lambda=10^{-5}$, $T=4380$
	Fire extinguisher hardware, $\lambda=10^{-5}$, $T=4380$
P_4	Fire alarm control, $\lambda=5\times10^{-5}$, $T=2190$
	Fire alarm hardware, $\lambda=10^{-5}$, $T=2190$

Solution:

Event C_0. The probability is $P_0(1-P_1)=(0.088)(1-0.02)=0.086$. The accompanying risk is (\$3000)(0.086)=\$258 (see Table 13.1).

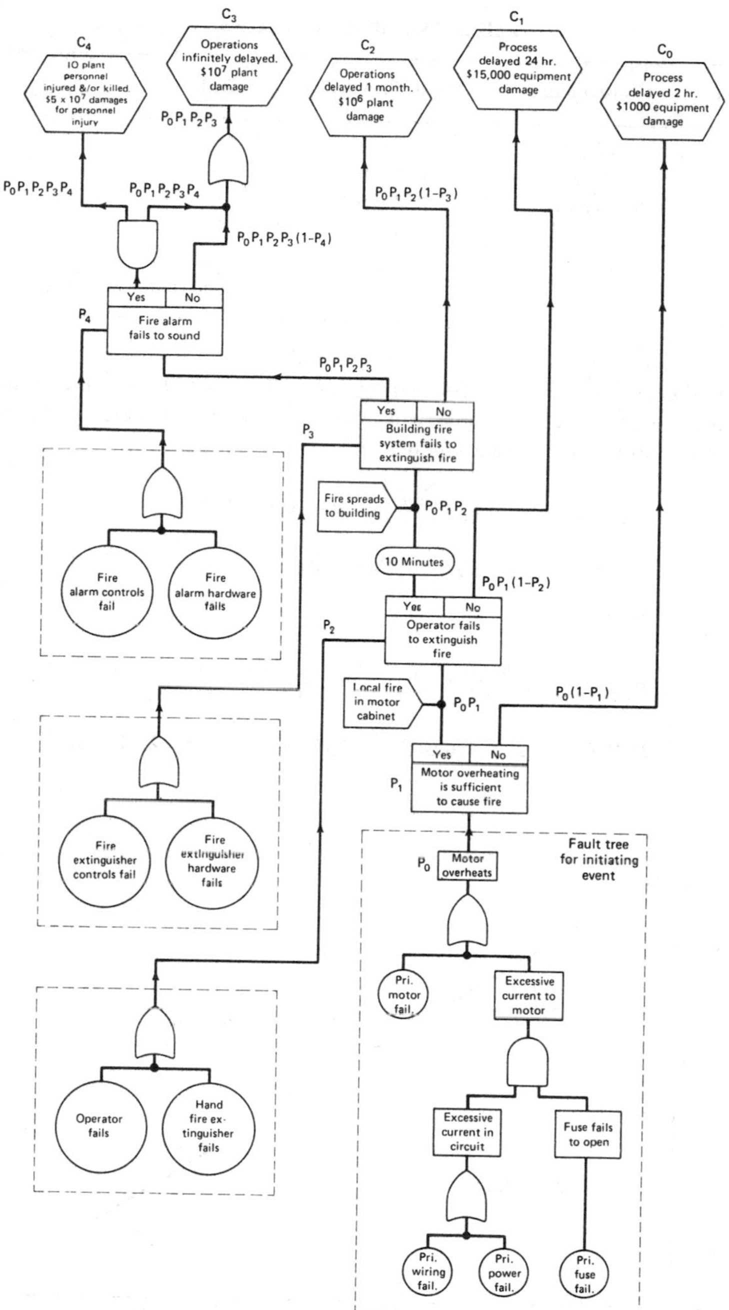

Figure 13.1. Sample system cause-consequence diagram.

TABLE 13.1 RISK CALCULATION

Event	Consequence	Event Probability	Risk
C_0	\$3,000	0.086	\$258
C_1	\$39,000	1.53×10^{-3}	60
C_2	$\$1.744 \times 10^6$	2.24×10^{-4}	391
C_3	$\$2 \times 10^7$	1.03×10^{-5}	206
C_4	$\$5 \times 10^7$	6.69×10^{-7}	33
			\$948/6 months = \$1896/year

Event C_1. Probability P_2 is the unavailability of the system in the lower left-hand dashed box. For the fire extinguisher, $Q = \lambda t = (10^{-4})(\frac{730}{2})$, assuming the failure is midway between test intervals. Since we have an OR gate, equation (7.18) of Chapter 7 provides an exact calculation.

$$Q_2 = p_2 = 0.1 + (10^{-4})\left(\frac{730}{2}\right)(1-0.1) = 0.133$$

$$\Pr\{C_1\} = P_0 P_1 (1 - P_2) = (0.088)(0.02)(1 - 0.133) = 1.53 \times 10^{-3}$$

Event C_2. Probability P_3 is

$$Q_3 = P_3 = (10^{-5})\left(\frac{4380}{2}\right) + \left[1 - (10^{-5})\left(\frac{4380}{2}\right)\right]\left[10^{-5}\left(\frac{4380}{2}\right)\right] = 0.043$$

$$\Pr\{C_2\} = P_0 P_1 P_2 (1 - P_3) = (0.088)(0.02)(0.133)(0.957) = 2.24 \times 10^{-4}$$

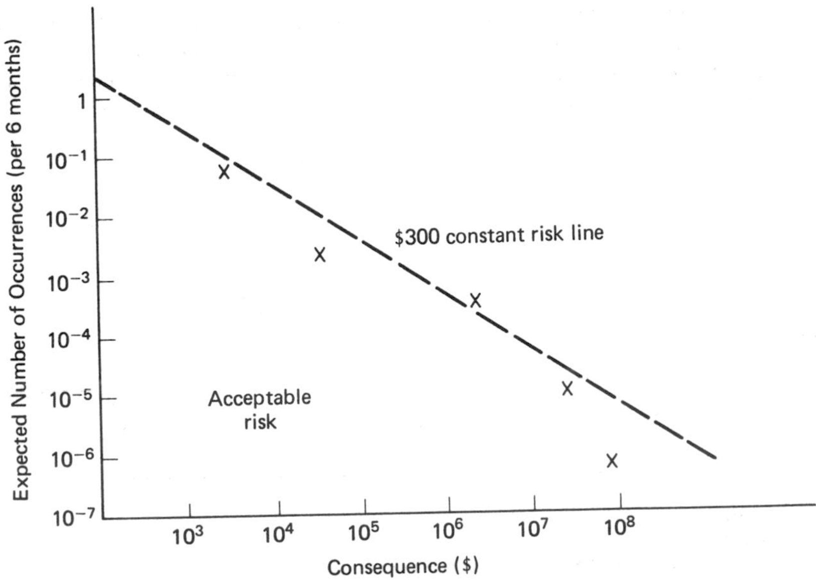

Figure 13.2. A farmer risk assessment curve.

Event C_3. $P_0P_1P_2P_3 = (0.088)(0.02)(0.133)(0.043) = 1.03 \times 10^{-5}$
Event C_4. Probability P_4 is

$$(10^{-5})\left(\frac{2190}{2}\right) + \left[1 - (10^{-5})\left(\frac{2190}{2}\right)\right]\left[(5 \times 10^{-5})\left(\frac{2190}{2}\right)\right] = 0.065$$

$$\Pr\{C_4\} = P_0P_1P_2P_3P_4 = (1.03 \times 10^{-5})(0.065) = 6.69 \times 10^{-7}$$

Figure 13.2 shows the Farmer risk curve, including the $300 risk line. This type of plot is useful for establishing design criteria for failure events, given their consequence and an acceptable level of risk.

Example 2. A Redundancy Study

In this example we use the short-cut calculation techniques of Section 7.7 of Chapter 7 to compare two strategies for increasing system reliability by adding redundancy.

A nuclear control rod shut-down system consists of the release mechanism activated by the radiation detector shown in Fig. 13.3. The component failure data are given in Table 13.2. The fault tree is shown as Fig. 13.4.

a) Calculate the number of hours per year the system is in a failed state.
b) Calculate the number of hours the system is in the failed state if a complete, duplicate (redundant) system is installed.
c) Repeat the calculation, assuming now that, instead of using a redundant system, we use redundant components.
d) How many inadvertant reactor shut-downs due to component failure are anticipated for the original system? For the redundant system?
e) How would a two-out-of-three redundant system perform in comparison with the two-fold redundant system, both in terms of expected number of failures and inadvertant trips?

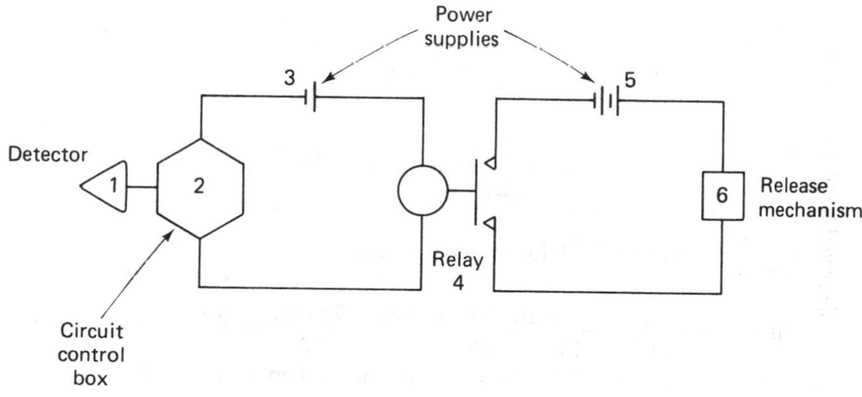

Figure 13.3. Reactor shut-down circuit.

TABLE 13.2 COMPONENT FAILURE DATA

Component	Mode	λ (1/hr)	τ=MTTR (hr)
1	Open signal	10^{-4}	10
	Closed signal	10^{-3}	10
2	Open circuit	10^{-4}	10
	Closed circuit	10^{-3}	10
3	Off	10^{-4}	24
	Surge	10^{-5}	24
4	Open contacts	10^{-4}	10
	Close contacts	10^{-3}	10
	Open coil circuit	10^{-5}	10
5	Off	10^{-5}	40
	Surge	10^{-6}	40
6	Inadvertent release	10^{-3}	28
	Fail to release	~0	28

Solution:

a) There are three, one-component cut sets. Thus, from Section 7.7.6,

Cut Set	λ_i^* (hr^{-1})	τ (hr)	$Q_i^*=(\lambda_i\tau_i)$	$w^*=\lambda^*[1-Q^*]$
Relay 4 closed	10^{-3}	10	10^{-2}	9.9×10^{-4}
Box 2 closed	10^{-3}	10	10^{-2}	9.9×10^{-4}
Sensor 1 defective	10^{-3}	10	10^{-2}	9.9×10^{-4}

For the system,

$$Q_s=\sum_{i=1}^{3} Q_i^*=3\times10^{-2}$$

$$\lambda_s=\sum_{i=1}^{3} \lambda_i^*=3\times10^{-3}\text{ hr}^{-1}$$

$$w_s=\sum_{i=1}^{3} w_i^*=2.97\times10^{-3}$$

The expected number of failures per year are

$$w_s t=(2.97\times10^{-3})(8760)=26.17\text{ failures/year}$$

Since the mean time to repair is 10 hr, the system is down $26.17\times10=262$ hr/year.

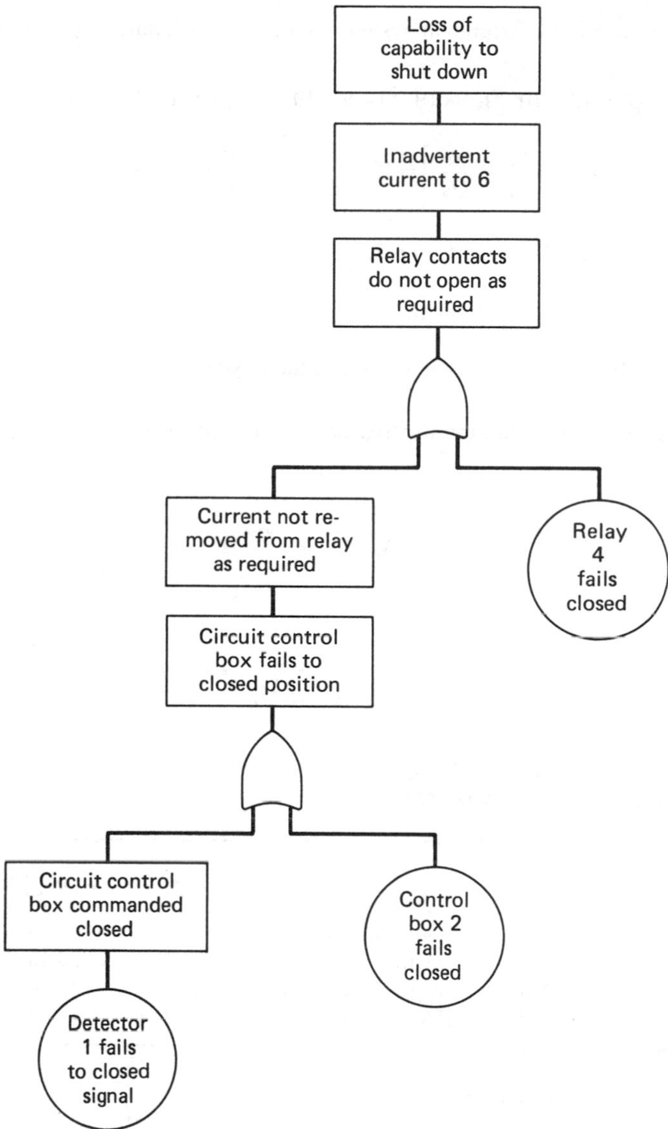

Figure 13.4. Fault tree for Fig. 13.3.

b) We now have a redundant system with identical failure parameters; thus:

$$Q_S = (3\times10^{-2})(3\times10^{-2}) = 9\times10^{-4}, \quad Q_i^* = 10^{-4} \quad i=1,\ldots,9$$

$$w_1^* = Q_1^*\left[\frac{\lambda_4}{Q_4} + \frac{\lambda_{4'}}{Q_{4'}}\right] = 10^{-4}\left[\frac{10^{-3}}{10^{-2}} + \frac{10^{-3}}{10^{-2}}\right]$$

$$= 2\times10^{-5} = w_i^*, \quad i=2,\ldots,9$$

$$w_s = \sum w_i^* = 9\times2\times10^{-5} = 1.8\times10^{-4}$$

$$\text{ENF} = (1.8\times10^{-4})(8760) = 1.6 \text{ failures/year}$$

If we assume constant failure and repair rates for the system, then, by short-cut calculation,

$$Q_s \cong \frac{\lambda_s}{\lambda_s + (1/\tau_s)} \cong \lambda_s\tau_s$$

Thus,

$$\tau_s \cong \frac{Q_s}{\lambda_s} = \frac{9\times10^{-4}}{1.8\times10^{-4}} = 5 \text{ hr}$$

The top event now exists: (1.6) (5) = 8 hr/year.

c) The system parameters are:

Cut Set	λ_i	τ_i	Q_i	Q_i^*	$w_i = Q_i^* \sum_{i=1}^{2} \lambda_i/Q_i$
4, 4′	$10^{-3}, 10^{-3}$	10, 10	$10^{-2}, 10^{-2}$	10^{-4}	2×10^{-5}
2, 2′	$10^{-3}, 10^{-3}$	10, 10	$10^{-2}, 10^{-2}$	10^{-4}	2×10^{-5}
1, 1′	$10^{-3}, 10^{-3}$	10, 10	$10^{-2}, 10^{-2}$	10^{-4}	2×10^{-5}

$$Q_s = \Sigma Q_i^* = 3\times10^{-4}$$

$$\lambda_s = 6\times10^{-5}\text{ hr}^{-1}, \quad w_s = [1-Q_s]\lambda_s = [1-3\times10^{-4}](6\times10^{-5})$$

$$\text{ENF} = (1-3\times10^{-4})(6\times10^{-5})(8760) = 0.53$$

$$\tau_s \cong \frac{Q_s}{\lambda_s} = \frac{3\times10^{-4}}{6\times10^{-5}} = 5 \text{ hr}$$

The top event exists only: (5) (0.53) = 2.6 hr/year. Although the calculations favor component redundancy, in practice, system redundancy is preferred because of the ease of testing and maintenance and less susceptibility to common-cause failures.

d) We can have an inadvertent trip in any of the following instances:

Cut Sets	λ	τ	$Q = Q^*$
Component 1 open	10^{-4}	10	10^{-3}
Component 2 open	10^{-4}	10	10^{-3}
Component 3 off	10^{-4}	24	2.4×10^{-3}
Component 4 open	10^{-4}	10	10^{-3}
Component 5 off	10^{-5}	40	4×10^{-4}
Component 6 release	10^{-3}	28	2.8×10^{-2}
	$\lambda_s = 1.41 \times 10^{-3}$		$Q_s = 3.38 \times 10^{-2}$

$$w_s = (1 - 3.38 \times 10^{-2})(1.4 \times 10^{-3})(8760) = 11.6/\text{year}$$

The average repair time for the system is

$$\tau_s = Q_s/\lambda_s = \frac{3.38 \times 10^{-2}}{1.41 \times 10^{-3}} = 24 \text{ hr}$$

Thus, the lost reactor hours add up to 278/year. If a redundant system is used, the lost hours are doubled, independent of component-by-component or complete system redundancy.

e) For a two-out-of-three system, when the top event is failure to trip, the system in a) can be regarded as a component. Since $Q_A = Q_B = Q_C = 3 \times 10^{-2}$ and $\lambda_A = \lambda_B = \lambda_C = 3 \times 10^{-3}$

Cut Sets	w^* (hr^{-1})
AB	$1.8 \times 10^{-4} = Q_A Q_B \left[\frac{\lambda_A}{Q_A} + \frac{\lambda_B}{Q_B} \right]$
AC	$1.8 \times 10^{-4} = Q_A Q_C \left[\frac{\lambda_A}{Q_A} + \frac{\lambda_C}{Q_C} \right]$
BC	$\frac{1.8 \times 10^{-4}}{w_s = 5.4 \times 10^{-4}} = Q_B Q_C \left[\frac{\lambda_B}{Q_B} + \frac{\lambda_C}{Q_C} \right]$

Thus, ENF$\cong$(5.4$\times$10^{-4})(8760)=4.7 failures/year. The number of inadvertent trips are obtained by regarding the system in d) as a component.

$$Q_A = Q_B = Q_C = 3.38\times 10^{-2} \quad \text{and} \quad \lambda_A = \lambda_B = \lambda_C = 1.41\times 10^{-3}$$

$$w^*_{AB} = Q_A Q_B\left[\frac{\lambda_A}{Q_A} + \frac{\lambda_B}{Q_B}\right] = 2(3.38\times 10^{-2})(1.41\times 10^{-3}) = 9.5\times 10^{-5}$$

$$w_s = \Sigma w^* = 2.85\times 10^{-4}$$

$$Q_s = 3(3.38\times 10^{-2})^2 = 3.43\times 10^{-3}$$

$$\lambda_s[1-Q_s] \cong w_s = 2.85\times 10^{-4}$$

$$\tau_s = \frac{Q_s}{\lambda_s} = \frac{3.43\times 10^{-3}}{2.85\times 10^{-4}} = 12 \text{ hr}$$

$$\text{ENF} = (2.85\times 10^{-4})(8760) = 2.5$$

The lost reactor hours are (2.5)(12)=30/year.

We see that this system results in a ten-fold decrease in lost reactor time due to spurious trips, but that the two-fold redundant system is considerably less failure-prone.

Example 3. Analysis of a Chemical Reactor Safety System

Figure 13.5(a) shows a reaction system in which the temperature increases with the feed rate of flow-controlled stream *A*. This system, when modified by Fig. 13.7, is identical to Fig. 2.25 in Chapter 2. Heat is removed by water circulation through a water-cooled exchanger. Normal reactor temperature is 200°F, but a catastrophic runaway will start if this temperature reaches 300°F. In view of this situation:

1. The reactor temperature is monitored (by TE/TT 714).
2. Rising temperature is alarmed at 225°F (see Horn).
3. An interlock shuts off stream *A* at 250°F, stopping the reaction (SV-1).
4. The operator can initiate the interlock by punching the panic switch (NC).

a) Construct the fault tree for this system.
b) Obtain the cut sets using MOCUS, and identify possible system upgrades based on the qualitative analysis.
c) Using the data of Table 13.5 obtain system failure characteristics.
d) Obtain component and cut set importances.

Solution:

a) Figure 13.6 shows the fault tree (constructed by Browning). He uses diamonds to designate both undeveloped failures and secondary failures due to external effects. Rectangles appearing under gates are usually command failures, i.e., those in which the component is “commanded” to failure by some in-system signal, force, or effect. The structure Browning tried to achieve is as shown in Fig. 13.5(b).

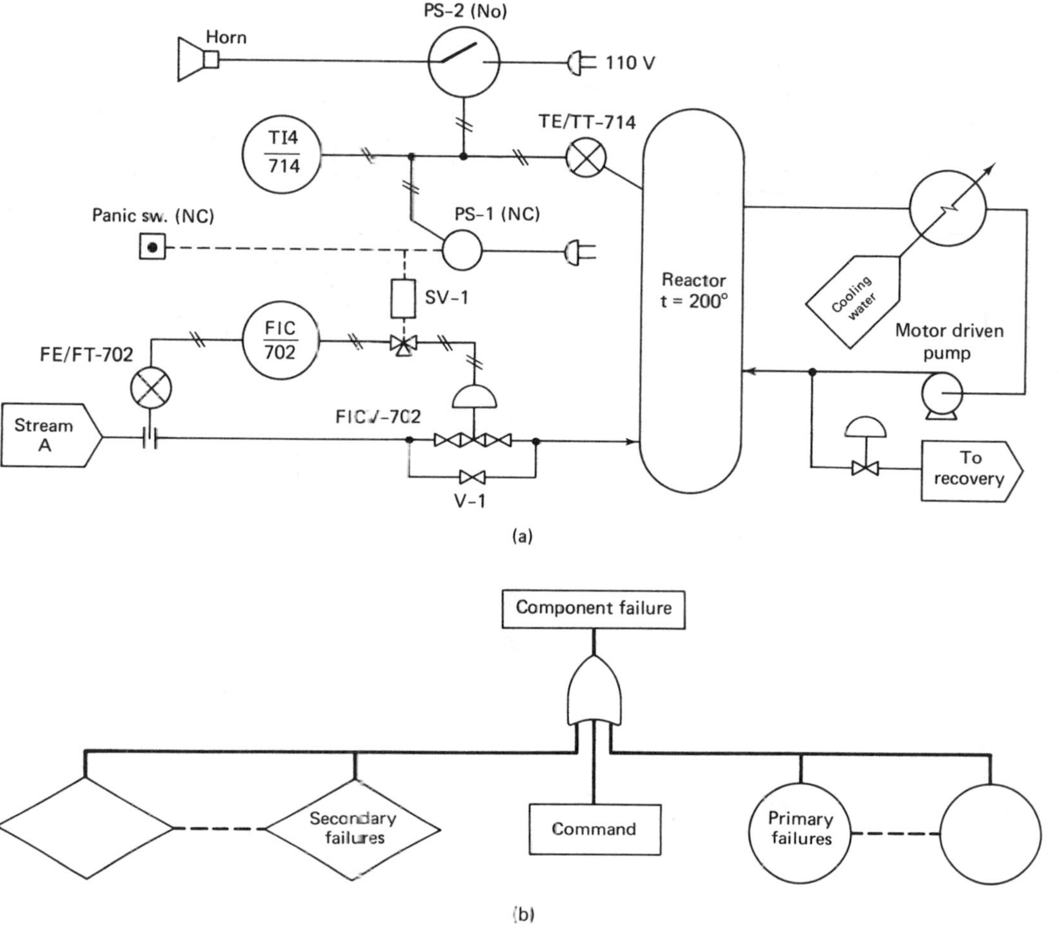

Figure 13.5 Example reaction system for Problem 3.

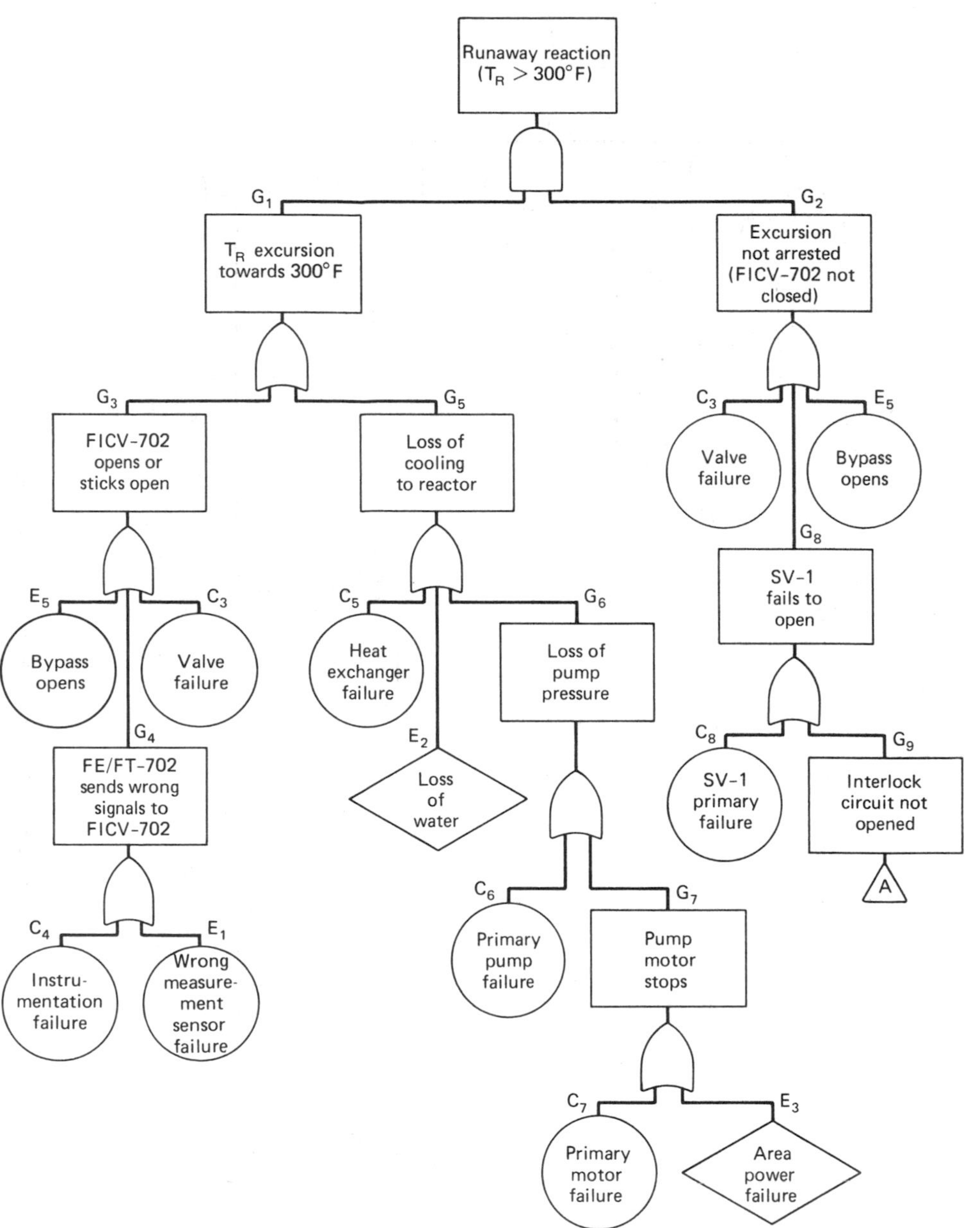

Figure 13.6. Fault tree for reactor system.

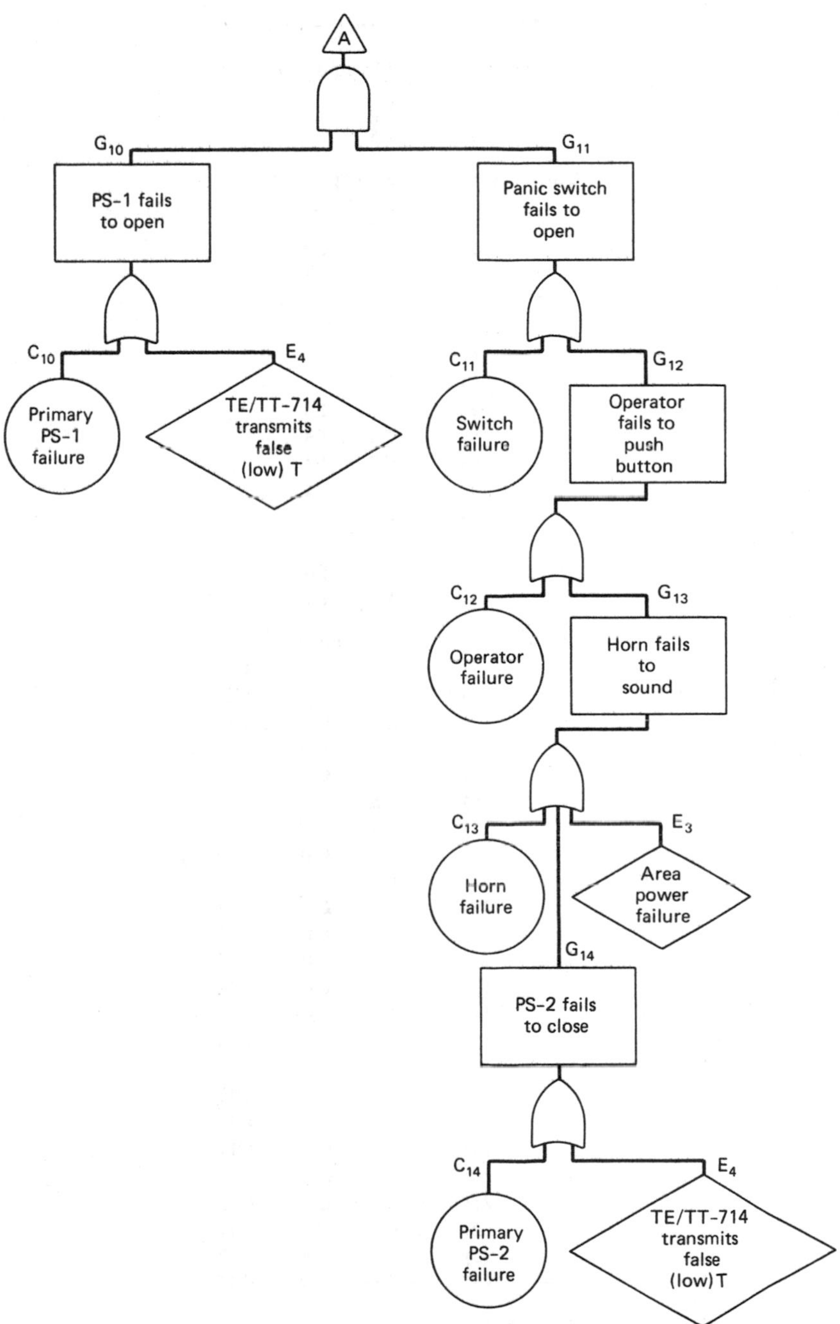

Figure 13.16. (Continued)

TABLE 13.3 Cut Sets for Reactor

Cut sets with	1 component:		
1)	C3		
2)	E5		

Cut sets with	2 components:	
1)	C4	C8
2)	C7	C8
3)	E1	C8
4)	C5	C8
5)	E2	C8
6)	C6	C8
7)	E3	C8
8)	C10	E3
9)	C4	E4
10)	C7	E4
11)	E1	E4
12)	C5	E4
13)	E2	E4
14)	C6	E4
15)	E4	E3

Cut sets with	3 components		
1)	C4	C10	C14
2)	C7	C10	C14
3)	E1	C10	C14
4)	C5	C10	C14
5)	E2	C10	C14
6)	C6	C10	C14
7)	C4	C10	C11
8)	C4	C10	C12
9)	C4	C10	C13
10)	C7	C10	C11
11)	C7	C10	C12
12)	C7	C10	C13
13)	E1	C10	C11
14)	E1	C10	C12
15)	E1	C10	C13
16)	C5	C10	C11
17)	C5	C10	C12
18)	C5	C10	C13
19)	E2	C10	C11
20)	E2	C10	C12
21)	E2	C10	C13
22)	C6	C10	C11
23)	C6	C10	C12
24)	C6	C10	C13

Total number of cut sets found was 41. All cut sets have been determined. The *W* array is dimensioned for 10,000 real*8 words.

b) The cut sets identified by MOCUS are shown in Table 13.3. The two single-event cut sets C_3 and E_5 are a serious potential problem which must be corrected. Figure 13.7 shows a reasonable modification. Here valve XV-714 is added in front of FICV-702, and SV-1 is relocated. Thus, XV-714 (C2) now forms a two-event cut set with C_3 and E_5. The value XV-714 corresponds to valve A of Figs. 2.25 and 2.26 in Chapter 2, and C_3 and E_5 to valve B and valve C.

With the fault tree modification shown in Fig. 13.7, the cut sets are as shown in Table 13.4.

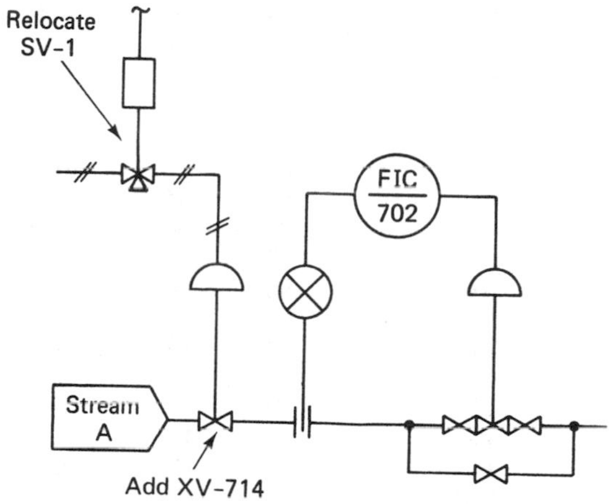

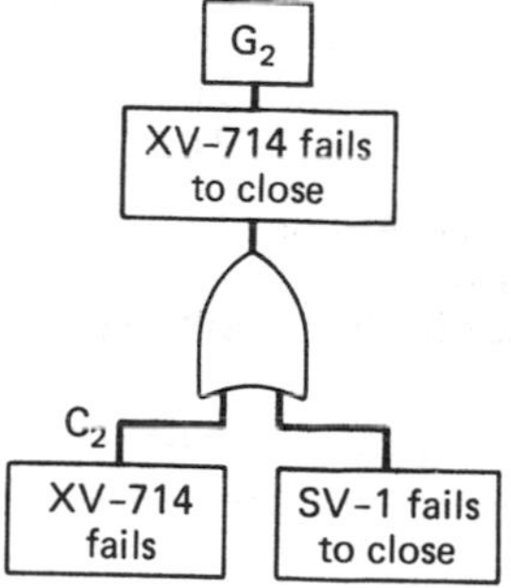

Figure 13.7. Flow sheet and fault tree modification.

c) The failure and repair data in Table 13.5 should not be taken literally. In general, the lambdas are low and are more typical of nuclear than chemical processes. Human operator, horn failures, and panic switch failures are postulated to occur with a constant probability, independent of time. This is correct for the operator, but not necessarily for the switch or horn unless they are not in the plant environment. Operator probabilities of failure are usually 0.01 for routine tasks,

TABLE 13.4 Cut Sets for Upgraded System

Cut sets with 1 component:

None exist

Cut sets with 2 components:

1)	C4	C2
2)	C4	C8
3)	C7	C2
4)	C7	C8
5)	C3	C2
6)	C3	C8
7)	E5	C2
8)	E5	C8
9)	E1	C2
10)	E1	C8
11)	C5	C2
12)	C5	C8
13)	E2	C2
14)	E2	C8
15)	C6	C2
16)	C6	C8
17)	E3	C2
18)	E3	C8
19)	C10	E3
20)	C4	E4
21)	C7	E4
22)	C3	E4
23)	E5	E4
24)	E1	E4
25)	C5	E4
26)	E2	E4
27)	C6	E4
28)	E4	E3

Cut sets with 3 components:

1)	C4	C10	C14
2)	C7	C10	C14
3)	C3	C10	C14
4)	E5	C10	C14
5)	E1	C10	C14
6)	C5	C10	C14
7)	E2	C10	C14
8)	C6	C10	C14
9)	C4	C10	C11
10)	C4	C10	C12
11)	C4	C10	C13
12)	C7	C10	C11
13)	C7	C10	C12
14)	C7	C10	C13
15)	C3	C10	C11
16)	C3	C10	C12
17)	C3	C10	C13
18)	E5	C10	C11
19)	E5	C10	C12
20)	E5	C10	C13
21)	E1	C10	C11
22)	E1	C10	C12
23)	E1	C10	C13
24)	C5	C10	C11
25)	C5	C10	C12
26)	C5	C10	C13
27)	E2	C10	C11
28)	E2	C10	C12
29)	E2	C10	C13
30)	C6	C10	C11
31)	C6	C10	C12
32)	C6	C10	C13

Total number of cut sets found was 60.
All cut sets have been determined.
The *W* array is dimensioned for 10,000 real*8 words.

TABLE 13.5 Failure Events and Probabilities

Computer Index No.	Event	Description	PR[a]	λ/hr	τ (hr)
15	C_3, C_2	Primary control valve failure (open)		7×10^{-5}	30
16	E_5	By-pass opens		1×10^{-6}	100
13	C_5	Heat exchanger failure		5×10^{-5}	50
6	C_4	Primary instrument failure, transmitters, controllers, recorders, etc.		5×10^{-5}	10

TABLE 13.5 (Continued)

Computer Index No.	Event	Description	PR[a]	λ/hr	τ (hr)
7	E_1	Sensor failure (low reading)		1.3×10^{-4}	20
14	E_2	Utility interruptions (water)		1×10^{-6}	300
12	C_6	Primary pump failure		3×10^{-6}	20
4	C_7	Primary motor failure		1×10^{-6}	20
5	E_3	Area power failure		3×10^{-7}	100
11	C_8	Primary switch failure		5×10^{-5}	20
3, 1	C_{10}, C_{14}	PS-1, PS-2 failure		4×10^{-5}	20
2	E_4	TE/TT-714 transmits or reads low temperature		4×10^{-5}	20
10	C_{11}	Panic switch fails (probability assumed constant because of limited use)	3×10^{-4}		
9	C_{12}	Operator failure	10^{-3}		
8	C_{13}	Horn fails to sound (probability assumed constant because of limited use)	3×10^{-4}		

[a]Probability assumed constant, independent of time.

0.001 for introducing a hazard, and 0.1 for emergency tasks under stress conditions. Portions of the KITT printouts from 0 to 100 hr are as shown in Table 13.6.†

Only the characteristics of the least and most reliable components (E1 = 7 and E3 = 5), and the inhibit conditions are given. For the repairable components, only WSUM = ENF increases after steady state is achieved (note that 5 is not at steady state at 100 hr). Some cut set values are as shown in Table 13.7.

TABLE 13.6

CHARACTERISTICS FOR COMPONENT NO. = 5

T (HOURS)	Q	W	L	WSUM
0.0000000	0.0000000	3.0000000–007	3.0000000–007	0.0000000
1.0000000+001	2.9999775–006	2.9999910–007	3.0000000–007	2.9999955–006
2.0000000+001	5.9999550–006	2.9999820–007	3.0000000–007	5.9999820–006
3.0000000+001	8.9999325–006	2.9999730–007	3.0000000–007	8.9999595–006
4.0000000+001	1.1999910–005	2.9999640–007	3.0000000–007	1.1999928–005
5.0000000+001	1.4999888–005	2.9999550–007	3.0000000–007	1.4999888–005
6.0000000+001	1.7999820–005	2.9999460–007	3.0000000–007	1.7999838–005
7.0000000+001	2.0999753–005	2.9999370–007	3.0000000–007	2.0999780–005
8.0000000+001	2.3999685–005	2.9999280–007	3.0000000–007	2.3999712–005
9.0000000+001	2.6999618–005	2.9999190–007	3.0000000–007	2.6999636–005
1.0000000+002	2.9999550–005	2.9999100–007	3.0000000–007	2.9999550–005

†Q = Unavailability, W = unconditional failure intensity, L = conditional failure intensity, WSUM = expected number of failures, FSUM = unreliability.

TABLE 13.6 (Continued)

CHARACTERISTICS FOR COMPONENT NO. = 7

T (HOURS)	Q	W	L	WSUM
0.0000000	0.0000000	1.3000000−004	1.3000000−004	0.0000000
1.0000000+001	1.2991554−003	1.2983111−004	1.3000000−004	1.2991555−003
2.0000000+001	2.5966229−003	1.2966244−004	1.3000000−004	2.5966233−003
3.0000000+001	2.5940938−003	1.2966277−004	1.3000000−004	3.8932493−003
4.0000000+001	2.5932546−003	1.2966288−004	1.3000000−004	5.1898776−003
5.0000000+001	2.5932572−003	1.2966288−004	1.3000000−004	6.4865063−003
6.0000000+001	2.5932575−003	1.2966288−004	1.3000000−004	7.7831351−003
7.0000000+001	2.5932575−003	1.2966288−004	1.3000000−004	9.0797638−003
8.0000000+001	2.5932575−003	1.2966288−004	1.3000000−004	1.0376393−002
9.0000000+001	2.5932575−003	1.2966288−004	1.3000000−004	1.1673021−002
1.0000000+002	2.5932575−003	1.2966288−004	1.3000000−004	1.2969650−002

COMPONENT 8 IS AN INHIBIT CONDITION WITH A PROBABILITY = 3.0000000−004
COMPONENT 9 IS AN INHIBIT CONDITION WITH A PROBABILITY = 1.0000000−003
COMPONENT 10 IS AN INHIBIT CONDITION WITH A PROBABILITY = 3.0000000−004

TABLE 13.7

MININIMAL SET INFORMATION

CHARACTERISTICS FOR SET NO. = 1 (COMPONENT 15)

T (HOURS)	Q	W	L	WSUM
0.0000000	0.0000000	7.0000000−005	7.0000000−005	0.0000000
1.0000000+001	6.9963263−004	6.9951026−005	7.0000000−005	6.9975513−004
2.0000000+001	1.3988981−003	6.9902077−005	7.0000000−005	1.3990206−003
3.0000000+001	2.0977965−003	6.9853154−005	7.0000000−005	2.0977968−003
4.0000000+001	2.0966961−003	6.9853231−005	7.0000000−005	2.7963287−003
5.0000000+001	2.0959632−003	6.9853283−005	7.0000000−005	3.4948613−003
6.0000000+001	2.0955977−003	6.9853308−005	7.0000000−005	4.1933943−003
7.0000000+001	2.0955986−003	6.9853308−005	7.0000000−005	4.8919273−003
8.0000000+001	2.0955991−003	6.9853308−005	7.0000000−005	5.5904604−003
9.0000000+001	2.0955992−003	6.9853308−005	7.0000000−005	6.2889935−003
1.0000000+002	2.0955992−003	6.9853308−005	7.0000000−005	6.9875266−003

CHARACTERISTICS FOR SET NO. = 2 (COMPONENT 16)

T (HOURS)	Q	W	L	WSUM
0.0000000	0.0000000	1.0000000−006	1.0000000−006	0.0000000
1.0000000+001	9.9997500−006	9.9999000−007	1.0000000−006	9.9999500−006
2.0000000+001	1.9999500−005	9.9998000−007	1.0000000−006	1.9999800−005
3.0000000+001	2.9999250−005	9.9997000−007	1.0000000−006	2.9999550−005
4.0000000+001	3.9999000−005	9.9996000−007	1.0000000−006	3.9999200−005
5.0000000+001	4.9998750−005	9.9995000−007	1.0000000−006	4.9998750−005
6.0000000+001	5.9998000−005	9.9994000−007	1.0000000−006	5.9998200−005
7.0000000+001	6.9997250−005	9.9993000−007	1.0000000−006	6.9997550−005
8.0000000+001	7.9996500−005	9.9992000−007	1.0000000−006	7.9996800−005
9.0000000+001	8.9995750−005	9.9991000−007	1.0000000−006	8.9995950−005
1.0000000+002	9.9995000−005	9.9990000−007	1.0000000−006	9.9995000−005

CHARACTERISTICS FOR SET NO. = 12 (COMPONENTS 4, 2)

T (HOURS)	Q	W	L	WSUM
0.0000000	0.0000000	0.0000000	0.0000000	0.0000000
1.0000000+001	3.9991801−009	7.9975404−010	7.9975405−010	3.9987702−009
2.0000000+001	1.5993442−008	1.5990164−009	1.5990164−009	1.5992622−008
3.0000000+001	1.5988525−008	1.5987707−009	1.5987708−009	3.1981558−008
4.0000000+001	1.5986889−008	1.5986890−009	1.5986890−009	4.7968856−008
5.0000000+001	1.5986890−008	1.5986890−009	1.5986891−009	6.3955746−008
6.0000000+001	1.5986890−008	1.5986890−009	1.5986891−009	7.9942637−008
7.0000000+001	1.5986890−008	1.5986890−009	1.5986891−009	9.5929527−008
8.0000000+001	1.5986890−008	1.5986890−009	1.5986891−009	1.1191642−007
9.0000000+001	1.5986890−008	1.5986890−009	1.5986891−009	1.2790331−007
1.0000000+002	1.5986890−008	1.5986890−009	1.5986891−009	1.4389020−007

TABLE 13.7 (Continued)

CHARACTERISTICS FOR SET NO. = 5 (COMPONENTS 13, 11)

T (HOURS)	Q	W	L	WSUM
0.0000000	0.0000000	0.0000000	0.0000000	0.0000000
1.0000000+001	6.4941532−007	1.2982463−007	1.2982471−007	6.4912313−007
2.0000000+001	2.5953250−006	2.5929901−007	2.5929968−007	2.5947413−006
3.0000000+001	2.5918252−006	2.5912448−007	2.5912515−007	5.1868587−006
4.0000000+001	2.5906635−006	2.5906652−007	2.5906719−007	7.7778137−006
5.0000000+001	2.5906665−006	2.5906667−007	2.5906734−007	1.0368480−005
6.0000000+001	2.5906669−006	2.5906669−007	2.5906736−007	1.2959146−005
7.0000000+001	2.5906669−006	2.5906669−007	2.5906736−007	1.5549813−005
8.0000000+001	2.5906669−006	2.5906669−007	2.5906736−007	1.8140480−005
9.0000000+001	2.5906669−006	2.5906669−007	2.5906736−007	2.0731147−005
1.0000000+002	2.5906669−006	2.5906669−007	2.5906736−007	2.3321814−005

CHARACTERISTICS FOR SET NO. = 31 (COMPONENTS 7, 3, 9)

T (HOURS)	Q	W	L	WSUM
0.0000000	0.0000000	0.0000000	0.0000000	0.0000000
1.0000000+001	5.1955823−010	1.0386749−010	1.0386749−010	5.1933746−010
2.0000000+001	2.0764676−009	2.0747033−010	2.0747033−010	2.0760266−009
3.0000000+001	2.0738230−009	2.0733843−010	2.0733843−010	4.1500704−009
4.0000000+001	2.0729451−009	2.0729464−010	2.0729464−010	6.2232357−009
5.0000000+001	2.0729474−009	2.0729475−010	2.0729475−010	8.2961827−009
6.0000000+001	2.0729477−009	2.0729477−010	2.0729477−010	1.0369130−008
7.0000000+001	2.0729477−009	2.0729477−010	2.0729477−010	1.2442078−008
8.0000000+001	2.0729477−009	2.0729477−010	2.0729477−010	1.4515026−008
9.0000000+001	2.0729477−009	2.0729477−010	2.0729477−010	1.6587973−008
1.0000000+002	2.0729477−009	2.0729477−010	2.0729477−010	1.8660921−008

The first two sets, 1 and 2, are for the single components. We see that their unavailabilities are from one to three orders of magnitude higher than that of the worst and best of the two-event sets, 5 and 12 (the other 13 two-event sets have intermediate Q's and are not shown). Set 31 is the worst of the 24, three-event sets and, as expected, includes the operator, component 9. The operator appears in none of the two-event cut sets, which is salutory. The system parameters are as shown in Table 13.8.

The average repair time for the system as obtained from the constant-rate steady-state assumption,

$$(1-Q_s)=\frac{1/\lambda_s}{1/\lambda_s+\tau_s}$$

is 31 hr. Thus, we are nearly at steady state in 100 hr. The safety system is unavailable for

$$Q_s t=(365)(2.2058\times 10^{-3})=0.81 \text{ days/year}$$

We can check this, by

$$t(w_s)\tau_s=(365)(7.1799\times 10^{-5})(31)=0.81 \text{ days/year}$$

This number is slightly optimistic because of the low λ's used. Note that in actual

TABLE 13.8

SYSTEM INFORMATION—UPPER BOUNDS

DIFFERENTIAL CHARACTERISTICS—UPPER BOUNDS

T (HOURS)	Q	W	L
0.00000000	0.00000000	7.10000000−005	7.10000000−005
1.00000000+001	7.11743018−004	7.13747516−005	7.14255883−005
2.00000000+001	1.42643144−003	7.17039407−005	7.18063676−005
3.00000000+001	2.13620508−003	7.17008478−005	7.18543434−005
4.00000000+001	2.14600318−003	7.17469750−005	7.19012754−005
5.00000000+001	2.15617123−003	7.17932003−005	7.19483333−005
6.00000000+001	2.16580806−003	7.17944218−005	7.19502523−005
7.00000000+001	2.17581120−003	7.17956176−005	7.19521720−005
8.00000000+001	2.18581452−003	7.17968416−005	7.19541200−005
9.00000000+001	2.19581802−003	7.17980937−005	7.19560962−005
1.00000000+002	2.20582139−003	7.17993458−005	7.19580724−005

INTEGRAL CHARACTERISTICS—UPPER BOUNDS

T (HOURS)	WSUM	FSUM
1.00000000+001	7.11873758−004	7.11874439−004
2.00000000+001	1.42726722−003	1.42726820−003
3.00000000+001	2.14429116−003	2.14428900−003
4.00000000+001	2.86153028−003	2.86126812−003
5.00000000+001	3.57923115−003	3.57820034−003
6.00000000+001	4.29716926−003	4.29486093−003
7.00000000+001	5.01511946−003	5.01102517−003
8.00000000+001	5.73308176−003	5.72669354−003
9.00000000+001	6.45105643−003	6.44186664−003
1.00000000+002	7.16904363−003	7.15654494−003

TABLE 13.9

SYSTEM INFORMATION—UPPER BOUNDS

DIFFERENTIAL CHARACTERISTICS—UPPER BOUNDS

T (HOURS)	Q	W	L
0.00000000	0.00000000	0.00000000	0.00000000
1.00000000+001	4.90004151−006	9.79794468−007	9.79799269−007
2.00000000+001	1.79906835−005	1.87801382−006	1.87804760−006
3.00000000+001	2.66687031−005	2.28597477−006	2.28603574−006
4.00000000+001	2.86890210−005	2.36876631−006	2.36883427−006
5.00000000+001	3.07172032−005	2.45192568−006	2.45200100−006
6.00000000+001	3.08012455−005	2.45540339−006	2.45547902−006
7.00000000+001	3.08896004−005	2.45905128−006	2.45912724−006
8.00000000+001	3.09791683−005	2.46274890−006	2.46282520−006
9.00000000+001	3.10699490−005	2.46649626−006	2.46657289−006
1.00000000+002	3.11607282−005	2.47024355−006	2.47032052−006

INTEGRAL CHARACTERISTICS—UPPER BOUNDS

T (HOURS)	WSUM	TSUM
1.00000000+001	4.89897234−006	4.89898435−006
2.00000000+001	1.91880138−005	1.91880466−005
3.00000000+001	4.00079567−005	4.00078471−005
4.00000000+001	6.32816621−005	6.32809951−005
5.00000000+001	8.73851221−005	8.73833556−005
6.00000000+001	1.11921767−004	1.11918311−004
7.00000000+001	1.36494041−004	1.36488290−004
8.00000000+001	1.61103042−004	1.61094390−004
9.00000000+001	1.85749268−004	1.85737107−004
1.00000000+002	2.10432967−004	2.10416684−004

practice the downtime due to maintenance would probably be greater than downtime due to failure for this situation. Common-mode failures have also not been analyzed.

For the upgraded system (Fig. 13.7) the upper-bound system parameters are as shown in Table 13.9.

We have achieved a two-fold order of magnitude increase in system availability by eliminating one-event cut sets.

d) For the original fault tree, we would expect the two, one-event cut set components to be dominantly important. With reference to the cut set tabulation in Table 13.3, we note that the structural importance ordering, based on the number of appearances in one- and two-event cut sets only, is as shown in Table 13.10.

TABLE 13.10

STRUCTURAL IMPORTANCE	NUMBER OF APPEARANCES IN SETS 14			
Rank	Component	One Event	Two Event	Three Event
1	C3-15	1	0	0
1	E5-16	1	0	0
4	C5-13	0	2	4
4	C4-6	0	2	4
4	E1-7	0	2	4
4	E2-14	0	2	4
4	C6-12	0	2	4
4	C7-4	0	2	4
3	E3-5	0	3	0
2	C8-11	0	7	0
5	C10-3	0	1	24
7	C14-1	0	0	6
2	E4-2	0	7	0
7	C11-10	0	0	6
7	C12-9	0	0	6
7	C13-8	0	0	6
		2	30	72

The general ranking shown in Table 13.10 was confirmed by nearly all of the importance measures computed. The Fussell-Vesely parameters, for example, are as shown in Table 13.11.

TABLE 13.11 FUSSELL-VESELY MEASURE OF BASIC-EVENT IMPORTANCE

PROB OF TOP EVENT=.606−03*

MISSION TIME=.100+02*

PROB OF TOP EVENT=.104−02*

MISSION TIME=.200+02*

PROB OF TOP EVENT=.209−02*

MISSION TIME=.100+03*

TABLE 13.11 (Continued)

RANK	BASIC EVENT	IMPORTANCE*
1	COMP15	.982+00*
2	COMP16	.157–01*
3	COMP7	.120–02*
4	COMP11	.119–02*
5	COMP2	.953–03*
6	COMP13	.530–03*
7	COMP6	.370–03*
8	COMP12	.276–04*
9	COMP14	.115–04*
10	COMP4	.920–05*
11	COMP5	.482–05*
12	COMP3	.331–05*
13	COMP1	.300–06*

RANK	BASIC EVENT	IMPORTANCE*
1	COMP15	.979+00*
2	COMP16	.174–01*
3	COMP11	.180–02*
4	COMP7	.179–02*
5	COMP2	.144–02*
6	COMP13	.899–03*
7	COMP6	.472–03*
8	COMP12	.414–04*
9	COMP14	.211–04*
10	COMP4	.138–04*
11	COMP5	.857–05*
12	COMP3	.567–05*
13	COMP1	.727–06*

RANK	BASIC EVENT	IMPORTANCE*
1	COMP15	.965+00*
2	COMP16	.302–01*
3	COMP11	.257–02*
4	COMP7	.220–02*
5	COMP2	.205–02*
6	COMP13	.184–02*
7	COMP6	.427–03*
8	COMP14	.726–04*
9	COMP12	.509–04*
10	COMP5	.234–04*
11	COMP4	.170–04*
12	COMP3	.121–04*
13	COMP1	.163–05*

An interesting point is that the IMPORTANCE code that was used to make these calculations does not rank events having constant probabilities. The rankings are a weak function of time. Component 7, for example is more important than 11 at 100 hr, but less important after 200 hr.

Example 4. System Trade-Off Analysis Using Short-Cut Calculation Methods

In the system of Fig. 13.8 the tank is filled in 10 min and empties in 50; thus, the cycle time is 1 hr. After the switch is closed, the timer is set to open the contacts in 10 min. If the mechanisms fail then the operator is instructed to open the switch to prevent a tank rupture due to overfilling.

a) Neglecting secondary failures and other effects, construct a fault tree and identify the cut sets which might lead to a tank rupture.
b) The failure and repair data of Table 13.12 have been collected. Using the data, compute, using short-cut methods, the expected number of failures per year due to tank rupture, assuming that the system is unrepairable (or that repair facilities are unavailable).

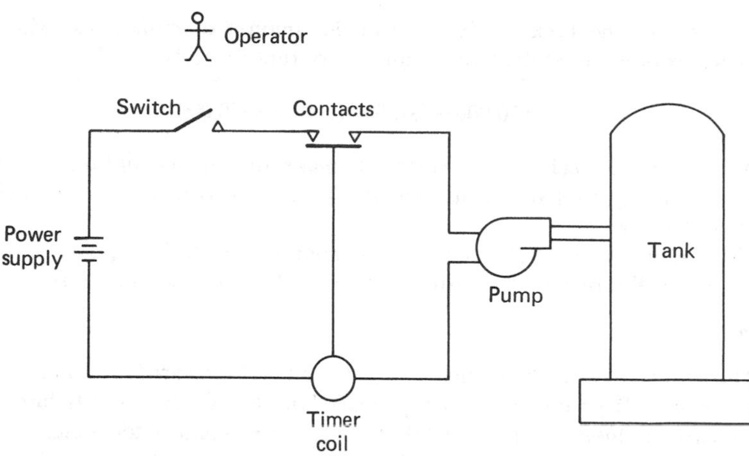

System Definitions:

Top = tank rupture
Initial conditions = switch and contacts closed
Not allowed events = wiring failures
external effects
tank draining faults
system restart during cycle after initial start

Figure 13.8. Schematic flow sheet, Example 4.

TABLE 13.12 RELIABILITY DATA

Component	Mode	$\lambda(\text{hr}^{-1})$	τ(hr)
Tank	Rupture	10^{-7}	500
Switch	Fails open	10^{-5}	10
	Transfers open	10^{-9}	10
	Fails closed	10^{-6}	10
	Transfers closed	10^{-9}	10
Timer coil and mechanism	Fails	2×10^{-5}	10
	Burns out	10^{-4}	10
Timer contacts	Fails open	10^{-4}	20
	Transfers open	10^{-9}	20
	Fails closed	10^{-5}	20
	Transfers closed	10^{-9}	20
Power supply	Surge	10^{-6}	50
	Off	10^{-3}	50
Pump	Off	10^{-3}	30
	Overheats	10^{-4}	30
	Shorts	10^{-4}	40
Operator unavailability = 10% = 0.1			

c) The cost of the tank is \$50,000 and the amount of money lost when the system is down is \$1200/hour. Thus, every rupture costs

$$\$50,000 + (\$1,200)(500) = \$650,000$$

At present, no mechanic is available to make any repairs, and the (prorated) cost of having a repair facility available is \$10,000/year. Is it worth having a repair facility?

d) Calculate the relative (Fussell-Vesely) importance of the components.

e) Show that the Birnbaum sensitivity rankings lead to a spurious result.

Solution:

a) The fault tree is shown as Figure 13.9. Note that hardware failures are always followed by an OR gate which shows primary failure and command failures. The state-of-system failures, such as "EMF through timer contact too long," can be followed by AND gates. Compared to Fig. 2.24 of Chapter 2, the present fault tree is simpler because the alarm failure is not involved.

The structure function, by inspection, is

$$\psi(Y) = Y_1 \vee (Y_3 \wedge Y_4) \vee (Y_3 \wedge Y_5) \vee (Y_2 \wedge Y_5) \vee (Y_2 \wedge Y_4)$$

The five cut sets are:

	λ (hr)	τ (hr)	Q
A-1. Tank ruptures (TR)	10^{-7}	500	
B-3. Timer coils fail closed (TC)	2×10^{-5}	10	
4. Operator failure (O)			0.1
C-3. Timer coils fail closed (TC)	2×10^{-5}	10	
5. Switch fails closed (S)	10^{-6}	10	
D-2. Timer contacts fail closed (C)	10^{-5}	20	
5. Switch fails closed (S)	10^{-6}	10	
E-2. Timer contacts fail closed (C)	10^{-5}	20	
4. Operator failure (O)			0.1

b) Since the components are considered non-repairable, the expected number of failures per year is Q_s(8760 hr), where Q_i^*, the n-event cut set unavailability, is obtained from

$$Q_i^* = \prod_{j=1}^{n} Q_j \simeq \prod_{j=1}^{n} (\lambda_j t)$$

and the system unavailability is

$$Q_s = \sum_{i=1}^{5} Q_i^*$$

If repair is considered for each component,

$$Q_j = \lambda\tau, \; Q_i^* = \prod_{j=1}^{n} Q_j, \quad \text{and} \quad w_i^* = Q_i^* \sum_{j=1}^{n} \frac{\lambda_i}{Q_i} \simeq \lambda_i^*(\text{hr}^{-1})$$

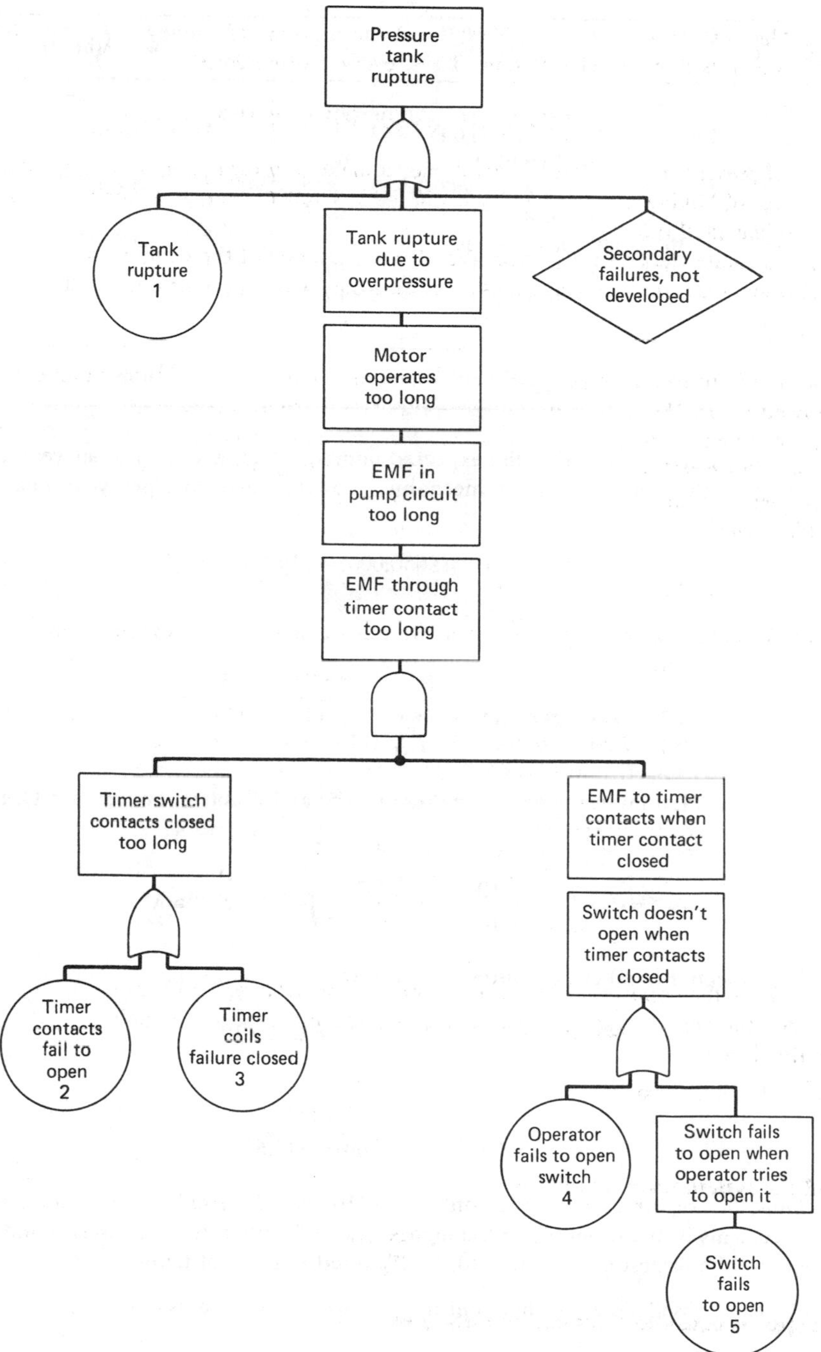

Figure 13.9. Fault tree.

Cut Set	λ_i	ENF/yr, W^*	$Q_j = \lambda\tau$	Q_i^*	$w^* \cong \lambda(\text{hr}^{-1})$
A-1	10^{-7}	8.76×10^{-4}	5×10^{-5}	5×10^{-5}	10^{-7}
B-3	2×10^{-5}	$(8760)(2\times10^{-5})(0.1)$	2×10^{-4}	2×10^{-5}	2×10^{-6}
4		$=1.75\times10^{-2}$	0.1		
C-3	2×10^{-5}	$(2\times10^{-5})(10^{-6})(8760)^2$	2×10^{-4}	2×10^{-9}	4×10^{-10}
5	10^{-6}	$=1.53\times10^{-3}$	10^{-5}		
D-2	10^{-5}	$(10^{-5})(10^{-6})(8760)^2$	2×10^{-4}	2×10^{-9}	3×10^{-10}
5	10^{-6}	$=7.67\times10^{-4}$	10^{-5}		
E-2	10^{-5}	$(10^{-5})(8760)(0.1)$	2×10^{-4}	2×10^{-5}	10^{-6}
4		$=8.76\times10^{-3}$	0.1		
		$\sum Q_i^* = 2.94\times10^{-2}$			$\sum \lambda_i^* = 3.1\times10^{-6}$

We see from column 3 that the expected number of system failures per year are 2.94×10^{-2}; thus, if there is no mechanic, the estimated cost per year due to breakdowns is

$$(2.94\times10^{-2})(\$650{,}000) = \$19{,}110$$

c) If there is a repair facility, then we must calculate the expected number of cut set failures from

$$w^* = Q_i^* \sum_{i=1}^{n} \frac{\lambda_i}{Q_i} \cong \lambda^*$$

These calculations are shown in columns 4, 5, and 6 above. For cut set D, for example,

$$w^* = (2\times10^{-9})\left(\frac{10^{-6}}{10^{-5}} + \frac{10^{-5}}{2\times10^{-4}}\right) = 3\times10^{-10} \cong \lambda^*$$

The expected number of failures per year are

$$8760\lambda_s = (8760)(3.1\times10^{-6}) = 2.72\times10^{-2}$$

with a financial risk of

$$(2.72\times10^{-2})(\$650{,}000) = \$17{,}655$$

Thus, the repair facility saves only ($19,110 − $17,655) = $1,455/yr and is not justified. This is due primarily to the high operator-failure rates in cut sets B and E, which are the largest contributors to the expected number of failures.

d) The Fussell-Vesely component importance is obtained as

$$\frac{g_i(Q(t))}{g(Q(t))}$$

Thus,

$$I_{TR}=\frac{8.76\times10^{-4}}{2.94\times10^{-2}}=2.97\times10^{-2}$$

$$I_{TC}=\frac{1.75\times10^{-2}+1.53\times10^{-3}}{2.94\times10^{-2}}=0.65$$

$$I_{O}=\frac{1.75\times10^{-2}+8.76\times10^{-3}}{2.94\times10^{-2}}=0.89$$

$$I_{S}=\frac{1.53\times10^{-3}+7.67\times10^{-4}}{2.94\times10^{-2}}=0.08$$

$$I_{C}=\frac{7.67\times10^{-4}+8.76\times10^{-3}}{2.94\times10^{-2}}=0.32$$

and

$$I_O>I_{TC}>I_C>I_S>I_{TR}$$
$$0.89>0.65>0.32>0.08>0.03$$

The operator is the weakest link, followed by the timer coil. The tank rupture, even though it is in a one-event cut set, contributes least to the probability of failure.

e) The Birnbaum importance $\delta Q_S/\delta Q_i$ can be obtained by differentiation.

$$Q_s=Q_{TR}+Q_{TC}Q_O+Q_{TC}Q_S+Q_CQ_s+Q_CQ_O$$

(upper bound unavailability).

For the non-repairable system,

$$I_{TR}=\frac{\delta Q_S}{\delta Q_{TR}}=1$$

$$I_{TC}=Q_O+Q_S=0.1+0.00876=0.1$$

$$I_O=Q_{TC}+Q_C=0.175+0.0876=0.26$$

$$I_s=Q_{TC}+Q_C=0.175+0.0876=0.26$$

$$I_C=Q_S+Q_O=0.175+0.1=0.275$$

$$I_{TR}>I_C>I_O=I_S>I_{TC}$$

This is not a reasonable result. Tank rupture, obviously, should not be ranked ahead of operator failure as a contributor to system failure.

Example 5. Pressure Tank Rupture

The pressure tank system and the fault tree (Fig. 2.27) are discussed in Chapter 2. Figure 13.10 is a condensed version of the tree, and the corresponding failure data are shown in Table 13.13. Note that the secondary failure of the tank or the S1 switch in Fig. 2.27 has two failure modes in Fig. 13.10.

TABLE 13.13 EVENT CODE AND FAILURE DATA

Basic Event i	Description of Basic Event	λ/hr	τ(hr)
1	Pressure tank failure	10^{-8}	—
2	Secondary failure of pressure tank due to improper selection	10^{-5}	50
3	Secondary failure of pressure tank due to out-of-tolerance conditions	10^{-5}	50
4	K2 relay contacts fail to open	10^{-5}	50
5	K2 relay secondary failure	10^{-5}	50
6	Pressure switch secondary failure	10^{-5}	50
7	Pressure switch contacts fail to open	10^{-5}	50
8	Excess pressure not sensed by pressure-actuated switch	10^{-5}	50
9	S1 switch secondary failure	10^{-5}	50
10	S1 switch contacts fail to open	10^{-5}	50
11	External reset actuation force remains on switch S1	10^{-5}	50
12	K1 relay contacts fail to open	10^{-5}	50
13	K1 relay secondary failure	10^{-5}	50
14	Timer does not "time off" due to improper setting	10^{-5}	50
15	Timer relay contacts fail to open		
16	Timer relay secondary failure	10^{-5}	50

a) Identify the cut sets by using the MOCUS program. Discuss possible system improvements.
b) Calculate the system failure parameters by using KITT-1.
c) Rank the components on the basis of the importance of their contribution to system failure at 20 and 1000 hr by using the IMPORTANCE code.

Solution:

a) The MOCUS inputs and outputs are as follows:

PRESSURE TANK PROBLEM—INPUT DATA

```
*DATA
7        BOTH   PRINT
END
*TREE
```

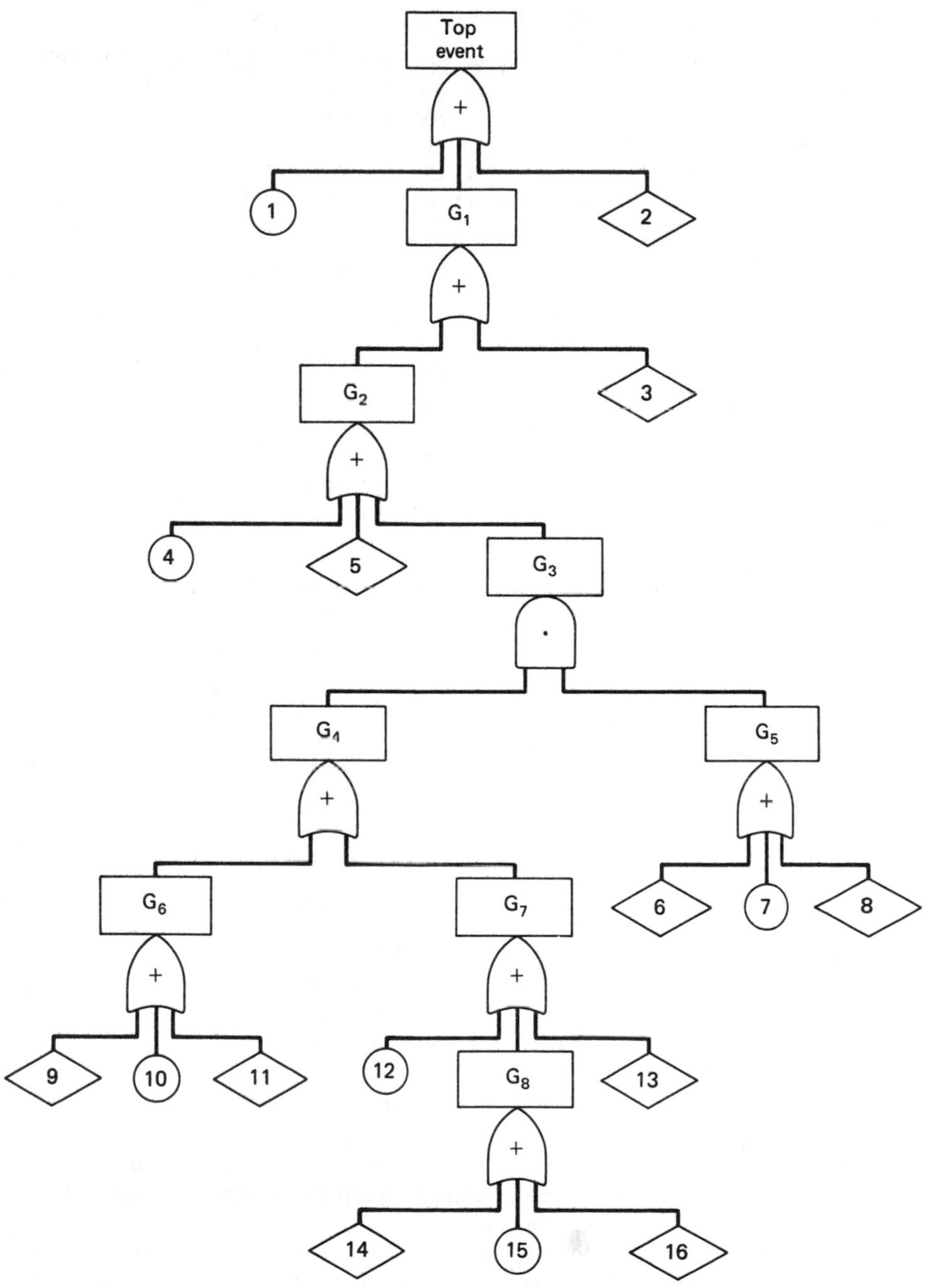

Figure 13.10. Pressure tank event tree.

TOP	OR	1	2	G1	C1	C2
G1	OR	1	1	G2	C3	
G2	OR	1	2	G3	C4	C5
G3	AND	2	0	G4	G5	
G4	OR	2	0	G6	G7	
G5	OR	0	3	C6	C7	C8
G6	OR	0	3	C9	C10	C11
G7	OR	1	2	G8	C12	C13
G8	OR	0	3	C14	C15	C16

MINIMAL CUT SETS FOR GATE TOP

CUT SETS WITH 1 COMPONENTS

1) C1
2) C2
3) C3
4) C4
5) C5

CUT SETS WITH TWO COMPONENTS

1)	C9	C6	13)	C10	C7
2)	C14	C6	14)	C10	C8
3)	C10	C6	15)	C11	C7
4)	C11	C6	16)	C11	C8
5)	C12	C6	17)	C12	C7
6)	C13	C6	18)	C12	C8
7)	C15	C6	19)	C13	C7
8)	C16	C6	20)	C13	C8
9)	C9	C7	21)	C15	C7
10)	C9	C8	22)	C15	C8
11)	C14	C7	23)	C16	C7
12)	C14	C8	24)	C16	C8

TOTAL NUMBER OF CUT SETS FOUND WAS 29

MINIMAL PATH SETS FOR GATE TOP

PATH SETS WITH 8 COMPONENTS

1) C6 C1 C2 C3 C4 C5 C7 C8

PATH SETS WITH 13 COMPONENTS

1) C9 C1 C2 C3 C4 C5 C14 C10
C11 C12 C13 C15 C16

TOTAL NUMBER OF PATH SETS FOUND WAS 2.

The program has found 29 cut sets, five of which are one-component sets, and 24 are two-event sets. Path sets are also found; note that these are much longer and, in a sense, contain less useful information.

We see that, in addition to possible direct tank failure modes, a primary failure of the K2 contacts could lead to system failure, a very uncomfortable situation. We should propose design modifications to include at least a relief valve on the pressure tank. We might also think of including the timer relay contacts in the pump motor circuit, thus eliminating a one-event system failure. Note, too, that safety devices should monitor directly and not monitor other safety devices as does the timer relay in this example.

b) The (truncated) KITT output with only upper-bound system values for $T=0$ to $T=80$ are shown in Table 13.14.

TABLE 13.14 KITT COMPUTER OUTPUT

CHARACTERISTICS FOR COMPONENT NO. = 1

T (HOURS)	Q	W	L	WSUM	FSUM
0.000	0.000	1.000–008	1.000–008	0.000	0.000
2.000+001	2.000–007	1.000–008	1.000–008	2.000–007	2.000–007
4.000+001	4.000–007	1.000–008	1.000–008	4.000–007	4.000–007
6.000+001	6.000–007	1.000–008	1.000–008	6.000–007	6.000–007
8.000+001	8.000–007	1.000–008	1.000–008	8.000–007	8.000–007

CHARACTERISTICS FOR COMPONENT NO. = 2→16

T (HOURS)	Q	W	L	WSUM	FSUM
0.000	0.000	1.000–005	1.000–005	0.000	0.000
2.000+001	2.000–004	9.998–006	1.000–005	2.000–004	2.000–004
4.000+001	3.999–004	9.996–006	1.000–005	3.999–004	3.999–004
6.000+001	4.998–004	9.995–006	1.000–005	5.998–004	5.998–004
8.000+001	4.998–004	9.995–006	1.000–005	7.997–004	7.997–004

MINIMAL SET INFORMATION
CHARACTERISTICS FOR SET NO. = 1

T (HOURS)	Q	W	L	WSUM	FSUM
0.000	0.000	1.000–008	1.000–008	0.000	0.000
2.000+001	2.000–007	1.000–008	1.000–008	2.000–007	2.000–007
4.000+001	4.000–007	1.000–008	1.000–008	4.000–007	4.000–007
6.000+001	6.000–007	1.000–008	1.000–008	6.000–007	6.000–007
8.000+001	8.000–007	1.000–008	1.000–008	8.000–007	8.000–007

CHARACTERISTICS FOR SET NO. = 2→5

T (HOURS)	Q	W	L	WSUM	FSUM
0.000	0.000	1.000–005	1.000–005	0.000	0.000
2.000+001	2.000–007	9.998–006	1.000–005	2.000–004	2.000–007
4.000+001	3.999–004	9.996–006	1.000–005	3.999–004	3.999–004
6.000+001	4.998–004	9.995–006	1.000–005	5.998–004	5.998–004
8.000+001	4.998–004	9.995–006	1.000–005	7.997–004	7.997–004

CHARACTERISTICS FOR SET NO. = 6→29

T (HOURS)	Q	W	L	WSUM	FSUM
0.000	0.000	0.000	0.000	0.000	0.000
2.000+001	3.999–008	3.999–009	3.999–009	3.999–008	3.999–008
4.000+001	1.599–007	7.995–009	7.995–009	1.599–007	1.599–007
6.000+001	2.498–007	9.992–009	9.992–009	3.398–007	3.398–007
8.000+001	2.498–007	9.991–009	9.991–009	5.396–007	5.396–007

SYSTEM INFORMATION—UPPER BOUNDS
DIFFERENTIAL CHARACTERISTICS—UPPER BOUNDS

T (HOURS)	Q	W	L
0.00000000	0.00000000	4.00100000–005	4.00100000–005
2.00000000+001	8.00818932–004	4.00979698–005	4.01301068–005
4.00000000+001	1.60292227–003	4.01858847–005	4.02504030–005
6.00000000+001	2.00443452–003	4.02297811–005	4.03105790–005
8.00000000+001	2.00438314–003	4.02297811–005	4.03105790–005

TABLE 13.14 (Continued)

INTEGRAL CHARACTERISTICS—UPPER BOUNDS

T (HOURS)	WSUM	FSUM
2.00000000+001	8.01079698−004	8.01080031−004
4.00000000+001	1.60391824−003	1.60391851−003
6.00000000+001	2.40807518−003	2.40791260−003
8.00000000+001	3.21267107−003	3.21185907−003

CONCLUSION OF OUTPUT FROM KITT-1

The failure characteristics for components 2–16 are identical, as are the characteristics of cut sets 2–5 (which are equal to the component values) and 6–29 (the two-event cut sets). An interesting aspect of this problem is that it consists of a mixture of non-repairable (event 1) and repairable (all other) failures. Thus, we expect steady-state values of Q for cut sets 2–29, but not for the system or cut set 1 (component 1).

c) This is an interesting test of the various importance criteria because some of the "common-sense" results which should be shown are:

1. Component 1, the only non-repairable component, should increase its ranking (or value) as time goes on.
2. Since all components (other than 1) have identical τ and λ values, their initial importances are clearly in the order given in Table 13.15 (component 1 with $\lambda = 10^{-8}$ should be intermediate between the one-event, $\lambda = 10^{-5}$, and two-event, $\lambda = 10^{-10}$, cut sets).

TABLE 13.15 STRUCTURAL IMPORTANCE

Component	APPEARANCE IN One-Event Cut Sets	Two-Event Cut Sets	Ranking
2, 3, 4, 5	1		Equal—1
1	1		Second
6		8	Third (tie)
7		8	
8		8	
9		3	Fourth (tie)
10		3	
11		3	
12		3	
13		3	
14		3	
15		3	
16		3	

On the basis of this we note that more than "common sense" is required to interpret the rankings in Table 13.16 given by the various computer outputs.

TABLE 13.16 Importance Code Output

BIRNBAUM'S MEASURE OF BASIC EVENT IMPORTANCE

MISSION TIME=.200+02*			MISSION TIME=.100+04*		
RANK	BASIC EVENT	IMPORTANCE*	RANK	BASIC EVENT	IMPORTANCE*
1	4	.100+01*	1	4	.998+00*
1	2	.100+01*	1	2	.998+00*
1	3	.100+01*	1	3	.998+00*
1	5	.100+01*	1	5	.998+00*
2	1	.999+00*	2	1	.998+00*
3	6	.132–02*	3	6	.398–02*
3	8	.132–02*	3	8	.398–02*
4	7	.115–02*	4	7	.349–02*
5	9	.659–03*	5	9	.199–02*
6	10	.494–03*	6	10	.150–02*
6	11	.494–03*	6	11	.150–02*
6	12	.494–03*	6	12	.150–02*
6	13	.494–03*	6	13	.150–02*
6	14	.494–03*	6	14	.150–02*
6	15	.494–03*	6	15	.150–02*
6	16	.494–03*	6	16	.150–02*

CRITICALITY BASIC EVENT IMPORTANCE

RANK	BASIC EVENT	IMPORTANCE*	RANK	BASIC EVENT	IMPORTANCE*
1	4	.250+00*	1	4	.248+00*
1	2	.250+00*	1	2	.248+00*
1	3	.250+00*	1	3	.248+00*
1	5	.250+00*	1	5	.248+00*
2	8	.329–03*	2	1	.496–02*
2	6	.329–03*	3	6	.989–03*
3	1	.303–03*	3	8	.989–03*
4	7	.288–03*	4	7	.865–03*
5	9	.165–03*	5	9	.495–03*
6	10	.123–03*	6	10	.371–03*
6	11	.123–03*	6	11	.371–03*
6	12	.123–03*	6	12	.371–03*
6	13	.123–03*	6	13	.371–03*
6	14	.123–03*	6	14	.371–03*
6	15	.123–03*	6	15	.371–03*
6	16	.123–03*	6	16	.371–03*

FUSSELL-VESELY MEASURE OF BASIC EVENT IMPORTANCE

MISSION TIME=.200+02*			MISSION TIME=.100+04*		
RANK	BASIC EVENT	IMPORTANCE*	RANK	BASIC EVENT	IMPORTANCE
1	4	.250+00*	1	4	.248+00*
1	2	.250+00*	1	2	.248+00*
1	3	.250+00*	1	3	.248+00*
1	5	.250+00*	1	5	.248+00*
2	8	.329–03*	2	1	.497–02*
2	6	.329–03*	3	6	.992–03*
3	1	.303–03*	3	8	.992–03*
4	7	.288–03*	4	7	.868–03*
5	9	.165–03*	5	9	.496–03*
6	10	.123–03*	6	10	.372–03*
6	11	.123–03*	6	11	.372–03*
6	12	.123–03*	6	12	.372–03*
6	13	.123–03*	6	13	.372–03*
6	14	.123–03*	6	14	.372–03*
6	15	.123–03*	6	15	.372–03*
6	16	.123–03*	6	16	.372–03*

TABLE 13.16 (Continued)

BARLOW-PROSCHAN MEASURE OF BASIC EVENT IMPORTANCE

MISSION TIME=.200+02*			MISSION TIME=.100+04*		
RANK	BASIC EVENT	IMPORTANCE	RANK	BASIC EVENT	IMPORTANCE
1	4	.250+00*	1	4	.249+00*
1	2	.250+00*	1	2	.249+00*
1	3	.250+00*	1	3	.249+00*
1	5	.250+00*	1	5	.249+00*
2	1	.250−03*	2	8	.940−03*
3	6	.178−03*	2	6	.940−03*
3	8	.178−03*	3	7	.823−03*
4	7	.155−03*	4	9	.471−03*
5	9	.888−04*	5	16	.353−03*
6	10	.666−04*	5	10	.353−03*
6	11	.666−04*	5	11	.353−03*
6	12	.666−04*	5	12	.353−03*
6	13	.666−04*	5	13	.353−03*
6	14	.666−04*	5	14	.353−03*
6	15	.666−04*	5	15	.353−03*
6	16	.666−04*	6	1	.249−03*

SEQUENTIAL CONTRIBUTORY BASIC EVENT IMPORTANCE

MISSION TIME=.200+02*			MISSION TIME=.100+04*		
RANK	BASIC EVENT	IMPORTANCE*	RANK	BASIC EVENT	IMPORTANCE*
1	8	.177−03*	1	8	.906−03*
1	6	.177−03*	1	6	.906−03*
2	7	.155−03*	2	7	.793−03*
3	9	.888−04*	3	9	.454−03*
4	12	.666−04*	4	12	.341−03*
4	10	.666−04*	4	10	.341−03*
4	11	.666−04*	4	11	.341−03*
4	16	.666−04*	4	15	.341−03*
4	15	.666−04*	4	13	.341−03*
4	13	.666−04*	4	14	.341−03*
4	14	.666−04*	5	16	.341−03*
5	2	.000 *	6	2	.000 *
5	3	.000 *	6	3	.000 *
5	4	.000 *	6	4	.000 *
5	5	.000 *	6	5	.000 *
5	1	.000 *	6	1	.000 *

BARLOW-PROSCHAN MEASURE OF CUT SET IMPORTANCE

MISSION TIME=.200+02*			MISSION TIME=.100+04		
RANK	BASIC EVENT	IMPORTANCE*	RANK	BASIC EVENT	IMPORTANCE*
1	3	.250+00*	1	4	.249+00*
1	2	.250+00*	1	2	.249+00*
1	3	.250+00*	1	3	.249+00*
1	5	.250+00*	1	5	.249+00*
2	1	.250−03*	2	9	.267−03*
3	28	.186−03*	2	11	.267−03*
4	22	.186−03*	3	10	.263−03*
4	26	.186−03*	4	12	.251−03*
4	24	.186−03*	5	1	.249−03*
4	20	.186−03*	6	16	.247−03*
4	18	.186−03*	7	13	.247−03*
5	23	.186−03*	8	14	.247−03*
5	25	.186−03*	9	15	.247−03*
5	29	.186−03*	10	7	.235−03*
5	27	.186−03*	10	8	.235−03*

TABLE 13.16 (Continued)

BARLOW-PROSCHAN MEASURE OF CUT SET IMPORTANCE

MISSION TIME=.200+02*			MISSION TIME=.100+04		
RANK	BASIC EVENT	IMPORTANCE*	RANK	BASIC EVENT	IMPORTANCE*
5	19	.186–03*	11	6	.235–03*
5	17	.186–03*	12	18	.112–04*
5	21	.186–03*	13	17	.112–04*
6	16	.444–04*	14	19	.112–04*
7	7	.444–04*	15	28	.108–04*
7	13	.444–04*	15	20	.108–04*
7	10	.444–04*	16	22	.108–04*
7	8	.444–04*	16	26	.108–04*
7	9	.444–04*	16	24	.108–04*
7	11	.444–04*	17	25	.108–04*
7	12	.444–04*	17	23	.108–04*
8	14	.444–04*	17	27	.108–04*
9	6	.444–04*	17	21	.108–04*
9	15	.444–04*	17	29	.108–04*

FUSSELL-VESELY MEASURE OF CUT SET IMPORTANCE

MISSION TIME=0.200+2*			MISSION TIME=.100+4*		
RANK	CUT SET NO.	IMPORTANCE	RANK	CUT SET NO.	IMPORTANCE
1	4	.250+00*	1	4	.248+00*
1	2	.250+00*	1	2	.248+00*
1	3	.250+00*	1	3	.248+00*
1	5	.250+00*	1	5	.248+00*
2	1	.303–03*	2	1	.497–02*
3	6	.412–04*	3	6	.124–03*
3	7	.412–04*	3	7	.124–03*
3	8	.412–04*	3	8	.124–03*
3	9	.412–04*	3	9	.124–03*
3	10	.412–04*	3	10	.124–03*
3	11	.412–04*	3	11	.124–03*
3	12	.412–04*	3	12	.124–03*
3	13	.412–04*	3	13	.124–03*
3	14	.412–04*	3	14	.124–03*
3	15	.412–04*	3	15	.124–03*
3	16	.412–04*	3	16	.124–03*
3	17	.412–04*	3	17	.124–03*
3	18	.412–04*	3	18	.124–03*
3	19	.412–04*	3	19	.124–03*
3	20	.412–04*	3	20	.124–03*
3	21	.412–04*	3	21	.124–03*
3	22	.412–04*	3	22	.124–03*
3	23	.412–04*	3	23	.124–03*
3	24	.412–04*	3	24	.124–03*
3	25	.412–04*	3	25	.124–03*
3	26	.412–04*	3	26	.124–03*
3	27	.412–04*	3	27	.124–03*
3	28	.412–04*	3	28	.124–03*
3	29	.412–04*	3	29	.124–03*

Example 6. Validity of the Constant Failure Rate Model

Consider a situation where a large chemical plant must be shut down if a refrigeration compressor fails. At the plant design stage, therefore, it is decided to install two compressors, one working and one spare. On the average, the compressor is expected to fail once per year and to require 30 days repair and 5 days for removal and reberthing.

a) Calculate the plant availability for the case of one- and two-fold compressor redundancy, assuming (1) a Poisson distribution and (2) steady-state availability approximations.

b) The actual plant failure data were as shown in Table 13.17. Compare the actual data with predicted values.

TABLE 13.17 FAILURE DATA

Compressor Failure No.	Time to Failure (days)
1	100
2	<1
3	1250
4	595
5	221
6	<1
7	<1
8	223

Solution:

a-1) *Poisson Analysis*. Assume a situation where a failed component is replaced by new components, one after another. As shown in Appendix 4.3 of Chapter 4, for components with a constant failure rate, the probability of having M or less failures in time t is, by the Poisson distribution,

$$F(M)=\sum_{i=0}^{M}\frac{(\lambda t)^i}{i!}\exp(-\lambda t) \tag{13.1}$$

where λ is the constant failure rate, i.e., the constant conditional failure intensity.

The probability of exactly M failures in time t is

$$\Pr(M\text{ failures})=\frac{(\lambda t)^M}{M!}\exp(-\lambda t) \tag{13.2}$$

and this is approximately equal to the probability of having M or more failures, provided that λt is sufficiently small.

$$\Pr(M\text{ or more failures})=\sum_{i=M}^{\infty}\frac{(\lambda t)^i}{i!}\exp(-\lambda t) \tag{13.3}$$

$$\simeq\frac{(\lambda t)^M}{M!}\exp(-\lambda t) \tag{13.4}$$

$$=\Pr(M\text{ failures})$$

In the calculations that follow, various higher-order terms are neglected, and events with probabilities nearly equal to one are assumed to always occur. We assume further that the first compressor fails on the 300th day. The Poisson distribution tells us that the probability of a second failure within 35 days is given by (13.2) with $M=1$,[†]

$$\frac{35}{300}\exp\left(-\frac{35}{300}\right)\cong\frac{1}{10}$$

i.e., a compressor fails approximately one in ten times; with this probability, we would expect a second failure before the failed compressor is restored. On average, this failure would occur about halfway through the repair period. Thus, we expect the mean percentage system unavailability to be about

$$\frac{1}{10}\times\frac{17.5}{300}\times 100\%=0.58\%$$

If this is judged to be too high, the simplest improvement is to buy another compressor to act as an uninstalled spare. It could be installed and ready to operate within 5 days of the first compressor failure. Analyzing this situation approximately, the refrigeration service would become unavailable if:

a) the second compressor failed within 5 days of the first failure, when the service would be down for 2.5 days on average, or
b) there were two failures within 35 days of the first failure, when the service would be down for about 15 days on the average.

From the Poisson distribution, the probability of a second failure within 5 days of the first failure is

$$\frac{5}{300}\exp\left(-\frac{5}{300}\right)\cong 0.0164$$

and the probability of two failures within 35 days is approximately

$$\frac{1}{2}\left(\frac{35}{300}\right)^2\exp\left(-\frac{35}{300}\right)\cong 0.0061$$

We therefore expect the mean percentage system unavailability to be about

$$\left(0.0164\times\frac{2.5}{300}+0.0061\times\frac{15}{300}\right)\times 100\%=0.049\%$$

a-2) Availability Analysis. Since MTTF$=300$ and MTTR$=35$, the availability of one compressor is $(300)(335)=0.896$. Unavailabilities for the two and three compressor cases are

$$\text{2 compressors:}\quad \bar{A}=(1-0.896)^2(100)=1.08\%$$

$$\text{3 compressors:}\quad \bar{A}=(1-0.896)^3(100)=0.11\%$$

[†]Note that a rigorous analysis of the standby system can be made by the Markov analysis of Chapter 8. However, it yields approximately the same results as the Poisson analysis here.

These unavailabilities are somewhat more conservative than those of a-1). The differences are due to the nature of the assumptions made: they are not exactly identical; i.e., the availability analysis assumes hot standby, whereas the Poisson analysis, assumes cold standby.

b) According to the data of Table 13.18, the total runtime is 2392 days, and the average time to failure is, therefore, about 300 days.

TABLE 13.18 SUMMARY OF PREDICTED AND ACTUAL UNAVAILABILITY

Predicted Plant Unavailability, Assuming Random Poisson Compressor Failure Model	Experienced Plant Unavailability
0.58% (no uninstalled spare)	4.2%
0.049% (1 uninstalled spare)	1.6%

If we imagine the system operating as a running compressor with only an installed spare, there would have been a 35-day plant outage after the second, sixth, and seventh compressor failures, corresponding to a percentage plant unavailability of

$$\frac{105\times 100}{2392+105}\% = 4.2\%$$

If we imagine the above system with an additional uninstalled spare compressor, there would have been a 5-day plant outage after the second compressor failure, and a 35-day plant outage after the seventh failure. The percentage plant unavailability would then have been

$$\frac{40\times 100}{2392+40}\% = 1.6\%$$

Table 13.11 summarizes the situation. There is a major divergence between the predicted and experienced unavailabilities.

What has gone wrong with the theory?

A re-examination of the recorded times to compressor failure shows three failures out of eight where the time to failure is less than 1 day. For a mean time to failure of 300 days and random failure model, the probability of a failure within 1 day of start-up is

$$\frac{1}{300}\exp\left(-\frac{1}{300}\right), \quad \text{which is approximately } \frac{1}{300}$$

This is another way of saying that on average 1 in 300 failures should occur within 1 day of start-up. The experienced frequency of failures of this type gives a strong indication that a random failure model is not appropriate for this particular compressor, and that a model where early failures predominate would be more relevant. The indication is confirmed by a Weibull analysis of the recorded data, which yields a best-estimate shape parameter B of 0.42 and a 90% confidence that it is, in fact, less than 0.60. This means that the Weibull distribution has a decreasing failure rate (such as may exist at the beginning of a bathtub curve: see Appendix 4.3).

This is not an isolated situation. Table 13.19 shows the recorded times to consecutive failure of a compressor of comparable complexity at another ICI plant.

TABLE 13.19 COMPRESSURE FAILURE

Failure No.	Time to Failure (days)
1	745
2	275
3	<1
4	670

A Weibull analysis of these data yields a best-estimate shape parameter of 0.58 and a 90% confidence that it is, in fact, less than 1.0.

On the basis of this evidence we might conclude that complex compressors, at least, can best be represented by a model which emphasizes early failure. This model could well apply also to other complex equipment. Experience has shown that universal, indiscriminate application of the random failure model can lead to seriously incorrect estimates of plant availability.

Example 7. Spares Inventory

At a certain site there are 100 valves of the same type. The maintenance policy is replacement with a spare on breakdown. The valves which have been removed are repaired. Generally, the repair period is 2 weeks. If the failure rate of this type of valve is 0.5 failure/year, how many spares should be held to provide a 90% chance of spare transmitters always being available?

Solution: For equipment at the constant failure rate stage, the probability of having r or less failures in a time t is given by the Poisson distribution, (13.1). 100 valves each with a failure rate of 0.5 failure/year will exhibit on average 50 failures/year.

$$\text{Take } \lambda = 50/\text{yr}$$

Since the valves are out of commission for 2 weeks for repair, there must be at least enough spares to give a 0.9 probability of coping with the breakdowns which may take place during a 2 week period.

$$\text{Take } t = 0.04 \text{ yr}, F(M) > 0.9$$

The answer can be obtained either from tables or by summing successive terms of the Poisson distribution. Using the latter method, we obtain

$$M = 3 \text{ gives } F(M) = 0.86$$
$$M = 4 \text{ gives } F(M) = 0.95$$

Hence, it appears that four spares will satisfy the requirement. This may not, in fact, be the correct answer, since after the first failure has taken place there will be few occasions when four spares are available on the shelf.

The spares problem can be tackled in other ways, by Monte Carlo simulation or rigorous Markov analysis. In this example, a Monte Carlo simulation of over 100 years of operation suggests that with $M=4$, the availability of the spares is just over 0.9. Thus, four spares should, in fact, be sufficient.

Example 8. Availability Study of White River Shale Oil Plant[1]

A shale oil plant is typical of raw material processing plants, since it consists of large stockpiles of materials and conveying, crushing and grinding, and screening units. A primary problem in the design of plants of this type is capacity-matching, particularly since frequent maintainance and breakdown of equipment occurs.

The flow sheet for the White River plant is shown as Fig. 13.11, and Fig. 13.12 is the corresponding block diagram. Table 13.20 is a listing of failure, repair, and storage data for the process units; the numbering of the units of Table 13.20 are keyed to Fig. 13.12. Units 7 and 16 are coarse and fine ore storage piles, which provide 36 and 7.35 hr of storage capacity at the rated production. The dashed lines of Fig. 13.12 and dummy units 14 and 15 show the *blocking* procedure used to combine the upstream units into a single, composite unit to obtain the system reliability parameters upstream of the storage tanks, as was done in Example 2, Section 11.1.5. The entries $\tau = -0.00001$ for blocks 14 and 15 are computer inputs which indicate blocked units, with a storage tank or single-series units to follow. The condensed block diagram is shown as Fig. 13.13, and the minimal cut sets are readily identified as:

1) (14): Primary crusher train failure.
2) (4): Collecting conveyor failure.
3) (4): Collecting conveyor failure (either conveyor failure results in system failure with respect to the rated throughput).
4) (5): Transport conveyor failure.
5) (6): Tripper conveyor failure.
6) (15): Secondary crusher train failure.

Note that the redundant units are actually a "train," or a series of units in parallel, and that these trains are treated as a separate subsystem to make the problem manageable.

The KITT-IT code stores the final overall reliability parameters for a subsystem with a time delay for future reference. The results are recalled as if the subsystem and time delay combined were a single component. Thus, addition of a "dummy" time delay, unit 14, groups units 1, 2, and 3 together as a subsystem with a time delay specified by component 14 input data. The subsystem, adjusted for the presence of a time delay, can then be treated as a single unit, unit 14, in future calculations.

[1]Ong, J., MS Thesis, Department of Chemical Engineering, University of Houston, June, 1978.

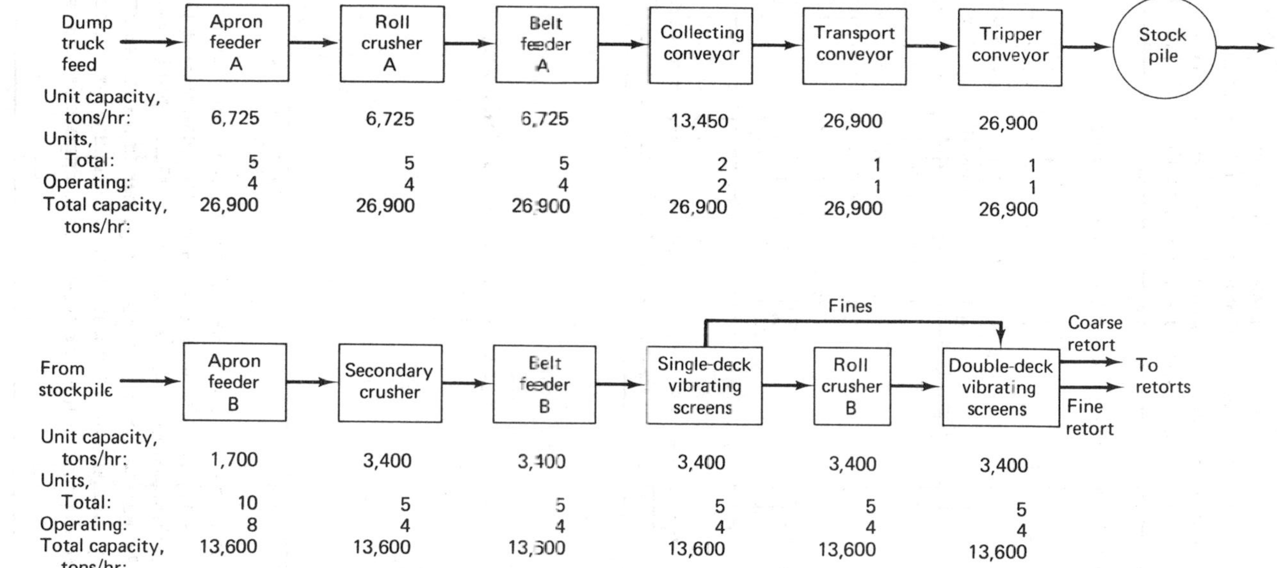

Figure 13.11. White River shale oil project flow sheet.

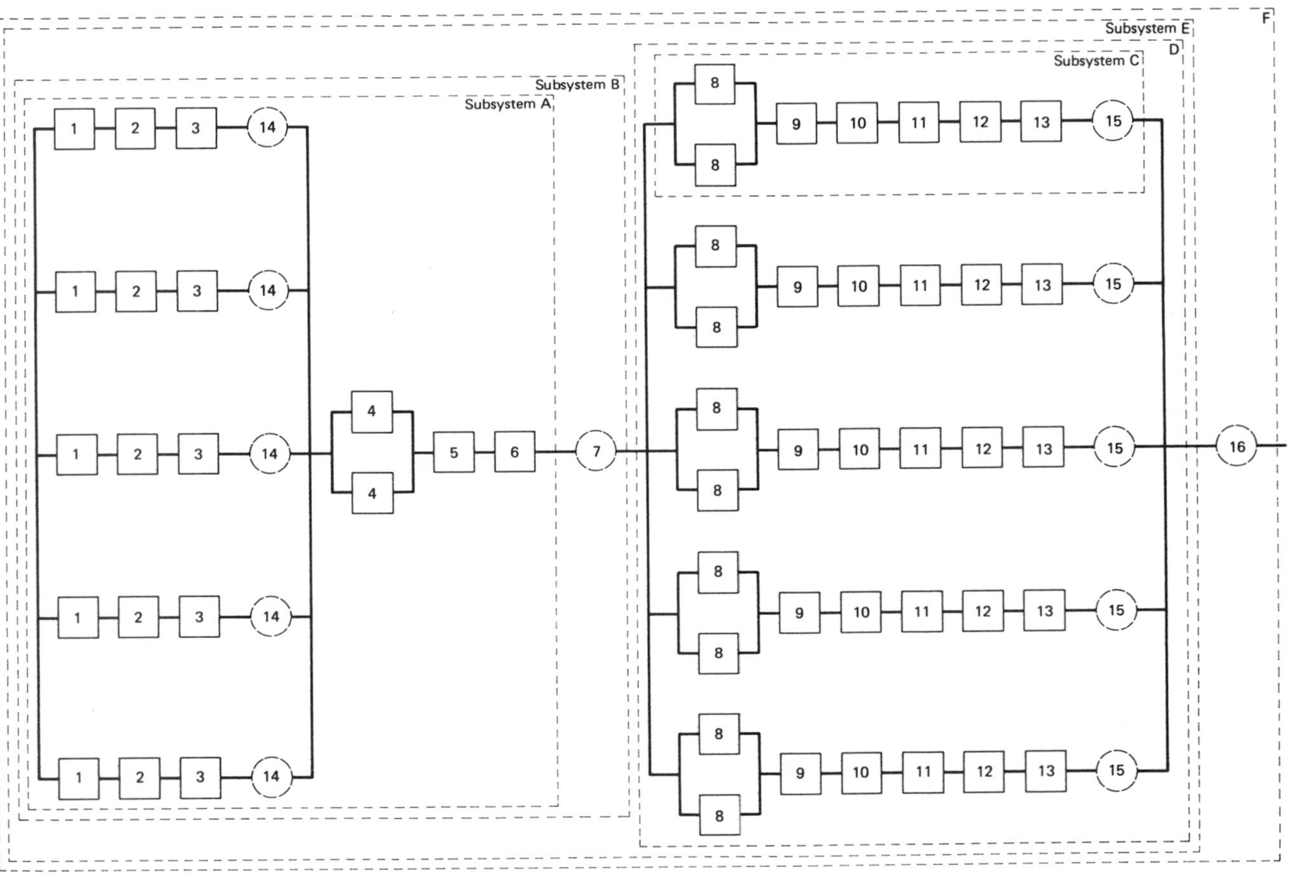

Figure 13.12. Block diagram for Fig. 13.11.

TABLE 13.20

Unit	Name	$\lambda(\text{hr}^{-1})$	τ(hr)
1	Apron feeder A	0.003	25
2	Roll crusher A	0.0015	50
3	Belt feeder A	0.003	25
4	Collecting conveyor	0.001	25
5	Transport conveyor	0.0005	25
6	Tripper conveyor	0.0005	25
7	Coarse ore stockpile	0	−36[a]
8	Apron feeder B	0.002	25
9	Secondary crusher	0.001	50
10	Belt feeder B	0.002	25
11	Single-deck vibrating screen	0.003	15
12	Roll crusher B	0.001	50
13	Double-deck vibrating screen	0.0003	15
14	Primary crusher train	0	−0.000001
15	Secondary crusher train	0	−0.000001
16	Shale fines storage	0	−7.35[b]

[a]500,000 tons/13,600 tons/hr = 36 hr.
[b]100,000 tons storage at 17,000 tons/hr.

The order of calculating subsystem reliability parameters is left to the process designer. In general, subsystems will normally be treated in a left to right sequence where system flow is from left to right. The subsystems, A,B,C,D,E,F are shown by the dashed lines in Fig. 13.12.

The unavailability Q of various portions of the process as obtained by computer are graphed in Fig. 13.14. Steady-state unavailability can be considered as existing at $t=100$ hr for all the subsystems shown. Curve E indicates that an onstream factor of 0.62 can be regarded as the best estimate of the average availability of the overall process. This availability exists for a capacity of $3400\times 4=13{,}600$ tons/hr (with four secondary crusher trains in operation).

To illustrate where process changes can be made to improve overall system availability, the bar graph representation in Fig. 13.15 may be instructive. According to equation (11.35), $C_A A_A$ must be greater than or equal to $C_D A_D$ to maintain storage, and this is true. However $C_D A_D$ is less than half of $C_A A_A$. This indicates that subsystem D may be increased in capacity if it is necessary to increase availability through the use of a storage tank.

Assuming that an overall system availability of 0.9 is desired at a capacity rating of 13,600 tons/hr implies that an average throughput of $13{,}600\times 0.9=12{,}240$ tons/hr is required. Keeping subsystem A the same, the availability of subsystem D can be estimated by

$$0.01 \geq Q_E \cong 1-(1-Q_B)(1-Q_D)$$

$$(1-Q_D) \geq \frac{0.9}{1-0.009}$$

$$Q_D \leq 0.092$$

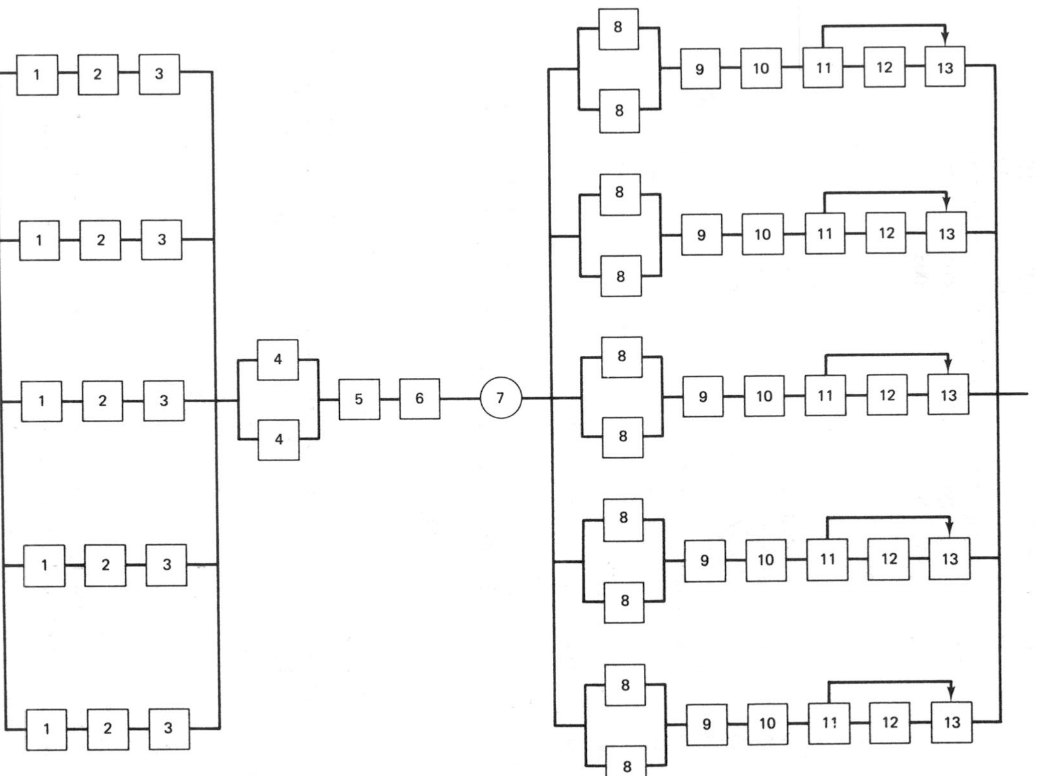

Figure 13.13. Condensed block diagram.

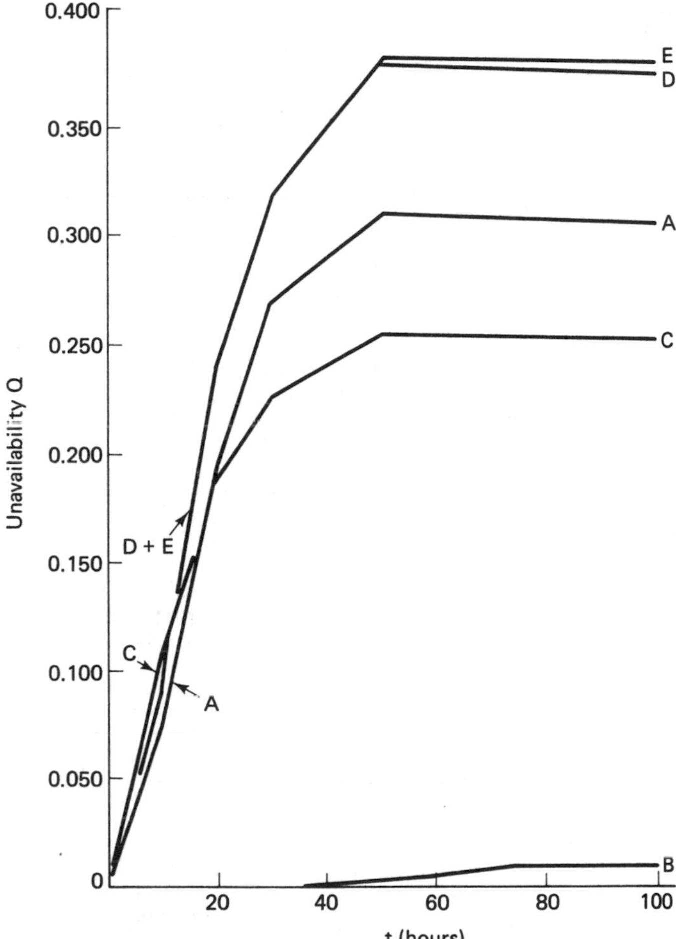

Figure 13.14. Unavailability of subsections.

If only additional secondary crusher trains were added to increase the availability of subsystem D, two standby trains would provide (see equation 11.66):

$$Q_D = 1 - \sum_{k=4}^{6} \frac{6}{k!(6-k)!}(1-0.245)^k (0.245)^{7-k}$$

$$= 0.162$$

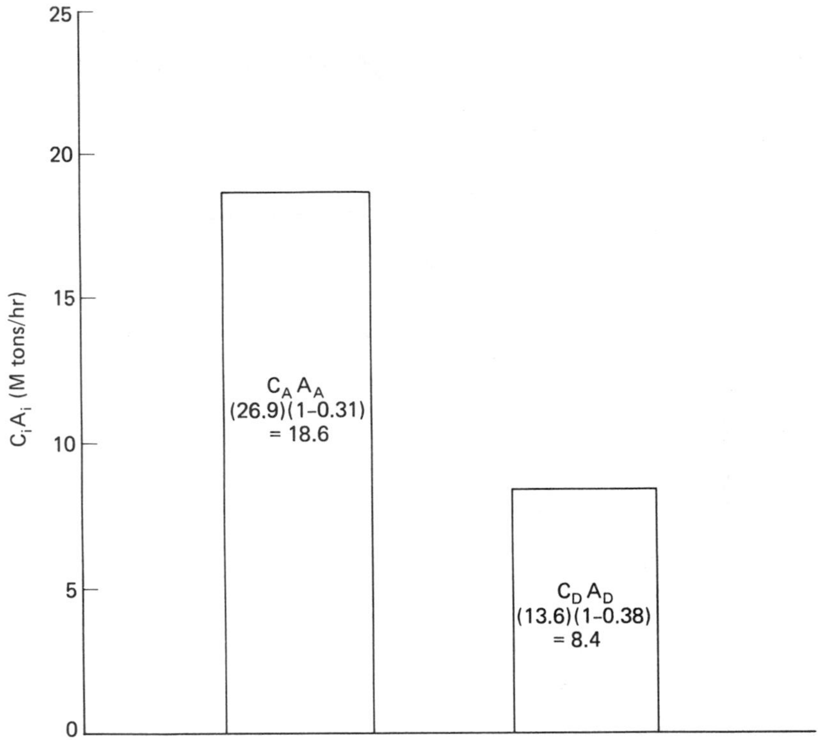

Figure 13.15. Throughputs of sub-units.

Three standby trains would provide

$$Q_D = 1 - \sum_{k=4}^{7} \frac{7!}{k!(7-k)!}(1-0.245)^k(0.245)^{8-k}$$
$$= 0.066$$

which gives

$$Q_E = 1 - (1-0.009)(1-0.066) = 0.074$$

or

$$A_E \cong 0.926$$

Thus, the design objective is met. The availability improvement for units E and D are shown in Fig. 13.16. Further calculations of this kind must, of course, be carried out to obtain a more optional configuration.

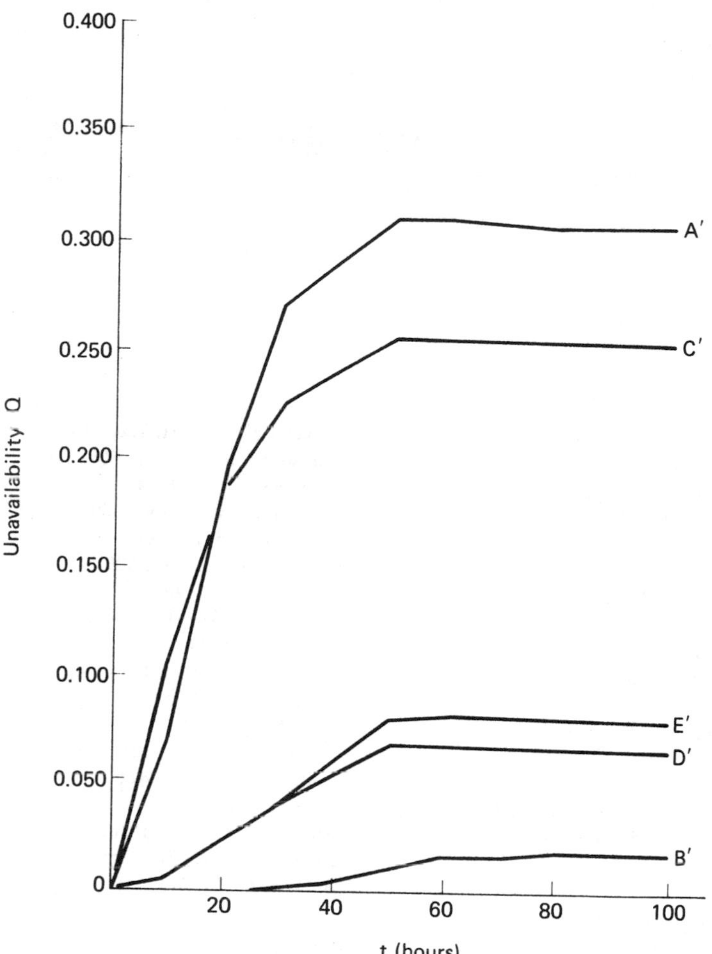

Figure 13.16. Availability improvement.

INDEX